GENERALIZED FUNCTIONS, VOLUME 1

PROPERTIES AND OPERATIONS

GENERALIZED FUNCTIONS, VOLUME 1

PROPERTIES AND OPERATIONS

I. M. GEL´FAND
G. E. SHILOV

AMS CHELSEA PUBLISHING
American Mathematical Society • Providence, Rhode Island

2010 *Mathematics Subject Classification*. Primary 46Fxx.

For additional information and updates on this book, visit
www.ams.org/bookpages/chel-377

Library of Congress Cataloging-in-Publication Data

Names: Gel'fand, I. M. (Izrail' Moiseevich) | Shilov, G. E. (Georgiĭ Evgen'evich)
Title: Generalized functions / I. M. Gel'fand, G. E. Shilov ; translated by Eugene Saletan.
Other titles: Obobshchennye funkt̄sii. English
Description: [2016 edition]. | Providence, Rhode Island : American Mathematical Society : AMS
    Chelsea Publishing, 2016- | Originally published in Russian in 1958. | Originally published in
    English as 5 volume set: New York : Academic Press, 1964-[1968]. | Includes bibliographical
    references and index.
Identifiers: LCCN 2015040021 | ISBN 9781470426583 (v. 1 : alk. paper) | ISBN 9781470426590
    (v. 2) | ISBN 9781470426613 (v. 3) | ISBN 9781470426620 (v. 4) | ISBN 9781470426637 (v. 5)
Subjects: LCSH: Theory of distributions (Functional analysis) | AMS: Functional analysis – Dis-
    tributions, generalized functions, distribution spaces – Distributions, generalized functions,
    distribution spaces. msc
Classification: LCC QA331.G373 2016 | DDC 515.7–dc23 LC record available at http://lccn.
    loc.gov/2015040021

Translator's Note

This book is a translation of the Russian edition of 1958. No attempt has been made to revise the original except for the correction of many errors, mostly typographical, and some formal changes. The translator welcomes further correction. References to the Western literature have been added where it was felt necessary. Some of the Russian Literature cited has, in addition, been translated into English since the preparation of the present volume was completed.

Appendix B appears in the Russian edition as the Appendix to Volume 5, from which it was moved at Professor Gel'fand's suggestion.

The term *generalized function*, being more general than *distribution*, has been used throughout. As with some other terms perhaps somewhat different from the usual English ones, this is in keeping with the terminology of the Russian authors.

EUGENE SALETAN

October 1963

Foreword
to the First Russian Edition

Generalized functions have of late been commanding constantly expanding interest in several different branches of mathematics. In somewhat nonrigorous form, they have already long been used in essence by physicists.

Important to the development of the theory have been the works of Hadamard dealing with divergent integrals occurring in elementary solutions of wave equations, as well as some work of M. Riesz. We shall not discuss here the even earlier mathematical work which could also be said to contain some groundwork for the future development of this theory.

The first to use generalized functions in the explicit and presently accepted form was S. L. Sobolev in 1936 in studying the uniqueness of solutions of the Cauchy problem for linear hyperbolic equations.

From another point of view Bochner's theory of the Fourier transforms of functions increasing as some power of their argument can also bring one to the theory of generalized functions. These Fourier transforms, in Bochner's work the formal derivatives of continuous functions, are in essence generalized functions.

In 1950-1951 there appeared Laurent Schwartz's monograph *Théorie des Distributions*. In this book Schwartz systematizes the theory of generalized functions, basing it on the theory of linear topological spaces, relates all the earlier approaches, and obtains many important and far reaching results. Unusually soon after the appearance of *Théorie des Distributions*, in fact literally within two or three years, generalized functions attained an extremely wide popularity. It is sufficient just to point out the great increase in the number of mathematical works containing the delta function.

In the volumes of the present series we will give a systematic development of the theory of generalized functions and of problems in analysis connected with it. On the one hand our aims do not include the colation of all material related in some way to generalized functions, and on the other hand many of the problems we shall consider can be treated without invoking them. However, the concept is a convenient link connecting many aspects of analysis, functional analysis, the theory of differential equations, the representation theory of locally compact Lie groups, and the theory of probability and statistics. It is perhaps for this reason that the title generalized functions is most appropriate for this series of volumes on functional analysis.

Let us briefly recount the contents of the first four volumes of the series.

The first volume is devoted essentially to algorithmic questions of the theory. Its first two chapters represent an elementary introduction to generalized functions. In this volume the reader will encounter many applications of generalized functions to various problems of analysis. Here and there throughout the book theorems are presented whose proofs· will be found in the second volume. Volume I makes wide use of Schwartz's book and of the article on homogeneous functions by Gel'fand and Z. Ya. Shapiro (*Uspekhi Matem. Nauk*, 1955). Shapiro has also written some of the paragraphs of the present volume.

The second volume develops the concepts introduced in the first, uses topological considerations to prove theorems left unproved in the latter, and constructs and studies a large number of specific generalized function spaces. The basis for all this is one of the most elementary and, for analysts, one of the most useful fields of the theory of linear topological spaces (developed in Chapter I of the second volume), namely the theory of countably normed spaces.

The third volume is devoted to some applications of generalized functions to the' theory of differential equations, in particular to constructing the uniqueness and consistency classes for solutions to the Cauchy problem in partial differential equations, and to expansions in eigenfunctions of differential operators. Here we make systematic use of the results obtained in the second volume.

In the fourth and fifth volumes we consider problems in probability theory related to generalized functions (generalized random processes) and the theory of the representation of Lie groups. The unifying concept here is that of harmonic analysis (the analog of Fourier integral theory) of generalized functions, in particular questions related to the representations of positive definite functions. In these volumes we present the kernel theorem of Schwartz.

Volumes I to III are written by G. E. Shilov and myself, while Volumes IV and V are written by N. Ya. Vilenkin and me.

The section entitled *Notes and References to the Literature* contains some historical remarks, citations of sources, and bibliography. In the text, however, no source references are made; references to the textbook literature are given in footnotes.

Of course all of this hardly exhausts the possibilities of application for generalized functions. The necessity for going deeper into the relations to differential equations is quite apparent (for instance with respect to boundary value problems, equations with variable coefficients, and many problems in quasi-linear equations). Further, the theory of generalized functions is the most convenient foundation on which to construct a general theory of the representation of Lie groups and, in particular, the general theory of spherical and generalized automorphic functions. We hope later to be able to shed some light also on these questions.

The authors of the first volume express their gratitude to their colleagues and students who have in one way or another taken part in the creation of this volume, in particular to V. A. Borovik, N. Ya. Vilenkin, M. I. Graev, and Z. Ya. Shapiro. The authors also express their gratitude to M. S. Agranovich, who edited the entire manuscript and introduced many improvements.

I. M. GEL'FAND

1958

Foreword
to the Second Russian Edition

The material in this second edition of the first volume has been some-what rearranged in order to make it easier to read. The first two chapters, Definition and Simplest Properties of Generalized Functions, and Fourier Transforms of Generalized Functions, can be recommended as an initial introduction; they contain the standard minimum which must be known by all mathematicians and physicists who have to deal with generalized functions.

The plan for further reading depends on the interests of the reader. Readers interested in the algorithmic aspects of the discussion can turn to Chapter III of this volume, which is devoted to special classes of generalized functions, namely to delta functions on surfaces of various dimension, generalized functions related to higher dimensional quadratic forms (of arbitrary signature), homogeneous functions, and functions equivalent to homogeneous functions. Such readers may also turn to Appendix B, which treats homogeneous generalized functions in the complex domain. To the reader interested in the general theory, we recommend that after reading the first two chapters of this volume he turn to the first three chapters of Volume II. Those contain, among other things, the necessary information from the theory of linear topological spaces. He may then turn to Chapter I of Volume IV, which describes nuclear spaces and measures in them. Those readers who wish to learn about the applications of generalized functions to the theory of partial differential equations may turn to Chapters II and III of Volume III, after first looking at the chapters on spaces of type S and W (Chapter IV of Volume II and Chapter I of Volume III). Spectral theory and its applications will be found in Chapter IV of Volume III and Chapter I of Volume IV; prerequisite for these are the first two chapters of Volume II. Other programs for reading are also possible; for instance questions concerned with the application of Fourier transforms of generalized functions are discussed after Chapter II of Volume I and Chapters III and IV of Volume II, the first three chapters of Volume III, and some of the chapters of Volume V. The application of generalized functions to the theory of group representations and of Fourier transforms over a group is described in Volume V, for which one need only have read the first two chapters of Volume I.

For convenience in using the first volume, we have placed at the end
of it a résumé of the basic definitions and formulas and a table of Fourier
transforms of generalized functions.

THE AUTHORS

1959

Contents

Chapter I

Chapter II

CHAPTER I

DEFINITION AND SIMPLEST PROPERTIES OF GENERALIZED FUNCTIONS

1. Test Functions and Generalized Functions

1.1. Introductory Remarks

Physicists have long been using so-called singular functions, although these cannot be properly defined within the framework of classical function theory.

The simplest of the singular functions is the delta function $\delta(x - x_0)$. As the physicists define it, this function is "equal to zero everywhere except at x_0, where it is infinite, and its integral is one." It is unnecessary to point out that according to the classical definition of a function and an integral these conditions are inconsistent.

One may, however, attempt to analyze the concept of a singular function in order to exhibit its actual content.

First of all, we remark that in solving any specific problems of mathematical physics, the delta function (and other singular functions) occur as a rule only in the intermediate stages. If the singular function occurs at all in the final result, it is only in an integrand where it is multiplied by some other sufficiently well-behaved function. There is therefore no actual necessity for answering the question of just what a singular function is per se; it is sufficient to know what is meant by the integral of a product of a singular function and a "good" function. For instance, rather than answer the question of what a delta function is, it is sufficient for our purposes to point out that for any sufficiently well-behaved function $\varphi(x)$ we have

$$\int_{-\infty}^{+\infty} \delta(x - x_0)\, \varphi(x)\, dx = \varphi(x_0).$$

In other words, to every singular function corresponds a functional which associates with every "sufficiently good" function some well-defined number. For instance, for the delta function the number corresponding to each "sufficiently good" function $\varphi(x)$ is $\varphi(x_0)$.

1

But if this is so, we need no longer be puzzled by the concept of a "singular function"; we shall identify it now with the functional actually under discussion. This will then be a perfectly good definition (so long as we have clearly specified also the class of "sufficiently good" functions on which this functional is defined).

Ordinary summable functions are obviously included in this concept, since given every such function $f(x)$ one can calculate the integral of the product of $f(x)$ with some "good" function. Thus the definition of generalized functions in terms of functionals will include both "singular" and ordinary functions.

Let us now proceed to formulate the exact definitions.

1.2. Test Functions

First of all we must define the set of those functions which we have conditionally called "sufficiently good," and on which our functionals will act.

As this set we shall choose the set K of all real functions[1] $\varphi(x)$ with continuous derivatives of all orders and with bounded support,[2] which means that the function vanishes outside of some bounded region [which may be different for each of the $\varphi(x)$].

We shall call these functions the *test functions*, and we shall call K the *space of test functions*.

The test functions can be added and multiplied by real numbers to yield new test functions, so that K is a linear space.

Further, we shall say that a sequence $\varphi_1(x)$, $\varphi_2(x)$, ..., $\varphi_\nu(x)$, ... of test functions *converges to zero in K* if all these functions vanish outside a certain fixed bounded region, the same for all of them, and converge uniformly to zero (in the usual sense) together with their derivatives of any order.

As an example of such a function which vanishes for

$$r \equiv |x| = \sqrt{\sum x_i^2} \geqslant a,$$

consider

$$\varphi(x, a) = \begin{cases} \exp\left(-\dfrac{a^2}{a^2 - r^2}\right) & \text{for } r < a, \\ 0 & \text{for } r \geqslant a. \end{cases}$$

---

[1] As a rule we shall let $x = \{x_1, x_2, ..., x_n\}$ denote a point in the n-dimensional space R_n. On first reading the reader may visualize x as a point on the line.

[2] The support of a continuous function $\varphi(x)$ is the closure of the set on which $\varphi(x) \neq 0$.

The sequence $\varphi_\nu(x) = \nu^{-1}\varphi(x, a)$ $(\nu = 1, 2, ...)$ converges in K. The sequence $\varphi_\nu(x) = \nu^{-1}\varphi(x/\nu, a)$ $(\nu = 1, 2, ...)$ converges to zero uniformly together with all its derivatives, but does not converge to zero in K, since there exists no common bounded region outside which all these functions vanish.

There exist many different kinds of functions in K. For instance (see Appendix 1.1 to this chapter), for a given continuous function $f(x)$ with bounded support there always exists a function $\varphi(x)$ in K arbitrarily close to it, i.e., such that for all x and for any $\epsilon > 0$,

$$|f(x) - \varphi(x)| < \epsilon.$$

1.3. Generalized Functions

We shall say that f is a *continuous linear functional* on K if there exists some rule according to which we can associate with every $\varphi(x)$ in K a real number (f, φ) satisfying the following conditions.

(a) For any two real numbers α_1 and α_2 and any two functions $\varphi_1(x)$ and $\varphi_2(x)$ in K we have $(f, \alpha_1\varphi_1 + \alpha_2\varphi_2) = \alpha_1(f, \varphi_1) + \alpha_2(f, \varphi_2)$ (linearity of f).

(b) If the sequence $\varphi_1, \varphi_2, ..., \varphi_\nu, ...$ converges to zero in K, then the sequence $(f, \varphi_1), (f, \varphi_2), ..., (f, \varphi_\nu), ...$ converges to zero (continuity of f).

For instance, let $f(x)$ be absolutely integrable in every bounded region of R_n (we shall call such functions *locally summable*). By means of such a function we can associate every $\varphi(x)$ in K with

$$(f, \varphi) = \int_{R_n} f(x)\,\varphi(x)\,dx, \tag{1}$$

where the integral is actually taken only over the bounded region in which $\varphi(x)$ fails to vanish. It is easily verified that conditions (a) and (b) are satisfied for the functional f. Condition (b) follows, in particular, from the possibility of passing to the limit under the integral sign when the functions in the integrand converge uniformly in a bounded region.

Equation (1) represents a very special kind of continuous linear functional on K. Other kinds of functionals are easily shown to exist. For instance, the functional which associates with every $\varphi(x)$ its value at $x_0 = 0$ is obviously linear and continuous. It is easily shown, however, that this functional cannot be written in the form of (1) with any locally summable function $f(x)$.

Indeed, let us assume that there exists some locally summable function $f(x)$ such that for every $\varphi(x)$ in K we have

$$\int_{R_n} f(x)\, \varphi(x)\, dx = \varphi(0).$$

In particular, for the function $\varphi(x, a)$ discussed in the previous subsection, we have

$$\int_{R_n} f(x)\, \varphi(x, a)\, dx = \varphi(0, a) = e^{-1}. \tag{2}$$

But as $a \to 0$ the integral on the left converges to zero, which contradicts Eq. (2).

We shall call the functional we are now discussing the delta function, in accordance with the established terminology (although this terminology is inaccurate, since the delta function is not a function in the classical sense of the word), and we shall denote it by $\delta(x)$. We thus write

$$(\delta(x), \varphi(x)) = \varphi(0).$$

One often has to deal with the "translated" delta function, or the functional $\delta(x - x_0)$ defined by

$$(\delta(x - x_0), \varphi(x)) = \varphi(x_0).$$

We now define a *generalized function* as any linear continuous functional defined on K. Those functionals which can be given by an equation such as (1) shall be called *regular*, and all others (including the delta function) will be called *singular*.

We shall call the regular generalized function f defined by[3]

$$(f, \varphi) = C \int \varphi(x)\, dx = \int C\varphi(x)\, dx$$

the constant C. For instance, the unit generalized function is defined by

$$(1, \varphi) = \int \varphi(x)\, dx.$$

It can be shown (see Volume II, Chapter I, Section 1.5) that if one

---

[3] We shall suppress the symbol R_n on the integral sign whenever the integral is taken over the entire space.

knows the value of a regular functional on all functions of K, the function $f(x)$ corresponding to it can be established everywhere except on a set of measure zero (almost everywhere). This means that to different functions $f_1(x)$ and $f_2(x)$ correspond different generalized functions (i.e., for some functions in K these functionals have different values). Thus the set of ordinary locally summable functions can be considered a subset of the set of all generalized functions.

For this reason, it is sometimes convenient to use the notation $f(x)$ for generalized functions, as in the case of the delta function, although we may no longer speak of the value of a generalized function at a given point (so that, rigorously speaking, the notation $f(x)$ is meaningless for a generalized function). In addition, we shall sometimes denote (f, φ) by $\int f(x)\, \varphi(x)\, dx$, although according to ordinary analysis such notation is meaningless. For instance, we will sometimes write $\int \delta(x)\, \varphi(x)\, dx$ instead of $(\delta(x), \varphi(x))$. Thus $\int \delta(x)\, \varphi(x)\, dx = \varphi(0)$.

We shall denote the set of all generalized functions by K'.

1.4. Local Properties of Generalized Functions

We have already seen that generalized functions cannot be assigned values at isolated points. One cannot, for instance, say that "a generalized function f is equal to zero at x_0." However, the statement that "a generalized function f is equal to zero in a neighborhood U of x_0" can be given a quite well-defined meaning. This will mean, namely, that for every $\varphi(x)$ in K with support in U, we have $(f, \varphi) = 0$. Thus, for instance, the generalized function f corresponding to an ordinary function $f(x)$ vanishes in a neighborhood U of x_0 if $f(x)$ itself vanishes (almost everywhere) in this neighborhood. The singular function $\delta(x - x_1)$ vanishes in a neighborhood of every point $x_0 \neq x_1$.

We shall now say that the generalized function f vanishes on some open set G if it vanishes in a neighborhood of every point in this set.

It can be proven (see Appendix 1) that the generalized function which vanishes in a neighborhood of every point vanishes also in the large, i.e., that for every $\varphi(x)$ in K we have

$$(f, \varphi) = 0.$$

If f is a generalized function which fails to vanish in any neighborhood of x_0, then x_0 is called an *essential point* of the functional f. Thus, for instance, the point $x_0 = 0$ on the line is an essential point of the functional $f(x) = x^2$ (although the function x^2 itself vanishes at this point!). It is also true, of course, that all the other points on the line are also essential points for this functional. The set of all essential points of a generalized function f is called its *support*. The support of the regular generalized function f corresponding to the continuous (or piecewise continuous) function $f(x)$ is the closure of the set on which $f(x) \neq 0$, i.e., the support of $f(x)$. The support of the generalized function $\delta(x - x_0)$ is the single point x_0. If F is a set which contains the support of a functional f, one says that f is *concentrated* in F.

The term "essential point" is justified by the following property (proved in Appendix 1.3). If $\varphi(x)$ in K is a function that vanishes in a neighborhood of the support of the functional f, then $(f, \varphi) = 0$. It follows that however φ may vary outside of the neighborhood of the support of f has no bearing on (f, φ). Indeed, any such variation is equivalent to adding to φ another test function ψ which vanishes in the neighborhood of the support of f, or such that $(f, \psi) = 0$; therefore $(f, \varphi + \psi) = (f, \varphi)$.

Let us now go on to a local comparison of two arbitrary generalized functions. We will say that the generalized functions f and g coincide on the open set G if the difference $f - g$ vanishes on this set. It can be shown that if f and g coincide in a neighborhood of every point, they coincide in the large, or $(f, \varphi) = (g, \varphi)$ for all φ. It follows from this that a generalized function f is determined uniquely by its local properties. More even than that, a generalized function can in fact be constructed if its local properties are known (Appendix 1.3).

In particular, we shall say that a generalized function f is regular in some region G if in this region it coincides with some ordinary locally summable function.

For instance, the delta function $\delta(x - x_0)$ is regular (and equal to zero) everywhere except at x_0.

One of the important problems in the theory of generalized functions is the following. Given an ordinary function $f(x)$, in general not locally summable, for instance $1/x$ on the line. One may then ask whether or not there exists a generalized function f which coincides with $f(x)$ at all points at which the latter is locally summable. Further, is it possible to establish the correspondence from $f(x)$ to f in a way which preserves the operations of addition, multiplication by a function, and differentiation (which we shall define below for generalized functions)? It is clear that the answers to these questions are quite important, since if this were possible one could include in the generalized functions those ordinary functions which have nonsummable singularities.

At present only partial answers exist to these questions; they will be discussed in Sections 1.7 and 3.

1.5. Addition and Multiplication by a Number and by a Function

Consider two generalized functions f and g. We define their sum $f + g$ as the functional on K defined by

$$(f + g, \varphi) = (f, \varphi) + (g, \varphi).$$

It is easily verified that according to this definition $f + g$ is also a continuous linear functional. In particular, if f and g are regular functionals corresponding to the functions $f(x)$ and $g(x)$, then $f + g$ is also a regular functional, and it corresponds to $f(x) + g(x)$. This shows how natural is this definition of the sum of generalized functions.

The product of a generalized function f by a number α is defined by

$$(\alpha f, \varphi) = \alpha(f, \varphi) = (f, \alpha\varphi).$$

Clearly, this functional is also continuous and linear. If f is a regular functional corresponding to the locally summable function $f(x)$, then this operation corresponds to multiplication of $f(x)$ by α.

There does not seem to be any natural way to define the product of two arbitrary generalized functions. Nevertheless, it is possible to define the product of any generalized function f by an infinitely differentiable function $a(x)$. We note first that the product of an infinitely differentiable function $a(x)$ and a function $\varphi(x)$ in K is a function $\psi(x) = a(x)\,\varphi(x)$ in K. Further, if the sequence of functions $\varphi_\nu(x)$ converges to zero in K, then the sequence $a(x)\,\varphi_\nu(x)$ also converges to zero in K. Now consider any generalized function f. We define the new functional af by

$$(af, \varphi) = (f, a\varphi).$$

Clearly, af is linear. It is also a continuous functional, as can be seen from the following. If $\varphi_\nu(x) \to 0$, then as was pointed out above, it follows that $a(x)\,\varphi_\nu(x) \to 0$, so that

$$(af, \varphi_\nu) = (f, a\varphi_\nu) \to 0.$$

For the regular functional f corresponding to the locally summable function $f(x)$, multiplication by $a(x)$ corresponds to multiplication of $f(x)$ by $a(x)$. Indeed, for this case we have

$$(af, \varphi) = (f, a\varphi) = \int f(x)\,[a(x)\,\varphi(x)]\,dx$$

$$= \int [a(x)\,f(x)]\,\varphi(x)\,dx,$$

which is the desired result.

1.6. Translations, Rotations, and Other Linear Transformations in the Space of the Independent Variables

For $h > 0$, the function $f(x - h)$ on the line is called in analysis the right translation of $f(x)$ through the distance h. We note, by the way, that the operation performed on the independent variable here is a left translation through the distance h, the opposite of the operation performed on the function.

This may be generalized in the following way to the case of functions of n variables. Let u be some nonsingular linear transformation in the n-dimensional space R_n and let u^{-1} be its inverse. Then the corresponding operation u on a function $f(x)$ is defined by

$$uf(x) = f(u^{-1}x). \tag{1}$$

This concept can be extended to generalized functions. We note first that if $\varphi(x)$ is in K, then so is $\varphi(ux)$. Let us further determine what equation for functionals corresponds to Eq. (1) for functions. Assuming $f(x)$ to be locally summable and letting $\varphi(x)$ be a function in K, we obtain

$$(uf(x), \varphi(x)) = (f(u^{-1}x), \varphi(x)) = \int f(u^{-1}x)\, \varphi(x)\, dx.$$

Now we make the substitution $u^{-1}x = y$, so that $x = uy$, and in the integral we may write $dx = |u|\, dy$, where $|u|$ is the absolute value of the determinant of the matrix of the transformation. Then

$$(uf, \varphi) = |u| \int f(y)\, \varphi(uy)\, dy = |u|\, (f(x), \varphi(ux)).$$

From this equation we define the operation u applied to an arbitrary generalized function in the following way. We define uf as the functional such that

$$(uf, \varphi) = |u|\, (f, \varphi(ux)). \tag{2}$$

It is also possible to denote uf by $f(u^{-1}x)$ (as with ordinary functions), a notation which sometimes clarifies the meaning of an equation. In that case we can use the previously established convention (see the end of Section 1.3) to write (2) in the form

$$\int f(u^{-1}x)\, \varphi(x)\, dx = |u| \int f(x)\, \varphi(ux)\, dx, \tag{2'}$$

where f is any generalized function.

A particularly simple equation is obtained for unimodular transformations (i.e., those with determinant one), in particular for rotations. For these we have

$$(f(u^{-1}x), \varphi(x)) \equiv (uf, \varphi) = (f, \varphi(ux)). \qquad (3)$$

Examples.

1. *Translation through the vector h*. Here $ux = x + h$. The translation of a generalized function f through the vector h is given by

$$(uf, \varphi) = (f, \varphi(ux)) = (f, \varphi(x + h)).$$

This formula can also be written in the form

$$(f(x - h), \varphi) = (f, \varphi(x + h)),$$

or in the form

$$\int f(x - h)\, \varphi(x)\, dx = \int f(x)\, \varphi(x + h)\, dx$$

(where f is a generalized function).

2. *Reflection in the origin*.   Here $ux = -x$. Reflection of a generalized function f is given by

$$(f(-x), \varphi(x)) \equiv (uf, \varphi) = (f, \varphi(-x)),$$

or

$$\int f(-x)\, \varphi(x)\, dx = \int f(x)\, \varphi(-x)\, dx.$$

3. *Similarity transformation*.   Here $ux = \alpha x$. The similarity transformation of a generalized function f is given by

$$(f(u^{-1}x), \varphi(x)) \equiv (uf, \varphi) = \alpha^n(f(x), \varphi(\alpha x)).$$

The generalized function f is naturally called invariant with respect to the operation u if $uf = f$.

For instance, a generalized function may be invariant with respect to reflection in the origin; i.e., it may satisfy the equation $(f, \varphi(-x)) = (f, \varphi)$. Such a function may be called centrally symmetric. Generalized functions invariant with respect to all rotations will be called spherically symmetric. Among such are all the regular functionals corresponding to functions depending only on $r = \sqrt{\Sigma x_j^2}$; another example is the δ function. A function invariant under translations through the vector h is called periodic with period h. It can be shown that a generalized function invariant with respect to all translations is a constant (see Section 2.6).

If $f(x)$ is a generalized function such that

$$f(\alpha x) = \alpha^\lambda f(x) \tag{4}$$

for all real $\alpha > 0$, then it is called homogeneous of degree λ. Equation (4) can also be written in the form

$$\alpha^{-n}\left(f(x), \varphi\left(\frac{x}{\alpha}\right)\right) = \alpha^\lambda (f(x), \varphi(x))$$

or, equivalently,

$$\left(f(x), \varphi\left(\frac{x}{\alpha}\right)\right) = \alpha^{\lambda+n}(f(x), \varphi(x)).$$

Thus, $\delta(x_1, ..., x_n)$ is a homogeneous generalized function of degree $-n$. It is clear that ordinary locally summable homogeneous functions are also homogeneous when considered as generalized functions, and of the same degree. We shall consider homogeneous generalized functions in more detail in Section 4 and then again in Chapter III.

1.7. Regularization of Divergent Integrals

Let $f(x)$ be a function locally summable everywhere except at x_0, where it has a nonsummable singularity [for instance, $f(x) = 1/x$ on the line]. Then in general the integral

$$\int f(x)\, \varphi(x)\, dx, \tag{1}$$

where $\varphi(x)$ is in K, will diverge. But this integral will converge if $\varphi(x)$ vanishes in a neighborhood of x_0. We may now ask whether it is possible to use this result to redefine a functional, that is to construct an $f \in K'$ such that for all $\varphi(x)$ in K vanishing in a neighborhood of x_0 the functional has the value given by (1). Any such functional f is called a *regularization* of the divergent integral of (1) [or a regularization of $f(x)$].

Thus, for instance, for $f(x) = 1/x$ we may set

$$(f, \varphi) = \int_{-\infty}^{-a} \frac{\varphi(x)}{x}\, dx + \int_{-a}^{b} \frac{\varphi(x) - \varphi(0)}{x}\, dx + \int_{b}^{+\infty} \frac{\varphi(x)}{x}\, dx \tag{2}$$

with any $a, b > 0$.

Regularization can also be defined in the following somewhat different way. The regularization of a function $f(x)$ is a continuous linear functional f which coincides with $f(x)$ everywhere except at x_0 (see Section 1.4).

Indeed, if there exists a functional f given by (1) for all $\varphi(x)$ in K which vanish in a neighborhood of x_0, then f coincides with $f(x)$ in some neighborhood of every point $x_1 \neq x_0$, in particular in every neighborhood that does not contain x_0 either in its interior or on its boundary. If, further, the functional f coincides with $f(x)$ everywhere except at x_0, Eq. (1) will hold for every φ in K which vanishes in a neighborhood of x_0, and therefore f represents a regularization of $f(x)$.

We now present some general propositions concerning the existence of regularizations. For simplicity we shall take $x_0 = 0$.

Proposition 1. If there exists an integer $m > 0$ such that $f(x) \cdot r^m$ is locally summable, the integral of (1) can be regularized.

For this case we can construct the regularization f in accordance with the equation

$$(f, \varphi) = \int f(x) \left\{ \varphi(x) - \left[\varphi(0) + \frac{\partial \varphi(0)}{\partial x_1} x_1 + \ldots + \frac{\partial^m \varphi(0)}{\partial x_n^m} \frac{x_n^m}{m!} \right] \theta(1 - r) \right\} dx \tag{3}$$

(where we subtract enough terms of the Taylor's series to leave a remainder of order greater than r^m). The function $\theta(1 - r)$ is unity for $r < 1$ and zero for $r > 1$.

Clearly the integral in (3) converges for all $\varphi(x)$ in K and is a continuous linear functional. If, further, $\varphi(x)$ vanishes in a neighborhood of the origin, (3) becomes

$$(f, \varphi) = \int f(x) \, \varphi(x) \, dx,$$

so that except at the origin f coincides with $f(x)$.

For this case, therefore, the regularization exists.

Proposition 2. If f_0 is a special solution to the regularization problem, i.e., if f_0 regularizes the integral of (1), the general solution f is obtained by adding to f_0 any functional concentrated on $x_0 = 0$.

Let f_0 be a regularization, and let g be a functional concentrated on the origin. Then for any $\varphi(x)$ in K which vanishes in a neighborhood of the origin,

$$(f_0 + g, \varphi) = (f_0, \varphi) + (g, \varphi) = (f_0, \varphi),$$

so that $f_0 + g$ is also a regularization. Conversely, if f_0 and f_1 are two different regularizations, then for all·such $\varphi(x)$ we have

$$(f_1 - f_0, \varphi) = (f_1, \varphi) - (f_0, \varphi) = 0,$$

so that $f_1 - f_0$ is concentrated on $f_0 = 0$.

For instance, if $f(x) = 1/x$, the difference of any two regularizations given by (2), as is easily verified, is simply $C\delta(x)$, where C is a constant.

In Section 3 we shall consider the problem of choosing from all the regularizations of a given function the most natural one.

Proposition 3.   If within some solid angle whose vertex is at the origin $f(x)$ satisfies the condition

$$f(x) \geqslant F(r), \tag{4}$$

where $F(r)$ increases monotonically faster than any power of $1/r$ as r approaches zero, the integral of (1) cannot be regularized.

For simplicity let us assume that the solid angle under discussion contains the region $H = (x_1 > 0, ..., x_n > 0)$. The proof of the general case is then quite similar to the one that follows. Consider a nonnegative function $\varphi(x)$ in K which vanishes for $|x| \geqslant 1$ and whose integral is equal to unity. Further, let $\epsilon_\nu > 0$ be any sequence of numbers approaching zero more rapidly than any power of $1/\nu$.

We now translate the function $\epsilon_\nu\varphi(\nu x)$ through the distance $\sqrt{n}/\nu$ along the line $x_1 = ... = x_n$. Call the function so obtained $\psi_\nu(x)$. Then $\psi_\nu(x)$ vanishes outside H and, for sufficiently large ν, outside the intersection of H with an arbitrarily small ball centered at the origin. It is easily seen further that the sequence of the $\psi_\nu(x)$ converges to zero in K as $\nu \to \infty$. Thus if there were to exist a functional f regularizing (1), it would follow that

$$(f, \psi_\nu) \to 0. \tag{5}$$

But, since $\psi_\nu(x)$ vanishes in a neighborhood of $x = 0$, and since except at the origin f coincides with the function $f(x)$ of (4), we may write

$$(f, \psi_\nu) = \int f(x)\,\psi_\nu(x)\,dx \geqslant \epsilon_\nu \int F(r)\,u\varphi(\nu x)\,dx,$$

where $u\varphi$ is the translated φ function. Now the integral of $\varphi(\nu x)$ is obviously $1/\nu^n$, and we thus arrive at

$$(f, \psi_\nu) > \frac{\epsilon_\nu}{\nu^n} F\left(\frac{2}{\nu}\sqrt{n}\right). \tag{6}$$

We have not yet picked the ϵ_ν. Let us write

$$\epsilon_\nu = \frac{\nu^n}{F\left(\frac{2}{\nu}\sqrt{n}\right)} ; \tag{7}$$

so that, since $F(r)$ increases more rapidly than any power of $1/r$, these ϵ_ν converge to zero more rapidly than any power of $1/\nu$, as we have required. But it is then seen from (6) that the (f, ψ_ν) are all greater than unity and cannot therefore converge to zero; this contradicts (5).

Thus in this case the integral in (1) cannot be regularized.

Remark.   This result does not mean that in generalized functional analysis one may never consider functions with singularities of infinite order. So far we have considered test functions of only a very special kind (namely those in K). In the second volume we shall discuss other test function spaces, and among them it is always possible to find those for which functions with singularities of any kind can be given meaning as functionals.

In conclusion, we remark that regularization can be defined similarly also if $f(x)$ has not one but several or even a countable number of isolated singular points, so long as the number of singular points in any finite interval is finite.

Any such function $f(x)$ can always be written in the form

$$f(x) = \sum f_k(x),$$

where each $f_k(x)$ has only one singular point (see Appendix 1.2). Therefore the case of a countable number of isolated singularities is essentially the same as the one just discussed.

1.8. Convergence of Generalized Function Sequences

A sequence $f_1, f_2, ..., f_\nu, ...$ of generalized functions is defined to converge to the generalized function f if for every $\varphi(x)$ in K

$$\lim_{\nu \to \infty} (f_\nu, \varphi) = (f, \varphi).$$

One may, of course, assume ν to vary over a continuous set. The definition of the limit of the sequence remains the same.

Similarly, a series $h_1 + h_2 + ... + h_\nu + ...$ of generalized functions is said to converge to the generalized function g if the sequence

$$g_1 = h_1, \qquad g_2 = h_1 + h_2, ... \qquad g_\nu = h_1 + h_2 + ... + h_\nu, ...$$

converges to g in the above sense.

It is easily verified that this definition of the limit of a convergent sequence of generalized functions is unambiguous and consistent.

Further, the operation of passing to a limit is linear. That is to say, if $f = \lim_{\nu \to \infty} f_\nu$, and α is a number (or an infinitely differentiable function), then $\lim_{\nu \to \infty} \alpha f_\nu$ exists and is equal to $\alpha \lim_{\nu \to \infty} f_\nu = \alpha f$. Further, if $f = \lim_{\nu \to \infty} f_\nu$, $g = \lim_{\nu \to \infty} g_\nu$, then $\lim_{\nu \to \infty} (f_\nu + g_\nu)$ exists and is equal to $f + g$.

If the sequence of locally summable functions $f_\nu(x)$ (with $\nu = 1, 2, ...$) converges uniformly to the locally summable function $f(x)$ in every bounded region, the corresponding functionals f_ν converge to the regular functional f. To see this, let $\varphi(x)$ be some function in K and let G be the support of $\varphi(x)$. Then as a result of a well-known theorem on the convergence of an integral, we have

$$(f_\nu, \varphi) = \int_G f_\nu(x)\, \varphi(x)\, dx \to \int_G f(x)\, \varphi(x)\, dx = (f, \varphi),$$

which is the required result.

We may, on the other hand, propose a weaker requirement than the uniform convergence of the $f_\nu(x)$ on bounded regions. The theorem on the limit of an integral will hold, for instance, under either of the following:

(a) $f_\nu(x) \to f(x)$ almost everywhere, and $|f_\nu(x)|$ bounded by a fixed constant (or even by a locally summable function);

(b) $f_\nu(x) \to f(x)$ monotonically increasing or decreasing, $f(x)$ locally summable.

Example. A sequence of regular functionals may converge to a singular functional. For instance, the functional

$$(f, \varphi) = \lim_{\epsilon \to 0} \int_{|x| > \epsilon} \frac{\varphi(x)}{x}\, dx$$

coincides with the ordinary function $1/x$ for $x \neq 0$. This function is not summable in any neighborhood of the origin, and the functional is therefore not regular. But it is seen from the construction that it is the limit as $\epsilon \to 0$ of the regular functionals corresponding to the ordinary functions defined as equal to $1/x$ for $|x| > \epsilon$ and to zero for $|x| < \epsilon$.

It can in general be shown (and will be shown in the second volume) that every singular functional is the limit of a sequence of regular functionals.

It is easily verified that every generalized function is the limit of a sequence of generalized functions concentrated in bounded sets.

To prove this, consider an infinitely differentiable function $g_\nu(x)$ equal to one in the ball $|x| \leqslant \nu$ and equal to zero outside the ball $|x| \leqslant 2\nu$. It is clear that for each φ in K there exists a sufficiently high ν such that $g_\nu\varphi = \varphi$. From this it follows that for any functional f the sequence $g_\nu f$ converges to f; for let φ be any function in K, so that $(g_\nu f, \varphi) = (f, g_\nu\varphi) = (f, \varphi)$ for sufficiently high ν, or $(g_\nu f, \varphi) \to (f, \varphi)$. Now the functional $g_\nu f$ is easily seen to vanish for $|x| > 2\nu$, so that it is concentrated on the bounded set $|x| \leqslant 2$. Thus f is the limit of the sequence $g_\nu f$ of generalized functions concentrated in bounded sets, as asserted.

One important property of the space of generalized functions is its completeness with respect to convergence as defined here. In other words, if the sequence $f_1, f_2, ..., f_\nu, ...$ is such that for every φ in K the number sequence (f_ν, φ) has a limit, this limit is again a continuous linear functional on K.

This theorem is proven in Appendix A at the end of this volume.

1.9. Complex Test Functions and Generalized Functions

So far we have assumed that both the $\varphi(x)$ and the (f, φ) take on only real values.

One may, however, also define complex generalized functions. To do this one goes over from the space of a real test function to the space of complex test functions (again infinitely differentiable functions with bounded support) using the previously defined operations.[4]

We now define *complex generalized functions* as continuous linear functionals taking on, in general, complex values on this new test-function space.

With every complex locally summable function $f(x)$ we associate the functional

$$(f, \varphi) = \int \overline{f(x)}\, \varphi(x)\, dx, \qquad (1)$$

where the bar denotes the complex conjugate.

---

[4] These are not generalized functions of a complex variable; see Appendix B.

As before, we shall denote the test-function space by K and the generalized-function space by K'.

Addition and multiplication by a (complex) number in K' is then given by

$$(f_1 + f_2, \varphi) = (f_1, \varphi) + (f_2, \varphi), \atop (\alpha f, \varphi) = \bar{\alpha}(f, \varphi) = (f, \bar{\alpha}\varphi). \quad\Bigg\} \tag{2}$$

Multiplication by a complex infinitely differentiable function $a(x)$ is defined by

$$(a(x) f, \varphi) = (f, \overline{a(x)}\, \varphi(x)). \tag{3}$$

For every generalized function f there exists the complex conjugate generalized function $\bar{f}$ defined by

$$(\bar{f}, \varphi) = \overline{(f, \bar{\varphi})}. \tag{4}$$

It can be shown that if an ordinary function $f(x)$ is treated as a generalized function, the newly defined operation of complex conjugation corresponds to the ordinary one. In the sequel we shall deal almost entirely with the real case. The results obtained, however, can for the most part be automatically extended to the complex case with obvious changes which follow from the above Eqs. (1)-(4).

1.10. Other Test-Function Spaces

It is often convenient to extend the functionals defined on K to a broader function space and then to study their behavior on this new space.

A space which very often occurs in applications is S, the space of infinitely differentiable functions which, together with their derivatives, approach zero more rapidly than any power of $1/|x|$ as $|x| \to \infty$ [for instance, $\exp(-x^2)$]. Thus on the line $-\infty < x < \infty$ the functions $\varphi(x) \in S$ satisfy inequalities of the form

$$|x^k \varphi^{(q)}(x)| \leqslant C_{kq} \tag{1}$$

for any $k, q = 0, 1, 2, \ldots$.

In the case of several variables, (1) is replaced by

$$\left| x_1^{k_1} \ldots x_n^{k_n} \frac{\partial^{q_1 + \ldots + q_n} \varphi(x)}{\partial x_1^{q_1} \ldots \partial x_n^{q_n}} \right| \leqslant C_{k_1, \ldots, k_n, q_1, \ldots, q_n}$$

$$(k_1, \ldots, q_n = 0, 1, 2, \ldots),$$

which we shall write symbolically in the form

$$| x^k D^q \varphi(x) | \leqslant C_{kq},$$

where $k = (k_1, ..., k_n)$, $q = (q_1, ..., q_n)$, $x_k = x_1^{k_1} ... x_n^{k_n}$, and

$$D^q = \frac{\partial^{q_1 + ... + q_n}}{\partial x_1^{q_1} ... \partial x_n^{q_n}}.$$

Convergence in S is defined as follows. A sequence $\varphi_\nu(x)$ is said to converge to $\varphi(x)$ if in every bounded region the derivatives of all orders of the $\varphi_\nu(x)$ converge uniformly to the corresponding derivatives of $\varphi(x)$ and if the constants C_{kq} in the conditions

$$| x^k D^q \varphi_\nu(x) | \leqslant C_{kq} \tag{2}$$

can be chosen independent of ν. Then the limit function $\varphi(x)$ also belongs to S, as can be seen by going to the limit as $\nu \to \infty$ in conditions (2).

Obviously every infinitely differentiable function with bounded support, i.e., every function in K, belongs also to S. Moreover, K is dense in S. To prove this we construct functions $e_\nu(x) \in K$ equal to one in the cubes $| x_j | \leqslant \nu$ and to zero outside the cubes $| x_j | \leqslant 2\nu$ and such that their derivatives of any fixed order all be bounded by the same number independent of ν (with $\nu = 1, 2, ...$).[5]

Now let $\varphi(x)$ be any function in S. Then the functions $\varphi_\nu(x) = \varphi(x) e_\nu(x)$ have bounded support and, as is easily seen, converge to $\varphi(x)$ in S, as asserted.

We note also that if $\varphi_\nu, \varphi \in K$, and $\varphi_\nu \to \varphi$ in K, then clearly $\varphi_\nu \to \varphi$ in S. In other words convergence in K is a stronger condition than, or includes, convergence in S.

The set of continuous linear functionals on S is denoted by S'. Clearly every continuous linear functional on S is also a continuous linear functional on K, so that $S' \subset K'$. But of course far from all the continuous linear functionals on K can be extended to S. (It may be noted in any case that if such an extension is possible for a given functional, it is unique. This follows from the fact that K is dense in S.) Generalized functions that can be extended to S are those with bounded support, regular functionals corresponding to slowly increasing functions, i.e. those that increase no more rapidly than any power of $| x |$ as $x \to \infty$, and some others.

It is clear that S' is a linear space, namely that $f, g \in S'$ implies,

---

[5] For instance, if $e_1(x)$ is constructed somehow, then one may set $e_\nu(x) = e_1(x/\nu)$.

$\alpha f + \beta g \in S'$ for all numbers α and β. In other words, therefore, S' is a linear subspace of K'.

Convergence may be defined in S' in analogy with the definition in K'. A sequence of functionals f_ν in S' converges to a functional f in S' if for every $\varphi(x) \in S$ we have

$$(f_\nu, \varphi) \to (f, \varphi).$$

Since S contains K it is clear that f_ν, f in S' and $f_\nu \to f$ in S' imply $f_\nu \to f$ in K'. Thus S' is contained in K' together with convergence as we have defined it.

In addition to S there are many other types of test-function spaces defined according to how the test functions behave at infinity. We shall deal with such spaces in the second volume.

In this volume the spaces S and S' will play only a secondary role. As a rule we shall deal only with K and K', and when we wish to refer to other spaces, we shall make special note of it.

2. Differentiation and Integration of Generalized Functions

2.1. Fundamental Definitions

It is well known that not all ordinary functions are differentiable, that there exist in fact many functions that have no derivative in the usual sense of the word. In contradistinction, as we shall show, generalized functions always have derivatives (and furthermore of all orders) which are also generalized functions.

In order to define the derivative of a generalized function, we shall first consider an ordinary function of a single variable.

Let $f(x)$ be a continuous function with a continuous first derivative (in the usual sense), and consider the functional

$$(f', \varphi) = \int_{-\infty}^{\infty} f'(x)\, \varphi(x)\, dx.$$

Integrating by parts and recalling that $\varphi(x)$ is in K so that outside some interval $[a, b]$ it vanishes, we arrive at

$$(f', \varphi) = f(x)\, \varphi(x) \Big|_{-\infty}^{\infty} - \int_{-\infty}^{\infty} f(x)\, \varphi'(x)\, dx = (f, -\varphi'). \tag{1}$$

We shall use this equation to define the derivative of a generalized function. Let f be any continuous linear functional on K. Then the functional g defined by

$$(g, \varphi) = (f, -\varphi'), \tag{2}$$

will be called *the derivative* of f and be denoted by f' or df/dx.

In accordance with our notational convention (Section 1.3) we may write f' also in the form

$$\int f'(x)\, \varphi(x)\, dx = - \int f(x)\, \varphi'(x)\, dx. \tag{2'}$$

To show the consistency of this definition we must prove that g is also a continuous linear functional on K. We do this as follows.

First, note that g is defined for all $\varphi(x)$ in K, since if $\varphi(x)$ is in K then so is $\varphi'(x)$. It is obvious that g is linear. We need show only that it is continuous.

Consider a sequence of functions $\varphi_\nu(x)$ (with $\nu = 1, 2, ...$) converging to zero in K. Then according to the definition of convergence, the sequence $-\varphi_\nu'(x)$ also converges to zero in K. The fact that f is continuous then implies that

$$(g, \varphi_\nu) = (f, -\varphi_\nu') \to 0,$$

as asserted.

Thus every generalized function has a derivative.

It can be shown that the ordinary rules of differentiation apply also to generalized functions. For instance, the derivative of a sum is the sum of the derivatives, and a constant can be commuted with the derivative operator.

For the product of an infinitely differentiable function $a(x)$ with a generalized function f, the product rule

$$(af)' = a'f + af' \tag{3}$$

remains valid. Indeed, we have

$$\begin{aligned}
((af)', \varphi) &= - (af, \varphi') = - (f, a\varphi') = - (f, (a\varphi)' - a'\varphi) \\
&= - (f, (a\varphi)') + (f, a'\varphi) = (f', a\varphi) + (a'f, \varphi) \\
&= (af', \varphi) + (a'f, \varphi) = (af' + a'f, \varphi),
\end{aligned}$$

as asserted.

Consider now the case of several variables. Then the partial derivative of a generalized function f with respect to each of the variables may be defined as

$$\left(\frac{\partial f}{\partial x_j}, \varphi \right) = \left(f, - \frac{\partial \varphi}{\partial x_j} \right) \qquad (j = 1, 2, ..., n). \tag{4}$$

As in the case of one variable, the consistency of this definition is easily verified. Further, if f is a regular functional corresponding to a function $f(x_1, x_2, ..., x_n)$ which is continuous and whose partial derivative with respect to x_j is continuous, we again find by integration by parts that the functional $\partial f/\partial x_j$ corresponds to $\partial f(x_1, ..., x_n)/\partial x_j$.

Since differentiation of a generalized function yields again a generalized function, the process may be continued to partial derivatives of any order such as $\partial^2 f/\partial x_i \partial x_j$, $\partial^3 f/\partial x_i \partial x_j \partial x_k$, etc.

Thus all generalized functions have derivatives of all orders.

In particular, every locally summable function has derivatives of all orders in the sense of generalized functions. (We remark, however, that although the function f may have an ordinary derivative, the functional defined by this ordinary derivative need not coincide with the derivative of f in the sense of generalized functions.)

In Volume II (Chapter II, Section 4) we shall prove the converse theorem, namely that every generalized function is the derivative of some order in the sense of generalized functions of some locally summable function (if desired, even a continuous function) or a finite sum of such derivatives.

Mixed derivatives of generalized functions are independent of the order of differentiation. For instance,

$$\frac{\partial^2 f}{\partial x_1 \, \partial x_2} = \frac{\partial^2 f}{\partial x_2 \, \partial x_1}.$$

This may be seen as follows:

$$\left(\frac{\partial^2 f}{\partial x_1 \, \partial x_2}, \varphi\right) = \left(\frac{df}{\partial x_2}, -\frac{\partial \varphi}{\partial x_1}\right) = \left(f, \frac{\partial^2 \varphi}{\partial x_2 \, \partial x_1}\right) = \left(f, \frac{\partial^2 \varphi}{\partial x_1 \, \partial x_2}\right)$$
$$= \left(\frac{\partial f}{\partial x_1}, -\frac{\partial \varphi}{\partial x_2}\right) = \left(\frac{\partial^2 f}{\partial x_2 \, \partial x_1}, \varphi\right).$$

Remark. One may define the derivative of a generalized function also as the limit of a certain ratio, a definition which then bears closer analogy to that for the ordinary derivative. We shall do this here, restricting our considerations for simplicity to the case of a single variable.

Recall that every generalized function f can be translated, say through the distance Δx, in accordance with the expression (see Section 1.6)

$$(f(x + \Delta x), \varphi(x)) = (f(x), \varphi(x - \Delta x)). \tag{5}$$

We now show that for any generalized function the limit (in the sense of generalized functions) of the ratio

$$\frac{f(x + \Delta x) - f(x)}{\Delta x} \tag{6}$$

exists as $\Delta x \to 0$, and that this limit is exactly the derivative df/dx defined above. Consider any function $\varphi(x)$ in K. Then applying (5) to (6) we have

$$\left(\frac{f(x + \Delta x) - f(x)}{\Delta x}, \varphi(x) \right) = \left(f(x), \frac{\varphi(x - \Delta x) - \varphi(x)}{\Delta x} \right).$$

Now the expression $[\varphi(x - \Delta x) - \varphi(x)]/\Delta x$ converges to a limit in K as $\Delta x \to 0$, and this limit is $-\varphi'(x)$. Since f is a continuous functional, it follows that

$$\left(\frac{f(x + \Delta x) - f(x)}{\Delta x}, \varphi(x) \right) \to (f(x), -\varphi'(x)) = (f', \varphi),$$

where f' is the previously defined derivative of the generalized function f. Thus the expression $[f(x + \Delta x) - f(x)]/\Delta x$ will indeed converge in K' to the functional $f'(x)$, as asserted.

2.2. Examples for the Case of a Single Variable

We have seen above that the functional f' corresponds to the function $f'(x)$ if both functions $f(x)$ and $f'(x)$ are continuous. It is easily shown that this is true if $f(x)$ is continuous and $f'(x)$ is merely piecewise continuous (even if it fails to exist at a finite number of points). In this case, in fact, Eq. (1) of Section 2.1 remains valid.

An even more general condition for the validity of this equation is that $f(x)$ be absolutely continuous; as is well known this implies that $f'(x)$ exists almost everywhere and is locally summable, and that the formula for integration by parts, namely Eq. (1) of Section 2.1, is valid.

Let us proceed to consider some examples.

Example 1. Consider the function

$$\theta(x) = \begin{cases} 0 & \text{for} \quad x < 0, \\ 1 & \text{for} \quad x > 0. \end{cases}$$

We shall denote the functional corresponding to this function also by $\theta(x)$. According to Eq. (2) of Section 2.1, given any $\varphi(x)$ in K, we have

$$(\theta'(x), \varphi(x)) = (\theta(x), -\varphi'(x)) = -\int_0^\infty \varphi'(x)\, dx = \varphi(0).$$

Then by Section 1.3 it follows that

$$\theta'(x) = \delta(x).$$

Similarly, it is a simple matter to show that

$$\theta'(x - h) = \delta(x - h).$$

Example 2. Now let $f(x)$ be a piecewise continuous function with piecewise continuous derivative $f'(x)$ and discontinuities of the first kind (jumps) h_1, h_2, ... at x_1, x_2, ... (Fig. 1). Let $f'(x)$ be defined everywhere except at these discontinuities, finite in number. Let us obtain the derivative of the functional f corresponding to $f(x)$.

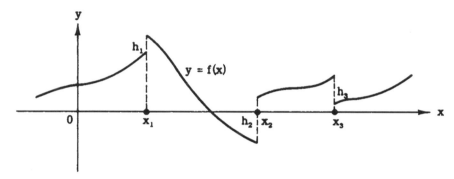

FIG. 1.

Consider the function

$$f_1(x) = f(x) - \sum_k h_k \theta(x - x_k). \tag{1}$$

This function is obviously everywhere continuous and has, except at a finite number of points, a derivative equal to $f'(x)$. The regular functional f_1 corresponding to $f_1(x)$ has, according to what has been said, a derivative which coincides with the regular functional corresponding to $f'(x)$. Hence differentiation of (1) gives

$$f_1' = f' - \sum_k h_k \delta(x - x_k),$$

from which we arrive at

$$f' = f_1' + \sum_k h_k \delta(x - x_k).$$

Thus we obtain the following result. Let $f(x)$ be a piecewise continuous function with piecewise continuous derivative. Then the discontinuity h_k occurring at the point x_k contributes the term $h_k \delta(x - x_k)$ to the derivative.

A somewhat different situation arises when $f(x)$ is an ordinary function whose derivative $f'(x)$ exists in the usual sense except possibly at isolated points, but is not locally summable [for instance, $f(x) = 1/x$ or $f(x) = \ln | x |$]. For this case the generalized function f' is formally given by the integral

$$(f', \varphi) = \int_{-\infty}^{\infty} f'(x)\, \varphi(x)\, dx. \qquad (2)$$

But this integral will in general diverge, and will thus not define a functional.

It will be shown in Appendix 1.4 that the desired functional f' must correspond to the function $f'(x)$ at all points at which $f'(x)$ is locally summable. Thus f' is one of the possible regularizations (see Section 1.7) of the integral appearing in (2), in particular the regularization defined by

$$\int_{-\infty}^{\infty} f'(x)\, \varphi(x)\, dx = (f', \varphi) = -(f, \varphi') = -\int_{-\infty}^{\infty} f(x)\, \varphi'(x)\, dx. \qquad (3)$$

In Section 1.7 we have seen that the regularization f of a function $f(x)$ with nonsummable singularities at isolated points is defined (if it exists) only up to an additive arbitrary functional concentrated on these points. Equation (3), however, defines a particular preferred regularization. It can be shown that this one is in a certain sense the natural regularization (see Sections 3.1 and 3.7).

It would be convenient if we could rewrite (3) in a form containing $\varphi(x)$ rather than its derivative. It is often possible to obtain this simplification by using the specific form of $f(x)$ to simplify the right-hand side of (3).

Example 3. Let us find the derivative of the generalized function

$$x_+^\lambda = \begin{cases} 0 & \text{for} \quad x \leqslant 0, \\ x^\lambda & \text{for} \quad x > 0, \end{cases} \qquad -1 < \lambda < 0.$$

This function is locally summable, but its ordinary derivative $\lambda x^{\lambda-1}$ is not, and we must thus regularize the divergent integral

$$\int_0^{\infty} \lambda x^{\lambda-1} \varphi(x)\, dx. \qquad (4)$$

According to the general rule for differentiating a generalized function, we have

$$((x_+^\lambda)', \varphi) = -(x_+^\lambda, \varphi') = -\int_0^\infty x^\lambda \varphi'(x)\, dx = -\lim_{\epsilon \to 0} \int_\epsilon^\infty x^\lambda \varphi'(x)\, dx.$$

Let us integrate by parts, setting $\varphi'(x)dx = du$, $x^\lambda = v$, $u = \varphi(x) + C$. We obtain

$$((x_+^\lambda)', \varphi) = -\lim_{\epsilon \to 0}\left[x^\lambda[\varphi(x) + C]\Big|_\epsilon^\infty - \int_\epsilon^\infty \lambda x^{\lambda-1}[\varphi(x) + C]\, dx\right].$$

The first term in the expression whose limit we are taking itself has limit zero as $\epsilon \to 0$ if we set C equal to $-\varphi(0)$. Let us choose C accordingly, thus obtaining

$$((x_+^\lambda)', \varphi) = \lim_{\epsilon \to 0} \int_\epsilon^\infty \lambda x^{\lambda-1}[\varphi(x) - \varphi(0)]\, dx$$

$$= \int_0^\infty [\varphi(x) - \varphi(0)]\, \lambda x^{\lambda-1}\, dx, \tag{5}$$

which is then the desired rule for assigning meaning to the integral in (4). In our case, as we see, the rule consists of replacing $\varphi(x)$ by $\varphi(x) - \varphi(0)$ in the integrand, which causes the integral to converge at $x = 0$ (without disturbing its convergence at infinity). The generalized function defined by (5) is conveniently denoted $\lambda x_+^{\lambda-1}$, so that we may write

$$(x_+^\lambda)' = \lambda x_+^{\lambda-1}.$$

Now $\lambda x_+^{\lambda-1}$ is no longer a regular functional. For $x \neq 0$, however, it nevertheless coincides with a regular functional, for according to (5) if $\varphi(x)$ is a function which vanishes at $x = 0$, the functional under discussion behaves like the ordinary function $\lambda x_+^{\lambda-1}$.

Example 4. It is somewhat more difficult to calculate the derivative of the generalized function

$$\ln x_+ = \begin{cases} 0 & \text{for} \quad x < 0, \\ \ln x & \text{for} \quad x > 0. \end{cases}$$

Here formal differentiation

$$((\ln x_+)', \varphi) = \int_0^\infty \frac{\varphi(x)}{x}\, dx \tag{6}$$

leads again to a divergent result. This divergent integral cannot be regularized by replacing $\varphi(x)$ by $\varphi(x) - \varphi(0)$, since this replacement causes the integral to diverge at infinity.

Proceeding, however, as in the preceding example, we obtain

$$((\ln x_+)', \varphi) = - (\ln x_+, \varphi') = - \int_0^\infty \ln x \, \varphi'(x) \, dx$$

$$= - \lim_{\epsilon \to 0} \int_\epsilon^\infty \ln x \, \varphi'(x) \, dx = - \lim_{\epsilon \to 0} \left[\ln x \, \varphi(x) \Big|_\epsilon^\infty - \int_\epsilon^\infty \frac{\varphi(x)}{x} \, dx \right]$$

$$= - \lim_{\epsilon \to 0} \left[- \varphi(\epsilon) \ln \epsilon - \int_\epsilon^\infty \frac{\varphi(x)}{x} \, dx \right].$$

Now we may replace $-\varphi(\epsilon) \ln \epsilon$ by $-\varphi(0) \ln \epsilon$, since the limit of the difference, namely the limit of $[\varphi(\epsilon) - \varphi(0)] \ln \epsilon$ is zero. Further,

$$-\varphi(0) \ln \epsilon = \int_\epsilon^1 \frac{\varphi(0)}{x} \, dx = \int_\epsilon^\infty \frac{\varphi(0) \, \theta(1 - x)}{x} \, dx,$$

where $\theta(x)$ is defined in Example 1. The result may thus be written in the form

$$((\ln x_+)', \varphi) = \lim_{\epsilon \to 0} \int_\epsilon^\infty \frac{\varphi(x) - \varphi(0) \, \theta(1 - x)}{x} \, dx$$

$$= \int_0^\infty \frac{\varphi(x) - \varphi(0) \, \theta(1 - x)}{x} \, dx,$$

which is then the desired rule for regularizing the integral of (6). We see that this rule leads to convergence of the integral at $x = 0$ without destroying the convergence at infinity.

As in the preceding example, the functional obtained is not regular throughout its region of definition. It nevertheless coincides with the ordinary functional $1/x$ for $x > 0$.

Example 5. Let us now find the derivative of the generalized function $\ln |x|$. To do this we must regularize the integral $\int_{-\infty}^\infty [\varphi(x)/x] \, dx$.

This problem could be solved by combining the rules we have already obtained for the derivatives of $\ln x_+$ and $\ln (-x)_+$. It is simpler, however, to proceed directly in the following way:

$$\left(\frac{d \ln |x|}{dx}, \varphi(x) \right) = (\ln |x|, -\varphi'(x))$$

$$= - \int_{-\infty}^\infty \ln |x| \, \varphi'(x) \, dx = - \lim_{\epsilon \to 0} \int_{|x| > \epsilon} \ln |x| \, \varphi'(x) \, dx$$

$$= - \lim_{\epsilon \to 0} \left\{ \ln |x| \, \varphi(x) \Big|_{-\infty}^{-\epsilon} + \ln |x| \, \varphi(x) \Big|_\epsilon^\infty - \int_{|x| > \epsilon} \frac{\varphi(x)}{x} \, dx \right\}$$

$$= \lim_{\epsilon \to 0} \int_{|x| > \epsilon} \frac{\varphi(x)}{x} \, dx.$$

What we have obtained is the well-known Cauchy principal value of the integral of $\varphi(x)/x$. We shall denote this merely by $\int_{-\infty}^{\infty} [\varphi(x)/x] \, dx$, and the corresponding generalized function simply by $1/x$. Thus

$$\frac{d \ln |x|}{dx} = \frac{1}{x}.$$

This newly defined functional $1/x$ is not regular; it nevertheless corresponds to the ordinary function $1/x$ for $x \neq 0$.

This same functional can also be defined otherwise by a formula which coincides in an obvious way with the above result, but contains only a convergent integral, namely

$$\left(\frac{d \ln |x|}{dx}, \varphi \right) = \int_0^\infty \frac{\varphi(x) - \varphi(-x)}{x} \, dx.$$

Example 6. Let us find the derivative of the generalized function $\ln (x + i0)$ defined by

$$\ln (x + i0) = \lim_{y \to +0} \ln (x + iy).$$

Writing $\ln (x + iy)$ in the form $\ln |x + iy| + i \arg (x + iy)$ and passing to the limit, we see that

$$\ln (x + i0) = \ln |x| + i\pi\theta(-x).$$

According to Example 2, $\theta'(x) = \delta(x)$. From the fact that $\theta(x) + \theta(-x) = 1$ one may derive the fact that $\theta'(-x) = -\delta(x)$. Therefore

$$\frac{d}{dx} \ln (x + i0) = \frac{d}{dx} \ln |x| + i\pi \frac{d}{dx} \theta(-x) = \frac{1}{x} - i\pi\delta(x),$$

where $1/x$ is the generalized function defined in Example 5. Compare this result also with that of Example 4 of Section 2.4.

Example 7. Finally, let us obtain the derivative of the delta function. In an obvious way we have

$$(\delta'(x - h), \varphi) = (\delta(x - h), -\varphi') = -\varphi'(h)$$

and in general

$$(\delta^{(k)}(x - h), \varphi) = (-1)^k \varphi^{(k)}(h) \qquad (k = 1, 2, \ldots),$$

so that in the notation introduced in Section 1.3,

$$\int \delta^{(k)}(x - h) \, \varphi(x) \, dx = (-1)^k \varphi^{(k)}(h).$$

2.3. Examples for the Case of Several Variables

We note first that if $f(x)$ is a continuous function with piecewise continuous partial derivatives, the derivatives of the regular functional corresponding to $f(x)$ are regular functionals corresponding to the appropriate derivatives.

Example 1. Let $f(x)$ be a function with continuous derivatives in a region G of the x_1, x_2 plane bounded by a piecewise smooth curve Γ. We shall assume that $f(x)$ vanishes outside of G and that it is discontinuous across Γ. Let us obtain an expression for the functional $\partial f / \partial x_1$. For any φ in K we have, from the general rule,

$$\left(\frac{\partial f}{\partial x_1}, \varphi \right) = \left(f, -\frac{\partial \varphi}{\partial x_1} \right) = -\iint_G f(x_1, x_2) \frac{\partial \varphi\,(x_1, x_2)}{\partial x_1}\, dx_1\, dx_2.$$

Integration over x_1 by parts gives

$$\iint_G f(x_1, x_2) \frac{\partial \varphi\,(x_1, x_2)}{\partial x_1}\, dx_1\, dx_2 = \int_{x_2} \left\{ f(x_1, x_2) \varphi(x_1, x_2) \big|_{x_1^0}^{x_1^1} \right.$$

$$\left. - \int_{x_1^0}^{x_1^1} \frac{\partial f\,(x_1, x_2)}{\partial x_1}\, \varphi(x_1, x_2)\, dx_1 \right\} dx_2$$

$$= \int_\Gamma f(x_1, x_2) \cos\,(n, x_2) \varphi(x_1, x_2)\, d\gamma$$

$$- \iint_G \frac{\partial f\,(x_1, x_2)}{\partial x_1}\, \varphi(x_1, x_2)\, dx_1\, dx_2,$$

where (n, x_2) is the angle between the normal and the x_2 axis at the point (x_1, x_2) on Γ. Thus $\partial f / \partial x_1$ is the sum of the regular functional corresponding to $\partial f(x_1, x_2)/\partial x_1$, and a singular one which arises from the discontinuity in $f(x_1, x_2)$ on crossing Γ. A similar result holds in a space of any dimensionality.

Green's well-known theorem

$$\iint_G f(x_1, x_2)\, \Delta\varphi\,(x_1, x_2)\, dx_1\, dx_2$$

$$= \iint_G \Delta f(x_1, x_2) \varphi(x_1, x_2)\, dx_1\, dx_2 + \int_\Gamma \left(f\frac{\partial \varphi}{\partial n} - \frac{\partial f}{\partial n}\, \varphi \right) d\gamma$$

(here Δ is the Laplacian) may be interpreted in the following way, using the equation $(\Delta f, \varphi) = (f, \Delta\varphi)$. Let f be a generalized function coinciding in a region G with an ordinary locally summable function $f(x_1, x_2)$ and vanishing elsewhere. Then the generalized function Δf is the sum of a regular functional corresponding to the function $\Delta f(x_1, x_2)$ in G and two singular functionals arising from the discontinuities in f and $\partial f / \partial n$ on crossing the boundary of G. A similar result holds in a space of any dimensionality.

In Chapter III, Section 1.4 we shall give an independent proof of this type of relation on the basis of certain generalized functions which we shall call $\delta(P)$ and which are similar to the ordinary delta function.

Example 2.   Consider the Laplacian in three-space applied to the regular functional corresponding to the function $1/r$ (where $r^2 = x_1^2 + x_2^2 + x_3^2$).

We note that $1/r$ is a harmonic function in any region which does not contain the origin, i.e., that $\Delta(1/r)$ vanishes (in the ordinary sense) for all $r \neq 0$. For the case of generalized functions we have

$$\left(\Delta \frac{1}{r}, \varphi\right) = \left(\frac{1}{r}, \Delta\varphi\right) = \int\int\int \frac{\Delta\varphi}{r}\, dv = \lim_{\epsilon \to 0} \int\int_{r \geqslant \epsilon}\int \frac{\Delta\varphi}{r}\, dv.$$

Now let us apply Green's theorem (Example 1) to this integral, choosing G to be the spherical shell $\epsilon \leqslant r \leqslant a$ where a is so large that for $r \leqslant a$ our function $\varphi(x)$ vanishes identically. Then

$$\int\int_{r \geqslant \epsilon}\int \frac{\Delta\varphi}{r}\, dv = \int\int_{r \geqslant \epsilon}\int \varphi \cdot \Delta \frac{1}{r}\, dv - \int\int_{r=\epsilon} \frac{\partial \varphi}{\partial r} \frac{1}{r}\, ds + \int\int_{r=\epsilon} \varphi \frac{\partial}{\partial r} \frac{1}{r}\, ds,$$

where ds is the element of area on the sphere $r = \epsilon$. Now

$$\int\int_{r \geqslant \epsilon}\int \varphi \cdot \Delta \frac{1}{r}\, dv = 0,$$

since outside the ball $r \leqslant \epsilon$ the function $1/r$ is harmonic. As for the other terms,[1]

$$\int\int_{r=\epsilon} \frac{\partial \varphi}{\partial r} \frac{1}{r}\, ds = \frac{1}{\epsilon}\int\int_{r=\epsilon} \frac{\partial\varphi}{\partial r}\, ds = O(\epsilon),$$

$$\int\int_{r=\epsilon} \varphi \frac{\partial}{\partial r} \frac{1}{r}\, ds = -\frac{1}{\epsilon^2}\int\int_{r=\epsilon} \varphi\, ds = -4\pi S_\epsilon(\varphi),$$

where $S_\epsilon(\varphi)$ is the mean value of $\varphi(x)$ on the sphere of radius ϵ. In the limit as $\epsilon \to 0$, of course, $S_\epsilon(\varphi) \to \varphi(0)$, so that

$$\left(\Delta \frac{1}{r}, \varphi\right) = \lim_{\epsilon \to 0} \int\int_{r \geqslant \epsilon}\int \frac{\Delta\varphi}{r}\, dv = -4\pi\varphi(0) = -4\pi(\delta(x), \varphi(x)).$$

[1] Here $y = O(x)$ means that y/x is bounded.

Hence we may write

$$\varDelta \frac{1}{r} = - 4\pi\delta(x).$$ (1)

A similar calculation for dimension $n \geqslant 3$ leads to the result

$$\varDelta \frac{1}{r^{n-2}} = - (n - 2)\varOmega_n\delta(x),$$

where $\varOmega_n$ is the hypersurface area of the unit sphere in n-space. For $n = 2$, however,

$$\varDelta \ln \frac{1}{r} = - 2\pi\delta(x)$$

Later (Chapter III, Section 3.3) we shall give some general rules for differentiating functions whose derivatives are not locally summable. These rules can be used to obtain formulas of the form of Eq. (1) in an automatic way [e.g., $\varDelta(1/r)$ can be calculated as the sum of second derivatives of $1/r$].

2.4. Differentiation as a Continuous Operation

Consider a sequence $f_1, f_2, ..., f_\nu, ...$ of generalized functions which converges to the generalized function f. We assert that the sequence of derivatives $\partial f_\nu/\partial x_j$ converges to $\partial f/\partial x_j$. This is immediately obvious, since for any $\varphi(x)$ in K we have

$$\left(\frac{\partial f_\nu}{\partial x_j}, \varphi\right) = \left(f_\nu, - \frac{\partial \varphi}{\partial x_j}\right) \rightarrow \left(f, - \frac{\partial \varphi}{\partial x_j}\right) = \left(\frac{\partial f}{\partial x_j}, \varphi\right),$$

as asserted.

Similarly, a series $h_1 + h_2 + ... + h_\nu + ...$ of generalized functions which converges to the generalized function g can be differentiated term by term. In other words, one may write

$$h_1' + h_2' + ... + h_\nu' + ... = g'.$$

In classical analysis such theorems do not hold, for the derivatives of a convergent sequence of differentiable functions will not in general converge. Consider, for instance, the sequence $f_\nu(x) = \nu^{-1} \sin \nu x$ on the real axis, which converges uniformly to zero. The derivatives $f_\nu'(x) = \cos \nu x$ of this sequence fail to converge in the classical sense;

in particular, they do not converge to the derivative of the limit. But in the sense of generalized functions, the $f'_\nu(x)$ converge, and furthermore to zero. We know this to be generally true, and in this particular case we may see this from the direct calculation

$$(f'_\nu, \varphi) = \int_{-a}^{a} \cos \nu x \, \varphi(x) \, dx = \frac{1}{\nu} \int_{-a}^{a} \sin \nu x \, \varphi'(x) \, dx \to 0,$$

where $[-a, a]$ is any interval containing the support of $\varphi(x)$. Moreover, the sequences $f''_\nu = -\nu \sin \nu x$, $f'''_\nu = -\nu^2 \cos \nu x$ $(\nu = 1, 2, ...)$, and higher derivatives also converge to zero in the sense of generalized functions.

Example 1. *Fourier series expansion of a periodic function.* Let $f(x)$ be a locally summable periodic function with period 2π, and let its (Fourier coefficients) c_n be defined by the classical formulas

$$c_n = \frac{1}{\pi} \int_{0}^{2\pi} f(x) e^{-inx} \, dx.$$

We then assert that the Fourier series $\sum_{-\infty}^{\infty} c_n e^{inx}$ converges (in the sense of generalized functions) to $f(x)$. Indeed, according to a well-known theorem of analysis[2] the formally integrated series

$$\sum_{-\infty}^{-1} \frac{c_n}{in} e^{inx} + c_0 x + \sum_{1}^{\infty} \frac{c_n}{in} e^{inx}$$

converges uniformly to an absolutely continuous function $F(x)$ whose derivative is equal almost everywhere to $f(x)$. Then the equation

$$F(x) = \sum_{-\infty}^{-1} \frac{c_n}{in} e^{inx} + c_0 x + \sum_{1}^{\infty} \frac{c_n}{in} e^{inx}$$

can be differentiated term by term to yield $f(x) = \sum_{-\infty}^{\infty} c_n e^{inx}$, as asserted.

Example 2. Any series of the form $\sum_{-\infty}^{\infty} a_n e^{inx}$ whose coefficients increase no faster than some power of n as $|n| \to \infty$ converges in the sense of generalized functions. This follows from the fact that the series can be obtained by a sufficient number of term-by-term differentiations of $\sum_{-\infty}^{\infty} [a_n/(in)^k] e^{inx}$, which is known to converge uniformly for sufficiently high k.

[2] A. Zygmund, "Trigonometrical Series," p. 27 (2.621). Chelsea, New York, 1952.

Example 3. It is a well-known fact that the series $\sin x + \frac{1}{2} \sin 2x + \frac{1}{3} \sin 3x + ...$ converges to a function equal to $(\pi - x)/2$ for $0 < x < 2\pi$ and periodic with period 2π for the rest of the real axis, and further that the finite partial sums of this series are uniformly bounded.[3] Thus this series converges also in the sense of generalized functions [see Section 1.8, requirement (a)]. By differentiating this series we obtain the following expressions:

$$\cos x + \cos 2x + ... + \cos nx + ... = -\frac{1}{2} + \pi \sum_{-\infty}^{\infty} \delta(x - 2\pi n),$$

$$\sin x + 2 \sin 2x + ... + n \sin nx + ... = -\pi \sum_{-\infty}^{\infty} \delta'(x - 2\pi n), \qquad (1)$$

$$\cos x + 4 \cos 2x + ... + n^2 \cos nx + ... = -\pi \sum_{-\infty}^{\infty} \delta''(x - 2\pi n),$$

. . . .

If we use Euler's formula to write out the cosine in the first of these equations, we obtain

$$1 + e^{ix} + e^{2ix} + ... + e^{-ix} + e^{-2ix} + ... = 2\pi \sum_{-\infty}^{\infty} \delta(x - 2\pi n). \qquad (2)$$

Now let us apply this to some function $\varphi(x)$ in K, recalling that[4]

$$(e^{inx}, \varphi(x)) = \int_{-\infty}^{\infty} \varphi(x) e^{-inx}\, dx = \psi(-n)$$

is the Fourier transform of $\varphi(x)$ at the point $-n$. This leads to the relation

$$\sum_{-\infty}^{\infty} \psi(n) = 2\pi \sum_{-\infty}^{\infty} \varphi(2\pi n),$$

which is called Poisson's formula.[5] We have proven this only for functions $\varphi(x)$ in K, but by using limiting operations we could extend this result to a wider class of functions, such as for instance those which, together with their first derivative, are absolutely integrable.

---

[3] Fikhtengolʹts (Fichtenholz), "A Course in Differential and Integral Calculus," Vol. III, p. 539 (in Russian). Gostekhizdat, 1949. Henceforth we shall refer to this book simply as Fikhtengolʹts, "Calculus." See also K. Knopp, "Theory and Application of Infinite Series," p. 375. Blackie & Son, London, 1928.

[4] Since e^{inx} is a complex function, we use Eq. (1) of Section 1.9.

[5] See, for instance, E. C. Titchmarsh, "Introduction to the Theory of Fourier Integrals," p. 60. Oxford Univ. Press, London and New York, 1937.

Similarly, the well-known relation[6]

$$\cos x + \frac{\cos 2x}{2} + \frac{\cos 3x}{3} + \dots = - \ln \left| 2 \sin \frac{x}{2} \right| \tag{3}$$

can be differentiated to obtain new trigonometric series. By formal differentiation we obtain

$$\sin x + \sin 2x + \sin 3x + \dots = \frac{1}{2} \cot \frac{x}{2},$$

$$\cos x + 2 \cos 2x + 3 \cos 3x + \dots = \frac{1}{2} \left(\cot \frac{x}{2} \right)' = -\frac{1}{4} \frac{1}{\sin^2 (x/2)}, \tag{4}$$

$$\sin x + 4 \sin 2x + 9 \sin 3x + \dots = \frac{1}{2} \left(\cot \frac{x}{2} \right)'' = \frac{1}{4} \frac{\cos (x/2)}{\sin^3 (x/2)},$$

· · · · · · · · · · · · · · · · · · · ·

However, $\cot (x/2)$ is not locally summable, and the integral

$$\int_{-\infty}^{\infty} \cot \frac{x}{2} \varphi(x) \, dx$$

will in general diverge. In Section 3.7 we shall find the generalized functions corresponding to the ordinary functions on the right-hand side of (4). We shall find that the one corresponding to $-\frac{1}{4}[\sin^2(x/2)]^{-1}$ will be the derivative of that corresponding to $\frac{1}{2} \cot (x/2)$, etc. In other words, Eqs. (4) will be given meaning in terms of generalized functions.

Equations (1) and (4) can be used to separate out the singularities of a trigonometric series of the form $\Sigma (a_n \cos nx + b_n \sin nx)$, whose coefficients are given, for instance, by expressions of the form

$$a_n = \alpha_k n^k + \alpha_{k-1} n^{k-1} + \dots + \alpha_0 + \frac{\alpha_{-1}}{n} + \frac{\gamma_n}{n^2},$$

$$b_n = \beta_j n^j + \beta_{j-1} n^{j-1} + \dots + \beta_0 + \frac{\beta_{-1}}{n} + \frac{\delta_n}{n^2},$$

where α_i and β_i are constants, and $\{\gamma_n\}$ and $\{\delta_n\}$ are bounded sequences of numbers.[7]

[6] Fikhtengol'ts, "Calculus," Vol. III, p. 550. See also K. Knopp, "Theory and Application of Infinite Series," p. 378. Blackie & Son, London, 1928.

[7] Cf. V. I. Smirnov, "Lehrgang der höheren Mathematik," p. 416. VEB Deutscher Verlag der Wissenschaften, Berlin, 1955. Henceforth we shall refer to this work as Smirnov, "Higher Math."

Example 4. Consider the functional f defined by

$$\ln (x + i0) = \begin{cases} \ln |x| + i\pi & \text{for } x < 0, \\ \ln x & \text{for } x > 0. \end{cases}$$

It is seen from the definition that this is the limit as $y \to 0$ of the function $\ln (x + iy)$, analytic in the upper half-plane. Let us verify that the relation

$$\ln (x + i0) = \lim_{y \to 0} \ln (x + iy)$$

is true also in the sense of convergence of generalized functions. For fixed $y > 0$ we have

$$\ln (x + iy) = \frac{1}{2} \ln (x^2 + y^2) + i \arctan \frac{y}{x};$$

which converges to $\ln (x + i0)$ as $y \to 0$ with the following two properties:

(a) The first term on the right converges, decreasing monotonically to $\ln |x|$.

(b) The second term has modulus bounded by π and converges to

$$h(x) = \begin{cases} i\pi & \text{for } x < 0, \\ 0 & \text{for } x > 0. \end{cases}$$

Thus, according to Section 1.8, $\ln (x + iy) \to \ln (x + i0)$ in the sense of generalized functions.

Now the derivative of $\ln |x|$ is $1/x$, and the derivative of $h(x)$ is $-i\pi\delta(x)$ (Section 2.2, Examples 5 and 2). This leads again to the result of Example 6, Section 2.2, namely,

$$\frac{d}{dx} \ln (x + i0) = \frac{1}{x} - i\pi\delta(x)$$

Since, on the other hand, differentiation is a continuous operation,

$$\frac{d}{dx} \ln (x + i0) = \lim_{y \to +0} \frac{d}{dx} \ln (x + iy) = \lim_{y \to +0} \frac{1}{x + iy}.$$

We thus arrive at the interesting result

$$\lim_{y \to +0} \frac{1}{x + iy} = \frac{1}{x} - i\pi\delta(x), \tag{5}$$

which may be interpreted as follows. Let $\varphi(x)$ be any function in K; then

$$\lim_{y \to +0} \int_{-\infty}^{\infty} \frac{\varphi(x)}{x + iy}\, dx = \int_{-\infty}^{\infty} \frac{\varphi(x)}{x}\, dx - i\pi\varphi(0),$$

where the integral on the right is understood as the Cauchy principal value (see Section 2.2, Example 5).

2.5. Delta-Convergent Sequences

There are many ways to construct a sequence of regular functions which converge to the δ function. All that is needed is that the corresponding ordinary functions $f_\nu(x)$ form what we shall call a *delta-convergent sequence*, which means that they must possess the following two properties.

(a) For any $M > 0$ and for $|\,a\,| \leqslant M$ and $|\,b\,| \leqslant M$, the quantities

$$\left| \int_a^b f_\nu(\xi)\, d\xi \right|$$

must be bounded by a constant independent of a, b, or ν (in other words, depending only on M).

(b) For any fixed nonvanishing a and b, we must have

$$\lim_{\nu \to \infty} \int_a^b f_\nu(\xi)\, d\xi = \begin{cases} 0 & \text{for } a < b < 0 \text{ and } 0 < a < b, \\ 1 & \text{for } a < 0 < b. \end{cases}$$

Let $f_\nu(x)$ be such a delta-convergent sequence. Consider also the sequence of primitive functions

$$F_\nu(x) = \int_{-1}^{x} f_\nu(\xi)\, d\xi.$$

It follows simply from the two properties of a delta-convergent sequence that as ν is allowed to increase the $F_\nu(x)$ converge to zero for $x < 0$, and to one for $x > 0$. Moreover, these functions are uniformly bounded (in ν) in every interval. This implies that the $F_\nu(x)$ converge in the sense of generalized functions to $\theta(x)$, which is equal to 0 for $x < 0$ and to 1 for $x > 0$. Then in the sense of generalized functions the sequence $f_\nu(x) = F_\nu'(x)$ converges to $\theta'(x) = \delta(x)$, as asserted.

Example 1.   Consider the function

$$f_\epsilon(x) = \frac{1}{\pi} \frac{\epsilon}{x^2 + \epsilon^2} \qquad (\epsilon > 0).$$

Figure 2 is a graph of this function for two values of ϵ. Now from the integral

$$\int_a^b f_\epsilon(x)\,dx = \frac{1}{\pi}\left[\arctan\frac{b}{\epsilon} - \arctan\frac{a}{\epsilon}\right]$$

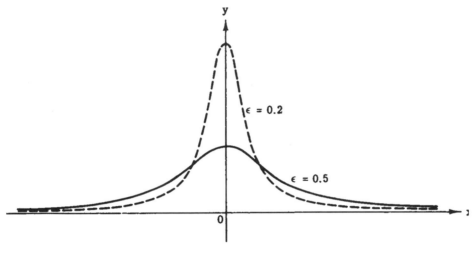

FIG. 2.

for $a \neq b$ it is easily verified that the $f_\epsilon(x)$ fulfill the two requirements of a delta-convergent sequence. Thus as $\epsilon \to 0$,

$$\frac{1}{\pi}\frac{\epsilon}{x^2 + \epsilon^2} \to \delta(x). \tag{1}$$

The same result is obtained by recalling that

$$\frac{1}{\pi}\frac{\epsilon}{x^2 + \epsilon^2} = -\frac{1}{\pi}\,\mathrm{Im}\,\frac{1}{x + i\epsilon};$$

so that from Eq. (5) of Section 2.4, we obtain

$$\frac{1}{\pi}\frac{\epsilon}{x^2 + \epsilon^2} = -\frac{1}{\pi}\,\mathrm{Im}\,\frac{1}{x + i\epsilon} \to \delta(x),$$

as above.

The derivatives with respect to x of a delta-convergent sequence give sequences which converge to the derivatives of the delta function. For instance, Eq. (1) implies that

$$-\frac{2}{\pi}\frac{\epsilon x}{(x^2+\epsilon^2)^2} \to \delta'(x) \tag{2}$$

and similar results. Figure 3 is a graph of the function on the left-hand side of Eq. (2) for two values of ϵ.

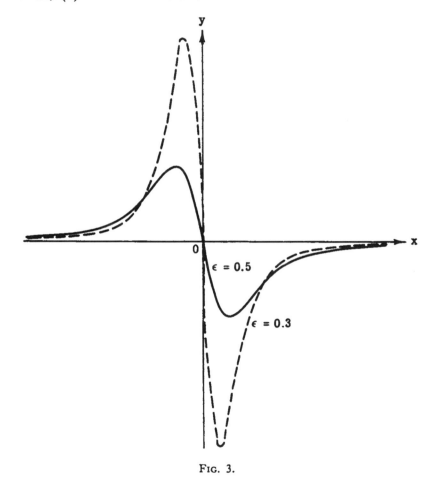

FIG. 3.

Example 2.　Consider the function

$$f_t(x) = \frac{1}{2\sqrt{\pi t}} \exp\left(-\frac{x^2}{4t}\right) \qquad (t > 0).$$

We shall show that as $t \to 0$ this function converges to the δ function.

We remark first that $f_t(x) > 0$, so that[8] for any a and b

$$\int_a^b f_t(x)dx \leqslant \frac{1}{2\sqrt{\pi t}} \int_{-\infty}^{\infty} \exp\left(-\frac{x^2}{4t}\right) dx = 1.$$

We now make the substitution of variables $x/\sqrt{t} = y$; then we see that for $a < 0 < b$,

$$\lim_{t \to 0} \int_a^b f_t(x)\, dx = \lim_{t \to 0} \frac{1}{2\sqrt{\pi}} \int_{a/\sqrt{t}}^{b/\sqrt{t}} \exp\left(-\frac{y^2}{4t}\right) dy = 1.$$

We note further that for any $b > 0$ the integrals

$$\frac{1}{2\sqrt{\pi t}} \int_b^{\infty} \exp\left(-\frac{x^2}{4t}\right) dx < \frac{1}{2\sqrt{\pi t}} \int_b^{\infty} \exp\left(-\frac{x^2}{4t}\right) \frac{x}{2t} \frac{2t}{b} dx = \frac{\sqrt{t}}{b\sqrt{\pi}} \exp\left(-\frac{b^2}{4t}\right)$$

converge to zero as $t \to 0$. A similar result holds for the integral over any segment $(-\infty, a)$ with $a < 0$. Thus the $f_t(x)$ are a delta-convergent sequence, so that

$$\frac{1}{2\sqrt{\pi t}} \exp\left(-\frac{x^2}{4t}\right) \to \delta(x). \tag{3}$$

Example 3. Consider the function

$$f_\nu(x) = \frac{1}{\pi} \frac{\sin \nu x}{x} \qquad (0 < \nu < \infty)$$

whose graph is shown in Fig. 4 for two values of ν. We shall show that as $\nu \to \infty$ this function converges to the δ function.

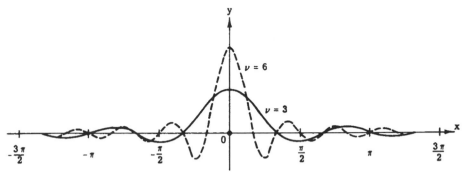

FIG. 4.

[8] Fikhtengolts, "Calculus," Vol. II, p. 624. See also R. Courant, "Differential and Integral Calculus," Vol. I, p. 496. Interscience, New York, 1959.

It is well known that[9]

$$\int_{-\infty}^{\infty} f_\nu(x)\, dx = \frac{1}{\pi} \int_{-\infty}^{\infty} \frac{\sin \nu x}{x}\, dx = 1.$$

Further, for any $b > a > 0$, the integrals

$$\int_a^b f_\nu(x)\, dx = \frac{1}{\pi} \int_a^b \frac{\sin \nu x}{x}\, dx = \frac{1}{\pi} \int_{a\nu}^{b\nu} \frac{\sin y}{y}\, dy$$

converge to zero as $\nu \to \infty$. A similar result will obviously hold true for the integral from a to b for $a < b < 0$. Further,

$$\left| \frac{1}{\pi} \int_a^b \frac{\sin \nu x}{x}\, dx \right| = \left| \frac{1}{\pi} \int_{a\nu}^{b\nu} \frac{\sin y}{y}\, dy \right|$$

is bounded uniformly in a and b for all ν. Hence the $f_\nu(x)$ are a delta-convergent sequence:

$$\lim_{\nu \to \infty} \frac{1}{\pi} \frac{\sin \nu x}{x} = \delta(x). \tag{4}$$

Now $\pi^{-1}[\sin(\nu x)/x]$ can be written as the integral of $(2\pi)^{-1}e^{i\xi x}$ over ξ from $-\nu$ to ν. Then we can formulate the above result in the following way: in the sense of convergence in K'

$$\lim_{\nu \to \infty} \int_{-\nu}^{\nu} e^{i\xi x}\, d\xi = 2\pi\delta(x). \tag{5}$$

The left-hand side of this equation is the *Fourier transform* of unity. (We shall discuss Fourier transforms in more detail from a different point of view in Chapter II. Compare the remark at the end of Section 3 of Chapter II.)

By differentiating this relation with respect to x we obtain the new and even more interesting results

$$\lim_{\nu \to \infty} \int_{-\nu}^{\nu} i\xi e^{i\xi x}\, d\xi = 2\pi\delta'(x), \tag{6}$$

$$\lim_{\nu \to \infty} \int_{-\nu}^{\nu} (i\xi)^2 e^{i\xi x}\, d\xi = 2\pi\delta''(x) \tag{7}$$

and others of higher orders. We may, in fact, assert quite generally that every locally summable function $f(x)$ which, as $|x| \to \infty$, increases

---

[9] Fikhtengol'ts, "Calculus," Vol. II, p. 625. See also R. Courant, "Differential and Integral Calculus," Vol. I, p. 450. Interscience, New York, 1959.

no faster than some power of $|x|$ has a Fourier transform in the sense of generalized functions, and this Fourier transform is given by the limit as $\nu \to \infty$ of

$$\int_{-\nu}^{\nu} f(\xi) e^{i\xi x} \, d\xi.$$

Indeed, we may always write $f(\xi)$ in the form $(\xi^2 + 1)^m f_0(\xi)$ where $f_0(\xi)$ is a summable function and has an ordinary Fourier transform, so that the limit

$$\lim_{\nu \to \infty} \int_{-\nu}^{\nu} f_0(\xi) e^{i\xi x} \, d\xi = G_0(x)$$

exists. Moreover, the convergence in this equation is uniform in x in any bounded region, so that the convergence holds also in K'. We now apply the operator $(-d^2/dx^2 + 1)^m$ to both sides, and recall that differentiation is a continuous operation. From this it follows that the limit

$$\lim_{\nu \to \infty} \int_{-\nu}^{\nu} f(\xi) e^{i\xi x} \, d\xi = \left(-\frac{d^2}{dx^2} + 1\right)^m G_0(x)$$

exists, as asserted.

2.6. Differential Equations for Generalized Functions

The operations that we have established for generalized functions, namely differentiation, multiplication by a function, and addition, can be used to write differential expressions of the form

$$a_0(x) y^{(n)}(x) + a_1(x) y^{(n-1)}(x) + \ldots + a_n(x) y(x) - b(x),$$

where $a_0(x)$, $a_1(x)$, ..., $a_n(x)$ are given infinitely differentiable functions, and $y(x)$ and $b(x)$ are generalized functions. If such an expression is set equal to zero, we obtain an ordinary linear differential equation of nth order for the generalized function $y(x)$. We then wish to obtain the set of all solutions of such an equation.

Let us start by considering the particularly simple equation

$$\frac{dy}{dx} = 0. \tag{1}$$

We shall show that this equation, treated as an equation for a generalized function, has the general solution $y = C$, a constant.

Equation (1) can be written in the form

$$(y', \varphi) = (y, -\varphi') = 0 \tag{2}$$

for any $\varphi(x)$ in K. But this then defines the functional y on the set Φ_0 of those functions of K that can be written as the derivatives of other functions in K. We need now to analyze the way in which this functional y can be extended from Φ_0 to all of K.

It is easily established that $\varphi_0(x)$ in K can be written as the derivative of some other function in K if and only if

$$\int_{-\infty}^{\infty} \varphi_0(x)\, dx = 0. \tag{3}$$

The proof follows. First, assume $\varphi_0(x) = \varphi_1'(x)$. Then it follows that

$$\int_{-\infty}^{\infty} \varphi_0(x)\, dx = \varphi_1(x)\Big|_{-\infty}^{\infty} = 0.$$

Now assume (3) to hold. Then we set

$$\varphi_1(x) = \int_{-\infty}^{x} \varphi_0(\xi)\, d\xi,$$

and we have only to show that $\varphi_1(x)$ is in K. But this is obvious, for the infinite differentiability of $\varphi_0(x)$ implies that of $\varphi_1(x)$, and Eq. (3) implies that it has bounded support.

Now let $\varphi_1(x)$ be some fixed function in K such that

$$\int_{-\infty}^{\infty} \varphi_1(x)\, dx = 1.$$

Then any $\varphi(x)$ in K can be written in the form

$$\varphi(x) = \varphi_1(x) \int_{-\infty}^{\infty} \varphi(x)\, dx + \varphi_0(x),$$

where $\varphi_0(x)$ obviously satisfies Eq. (3). This means that if the desired functional y us defined only on $\varphi_1(x)$, its value on any $\varphi(x)$ in K will be automatically defined by

$$(y, \varphi) = (y, \varphi_1) \int_{-\infty}^{\infty} \varphi(x)\, dx. \tag{4}$$

For instance, let $(y, \varphi_1) = C_1$ be any constant. Then from Eq. (4) we have

$$(y, \varphi) = C_1 \int_{-\infty}^{\infty} \varphi(x)\, dx = \int_{-\infty}^{\infty} C_1 \varphi(x)\, dx,$$

which means that y is the constant generalized function C_1 as asserted. We thus see that the differential equation (1) has no solutions in the class of generalized functions that it does not have also in the classical functions.

Example.   We wish to show that the generalized function f invariant with respect to translation on the line is a constant. For this case we have

$$f(x + \Delta x) - f(x) = 0,$$

whence

$$f'(x) = \lim_{\Delta x \to 0} \frac{f(x + \Delta x) - f(x)}{\Delta x} = 0.$$

But we have seen above that then $f(x) = $ const, as asserted.

It can be shown that any system of homogeneous equations of the form

$$\frac{dy_1}{dx} = a_{11}y_1 + ... + a_{1m}y_m,$$

$$\cdots \cdots \cdots \cdots \cdots \cdots$$ (5)

$$\frac{dy_m}{dx} = a_{m1}y_1 + ... + a_{mm}y_m,$$

where the a_{ij} are infinitely differentiable functions of x, also has no solutions in generalized functions other than the classical ones. A similar situation holds also for a single higher order equation of the form

$$y^{(n)} + a_1(x)y^{(n-1)} + ... + a_n(x)y = 0$$

with infinitely differentiable coefficients, as well as for any system of such equations.

These assertions may be proved by the method sketched below: For convenience we write Eqs. (5) in the vector form[10]

$$\frac{dy}{dx} = Ay, \qquad A = \| a_{ij} \|.$$

Let $U = \| u_j^k(x) \|$ be the matrix formed by a fundamental set of (ordinary) solutions of (5). Then U is nonsingular. Now let us transform from the variable y to z according to the equation $y = Uz$. Inserting this into (5), we have

$$\frac{dU}{dx} z + U \frac{dz}{dx} = AUz$$

[10] See F. R. Gantmakher, "Applications of the Theory of Matrices," Chapter IV. Interscience, New York, 1959.

or, since $dU/dx = AU$,

$$U \frac{dz}{dx} = 0.$$

Multiplying on the left by U^{-1}, we obtain the separated system

$$\frac{dz}{dx} = 0.$$

We have seen that this implies $z = \mathrm{const}$. This means that $y = Uz$ is a linear combination of the solutions that make up U, which thus form a complete set.

The remaining assertions follow from this, since any equation of higher order as well as any system of such equations can be replaced by an equivalent system of first order.

Remark. As opposed to what we have been saying, if the coefficients of the equations have singularities, new solutions may occur in generalized functions, while classical solutions may disappear.

Example 1. Consider the first-order equation

$$x \frac{dy}{dx} = 0.$$

Its solution must be a constant for $x < 0$ and $x > 0$. Thus it has the two linearly independent solutions

$$y_1 = 1, \qquad y_2 = \theta(x).$$

Example 2. The equation

$$- 2x^3 y' = y$$

has only the single solution

$$y = 0$$

in generalized functions. This is because for $x \neq 0$ the generalized solution must coincide with the classical one $y = C \exp(x^{-2})$ with either $C \neq 0$ or $C = 0$. The first of these alternatives is impossible, however, since according to Section 1.7 the integral

$$\int \exp(x^{-2}) \varphi(x)\, dx$$

cannot be regularized.

EXISTENCE OF THE PRIMITIVE FUNCTION

Consider the very simple inhomogeneous equation

$$\frac{dg}{dx} = f, \tag{6}$$

where f is a given generalized function and we wish to find g.

We shall show that for arbitrary f Eq. (6) has a solution in the class of generalized functions. This solution is called, quite naturally, the primitive or the indefinite integral of the generalized function f, and we then write

$$g = \int f \, dx.$$

Equation (6) may be written

$$(g, -\varphi') = (f, \varphi)$$

for every φ in K. But this then means that the functional g is defined for every ψ in K which is the derivative of some other function φ in K; in other words, g is defined on the set Φ_0 defined earlier. We now wish to extend g to all of K. This can be done, for instance, by considering first some $\varphi_1(x)$ in K such that $\int_{-\infty}^{\infty} \varphi_1(x) \, dx = 1$ and writing any φ in K in the form

$$\varphi = \varphi_1 \int_{-\infty}^{\infty} \varphi \, dx + \varphi_0,$$

where φ_0 is in Φ_0. In this way with every φ we associate uniquely its "projection" φ_0 onto Φ_0. We now define

$$(g_0, \varphi) = (g, \varphi_0) \tag{7}$$

It is easily shown that this new functional g_0 is linear and continuous. The general solution of (6) is obtained by adding to the particular solution we have found a general solution of the homogeneous equation, and it follows from the previous results that this latter is $g_1 = C = \text{const.}$

Thus any solution of (6) can be written

$$g = g_0 + C,$$

where g_0 is defined by (7).

The general solution of the inhomogeneous system

$$\frac{dy_i}{dx} + \sum_{j=1}^{m} a_{ij} y_j = f_i \qquad (i = 1, 2, ..., m), \tag{8}$$

where the f_i are generalized functions and the a_{ij} are ordinary infinitely differentiable functions, can be reduced to the solution of an equation of the form of (6).

Indeed, again writing $y = Uz$, where U is the matrix formed by a fundamental set of solutions of the corresponding homogeneous system ($f_i = 0$), we arrive at $U\,dz/dx = f$ or $dz/dx = U^{-1}f$. In this new set of equations the variables are separated: each of the equations is of the form of (6).

Finally, the inhomogeneous equation of higher order

$$y^{(m)} + a_1 y^{(m-1)} + \dots + a_m y = f \qquad (9)$$

(where the a_i are infinitely differentiable functions and f is any generalized function) reduces to a system of the same form as (8) when one writes

$$y_1 = y, \qquad y_2 = \frac{dy_1}{dx}, \dots, y_{m-1} = \frac{dy_{m-2}}{dx}.$$

Thus again the problem can be reduced to solving an equation of the form of (6).

2.7. Differentiation in S

In Section 1.10 we defined the test-function space S. Recall that S is the space of infinitely differentiable functions $\varphi(x)$ satisfying inequalities of the form

$$|\,x^k \varphi^{(q)}(x)\,| \leqslant C_{kq} \qquad (k, q = 0, 1, 2, \dots).$$

We have seen that the space S' of continuous linear functionals on S is a subspace of K', the continuous linear functionals on K.

We wish to show that $f \in S'$ implies that the derivative of f is in S'. For simplicity we consider the case of only a single variable. We note first that $\varphi(x) \in S$ implied that the derivative $\varphi'(x) \in S$; further, $\varphi_\nu \to 0$ in S implies $\varphi'_\nu(x) \to 0$ in S. Thus the functional f' defined by

$$(f', \varphi) = -(f, \varphi')$$

is again a continuous linear functional in S.

But it is clear that when restricted to K this functional coincides with the derivative of f. In other words, the derivative f' of f restricted to K can be extended to S together with f, which is what was asserted. The generalization of this result to higher derivatives and to several variables presents no difficulties.

We have seen before how all regular functionals corresponding to functions increasing no faster than some power of the variable (which we shall call slowly increasing functions) can be extended from K to S. We now see that this is true also of the derivatives of such functionals. We shall see in Volume II (Chapter II, Section 4) that every continuous linear functional on S can be obtained by applying some differential operator to a slowly increasing function.

3. Regularization of Functions with Algebraic Singularities

3.1. Statement of the Problem

Of the functions with nonsummable singularities at isolated points, the most important are those with *algebraic singularities*. These are functions which, as x approaches the singular point x_0, increase no faster than some power of $1/|x - x_0|$. In this section we shall construct the generalized functions corresponding to a rather broad class of such functions.

Recall the definition of regularization, as given in Section 1.7. The regularization of an integral

$$\int f(x)\varphi(x)dx \tag{1}$$

or of a function $f(x)$ with, in general, locally nonsummable singular points was defined as the functional f given by (1) for those $\varphi(x)$ in K which vanish in a neighborhood of the singular points of $f(x)$. It was shown in Section 1.7 that every $f(x)$ with algebraic singularities (but with at most a finite number of them in any finite region) can be regularized. Moreover, this regularization is determined only up to an additive functional concentrated on the singularities of $f(x)$.

From this point of view the content of most of the present section may be summarized as follows. For a large class of functions of a single variable with algebraic singularities we obtain the regularization which is *natural* in the sense that the sum of two ordinary functions corresponds to the sum of their regularizations, the ordinary derivative of a function to the derivative of its regularization, and the product of a function with an infinitely differentiable function $h(x)$ to the product of its regularization with $h(x)$.

We shall start, however, with certain specific particularly important functions, putting off until Section 3.7 the more general considerations and the more complete proofs of the properties we shall be discussing.

As an example of a function with the type of singularity under discussion, consider

$$x_+^{-\frac{3}{2}} = \begin{cases} 0 & \text{for} \quad x \leqslant 0, \\ x^{-\frac{3}{2}} & \text{for} \quad x > 0. \end{cases}$$

The corresponding generalized function has already been constructed in Example 3 of Section 2.2, namely

$$(x_+^{-\frac{3}{2}}, \varphi) = \int_0^\infty \frac{\varphi(x) - \varphi(0)}{x^{\frac{3}{2}}} \, dx. \tag{2}$$

The basis for this construction was the fact that in the sense of generalized functions $-\frac{1}{2}x_+^{-\frac{3}{2}}$ should be the derivative of $x_+^{-\frac{1}{2}}$, that is, of the regular functional corresponding to the ordinary function

$$x_+^{-\frac{1}{2}} = \begin{cases} 0 & \text{for} \quad x \leqslant 0, \\ x^{-\frac{1}{2}} & \text{for} \quad x > 0. \end{cases}$$

We saw from other examples in Section 2 that similar considerations often make it possible to construct a generalized function corresponding to a given ordinary function with an algebraic singularity.

Another way to obtain the result of (2) is by analytic continuation. It is essentially this method that we shall use. Before explaining its underlying principle, we make the following definition. Consider a generalized function f_λ depending on a parameter λ running over some open region Λ in the complex λ plane. Then f_λ is called an *analytic function* of λ in Λ if (f_λ, φ) is an analytic function of λ for all φ in K.

The properties of generalized analytic functions are similar to those of ordinary ones. For instance, if the derivative $df_\lambda/d\lambda$ of f_λ is defined as

$$\lim_{\varDelta\lambda\to 0} \frac{f_{\lambda+\varDelta\lambda} - f_\lambda}{\varDelta\lambda}$$

(in the sense of Section 1.8) it follows that f_λ is analytic with respect to λ in Λ if and only if $df_\lambda/d\lambda$ exists at every point of Λ. Moreover, the analogs of the classical theorems on the Taylor's and Laurent series, on analytic continuation, etc., remain valid. These theorems will be found in Appendix 2 at the end of this chapter.

The analytic continuation method is the following. Let $f_\lambda(x)$ be a locally summable function (of x) when λ is in some region Λ of the complex plane, but not in general otherwise. Further, for $\lambda \in \Lambda$ let (f_λ, φ) be analytic for every $\varphi(x)$ in K, and assume that it can be extended analytically to a wider region Λ_1 independent of $\varphi(x)$. Then with the function $f_{\lambda_0}(x)$ for $\lambda_0 \in \Lambda_1 - \Lambda$ we may associate the functional (f_{λ_0}, φ)

obtained by analytic continuation of (f_λ, φ) out of Λ. In other words we shall write

$$\int f_{\lambda_0}(x)\varphi(x)\,dx = \operatorname*{anal\,cont}_{\lambda \to \lambda_0} \int f_\lambda(x)\varphi(x)\,dx.$$

For instance, to define the generalized function $x_+^{-\frac{3}{2}}$ we shall consider

$$x_+^\lambda = \begin{cases} 0 & \text{for} \quad x \leqslant 0, \\ x^\lambda & \text{for} \quad x > 0. \end{cases}$$

For $\operatorname{Re} \lambda > -1$ this is the regular functional given by

$$(x_+^\lambda, \varphi) = \int_0^\infty x^\lambda \varphi(x)\,dx. \tag{3}$$

Now (3) is a function which is obviously analytic in λ, for its derivative with respect to λ is

$$\int_0^\infty x^\lambda \ln x\, \varphi(x)\,dx.$$

Let us rewrite the right-hand side of (3) in the form

$$\int_0^1 x^\lambda[\varphi(x) - \varphi(0)]\,dx + \int_1^\infty x^\lambda\varphi(x)\,dx + \frac{\varphi(0)}{\lambda + 1}.$$

The first term here is defined for $\operatorname{Re} \lambda > -2$, the second for all λ, and the third for $\lambda \neq -1$. Thus the functional defined in (3) can be analytically continued to $\operatorname{Re} \lambda > -2$, $\lambda \neq -1$. In particular, for $\lambda = -\frac{3}{2}$ we have

$$(x_+^{-\frac{3}{2}}, \varphi(x)) = \int_0^1 x^{-\frac{3}{2}}[\varphi(x) - \varphi(0)]\,dx + \int_1^\infty x^{-\frac{3}{2}}\varphi(x)\,dx - 2\varphi(0). \tag{4}$$

The right-hand side of (4) agrees with that of (2), since in general

$$-\frac{1}{\lambda + 1} = \int_1^\infty x^\lambda\,dx.$$

It is, of course, possible to write the dependence on the parameter in a different way and by analytic continuation to obtain an entirely different result. For instance,

$$x_+^\lambda + \frac{1}{\pi}\frac{(\lambda + \frac{3}{2})}{x^2 + (\lambda + \frac{3}{2})} \to x_+^{-\frac{3}{2}} + \delta(x) \qquad \text{as} \quad \lambda \to -\frac{3}{2}$$

(see Section 2.5).

It should be emphasized that the specific method used to arrive at definitions such as (2) and (4) is actually only of secondary interest; it plays only an auxiliary role. The method is only the *means*, whereas the *end* is the definition itself, and as in (2) and (4) this definition has meaning in and of itself, independent of the method used to arrive at it.

3.2. The Generalized Functions x_+^λ and x_-^λ

Consider the function x_+^λ equal to x^λ for $x > 0$, and to zero for $x \leqslant 0$. We wish to construct and study the generalized function corresponding to it. As has already been described, the regular functional

$$(x_+^\lambda, \varphi) = \int_0^\infty x^\lambda \varphi(x)\, dx, \tag{1}$$

defined by x_+^λ for Re $\lambda > -1$ can be analytically continued to Re $\lambda > -2$ $\lambda \neq -1$ by means of the identity

$$\int_0^\infty x^\lambda \varphi(x)\, dx = \int_0^1 x^\lambda [\varphi(x) - \varphi(0)]\, dx + \int_1^\infty x^\lambda \varphi(x)\, dx + \frac{\varphi(0)}{\lambda + 1} \tag{2}$$

valid for Re $\lambda > -1$. Specifically, for Re $\lambda > -2$, $\lambda \neq -1$, the right-hand side exists and defines a regularization of the integral on the left. In other words, if $-2 < \text{Re } \lambda \leqslant -1$, $\lambda \neq -1$ and if the test function vanishes in a neighborhood of the identity, what remains on the right is $\int_0^\infty x^\lambda \varphi(x)\, dx$.

We may proceed similarly and continue x_+^λ into the region Re $\lambda > -n - 1$, $\lambda \neq -1, -2, ..., -n$ to obtain

$$\int_0^\infty x^\lambda \varphi(x)\, dx = \int_0^1 x^\lambda \Big[\varphi(x) - \varphi(0) - x\varphi'(0) - ... - \frac{x^{n-1}}{(n-1)!} \varphi^{(n-1)}(0)\Big]\, dx$$

$$+ \int_1^\infty x^\lambda \varphi(x)\, dx + \sum_{k=1}^n \frac{\varphi^{(k-1)}(0)}{(k-1)!\,(\lambda + k)}. \tag{3}$$

Here again the right-hand side regularizes the integral on the left. This defines the generalized function x_+^λ for all $\lambda \neq -1, -2, ...$.

In any strip of the form $-n - 1 < \text{Re } \lambda < -n$, Eq. (3) can be written in the simpler form

$$(x_+^\lambda, \varphi) = \int_0^\infty x^\lambda \Big[\varphi(x) - \varphi(0) - x\varphi'(0) - ... - \frac{x^{n-1}}{(n-1)!} \varphi^{(n-1)}(0)\Big]\, dx, \tag{4}$$

as follows from the fact that for $1 \leqslant k \leqslant n$

$$\int_1^\infty x^{\lambda+k-1}\, dx = -\frac{1}{\lambda+k}. \tag{5}$$

Equation (3) shows that when we treat (x_+^λ, φ) as a function of λ, it has simple poles at $\lambda = -1, -2, \ldots$, and that its residue at $\lambda = -k$ is $\varphi^{(k-1)}(0)/(k-1)!$. Since $\varphi^{(k-1)}(0) = (-1)^{(k-1)}(\delta^{(k-1)}(x), \varphi(x))$, we may say that the functional x_+^λ itself has a simple pole at $\lambda = -k$, and that the residue there is

$$\frac{(-1)^{k-1}}{(k-1)!}\, \delta^{(k-1)}(x) \qquad (k = 1, 2, \ldots).$$

Let us now calculate dx_+^λ / dx. For $\operatorname{Re}\lambda > 0$, we have the obvious relation $dx_+^\lambda / dx = \lambda x_+^{\lambda-1}$, or $(x_+^\lambda, \varphi'(x)) = -(\lambda x_+^{\lambda-1}, \varphi(x))$. Since both sides of this last equation can be analytically continued to the entire plane (except for the points $-1, -2, \ldots$), uniqueness of analytic continuation implies that the equation will hold in the entire plane. Thus

$$\frac{dx_+^\lambda}{dx} = \lambda x_+^{\lambda-1} \qquad (\lambda \neq -1, -2, \ldots).$$

For instance, for $-1 < \lambda < 0$ we have

$$\left(\frac{d}{dx}x_+^\lambda, \varphi\right) = (\lambda x_+^{\lambda-1}, \varphi) = \int_0^\infty \lambda x^{\lambda-1}[\varphi(x) - \varphi(0)]\, dx.$$

We have already derived this formula in a different way in Section 2.2.

In Section 4 we shall expand x_+^λ in a Taylor's series in the neighborhood of a regular point, and in a Laurent series in the neighborhood of a pole.

Let us now consider the generalized function corresponding to

$$x_-^\lambda = \begin{cases} |x|^\lambda & \text{for} \quad x < 0, \\ 0 & \text{for} \quad x \geqslant 0. \end{cases}$$

For $\operatorname{Re}\lambda > -1$, this function defines a regular functional given by

$$(x_-^\lambda, \varphi) = \int_{-\infty}^0 |x|^\lambda \varphi(x)\, dx. \tag{6}$$

This functional can be continued to the half-plane $\operatorname{Re}\lambda \leqslant -1$ in the same way as was x_+^λ. It is simplest to replace x by $-x$, writing (x_-^λ, φ) in the form

$$(x_-^\lambda, \varphi(x)) = \int_0^\infty x^\lambda \varphi(-x)\, dx = (x_+^\lambda, \varphi(-x)).$$

With this equation we can immediately transfer all the results obtained for x_+^λ to this new functional x_-^λ merely by replacing $\varphi(x)$ in the appropriate expressions by $\varphi(-x)$. Then where $\varphi^{(j)}(0)$ appears in these expressions, we replace it by $(-1)^j \varphi^{(j)}(0)$. We thus see that the generalized function x_-^λ, like x_+^λ, exists and is analytic in the entire λ plane except for $\lambda = -1, -2, \ldots$ and that at $\lambda = -k$ it has a simple pole with residue $\delta^{(k-1)}(x)/(k-1)!$.

The equation regularizing the integral

$$(x_-^\lambda, \varphi) = \int_{-\infty}^{0} |x|^\lambda \varphi(x) \, dx$$

in the strip $-n - 1 < \operatorname{Re} \lambda < -n$ can be written

$$\int_{-\infty}^{0} |x|^\lambda \varphi(x) \, dx = \int_{0}^{\infty} x^\lambda \varphi(-x) \, dx = \int_{0}^{\infty} x^\lambda \Big[\varphi(-x) - \varphi(0) $$

$$ + x\varphi'(0) - \ldots - \frac{(-1)^{n-1} x^{n-1}}{(n-1)!} \varphi^{(n-1)}(0) \Big] \, dx. \qquad (7)$$

In Section 4 we will give the Taylor's and Laurent series for x^λ.

3.3. Even and Odd Combinations of x_+^λ and x_-^λ

A generalized function f is called *even* if

$$(f(x), \varphi(-x)) = (f(x), \varphi(x)),$$

and *odd* if

$$(f(x), \varphi(-x)) = -(f(x), \varphi(x)).$$

We can use the functions we have been discussing in the preceding section to construct the even and odd generalized functions

$$|x|^\lambda = x_+^\lambda + x_-^\lambda, \qquad (1)$$

$$|x|^\lambda \operatorname{sgn} x = x_+^\lambda - x_-^\lambda. \qquad (2)$$

Let us find the singularities of these new generalized functions. Recall that at $\lambda = -k$ both x_+^λ and x_-^λ have poles, the first with residue

$$\frac{(-1)^{(k-1)}}{(k-1)!} \delta^{(k-1)}(x)$$

and the second with residue $\delta^{(k-1)}(x)/(k-1)!$. Thus $|x|^\lambda$ has poles only at $\lambda = -1, -3, -5, \ldots$. The residue of this function at $\lambda = -2m - 1$ is $2\delta^{(2m)}(x)/(2m)!$. At $\lambda = -2m$ $(m = 1, 2, \ldots)$, the generalized function $|x|^\lambda$ is well defined, and for these points we shall write, in a natural way, x^{-2m} instead of $|x|^{-2m}$.

Similarly, $|x|^\lambda \operatorname{sgn} x$ has poles at $\lambda = -2, -4, \ldots$, and the residue at $\lambda = -2m$ is $-2\delta^{(2m-1)}(x)/(2m - 1)!$. At $\lambda = -2m - 1$ $(m = 1, 2, \ldots)$, this function is well defined and we shall write, in a natural way, x^{-2m-1} rather than $|x|^{-2m-1} \operatorname{sgn} x$. Thus the generalized functions x^{-n} are now defined for all $n = 1, 2, \ldots$.

We wish to give explicit expressions for the generalized functions $|x|^\lambda$ and $|x|^\lambda \operatorname{sgn} x$. We do this by using the regularizations we have obtained for the integrals $\int_0^\infty x^\lambda \varphi(x)\, dx$ and $\int_{-\infty}^0 |x|^\lambda \varphi(x)\, dx$ in Eqs. (4) and (7) of Section 3.2. In the strip given by $-n - 1 < \operatorname{Re}\lambda < -n$,

$$(x_+^\lambda, \varphi) = \int_0^\infty x^\lambda \left[\varphi(x) - \varphi(0) - x\varphi'(0) - \cdots - \frac{x^{n-1}}{(n-1)!}\varphi^{(n-1)}(0)\right] dx,$$

$$(x_-^\lambda, \varphi) = \int_0^\infty x^\lambda \left[\varphi(-x) - \varphi(0) + x\varphi'(0)\right.$$

$$\left. - \cdots - \frac{(-1)^{n-1}x^{n-1}}{(n-1)!}\varphi^{(n-1)}(0)\right] dx.$$

Replacing n by $2m$ and adding and subtracting, we arrive at

$$(|x|^\lambda, \varphi) = \int_0^\infty x^\lambda \left\{\varphi(x) + \varphi(-x)\right.$$

$$\left. - 2\left[\varphi(0) + \frac{x^2}{2!}\varphi''(0) + \cdots + \frac{x^{2m-2}}{(2m-2)!}\varphi^{(2m-2)}(0)\right]\right\} dx; \qquad (3)$$

$$(|x|^\lambda \operatorname{sgn} x, \varphi) = \int_0^\infty x^\lambda \left\{\varphi(x) - \varphi(-x)\right.$$

$$\left. - 2\left[x\varphi'(0) + \frac{x^3}{3!}\varphi'''(0) + \cdots + \frac{x^{2m-1}}{(2m-1)!}\varphi^{(2m-1)}(0)\right]\right\} dx. \qquad (4)$$

The first expression converges for $-2m - 1 < \operatorname{Re}\lambda < -2m + 1$, and the second for $-2m - 2 < \operatorname{Re}\lambda < -2m$. In particular,

$$(x^{-2m}, \varphi) = \int_0^\infty x^{-2m} \left\{\varphi(x) + \varphi(-x)\right.$$

$$\left. - 2\left[\varphi(0) + \frac{x^2}{2!}\varphi''(0) + \cdots + \frac{x^{2m-2}}{(2m-2)!}\varphi^{(2m-2)}(0)\right]\right\} dx, \qquad (5)$$

$$(x^{-2m-1}, \varphi) = \int_0^\infty x^{-2m-1} \left\{\varphi(x) - \varphi(-x)\right.$$

$$\left. - 2\left[x\varphi'(0) + \frac{x^3}{3!}\varphi'''(0) + \cdots + \frac{x^{2m-1}}{(2m-1)!}\varphi^{(2m-1)}(0)\right]\right\} dx, \qquad (6)$$

so that, for instance,

$$(x^{-2}, \varphi) = \int_0^\infty \frac{\varphi(x) + \varphi(-x) - 2\varphi(0)}{x^2} \, dx, \qquad (7)$$

$$(x^{-1}, \varphi) = \int_0^\infty \frac{\varphi(x) - \varphi(-x)}{x} \, dx. \qquad (8)$$

This last expression is the same as that for the Cauchy principal value of the integral of $\varphi(x)/x$, namely

$$\int_{-\infty}^\infty \frac{\varphi(x)}{x} \, dx = \lim_{\epsilon \to 0} \left\{ \int_{-\infty}^{-\epsilon} \frac{\varphi(x)}{x} \, dx + \int_\epsilon^\infty \frac{\varphi(x)}{x} \, dx \right\}.$$

On differentiating $|x|^\lambda$ and $|x|^\lambda \operatorname{sgn} x$, we obtain

$$\frac{d}{dx} |x|^\lambda = \frac{d}{dx} (x_+^\lambda + x_-^\lambda) = \lambda x_+^{\lambda-1} - \lambda x_-^{\lambda-1} = \lambda |x|^{\lambda-1} \operatorname{sgn} x; \qquad (9)$$

$$\frac{d}{dx} |x|^\lambda \operatorname{sgn} x = \frac{d}{dx} (x_+^\lambda - x_-^\lambda) = \lambda x_+^{\lambda-1} + \lambda x_-^{\lambda-1} = \lambda |x|^{\lambda-1}. \qquad (10)$$

In particular, for $\lambda = -n$, we have

$$\frac{d}{dx} x^{-n} = -nx^{-n-1}. \qquad (11)$$

In Section 4 we will give the Taylor and Laurent expansions of $|x|^\lambda$ and $|x|^\lambda \operatorname{sgn} x$.

We note in conclusion that for Re $\lambda > -1$ the functionals x_+^λ, x_-^λ, $|x|^\lambda$, and $|x|^\lambda \operatorname{sgn} x$, being regular functionals corresponding to slowly increasing functions, can be extended onto S (see Section 1.10). Moreover, the functionals obtained by their analytic continuation to other values of λ can also be extended onto S. This follows both from the formulas for the analytic continuations themselves and from the differential formulas $dx_+^\lambda/dx = \lambda x_+^{\lambda-1}$ and Section 2.7, in which we discuss differentiation of functionals on S. Thus the functionals we have been discussing and the formulas obtained for them can be applied not only to functions of bounded support, but to all functions of S [for instance, $\exp(-x^2)$].

Example 1.   The gamma function is defined by the integral

$$\Gamma(\lambda) = \int_0^\infty x^{\lambda-1} e^{-x} \, dx,$$

which converges for Re $\lambda > -1$. This integral may be thought of as representing the application of $x_+^{\lambda-1}$ to the test function equal to e^{-x} for $0 \leqslant x < \infty$ (which is known to be in S). Using the regularization formulas of Section 3.2, we obtain expressions for the gamma function for $\mathrm{Re}\ \lambda \leqslant -1$. First, for $\mathrm{Re}\ \lambda > -n-1$, $\lambda \neq -1, ..., -n$, we have

$$\Gamma(\lambda) = \int_0^1 x^{\lambda-1} \left[e^{-x} - \sum_{k=0}^n (-1)^k \frac{x^k}{k!} \right] dx + \int_1^\infty x^{\lambda-1} e^{-x}\, dx + \sum_{k=0}^n \frac{(-1)^k}{k!(k+\lambda)} .$$

For $-n-1 < \mathrm{Re}\ \lambda < -n$, we have

$$\Gamma(\lambda) = \int_0^\infty x^{\lambda-1} \left[e^{-x} - \sum_{k=0}^n (-1)^k \frac{x^k}{k!} \right] dx.$$

Example 2. Consider the integral

$$\int_0^\infty x^\lambda [e^{-ax} - e^{-bx}]\, dx.$$

Note that this integral converges for $\mathrm{Re}\ \lambda > -2$. It may be thought of as the result of applying the functional x_+^λ to the test function $e^{-ax} - e^{-bx}$. Thus

$$\int_0^\infty x^\lambda [e^{-ax} - e^{-bx}]\, dx = \int_0^\infty x^\lambda e^{-ax}\, dx - \int_0^\infty x^\lambda e^{-bx}\, dx,$$

where each of the terms on the right is the result of applying this functional to the appropriate test function. But for $\mathrm{Re}\ \lambda > -1$, we may make a substitution of variables of the form $ax = \xi$, from which we obtain

$$\int_0^\infty x^\lambda e^{-ax}\, dx = \int_0^\infty \left(\frac{\xi}{a} \right)^\lambda e^{-\xi} \frac{d\xi}{a} = \frac{\Gamma(\lambda+1)}{a^{\lambda+1}},$$

an expression valid for all λ by the uniqueness of analytic continuation. As a result we arrive at

$$\int_0^\infty x^\lambda [e^{-ax} - e^{-bx}]\, dx = \left(\frac{1}{a^{\lambda+1}} - \frac{1}{b^{\lambda+1}} \right) \Gamma(\lambda+1).$$

This formula gives the value of the convergent integral on the left for $\mathrm{Re}\ \lambda > -2$. It is interesting that we have obtained it with the aid of divergent integrals. It would have been possible, of course, to arrive at this result in a more pedestrian way (for instance, by differentiating with respect to a and b).

3.4. Indefinite Integrals of x_+^λ, x_-^λ, $|x|^\lambda$, $|x|^\lambda \operatorname{sgn} x$

Since the indefinite integral is the inverse of the derivative (see Section 2.6), for $\lambda \neq -1, -2, \ldots$ we have

$$\int x_+^\lambda \, dx = \frac{x_+^{\lambda+1}}{\lambda+1} + C_1(\lambda);$$

$$\int x_-^\lambda \, dx = -\frac{x_-^{\lambda+1}}{\lambda+1} + C_2(\lambda);$$

and for $\lambda \neq -1, -3, -5, \ldots$ we have

$$\int |x|^\lambda \, dx = \frac{|x|^{\lambda+1} \operatorname{sgn} x}{\lambda+1} + C_3(\lambda),$$

while for $\lambda \neq -2, -4, -6, \ldots$ (and for $\lambda \neq -1$, since we have no derivative formula for this case) we have

$$\int |x|^\lambda \operatorname{sgn} x \, dx = \frac{|x|^{\lambda+1}}{\lambda+1} + C_4(\lambda). \tag{1}$$

Here $C_1(\lambda), \ldots, C_4(\lambda)$ are functions chosen arbitrarily.

The freedom of choice for these functions can be used to find the indefinite integral of $x^{-1} = |x|^{-1} \operatorname{sgn} x$ by a limiting operation on Eq. (1). At $\lambda = -1$ the first term on the right-hand side of that equation has a pole with residue one. Let us write $C_4(\lambda) = -(\lambda+1)^{-1} + C$ with which we can continue the right-hand side analytically to $\lambda = -1$, using continuity to give the integral meaning in the limit. Thus writing

$$\int |x|^\lambda \operatorname{sgn} x \, dx = \frac{|x|^{\lambda+1} - 1}{\lambda+1} + C,$$

we find that as $\lambda \to -1$

$$\lim_{\lambda \to -1} \frac{|x|^{\lambda+1} - 1}{\lambda+1} = \ln|x|$$

so that

$$\int x^{-1} \, dx = \ln|x| + C. \tag{2}$$

It is possible also to calculate multiple integrals of these functions. Thus, the q-fold integral of $|x|^\lambda$ is

$$\underbrace{\int \cdots \int}_{q} |x|^\lambda \, d^q x = \frac{|x|^{\lambda+q} (\text{sgn } x)^q}{(\lambda+1) \ldots (\lambda+q)} + Q_\lambda(x), \tag{3}$$

where $Q_\lambda(x)$ is any polynomial in x of degree less than q.

The integrand on the left has poles at $\lambda = -1, -3, \ldots$. On the right

$$\frac{|x|^{\lambda+q} (\text{sgn } x)^q}{(\lambda+1) \ldots (\lambda+q)}$$

has, in addition, poles at $\lambda = -2, -4, \ldots$ for $|\lambda| \leqslant q$. These poles can be eliminated in a natural way by special choice of $Q_\lambda(x)$. Since at $\lambda = -2k$ (for $2k < q$) the first term on the right has residue

$$\frac{|x|^{-2k+q} (\text{sgn } x)^q}{(-2k+1)(-2k+2) \ldots (-1) \cdot 1 \cdot 2 \ldots (-2k+q)} = \frac{-x^{q-2k}}{(2k-1)! \, (q-2k)!},$$

we may write

$$Q_\lambda(x) = \sum_{k=1}^{[q/2]} \frac{x^{-2k+q}}{(2k-1)! \, (q-2k)!} \frac{1}{\lambda+2k}$$

[and we can add to $Q_\lambda(x)$ an arbitrary polynomial in x of degree less than q and analytic in λ]. Thus

$$\underbrace{\int \cdots \int}_{q} |x|^\lambda \, d^q x = \frac{|x|^{\lambda+q} (\text{sgn } x)^q}{(\lambda+1) \ldots (\lambda+q)} + \sum_{k=1}^{[q/2]} \frac{x^{-2k+q}}{(2k-1)! \, (q-2k)!} \frac{1}{\lambda+2k}. \tag{4}$$

In particular, the double integral is

$$\iint |x|^\lambda \, d^2 x = \frac{|x|^{\lambda+2}}{(\lambda+1)(\lambda+2)} + \frac{1}{\lambda+2}.$$

3.5. Normalization of x_+^λ, x_-^λ, $|x|^\lambda$, and $|x|^\lambda \text{ sgn } x$

As we have seen, x_+^λ, x_-^λ, $|x|^\lambda$, and $|x|^\lambda \text{ sgn } x$ have simple poles in the λ plane. It is natural to try to eliminate these by dividing each of the generalized functions by an ordinary function of λ with simple poles at the same points. Such a function is easiest to obtain by applying the generalized function under discussion to a fixed $\varphi_0(x)$ in S (that we may work in S is established just before the examples at the end of Section 3.3).

Consider first x_+^λ. This has a pole with residue $(-1)^{n-1}\delta^{(n-1)}(x)/(n-1)!$ at $\lambda = -n$. The function $\varphi_0(x)$ we choose in S must be such that $(x_+^\lambda, \varphi_0(x))$ have nonzero residue at the same points. This means that all the derivatives of $\varphi_0(x)$ at $x = 0$ must be nonzero.

It is convenient to choose e^{-x} for this function, which leads to the normalizing factor whose denominator is

$$(x_+^\lambda, e^{-x}) = \int_0^\infty x^\lambda e^{-x}\, dx = \Gamma(\lambda + 1).$$

Normalizing denominators can be chosen for the remaining three functions in a similar way. For x_-^λ we must again choose $\varphi_0(x)$ so that its derivatives at $x = 0$ fail to vanish. This time it is convenient to choose $\varphi_0(x) = e^x$, which gives

$$(x_-^\lambda, e^x) = \int_{-\infty}^0 |x|^\lambda e^x\, dx = \int_0^\infty x^\lambda e^{-x}\, dx = \Gamma(\lambda + 1).$$

The poles of $|x|^\lambda$ occur at $-1, -3, \ldots$, and the residue at $\lambda = -2m - 1$ is $2\delta^{(2m)}(x)/(2m)!$. Thus $\varphi_0(x)$ should in this case be chosen so that its derivatives of even order fail to vanish at $x = 0$. This time it is convenient to write $\varphi_0(x) = \exp(-x^2)$, so that the normalizing denominator becomes

$$(|x|^\lambda, \exp\{-x^2\}) = \int_{-\infty}^\infty |x|^\lambda \exp(-x^2)\, dx = 2\int_0^\infty x^\lambda \exp(-x^2)\, dx$$

$$= \int_0^\infty t^{\frac{1}{2}(\lambda-1)} e^{-t}\, dt = \Gamma\left(\frac{\lambda + 1}{2}\right).$$

Finally, $|x|^\lambda \operatorname{sgn} x$ has poles at $-2, -4, \ldots$, and its residue at $\lambda = -2m$ is $-2\delta^{(2m-1)}(x)/(2m - 1)!$. Thus in this case $\varphi_0(x)$ should be chosen so that its derivatives of odd order fail to vanish at $x = 0$. We may choose $\varphi_0(x) = x \exp(-x^2)$, with which the normalizing denominator is then given by

$$(|x|^\lambda \operatorname{sgn} x, x \exp\{-x^2\}) = 2\int_0^\infty x^{\lambda+1} \exp(-x^2)\, dx = \Gamma\left(\frac{\lambda + 2}{2}\right).$$

We may thus construct the following entire functions of λ:

$$\frac{x_+^\lambda}{\Gamma(\lambda + 1)}, \quad \frac{x_-^\lambda}{\Gamma(\lambda + 1)}, \quad \frac{|x|^\lambda}{\Gamma\left(\frac{\lambda + 1}{2}\right)}, \quad \frac{|x|^\lambda \operatorname{sgn} x}{\Gamma\left(\frac{\lambda + 2}{2}\right)}.$$

The values of these generalized functions at the singular points of the numerator and denominator can be obtained by taking the ratios of the corresponding residues. On doing this, we arrive at

$$\left.\frac{x_+^\lambda}{\Gamma(\lambda+1)}\right|_{\lambda=-n} = \frac{\operatorname*{res}_{\lambda=-n} x_+^\lambda}{\operatorname*{res}_{\lambda=-n}(x_+^\lambda, e^{-x})}$$

$$= \frac{(-1)^{n-1}\delta^{(n-1)}(x)\,(n-1)!}{(-1)^{n-1}(\delta^{(n-1)}(x), e^{-x})\,(n-1)!} = \delta^{(n-1)}(x); \tag{1}$$

$$\left.\frac{x_-^\lambda}{\Gamma(\lambda+1)}\right|_{\lambda=-n} = \frac{\operatorname*{res}_{\lambda=-n} x_-^\lambda}{\operatorname*{res}_{\lambda=-n}(x_-^\lambda, e^{x})}$$

$$= \frac{\left[\dfrac{\delta^{(n-1)}(x)}{(n-1)!}\right]}{\left[\dfrac{(\delta^{(n-1)}(x), e^{x})}{(n-1)!}\right]} = (-1)^{n-1}\delta^{(n-1)}(x); \tag{2}$$

$$\left.\frac{|x|^\lambda}{\Gamma\!\left(\dfrac{\lambda+1}{2}\right)}\right|_{\lambda=-2m-1} = \frac{\operatorname*{res}_{\lambda=-2m-1} |x|^\lambda}{\operatorname*{res}_{\lambda=-2m-1}(|x|^\lambda, \exp\{-x^2\})} = \frac{2\delta^{(2m)}(x)/(2m)!}{2(\delta^{(2m)}(x), \exp\{-x^2\})/(2m)!}.$$

In the last expression we note that applying $\delta^{(2m)}(x)$ to $\exp(-x^2)$ gives $(\exp\{-x^2\})^{(2m)}|_{x=0}$. But

$$\exp(-x^2) = 1 - x^2 + \frac{x^4}{2!} - \frac{x^6}{3!} + \dots + (-1)^m \frac{x^{2m}}{m!} + \dots$$

$$= \sum_{m=0}^{\infty} (\exp\{-x^2\})^{(2m)}|_{x=0}\, \frac{x^{2m}}{(2m)!}$$

so that

$$(\exp\{-x^2\})^{(2m)}|_{x=0} = \frac{(-1)^m\,(2m)!}{m!}.$$

We may thus write

$$\frac{|x|^\lambda}{\Gamma\!\left(\dfrac{\lambda+1}{2}\right)}_{\lambda=-2m-1} = \frac{(-1)^m\,\delta^{(2m)}(x)m!}{(2m)!}. \tag{3}$$

Finally, consider

$$\left.\frac{|x|^\lambda \operatorname{sgn} x}{\Gamma\!\left(\dfrac{\lambda+2}{2}\right)}\right|_{\lambda=-2m} = \frac{\operatorname*{res}_{\lambda=-2m} |x|^\lambda \operatorname{sgn} x}{\operatorname*{res}_{\lambda=-2m}(|x|^\lambda \operatorname{sgn} x, x\exp\{-x^2\})}$$

$$= \frac{-2\delta^{(2m-1)}(x)/(2m-1)!}{-2(\delta^{(2m-1)}(x), x\exp\{-x^2\})/(2m-1)!}.$$

On applying $\delta^{(2m-1)}(x)$ to $x \exp(-x^2)$ we obtain $-(x \exp\{-x^2\})^{(2m-1)}|_{x=0}$ and since

$$x \exp\{-x^2\} = x - x^3 + \frac{x^5}{2!} - \frac{x^7}{3!} + \cdots + (-1)^{m-1} \frac{x^{2m-1}}{(m-1)!} + \cdots$$

$$= \sum_{m=1}^{\infty} (x \exp\{-x^2\})^{(2m-1)}|_{x=0} \frac{x^{2m-1}}{(2m-1)!},$$

we arrive at

$$(\exp\{-x^2\})^{(2m-1)}|_{x=0} = (-1)^{m-1} \frac{(2m-1)!}{(m-1)!};$$

Thus

$$\left. \frac{|x|^\lambda \operatorname{sgn} x}{\Gamma\left(\frac{\lambda+2}{2}\right)} \right|_{\lambda=-2m} = (-1)^m \frac{\delta^{(2m-1)}(x)\,(m-1)!}{(2m-1)!}. \tag{4}$$

Equations (1)–(4) could have been obtained differently by using the known residues of the gamma functions at the points in question.

For the functional

$$f_+^\lambda = \frac{x_+^\lambda}{\Gamma(\lambda+1)}$$

the derivative formula is simpler than that for x_+^λ. Indeed,

$$\frac{d}{dx} f_+^\lambda = \frac{d}{dx} \frac{x_+^\lambda}{\Gamma(\lambda+1)} = \frac{\lambda x_+^{\lambda-1}}{\Gamma(\lambda+1)} = \frac{x_+^{\lambda-1}}{\Gamma(\lambda)} = f_+^{\lambda-1}; \tag{5}$$

so that differentiation of f_+^λ is equivalent to reducing the index λ by one. Similarly, the derivative of

$$f_-^\lambda = \frac{x_-^\lambda}{\Gamma(\lambda+1)}$$

is given by

$$\frac{d}{dx} f_-^\lambda = -f_-^{\lambda-1} \tag{6}$$

so that taking its derivative is equivalent to reducing the index λ by one and changing its sign.

As for $\dfrac{|x|^\lambda}{\Gamma\left(\frac{\lambda+1}{2}\right)}$, its derivative is $\dfrac{|x|^\lambda \operatorname{sgn} x}{\Gamma\left(\frac{\lambda+2}{2}\right)}$ with index reduced by one and multiplied by a certain numerical factor. Taking its derivative a second time reduces its index by two, namely

$$\frac{d^2}{dx^2} \frac{|x|^\lambda}{\Gamma\left(\frac{\lambda+1}{2}\right)} = 2\lambda \frac{|x|^{\lambda-2}}{\Gamma\left(\frac{\lambda-1}{2}\right)}. \tag{7}$$

In fact the formula

$$\frac{|x|^{\lambda-2}}{\Gamma\left(\frac{\lambda-1}{2}\right)} = \frac{1}{2\lambda} \frac{d^2}{dx^2} \frac{|x|^\lambda}{\Gamma\left(\frac{\lambda+1}{2}\right)} \tag{8}$$

could have been used to obtain the analytic continuation of $\dfrac{|x|^\lambda}{\Gamma\left(\frac{\lambda+1}{2}\right)}$.

3.6. The Generalized Functions $(x + i0)^\lambda$ and $(x - i0)^\lambda$

We shall now define two new generalized functions, namely $(x + i0)^\lambda$ and $(x - i0)^\lambda$ which, unlike those introduced in Sections 3.2 and 3.3, will need no normalization. They will be, in fact, entire analytic functions of λ. Other nice properties of these new functions will also become evident in the chapter on Fourier series (Chapter II).

As is known, the expression $(x + iy)^\lambda$ is defined by

$$(x + iy)^\lambda = \bm{*} \exp\left[\lambda \operatorname{Ln}(x + iy)\right] = \exp\left\{\lambda\left[\ln|x + iy| + i \operatorname{Arg}(x + iy)\right]\right\}.$$

We shall choose

$$\operatorname{Arg}(x + iy) = \arg(x + iy)$$

where $-\pi < \arg z < \pi$. Then $(x + iy)^\lambda$ is a single-valued analytic function of the complex variable $z = x + iy$ in the upper half-plane $(y > 0)$. In exactly the same way $(x + iy)^\lambda$ is a single-valued analytic function in the lower half-plane $(y < 0)$. We shall be interested in the limits of these two functions as we approach the real axis from above and below. These are easily calculated; in fact,

$$(x + i0)^\lambda = \lim_{y\to+0} (x^2 + y^2)^{\lambda/2} \exp\left[i\lambda \arg(x + iy)\right] = \begin{cases} e^{i\lambda\pi} |x|^\lambda & \text{for } x < 0 \\ x^\lambda & \text{for } x > 0; \end{cases} \tag{1}$$

$$(x - i0)^\lambda = \lim_{y\to-0} (x^2 + y^2)^{\lambda/2} \exp\left[i\lambda \arg(x + iy)\right] = \begin{cases} e^{-i\lambda\pi} |x|^\lambda & \text{for } x < 0, \\ x^\lambda & \text{for } x > 0. \end{cases} \tag{2}$$

These functions are defined for all complex λ.

The problem now is to find generalized functions corresponding to these ordinary functions. We shall denote these generalized functions also by $(x + i0)^\lambda$ and $(x - i0)^\lambda$.

In terms of the already familiar functions x_+^λ and x_-^λ, for Re $\lambda > -1$ we may write

$$(x + i0)^\lambda = x_+^\lambda + e^{i\lambda\pi}x_-^\lambda, \tag{3}$$

$$(x - i0)^\lambda = x_+^\lambda + e^{-i\lambda\pi}x_-^\lambda. \tag{4}$$

These equations imply for any $\lambda \neq -1, -2, ...$, that with the (ordinary) functions (1) and (2) we can associate the generalized functions on the right-hand sides of (3) and (4) as defined in Section 3.2. Thus the generalized functions $(x + i0)^\lambda$ and $(x - i0)^\lambda$ are defined for all $\lambda \neq -1, -2, ...$.

If, moreover, we compare the residues of x_+^λ and x_-^λ at $\lambda = -n$, we see immediately that the singularities on the right-hand sides of (3) and (4) cancel out. Thus our new functions are entire functions of λ. In Section 4 we shall obtain the specific expressions

$$(x + i0)^{-n} = x^{-n} - \frac{i\pi(-1)^{n-1}}{(n-1)!}\delta^{(n-1)}(x),$$

$$(x - i0)^{-n} = x^{-n} + \frac{i\pi(-1)^{n-1}}{(n-1)!}\delta^{(n-1)}(x).$$

A clearer understanding of the reasons for the absence of singularities in these functions may be obtained in the following way. Let $\varphi(x)$ in S be continued analytically to complex values of the variable in a neighborhood of the real axis. Consider the integral

$$\int_{L_\epsilon} z^\lambda\varphi(z)\,dz \qquad (z = x + iy)$$

along the contour L_ϵ consisting of the real axis from $-\infty$ to $-\epsilon$ (with $\epsilon > 0$), a semicircle of radius ϵ in the upper half-plane [we shall consider only $(x + i0)^\lambda$] centered at the origin, and the real axis from ϵ to $+\infty$. According to Cauchy's theorem this integral is independent of ϵ. Since the integrand has no singularities on L_ϵ and since $\varphi(x)$ decreases more rapidly than any power of x as $|x| \to \infty$, the integral exists for all λ and, as is easily seen, is an analytic function of λ. For Re $\lambda > -1$ this function is clearly equal to

$$\int_{-\infty}^{\infty} (x + i0)^\lambda\varphi(x)\,dx$$

and therefore the analytic continuation has no singularities.

3.7. Canonical Regularization

In the above discussion we have been finding generalized functions corresponding to certain specific functions with algebraic singularities (and have in the process learned how to calculate many divergent integrals). We shall now consider a particular type of regularization that can be used for functions of a rather large subclass of the class of functions with algebraic singularities. For convenience we shall call this regularization *canonical* and shall temporarily introduce the notation

$$f = \mathrm{CR}\, f(x)$$

[where f is the functional which regularizes the function $f(x)$]. We shall show that this canonical regularization is most natural in the sense that the following two conditions are fulfilled:

(A) $\mathrm{CR}\, [\alpha_1 f_1(x) + \alpha_2 f_2(x)] = \alpha_1\, \mathrm{CR}\, f_1(x) + \alpha_2\, \mathrm{CR}\, f_2(x).$

(B) $\mathrm{CR}\, \left[\dfrac{d}{dx} f(x)\right] = \dfrac{d}{dx}\, [\mathrm{CR}\, f(x)].$

Here d/dx represents the ordinary derivative on the left-hand side and the derivative in the sense of generalized functions on the right-hand side of the equations.

(C) $\mathrm{CR}\, [h(x) f(x)] = h(x) \cdot \mathrm{CR}\, f(x)$

for every infinitely differentiable function $h(x)$.

Let us first consider functions with nonsummable singularities only at the single point $x = 0$.

We shall further restrict our considerations to functions that can be written in the form

$$f(x) = \sum p_i(x) q_i(x), \tag{1}$$

where the $p_i(x)$ are infinitely differentiable functions, and the $q_i(x)$ are chosen from among x_+^λ, x_-^λ, and x^{-n}, in which λ may not take on the values $-1, -2, \ldots$. In Sections 3.2 and 3.3 we obtained the generalized functions corresponding to these functions, and it was shown that the ordinary equations for the derivatives with respect to x remain valid. We note that the operations of addition, multiplication by an infinitely differentiable function, and ordinary differentiation do not take us out of the class defined by (1).

The desired rule for regularizing $f(x)$ is now immediately evident. Since multiplication by an infinitely differentiable function and addition are defined for generalized functions, we need only replace the $q_i(x)$ on the right-hand side of (1) by the corresponding generalized functions.

We shall henceforth consider the canonical regularization of x_+^λ, x_-^λ, and x^{-n} to be given by the definitions of Sections 3.2 and 3.3, and the canonical regularization of $f(x)$ then to be defined by

$$\text{CR} f(x) = \sum p_i(x) \cdot \text{CR} \ q_i(x). \tag{2}$$

Since, in particular, $| x |^0 = 1$, the canonical regularization of an infinitely differentiable function will be simply the regular functional corresponding to it.

Conditions A and C are obviously fulfilled. Let us verify B. We need do this only for a single term, i.e., for

$$f(x) = p(x)q(x).$$

For convenience we shall denote differentiation in the sense of generalized functions by d/dx and ordinary differentiation by a prime. From Eq. (3) of Section 2.1 we have

$$\frac{d}{dx} \text{CR} f(x) = p'(x) \cdot \text{CR} \ q(x) + p(x) \ \frac{d}{dx} \text{CR} \ q(x).$$

But, as already mentioned,

$$\frac{d}{dx} \text{CR} \ q(x) = \text{CR} \ q'(x),$$

so that using conditions A and C we arrive at

$$\frac{d}{dx} \text{CR} \ f(x) = \text{CR} \ [p'(x)q(x) + p(x)q'(x)] = \text{CR} \ f'(x).$$

One may still ask whether our definition of canonical regularization is unique. That it is can be seen from the following. Note first that by expanding the $p_i(x)$ in a Taylor's series we can write $f(x)$ as the finite sum

$$f(x) = \sum c_i r_i(x) + h(x), \tag{1'}$$

where the $r_i(x)$ are again functions chosen from x_+^λ, x_-^λ, and x^{-n} (except that λ will now in general take on new values), and the c_i are constants, while $h(x)$ is a locally summable function. Let us assume that similar terms in $\sum c_i r_i(x)$ are grouped and that only those of the possible $r_i(x)$ occur in it for which Re $\lambda \leqslant -1$, while the rest are all in $h(x)$.

Then (1') is a unique representation of $f(x)$, since the different terms are all of different order as $x \to 0$. Then according to (1') we would write the regularization in the form

$$\sum c_i \cdot \text{CR } r_i(x) + h(x). \tag{2'}$$

We shall prove the uniqueness of our definition of canonical regularization by showing that (2) and (2') are the same.

We may without loss of generality assume that the sum in (1) consists of the single term $p(x)\,q(x)$, since both (2) and (2') satisfy condition A. Let us treat, specifically, the case $q(x) = x_+^\lambda$; the other two can be treated analogously. Let $-m - 2 < \text{Re } \lambda < -m - 1$. Expanding $p(x)$ in a Taylor's series we have

$$p(x) = \sum_{j=0}^{m} \frac{x^j}{j!}\, p^{(j)}(0) + x^{m+1} s(x).$$

We wish to show that for every test function $\varphi(x)$

$$(x_+^\lambda, p(x)\varphi(x)) = \sum_{j=0}^{m} \frac{p^{(j)}(0)}{j!} (x_+^{\lambda+j}, \varphi(x)) + \int_0^\infty x^{\lambda+m+1} s(x)\varphi(x)\, dx. \tag{3}$$

To do this we merely calculate the left-hand side:

$$(x_+^\lambda, p(x)\varphi(x)) = \int_0^\infty x^\lambda \left\{ p(x)\varphi(x) - \sum_{k=0}^{m} \frac{x^k}{k!} [p(x)\varphi(x)]^{(k)} \big|_{x=0} \right\} dx$$

$$= \int_0^\infty x^\lambda \left\{ \varphi(x) \left[\sum_{j=0}^{m} \frac{x^j}{j!} p^{(j)}(0) + x^{m+1} s(x) \right] - \sum_{i=0}^{m} \frac{x^i}{i!} \varphi^{(i)}(0) \cdot \sum_{j=0}^{m} \frac{x^j}{j!} p^{(j)}(0) \right\} dx$$

$$= \sum_{j=0}^{m} \frac{p^{(j)}(0)}{j!} \int_0^\infty x^{\lambda+j} \left[\varphi(x) - \sum_{i=0}^{m} \frac{x^i}{i!} \varphi^{(i)}(0) \right] + \int_0^\infty x^{\lambda+m+1} s(x)\varphi(x)\, dx,$$

which is the right-hand side of (3).

Note that Eq. (2') could have been used as the definition of the canonical regularization. In any specific case, of course, either (2) or (2') may be used.

Now consider the case in which $f(x)$ has not one, but several singularities of the same type as in (1), or even a countable number of such singularities so long as any finite interval contains only a finite number. For this case we will make use of the fact, to be proven in Appendix 1.2, that unity may be partitioned according to

$$1 = \sum_{i=1}^{\infty} e_i(x),$$

where the $e_i(x)$ are infinitely differentiable functions such that every interval on the real line intersects only a finite number of the supports of the $e_i(x)$ and such that each support contains only one singularity of $f(x)$. If we multiply this equation by $f(x)$ we see that we may write

$$f(x) = \sum_{i=1}^{\infty} p_i(x) q_i(x), \tag{4}$$

where the $q_i(x)$ are translations of the same functions x_+^{λ}, x_-^{λ}, and x^{-n} that we have been discussing, and the $p_i(x)$ are infinitely differentiable functions such that on any finite interval Eq. (4) represents a finite sum. Then again we may write

$$\text{CR } f(x) = \sum_{i=1}^{\infty} p_i(x) \cdot \text{CR } q_i(x), \tag{5}$$

or

$$(\text{CR } f(x), \varphi(x)) = \sum_{i=1}^{\infty} (\text{CR } q_i(x), p_i(x)\varphi(x)). \tag{6}$$

In this expression the series will always be finite, since the test functions all have bounded support. The operations of addition, multiplication by an infinitely differentiable function, and differentiation on these functions will again leave them in the class defined by (4). It is true, of course, that Eq. (4) is not a unique representation of $f(x)$, just as (1) was not unique, but it can be shown that the definition in (5) is independent of this representation.

Conditions A, B, and C are shown quite simply to be satisfied.

We shall no longer make use of the notation CR $f(x)$, merely denoting the canonical regularization of a function $f(x)$ [or more properly the generalized function corresponding to the ordinary function $f(x)$] by the same symbol $f(x)$, just as in Sections 3.2 and 3.3 we have done for x_+^{λ} and the other functions we discussed.

In particular, in accordance with the notation of Section 1.3, if $f(x)$ is an ordinary function which has a canonical regularization, we shall understand $\int f(x)\,\varphi(s)\,dx$ to be the result of applying this canonical regularization to $\varphi(x)$.

Example. In Section 2.4 we dealt with the function cot $(x/2)$. For $|x| < \pi/2$ this function can be written in the form

$$\cot \frac{x}{2} = x \frac{\cos (x/2)}{\sin (x/2)} \frac{1}{x} = \frac{p(x)}{x}.$$

where $p(x)$ is an infinitely differentiable function. Moreover $\cot(x/2)$ can be written in the same way in the neighborhood of $k\pi$, for $k = \pm 1, \pm 2, \dots$. We thus see that $\cot(x/2)$ has a canonical regularization. For functions in K with support in a small neighborhood of the origin, this regularization can be written

$$\left(\cot\frac{x}{2}, \varphi(x)\right) = \left(\frac{p(x)}{x}, \varphi(x)\right) = \left(\frac{1}{x}, p(x)\varphi(x)\right)$$

$$= \int_0^\infty \frac{p(x)\varphi(x) - p(-x)\varphi(-x)}{x}\, dx$$

$$= \int_0^\infty [\varphi(x) - \varphi(-x)] \cot\frac{x}{2}\, dx.$$

Because of property B, the derivative of this functional will be the canonical regularization of

$$\frac{d}{dx}\cot\frac{x}{2} = -\frac{1}{2}\frac{1}{\sin^2(x/2)},$$

or, using the definition of x^{-2}, the functional defined by

$$\left(-\frac{1}{2}\frac{1}{\sin^2(x/2)}, \varphi\right) = -\frac{1}{2}\int_0^\infty\left[\frac{\varphi(x) + \varphi(-x)}{\sin^2(x/2)} - \frac{4\varphi(0)}{x^2}\right] dx$$

[again assuming that $\varphi(x) = 0$ outside a small neighborhood of the origin]. Higher derivatives of the generalized function $\cot(x/2)$ can be calculated in a similar way.

3.8 Regularization of Other Integrals

We have been discussing how the theory of generalized functions interprets a (classically divergent) integral of the form

$$\int_{-\infty}^{\infty} f(x)\varphi(x)\, dx,$$

where $f(x)$ is a certain fixed function with relatively general conditions placed on its singularities, and $\varphi(x)$ is an arbitrary test function. In particular, by fixing the test function we have been able to calculate certain specific "divergent" integrals.

We have, however, not considered such simple but important integrals as $\int_a^b x^\lambda\, dx$ and $\int_0^\infty x^\lambda\, dx$ (the first of which diverges for Re $\lambda \leqslant -1$, while the second diverges for all λ), as well as many others. In this section we shall try to give meaning to such integrals.

REGULARIZATION IN A FINITE INTERVAL

We have been discussing the functionals corresponding to the functions x_+^λ, x_-^λ, and x^{-n} on the real line and on the positive real line. We might also consider functionals defined by integrals over a finite interval, for instance,

$$(x_{0 \leqslant x \leqslant b}^\lambda, \varphi) = \int_0^b x^\lambda \varphi(x) \, dx. \tag{1}$$

We have chosen the interval $0 \leqslant x \leqslant b$ as being particularly typical in that it has a single point of possible divergence which is, further, one of the end points ($x = 0$). The interval $a \leqslant x \leqslant b$ without any singular points is of no interest, and an interval such that a singular point lies in its interior can be broken up into two intervals each of which has a singular point at an end.

Consider, therefore, Eq. (1). The integral converges for Re $\lambda > -1$ and is an analytic function of λ. Writing

$$\int_0^b x^\lambda \varphi(x) \, dx = \int_0^b x^\lambda \left[\varphi(x) - \varphi(0) - x\varphi'(0) - \dots - \frac{x^{n-1}}{(n-1)!} \varphi^{(n-1)}(0) \right] dx$$

$$+ \varphi(0) \frac{b^{\lambda+1}}{\lambda+1} + \varphi'(0) \frac{b^{\lambda+2}}{(\lambda+2)} + \dots + \varphi^{(n-1)}(0) \frac{b^{\lambda+n}}{(n-1)!\,(\lambda+n)}, \tag{2}$$

we see that as a function of λ the integral can be analytically continued into the entire λ plane except for the points $\lambda = -1, -2, \dots, -n, \dots$, where it has simple poles. We will use (2) to define the regularization of the integral for $\lambda \neq -1, -2, \dots$. We shall denote the corresponding functional by $x_{0 \leqslant x \leqslant b}^\lambda$.

We can here choose $\varphi(x)$ to be in fact any function that is infinitely differentiable in $0 \leqslant x \leqslant b$, since such a function can always be continued past 0 and b so as to yield a function in K.

Let us take, in particular, $\varphi(x) \equiv 1$ for $0 \leqslant x \leqslant b$. This leads to the equation

$$\int_0^b x^\lambda dx = \frac{b^{\lambda+1}}{\lambda+1}, \tag{3}$$

valid for all $\lambda \neq -1, -2, \dots$. It is needless to point out that (for Re $\lambda \leqslant -1$) the integral on the left-hand side of this equation is here understood not in the ordinary sense, but as the regularization according to our rules.[1]

---

[1] In the present case the right-hand side, and therefore also the left-hand side, has only the one singularity at $\lambda = -1$.

Now consider any $f(x)$ of the form

$$f(x) = \begin{cases} x^\lambda p(x) & \text{for} \quad 0 < x \leqslant b, \\ 0 & \text{for} \quad \text{all other } x \end{cases} \qquad (4)$$

where $p(x)$ is an infinitely differentiable function. In analogy with Section 3.7, we shall consider the regularization of $f(x)$ to be defined by

$$(\text{Reg } f(x), \varphi(x)) = (x^\lambda_{0 \leqslant x \leqslant b}, p(x)\varphi(x)). \qquad (5)$$

The following remark will be of help in what follows. The equation

$$\int_0^b x^\lambda \varphi(x) \, dx = \int_0^c x^\lambda \varphi(x) \, dx + \int_c^b x^\lambda \varphi(x) \, dx \qquad (6)$$

(where $0 < c < b$), valid for $\text{Re } \lambda > -1$, remains valid for all $\lambda \neq -1, -2, \dots$ if the first two integrals are understood in the terms we have just discussed. Indeed, all three integrals can be analytically continued independently into the entire λ plane except for the points $-1, -2, \dots$ (and moreover the third of the integrals exists in the ordinary sense for all λ), so that (6) remains valid in view of the uniqueness of analytic continuation.

Let us now consider a function $f(x)$ with algebraic singularities at a and b, so that in the neighborhood of a we may write

$$f(x) = (x - a)^\lambda p_a(x),$$

and in the neighborhood of b

$$f(x) = (b - x)^\mu p_b(x),$$

where $p_a(x)$ and $p_b(x)$ are infinitely differentiable functions in the intervals $a \leqslant x < b$ and $a < x \leqslant b$, respectively. We now define the regularization of $f(x)$, that is, of the integral

$$\int_a^b f(x)\varphi(x) \, dx,$$

by the equation

$$\int_a^b f(x)\varphi(x) \, dx = \int_a^c f(x)\varphi(x) \, dx + \int_c^b f(x)\varphi(x) \, dx,$$

where c is a point between a and b, and the integrals on the right-hand side stand for the appropriate regularizations.

Note first that the result is independent of the choice of c. Indeed, if $c' > c$ be any other point on $[a, b]$, according to the previous discussion we have

$$\int_a^{c'} f(x)\varphi(x)\, dx = \int_a^c f(x)\varphi(x)\, dx + \int_c^{c'} f(x)\varphi(x)\, dx,$$

$$\int_c^b f(x)\varphi(x)\, dx = \int_c^{c'} f(x)\varphi(x)\, dx + \int_{c'}^b f(x)\varphi(x)\, dx,$$

so that

$$\int_a^{c'} f(x)\varphi(x)\, dx + \int_{c'}^b f(x)\varphi(x)\, dx = \int_a^c + \int_c^{c'} + \int_{c'}^b = \int_a^c + \int_c^b,$$

as stated.

For example, consider the beta function

$$B(\lambda, \mu) = \int_0^1 x^{\lambda-1}(1-x)^{\mu-1}\, dx.$$

For Re $\lambda > 0$ and Re $\mu > 0$ this integral converges in the usual sense, and from what we have proven it can be analytically continued to both the λ and μ planes except for the points $\lambda = 0, -1, -2, \dots$ and $\mu = 0, -1, -2, \dots$. The regularization formula for Re $\lambda > -k$ and Re $\mu > -s$ is

$$B(\lambda, \mu) = \int_0^{1/2} x^{\lambda-1}\left[(1-x)^{\mu-1} - \sum_{r=0}^{k-1}(-1)^r \frac{\Gamma(\mu)x^r}{r!\Gamma(\mu-r)}\right] dx$$

$$+ \int_{1/2}^1 (1-x)^{\mu-1}\left[x^{\lambda-1} - \sum_{r=0}^{s-1}(-1)^r \frac{\Gamma(\lambda)(1-x)^r}{r!\Gamma(\lambda-r)}\right] dx$$

$$+ \sum_{r=0}^{k-1} \frac{(-1)^r\Gamma(\mu)}{2^{r+\lambda}r!\Gamma(\mu-r)(r+\lambda)} + \sum_{r=0}^{s-1} \frac{(-1)^r\Gamma(\lambda)}{2^{r+\mu}r!\Gamma(\lambda-r)(r+\mu)}.$$

Regularization at Infinity

We have now interpreted the integral

$$\int_0^b x^\lambda\, dx$$

for all $\lambda \neq -1$ as the result of applying the generalized function $x_{0 \leqslant x \leqslant b}^\lambda$ to the test function $\varphi(x)$ equal to one on $0 \leqslant x \leqslant b$. We would like to be able to interpret the integral

$$\int_b^\infty x^\lambda\, dx \qquad (b > 0)$$

over the infinite interval in a similar way. But unfortunately there exists no test function equal to one from b to ∞. We shall therefore proceed in the following way. Consider the class $K(b, \infty)$ of all functions $\varphi(x)$ defined and infinitely differentiable for all $x \geqslant b$ and such that the operation of inversion $\varphi(1/x) = \psi(x)$ transforms them into functions in K with support contained in $(0, 1/b)$, or more accurately into a function which on this interval is equal to some function in K. We then define the functional

$$\int_b^\infty x^\lambda \varphi(x)\, dx$$

by making the substitution $1/x = y$ and writing

$$\int_b^\infty x^\lambda \varphi(x)\, dx = \int_0^{1/b} y^{-\lambda-2} \varphi\left(\frac{1}{y}\right) dy,$$

in which the integral, if necessary, is understood in the sense of the regularization discussed above. Accordingly, this integral exists for $-\lambda - 2 \neq -1, -2, \ldots$ or $\lambda \neq -1, 0, +1, \ldots$.

Let $F(x)$ be a function defined on $[b, \infty)$ such that $F(x) = x^\lambda f(x)$, where $f(x)$ is in $K(b, \infty)$. We define the regularization of this function by

$$(\text{Reg } F, \varphi) = \int_b^\infty F(x)\varphi(x)\, dx = \int_b^\infty x^\lambda [f(x)\varphi(x)]\, dx.$$

It is clear that this definition of $\text{Reg } F$ on $K(b, \infty)$ satisfies the linearity conditions

$$\text{Reg } [\alpha_1 F_1 + \alpha_2 F_2] = \alpha_1 \cdot \text{Reg } F_1 + \alpha_2 \cdot \text{Reg } F_2,$$

$$\text{Reg } [g(x)F(x)] = g(x) \cdot \text{Reg } F(x),$$

where $g(x)$ is any infinitely differentiable function.

Let us now consider the most general case, in which $F(x)$ has an arbitrary finite number of algebraic singularities. Let these singularities be ordered according to increasing argument, so that $-\infty < b_1 < \ldots < b_n < \infty$ and consider the finite number of intervals

$$(-\infty, a_1), (a_1, b_1), (b_1, a_2), (a_2, b_2), \ldots, (b_n, a_{n+1}), (a_{n+1}, \infty),$$

each one of which contains an isolated singularity on one of its ends. In each of these intervals we use the regularization formulas developed above, and add them up. As above, it is easily shown that this result is independent of the choice of the a_i.

We may note further that this regularization of $F(x)$ on the entire line satisfies the above linearity conditions.

Example 1. Consider $\int_0^\infty x^\lambda \, dx$.

As described above, for $b > 0$ we have

$$\int_0^b x^\lambda \, dx = \frac{b^{\lambda+1}}{\lambda+1} \qquad (\lambda \neq -1).$$

Similarly, according to the above treatment,

$$\int_b^\infty x^\lambda \, dx = -\frac{b^{\lambda+1}}{\lambda+1} \qquad (\lambda \neq -1).$$

Thus finally

$$\int_0^\infty x^\lambda \, dx = \int_0^b x^\lambda \, dx + \int_b^\infty x^\lambda \, dx = 0 \qquad (\lambda \neq -1).$$

Since this result is an analytic function of λ it is valid also at the single excluded point $\lambda = -1$.

Example 2. The integral

$$\int_0^\infty x^{\lambda-1}(1 + x)^{-\lambda-\mu} \, dx$$

converges in the usual sense for Re $\lambda > 0$ and Re $\mu > 0$. By writing $x/(1 + x) = y$, we see immediately that for these values it is equal to $B(\lambda, \mu)$. This is therefore true also of its analytic continuation, i.e., everywhere except at $\lambda, \mu = -1, -2, \ldots$.

Example 3. Consider

$$I = \int_0^\infty [(x + 1)^\lambda - x^\mu]^n \, dx.$$

To evaluate this integral, expand the integrand by the binomial theorem and use the first property of canonical regularization. This gives

$$I = \sum_{k=0}^n (-1)^k C_n^k \int_0^\infty (x + 1)^{(n-k)\lambda} x^{k\mu} \, dx$$

$$= \sum_{k=0}^n (-1)^k C_n^k B(k\mu + 1, k[\lambda - \mu] - n\lambda - 1),$$

where we have used the preceding result.

Note that I converges in the usual sense for $\lambda = \mu$, $-1/n < \lambda < 1 - 1/n$, which leads to the "classical" result

$$\int_0^\infty [(x + 1)^\lambda - x^\lambda]^n \, dx = \sum_{k=0}^n (-1)^k C_n^k B(k\lambda + 1, -n\lambda - 1)$$

$$(-1/n < \lambda < 1 - 1/n).$$

3.9. The Generalized Function r^λ

Let $r = \sqrt{x_1^2 + \ldots + x_n^2}$ and consider the functional r^λ defined by

$$(r^\lambda, \varphi) = \int_{R_n} r^\lambda \varphi(x)\, dx \qquad (1)$$

for Re $\lambda > -n$. Because the derivative

$$\frac{\partial}{\partial \lambda} (r^\lambda, \varphi) = \int r^\lambda \ln r\, \varphi(x)\, dx$$

exists, the functional r^λ is an analytic function of λ for Re $\lambda > -n$. For Re $\lambda \leqslant -n$, further, r^λ is a locally summable function of the x_i, so that we may define the functional r^λ by analytic continuation. This may be done by generalizing the preceding considerations [an approach we shall use later in application to the wider class of functions of the form $P^\lambda(x)$, where $P(x)$ is a homogeneous positive function; see Chapter III, Section 3.2]. Here we shall do this in a relatively simple way involving a relation between r^λ and x_+^λ.

Let us go over to spherical coordinates in (1), writing it in the form

$$(r^\lambda, \varphi) = \int_0^\infty r^\lambda \left\{ \int_\Omega \varphi(r\omega)\, d\omega \right\} r^{n-1}\, dr,$$

where $d\omega$ is the hypersurface element on the unit sphere. The integral appearing in the integrand can be written in the form

$$\int_\Omega \varphi(r\omega) d\omega = \Omega_n S_\varphi(r),$$

where Ω_n is the hypersurface area of the unit sphere imbedded in Euclidean space of n dimensions, and $S_\varphi(r)$ is the mean value of $\varphi(x)$ on the sphere of radius r. Thus we arrive at

$$(r^\lambda, \varphi) = \Omega_n \int_0^\infty r^{\lambda+n-1} S_\varphi(r) dr. \qquad (2)$$

Let us consider some of the properties of $S_\varphi(r)$. We assert that $S_\varphi(r)$ (defined for $r \geqslant 0$) has bounded support and is infinitely differentiable, and that all of its derivatives of odd order vanish at $r = 0$.

For sufficiently large r, of course, $\varphi(x)$ vanishes, so that its average $S_\varphi(r)$ has bounded support.

That $S_\varphi(r)$ is infinitely differentiable for $r > 0$ is obvious.

To show that all the derivatives of $S_\varphi(r)$ exist at $r = 0$, expand $\varphi(x)$ in a Taylor's series. We then have

$$\Omega_n S_\varphi(r) = \int_\Omega \left[\varphi(0) + \sum \frac{\partial \varphi(0)}{\partial x_j} x_j + \frac{1}{2} \sum \frac{\partial^2 \varphi(0)}{\partial x_i \partial x_j} x_i x_j \right.$$
$$\left. + \frac{1}{3!} \sum \frac{\partial^3 \varphi(0)}{\partial x_i \partial x_j \partial x_k} x_i x_j x_k + ... \right] d\omega.$$

Clearly every term in the integrand (other than the remainder) with an odd number of factors of the x_j fails to contribute to the integral. The sum of terms with $2m$ factors of the x_j contributes an expression of the form $a_m r^{2m}$. We thus arrive at[2]

$$S_\varphi(r) = \varphi(0) + a_1 r^2 + a_2 r^4 + ... + a_n r^{2k} + o(r^{2k}) \tag{3}$$

This expression shows that $S_\varphi(r)$ has all derivatives up to order $2k$ at $r = 0$, and further that the odd derivatives vanish. Since k may be chosen arbitrarily, $S_\varphi(r)$ is infinitely differentiable at $r = 0$, where all of its derivatives of odd order vanish.

All of this implies that $S_\varphi(r)$ is an even function of the single variable r in K. Then the integral of (2) represents the application of $\Omega_n x_+^\mu$ (with $\mu = \lambda + n - 1$) to $S_\varphi(x)$. But we have already established that x_+^μ, analytic for $\mathrm{Re}\,\mu > -1$ (or for $\mathrm{Re}\,\lambda > -n$), can be analytically continued to the entire μ (or λ) plane except for the points $\mu = -1, -2, ...$ ($\lambda = -n, -n + 1, ...$), where it has simple poles. Further, the residue at $\mu = -m$ (or $\lambda = -n - m + 1$) is

$$\frac{((-1)^{m-1} \delta^{(m-1)}(x), S_\varphi(x))}{(m-1)!} = \frac{S_\varphi^{(m-1)}(0)}{(m-1)!}.$$

But since the derivatives of odd order of $S_\varphi(x)$ vanish at $x = 0$, the poles corresponding to even values of m do not in fact exist. This leaves us with the poles corresponding to $m = 1, 3, 5, ...$ or equivalently $\lambda = -n, -n - 2, -n - 4, ...$

Accordingly, the residue of $(r^\lambda, \varphi(x))$ at $\lambda = -n - 2k$ for non-negative integral k is given by

$$\Omega_n \frac{(\delta^{(2k)}(x), S_\varphi(x))}{(2k)!} = \Omega_n \frac{S_\varphi^{(2k)}(0)}{(2k)!}. \tag{4}$$

In particular, at $\lambda = -n$ this funcion has a simple pole whose residue

---

[2] By $y(x) = o(x)$ is meant that y/x converges to zero.

is $\Omega_n S_\varphi(0) = \Omega_n \varphi(0)$. Thus means that at $\lambda = -n$ the generalized function r^λ has a simple pole whose residue is $\Omega_n \delta(x)$.

The $S_\varphi^{(2k)}(0)$ can be written directly in terms of φ, not involving its average.

In order to obtain such an expression we shall derive another expression for the residue of r^λ at $\lambda = -n - 2k$, using the formula

$$\Delta(r^{\lambda+2}) = (\lambda + 2)(\lambda + n)r^\lambda,$$

where Δ is the Laplacian operator. (For Re $\lambda > 0$, this formula can be obtained simply by calculating the left-hand side; for other values of λ it follows from analytic continuation.) By iteration we find for any integer k that

$$r^\lambda = \frac{\Delta^k r^{\lambda+2k}}{(\lambda + 2)(\lambda + 4)\dots(\lambda + 2k)(\lambda + n)(\lambda + n + 2)\dots(\lambda + n + 2k - 2)}.$$

The residue of r^λ for $\lambda = -n - 2k$ can now be calculated by obtaining it for the right-hand side in this expression. Since at $\lambda = -n - 2k$ the denominator does not vanish, we need find only the residue of the numerator. But for any $\varphi(x)$ in K we have

$$(\Delta^k r^{\lambda+2k}, \varphi(x)) = (r^{\lambda+2k}, \Delta^k \varphi(x)),$$

so that we must calculate the residue of $(r^\mu, \Delta^k \varphi(x))$ at $\mu = -n$. We have already performed a similar calculation for any function in K, and found the residue to be the value of the function at $x = 0$ multiplied by Ω_n. Thus in our specific case we obtain $\Omega_n \Delta^k \varphi(0)$. This means that the residue of (r^λ, φ) at $\lambda = -n - 2k$ is

$$\frac{\Omega_n \Delta^k \varphi(0)}{(-2k - n + 2)\dots(-n)(-2k)\dots(-2)} = \frac{\Omega_n(\Delta^k \delta(x), \varphi(x))}{2^k k! \, n(n + 2)\dots(n + 2k - 2)}, \tag{5}$$

and that the residue of the generalized function r^λ for the same value of λ is

$$\frac{\Omega_n \Delta^k \delta(x)}{2^k k! \, n(n + 2)\dots(n + 2k - 2)}. \tag{5'}$$

Comparing (5) with (4), we see that

$$S_\varphi^{(2k)}(0) = \frac{(2k)! \, \Delta^k \varphi(0)}{2^k k! \, n(n + 2)\dots(n + 2k - 2)}. \tag{6}$$

This result can be used to write out the Taylor's series for $S_\varphi(r)$, namely,

$$
\begin{aligned}
S_\varphi(r) &= \varphi(0) + \frac{1}{2!} S''_\varphi(0) r^2 + \dots + \frac{1}{(2k)!} S^{(2k)}_\varphi(0) r^{2k} + \dots \\
&= \sum_{k=0}^{m} \frac{\Delta^k \varphi(0) r^{2k}}{2^k k! \, n(n+2) \dots (n+2k-2)} + \dots
\end{aligned}
\tag{7}
$$

(Pizetti's formula[3]).

Later we shall find it convenient to normalize the generalized function r^λ as we did the powers of x on the line. To do this we divide r^λ by

$$
(r^\lambda, \exp\{-r^2\}) = \int r^\lambda \exp(-r^2) \, r^{n-1} \, d\omega \, dr = \frac{\Omega_n}{2} \Gamma\left(\frac{\lambda+n}{2}\right).
$$

Now

$$
\frac{2 r^\lambda}{\Omega_n \Gamma(\tfrac{1}{2}\lambda + \tfrac{1}{2}n)}
$$

is obviously an entire analytic function of λ. Its values at the singular points of both the numerator and the denominator can be obtained from the ratio of the corresponding residues. The result is

$$
\frac{2}{\Omega_n} \frac{r^\lambda}{\Gamma(\tfrac{1}{2}\lambda + \tfrac{1}{2}n)} \bigg|_{\lambda=-n-2k} = \frac{\operatorname*{res}_{\lambda=-n-2k} r^\lambda}{\operatorname*{res}_{\lambda=-n-2k} (r^\lambda, \exp\{-r^2\})}
$$

$$
= \frac{\dfrac{\Omega_n \Delta^k \delta(x)}{2^k k! \, n(n+2) \dots (n+2k-2)}}{\dfrac{(\Omega_n \delta^{(2k)}(x), \exp\{-x^2\})}{(2k)!}} = (-1)^k \frac{\Delta^k \delta(x)}{2^k n \dots (n+2k-2)}.
\tag{8}
$$

In particular, for $k = 0$ this gives

$$
\frac{2}{\Omega_n} \frac{r^\lambda}{\Gamma(\tfrac{1}{2}\lambda + \tfrac{1}{2}n)} \bigg|_{\lambda=-n} = \delta(x).
\tag{9}
$$

3.10. Plane-Wave Expansion of r^λ

Let $\omega = (\omega_1, \omega_2, \dots, \omega_n)$ be a point on the unit sphere in R_n, and let $\operatorname{Re} \lambda > -1$. Consider the generalized function $F_\lambda(\omega_1 x_1 + \dots + \omega_n x_n)$ defined by

$$
(F_\lambda(\omega_1 x_1 + \dots + \omega_n x_n), \varphi(x)) = \int \frac{|\omega_1 x_1 + \dots + \omega_n x_n|^\lambda}{\Gamma(\tfrac{1}{2}\lambda + \tfrac{1}{2})} \varphi(x) \, dx.
\tag{1}
$$

[3] R. Courant and D. Hilbert, "Methods of Mathematical Physics," Vol. II, p. 287. Interscience, New York, 1962.

First we rotate the axes, writing $y = ux$, so that ω takes on the coordinates $(1, 0, ..., 0)$. Writing $\psi(y) = \varphi(u^{-1}y) = \varphi(x)$, we transform the integral to the form

$$\int \frac{|y_1|^\lambda}{\Gamma(\frac{1}{2}\lambda + \frac{1}{2})} \psi(y) \, dy$$

$$= \int_{y_1=-\infty}^{\infty} \frac{|y_1|^\lambda}{\Gamma(\frac{1}{2}\lambda + \frac{1}{2})} \left\{ \int_{y_2,...,y_n} \psi(y) \, dy_2 \, ... \, dy_n \right\} dy_1. \qquad (2)$$

Let us call the expression in brackets $\varphi_0(y_1)$. This is an infinitely differentiable function with bounded support, and thus belongs to K (in the variable y_1). But then (2) can be analytically continued to the entire λ plane. This will define $F_\lambda(\omega_1 x_1 + ... + \omega_n x_n)$ for all λ.

It is easily seen that $F_\lambda(\omega_1 x_1 + ... + \omega_n x_n)$ depends continuously on ω. Indeed, continuous variation of ω induces continuous variation of the subspace orthogonal to ω, which means that the integral of $\psi(y)$ over this subspace varies continuously, and therefore that $\varphi_0(y_1)$ also varies continuously. Moreover, this variation is continuous in the sense of convergence in K. For each fixed λ, therefore, $(F_\lambda(\omega_1 x_1 + ... + \omega_n x_n), \varphi(x))$ varies continuously, which means that F_λ is a functional depending continuously on the parameter ω (cf. Section 1.8).

We can therefore integrate F_λ in ω over the unit sphere Ω, obtaining a new functional G_λ such that for every $\varphi(x)$ in K

$$(G_\lambda, \varphi(x)) = \int_\Omega (F_\lambda, \varphi(x)) \, d\omega$$

(for a detailed discussion of integration over a parameter, see Appendix 2). We first calculate the integral for Re $\lambda > -1$. In this case, as is easily seen,

$$\int_\Omega |\omega_1 x_1 + ... + \omega_n x_n|^\lambda \, d\omega$$

is a spherically symmetric function of $x_1, x_2, ..., x_n$, homogeneous of degree λ, and equal to r^λ to within a factor. In other words we may write

$$\int_\Omega |\omega_1 x_1 + ... + \omega_n x_n|^\lambda \, d\omega = C(n, \lambda) r^\lambda. \qquad (3)$$

To find $C(n, \lambda)$, we set $x_n = 1$, $x_i = 0$ for $i \neq n$. We then obtain[4]

$$C(n, \lambda) = \int_{\Omega} |\omega_n|^{\lambda} \, d\omega = 2\pi^{\frac{1}{2}(n-1)} \frac{\Gamma(\frac{1}{2}\lambda + \frac{1}{2})}{\Gamma(\frac{1}{2}\lambda + \frac{1}{2}n)}.$$

Inserting this expression for $C(n, \lambda)$ into (3) and dividing by $\pi^{\frac{1}{2}(n-1)} \Gamma(\frac{1}{2}\lambda + \frac{1}{2})$, we arrive at

$$\frac{1}{\pi^{\frac{1}{2}(n-1)} \Gamma(\frac{1}{2}\lambda + \frac{1}{2})} \int_{\Omega} |\omega_1 x_1 + \ldots + \omega_n x_n|^{\lambda} \, d\omega = \frac{2r^{\lambda}}{\Gamma(\frac{1}{2}\lambda + \frac{1}{2}n)}, \qquad (4)$$

which we will find quite important in what follows. Although this formula was obtained for Re $\lambda > -1$, its validity can be extended by analytic continuation to the rest of the λ plane (see Appendix 2). It gives the so-called *plane-wave expansion* of

$$\frac{2r^{\lambda}}{\Gamma(\frac{1}{2}\lambda + \frac{1}{2}n)},$$

which can often be used, as we shall show below, to reduce problems of higher dimension to plane and one-dimensional problems.

Consider this formula for the special case $\lambda = -n$. If n is odd, according to Eq. (3) of Section 3.5 we have

$$\frac{|x|^{\lambda}}{\Gamma(\frac{1}{2}\lambda + \frac{1}{2})}\bigg|_{\lambda=-n} = (-1)^{\frac{1}{2}(n-1)} \frac{(\frac{1}{2}n - \frac{1}{2})!}{(n-1)!} \delta^{(n-1)}(x).$$

On the other hand, according to Eq. (9) of Section 3.9 we have

$$\frac{2r^{\lambda}}{\Gamma(\frac{1}{2}\lambda + \frac{1}{2}n)}\bigg|_{\lambda=-n} = \Omega_n \delta(x_1, x_2, \ldots, x_n),$$

---

[4] To perform the calculation we go over to the spherical coordinates $\theta_1, \theta_2, \ldots, \theta_{n-1}$. Then

$$\omega_n = \cos\theta_{n-1}, \quad \text{and} \quad d\omega = \sin^{n-2}\theta_{n-1} \, d\omega_{n-1},$$

where $d\omega_{n-1}$ is the element of (hypersurface) area on the unit sphere in $(n-1)$ dimensions. The integral becomes

$$2\Omega_{n-1} \int_0^{\pi/2} \sin^{n-2}\theta \cos^{\lambda}\theta \, d\theta = \frac{2\pi^{\frac{1}{2}(n-1)} \Gamma(\frac{1}{2}\lambda + \frac{1}{2})}{\Gamma(\frac{1}{2}\lambda + \frac{1}{2}n)},$$

where we have used

$$\Omega_{n-1} = \frac{2\pi^{\frac{1}{2}(n-1)}}{\Gamma(\frac{1}{2}n - \frac{1}{2})} \quad \text{and} \quad \int_0^{\pi/2} \sin^{n-2}\theta \cos^{\lambda}\theta \, d\theta = \frac{1}{2} B\left(\frac{n-1}{2}, \frac{\lambda+1}{2}\right).$$

3.10 Functions with Algebraic Singularities 77

where Ω_n is the area of the unit sphere. Inserting these expressions into (4) and dividing both sides by Ω_n, we arrive at

$$\delta(x_1, ..., x_n) = \frac{(-1)^{\frac{1}{2}(n-1)}(\frac{1}{2}n - \frac{1}{2})!}{\pi^{\frac{1}{2}(n-1)}(n-1)!\Omega_n} \int_\Omega \delta^{(n-1)}(\omega_1 x_1 + ... + \omega_n x_n)\,d\omega.$$

Now if the dimension is odd, then

$$\Omega_n = \frac{2\pi^{\frac{1}{2}n}}{\Gamma(\frac{1}{2}n)} = \frac{2(2\pi)^{\frac{1}{2}(n-1)}}{1 \cdot 3 \cdot 5 ... (n-2)}.$$

With this expression and some elementary manipulations, we finally obtain

$$\delta(x_1, ..., x_n) = \frac{(-1)^{\frac{1}{2}(n-1)}}{2(2\pi)^{n-1}} \int_\Omega \delta^{(n-1)}(\omega_1 x_1 + ... + \omega_n x_n)\,d\omega. \qquad (5)$$

Let us now derive the similar *plane-wave expansion of the δ function* for a space of even dimension. As for odd dimension, the right-hand side of (4) becomes $\Omega_n \delta(x_1, ..., x_n)$ when $\lambda = -n$. But for even n we have

$$\frac{|x|^\lambda}{\Gamma(\frac{1}{2}\lambda + \frac{1}{2})}\bigg|_{\lambda=-n} = \frac{1}{\Gamma(-\frac{1}{2}n + \frac{1}{2})}x^{-n},$$

where x^{-n} is defined by Eq. (5) of Section 3.3. Inserting these expressions into (4), we find that

$$\delta(x_1, ..., x_n) = \frac{1}{\Gamma(-\frac{1}{2}n + \frac{1}{2})\pi^{\frac{1}{2}(n-1)}\Omega_n} \int_\Omega (\omega_1 x_1 + ... + \omega_n x_n)^{-n}\,d\omega$$

We now again use the fact that

$$\Omega_n = \frac{2\pi^{\frac{1}{2}n}}{\Gamma(\frac{1}{2}n)}$$

to obtain the following expression[5] for even n:

$$\delta(x_1, ..., x_n) = \frac{(-1)^{\frac{1}{2}n}(n-1)!}{(2\pi)^n} \int_\Omega (\omega_1 x_1 + ... + \omega_n x_n)^{-n}\,d\omega \qquad (6)$$

---

[5] The cases of even and odd n can be combined by writing

$$\delta(x_1, ..., x_n) = \frac{(n-1)!}{(2\pi i)^n} \int_\Omega (\omega_1 x_1 + ... + \omega_n x_n - i0)^{-n}\,d\omega,$$

where $(x - i0)^{-n}$ is defined in Section 3.6.

In order to exhibit the meaning of (5), let us apply both sides to some function $\varphi(x_1, x_2, ..., x_n)$. We then have

$$\varphi(0, ..., 0) = \frac{(-1)^{\frac{1}{2}(n-1)}}{2(2\pi)^{n-1}} \int_\Omega d\omega \int_{\Sigma \omega_k x_k = 0} \frac{\partial^{n-1}\varphi(x_1, ..., x_n)}{\partial\nu^{n-1}} d\sigma_0, \qquad (7)$$

where $d\sigma_0$ is the element of area on the hyperplane $\Sigma\omega_k x_k = 0$, and $\partial/\partial\nu$ is the derivative along the normal ω to the plane.

Equation (6) can be handled similarly. Let us write

$$\psi(\xi, \omega) \equiv \psi(\xi) = \int_{\Sigma \omega_k x_k = \xi} \varphi(x_1, ..., x_n) d\sigma_\xi,$$

for any $\varphi(x)$ in K, where $d\sigma_\xi$ is the element of area of the hyperplane $\Sigma\omega_k x_k = \xi$ (the integral depends on the direction of the normal ω). Obviously $\psi(\xi)$ is an infinitely differentiable function with bounded support. Then according to Eq. (5) of Section 3.3

$$(\xi^{-n}, \psi(\xi)) = \int_0^\infty \xi^{-n} \left\{ \psi(\xi) + \psi(-\xi) - 2\left[\psi(0) + \frac{\xi^2}{2!} \psi''(0) + ... \right. \right.$$

$$\left. \left. ... + \frac{\xi^{n-2}}{(n-2)!} \psi^{(n-2)}(0)\right] \right\} d\xi.$$

We can now apply Eq. (6) to a $\varphi(x)$ in K, obtaining

$$\varphi(0, ..., 0) = \frac{(-1)^{\frac{1}{2}n}(n-1)!}{(2\pi)^n}$$

$$\times \int_\Omega ((\omega_1 x_1 + ... + \omega_n x_n)^{-n}, \psi(\omega_1 x_1 + ... + \omega_n x_n)) d\omega \qquad (8)$$

Equations (7) and (8) represent a solution of the so-called *problem of Radon*, namely, to find $\varphi(x)$ [or $\psi(x)$] when its integrals over the hyperplanes $(\omega, x) = C$ are known.

3.11. Homogeneous Functions

We wish now to return to the generalized functions x_+^λ, x_-^λ, etc., but now to study them as examples of homogeneous functions. In Section 1.6 we defined a homogeneous generalized function of degree λ by the equation

$$f(\alpha x) = \alpha^\lambda f(x), \qquad (1)$$

or, equivalently,

$$\left(f(x),\ \varphi\left(\frac{x}{\alpha}\right) \right) = \alpha^{\lambda+n}(f(x),\ \varphi(x))$$

for any $\varphi(x)$ in K and for any positive α.

We wish first to establish some simple properties of generalized homogeneous functions.

1. The sum of two homogeneous functions of degree λ is a homogeneous function of degree λ.

2. The product of a homogeneous generalized function f of degree λ with an infinitely differentiable homogeneous function $a(x)$ of degree μ is a homogeneous generalized function of degree $\lambda + \mu$.

The proofs of these assertions follow immediately from the definition of Eq. (1).

3. The derivative with respect to x_j of a homogeneous generalized function f of degree λ is a homogeneous generalized function of degree $\lambda - 1$.

Indeed, we have

$$\left(\frac{\partial f}{\partial x_j},\ \varphi\left(\frac{x}{\alpha}\right) \right) = -\left(f, \frac{\partial}{\partial x_j}\, \varphi\left(\frac{x}{\alpha}\right) \right) = -\frac{1}{\alpha}\, \alpha^{\lambda+n}\left(f, \frac{\partial \varphi}{\partial x_j} \right) = \alpha^{\lambda-1+n}\left(\frac{\partial f}{\partial x_j}, \varphi \right).$$

4. Homogeneous functions of different degrees are linearly independent. Assume that

$$c_1 f_1(x) + \ldots + c_m f_m(x) = 0,$$

where $f_k(x)$ is a homogeneous generalized function of degree λ_k, and that all the λ_k are different. By definition, for any positive α and any $\varphi(x)$ in K we have

$$c_1 \alpha^{\lambda_1}\, (f_1, \varphi) + c_2 \alpha^{\lambda_2}(f_2, \varphi) + \ldots + c_m \alpha^{\lambda_m}(f_m, \varphi) = 0.$$

The fact that all the λ_k differ by assumption implies that $c_k(f_k, \varphi) = 0$ for all k and all φ. If $f_k \neq 0$, we may choose $\varphi(x)$ such that $(f_k, \varphi) \neq 0$. Hence $c_k = 0$ for all k, which was to be proven.

5. Let f_λ be a homogeneous generalized function of degree λ analytic in λ in some region Λ. Further, let f_λ have an analytic continuation in λ into some larger region $\Lambda_1 \supset \Lambda$. Then in Λ_1 the functional f_λ is a homogeneous function of degree λ.

Indeed, for any fixed λ and any $\varphi(x)$ in K, the equation

$$\left(f_\lambda, \varphi\left(\frac{x}{\alpha}\right)\right) = \alpha^{\lambda+n}(f_\lambda, \varphi)$$

is valid in Λ. In this region each side of this equation is, moreover, an analytic function of λ. Then from the uniqueness of analytic continuation it follows that the equation holds also in Λ_1.

The generalized function x_+^λ that we have treated in previous sections is a homogeneous function of degree λ for Re $\lambda > -1$. We know further that x_+^λ can be analytically continued to the entire λ plane except for the points $-1, -2, \ldots$. But the fifth property we have listed for homogeneous functions implies that x_+^λ is then also a homogeneous function of degree λ for all complex $\lambda \neq -1, -2, \ldots, -n, \ldots$. The same may be said of x_-^λ. Since $|x|^\lambda$, $|x|^\lambda$ sgn x, $(x+i0)^\lambda$, and $(x-i0)$ are linear combinations of x_+^λ and x_-^λ, they are also homogeneous functions of degree λ, each throughout the region of its existence: in particular, $|x|^\lambda$ is homogeneous for $\lambda \neq -1, -3, -5, \ldots$; $|x|^\lambda$ sgn x for $\lambda \neq -2, -4, \ldots$; $(x+i0)^\lambda$ and $(x-i0)^\lambda$ throughout the entire λ plane without exception. For instance, x^{-m} is homogeneous of degree $-m$ for every integer m. There exists, however, another homogeneous function of degree $-m$, namely $\delta^{(m-1)}(x)$. To see this, consider

$$\left(\delta^{(m-1)}(x), \varphi\left(\frac{x}{\alpha}\right)\right) = (-1)^{m-1}\frac{\varphi^{(m-1)}(0)}{\alpha^{m-1}} = \alpha^{-m+1}(\delta^{(m-1)}(x), \varphi(x)).$$

Let us now find all the homogeneous generalized functions of degree λ on the line. By definition such functions satisfy the equation

$$f(\alpha x) = \alpha^\lambda f(x) \tag{2}$$

for every $\alpha > 0$. Differentiating this relation with respect to α and then setting $\alpha = 1$, we obtain

$$xf'(x) = \lambda f(x). \tag{3}$$

Let us now find all solutions of this equation. For $x \neq 0$ it can be integrated in the usual way. This leads to the conclusion that all the desired generalized functions $f(x)$ must coincide with $C_1 x^\lambda$ for $x > 0$ and $C_2 |x|^\lambda$ for $x < 0$. We have already studied such generalized functions, namely, $C_1 x_+^\lambda$ and $C_2 x_-^\lambda$ for $\lambda \neq -n, n = 1, 2, \ldots$. Assuming then, that λ is not a negative integer, consider the generalized function $f_0(x) = f(x) - C_1 x_+^\lambda - C_2 x_-^\lambda$. Each term in this expression, and therefore the expression as a whole, satisfies Eq. (2). Further, $f_0(x)$ vanishes for $x \neq 0$, and is therefore concentrated at a single point.

In Volume II we shall prove the following theorem, which we now accept without proof. Any generalized function f_0 concentrated at a point, say at $x_0 = 0$, is of the form

$$f_0(x) = \sum_{k=0}^{m} c_k \delta^{(k)}(x)$$

with a finite m.

This means that

$$f(x) - C_1 x_+^\lambda - C_2 x_-^\lambda - \sum_{k=0}^{m} c_k \delta^{(k)}(x) = 0.$$

Using the fourth property of homogeneous functions (linear independence for different degree), we see that $c_0 = c_1 = \ldots = c_m = 0$, so that

$$f(x) = C_1 x_+^\lambda + C_2 x_-^\lambda$$

is the general solution of the problem for $\lambda \neq -n$.

Now let $\lambda = -n$ be some negative integer. We shall assume that $f(x)$ is even for even n and odd for odd n. For $x \neq 0$, $f(x)$ must coincide with the ordinary function Cx^{-n}. A generalized homogeneous function with this property is already known, namely Cx^{-n}. As above, $f_0(x) = f(x) - Cx^{-n}$ is concentrated at $x = 0$ and is a linear combination of the δ function and its derivatives. Again applying the fourth property, we find that the general solution of Eq. (2) for $\lambda = -n$ is of the form

$$f(x) = Cx^{-n} + C_1 \delta^{(n-1)}(x).$$

In all cases we obtain two linearly independent homogeneous generalized functions of degree λ.

In n dimensions r^λ is obviously a homogeneous function of degree λ for Re $\lambda > -n$. By the fifth property (analyticity), this function remains homogeneous throughout its domain of analyticity, namely throughout the λ plane except for the points $\lambda = -n, -n - 2, \ldots$. The function $\delta(x_1, \ldots, x_n)$ is homogeneous of degree $-n$, as can be seen directly from its definition or from Eq. (9) of Section 3.9.

The homogeneity of $\delta^{(n-1)}(x)$ and x^{-n} can be used to transform the plane-wave expansion of the δ function [Eqs. (5) and (6) of Section 3.10] to *affine invariant* form. Consider any hypersurface S such that every radius vector from the origin crosses it exactly once. Then in terms of the unit sphere, S can be described by multiplying every radius vector $\omega = (\omega_1, \ldots \omega_n)$ by a positive number $f(\omega)$. Let us write $\rho_k = f(\omega)\,\omega_k$. Because $\delta^{(n-1)}(x)$ is homogeneous, we may write

$$\delta^{(n-1)}(\rho_1 x_1 + \ldots + \rho_n x_n) = [f(\omega)]^{-n} \delta^{(n-1)}(\omega_1 x_1 + \ldots + \omega_n x_n)$$

But $[f(\omega)]^n\, d\omega$ is the element ds of area on S, so that for odd n we have

$$\delta(x_1, \ldots, x_n) = c_n \int_\Omega \delta^{(n-1)}(\omega_1 x_1 + \ldots + \omega_n x_n)d\omega$$

$$= c_n \int_S \delta^{(n-1)}(\rho_1 x_1 + \ldots + \rho_n x_n)ds. \tag{4}$$

Similarly, for even n we have

$$\delta(x_1, \ldots, x_n) = c_n \int_{\Omega^-} (\omega_1 x_1 + \ldots + \omega_n x_n)^{-n}d\omega$$

$$= c_n \int_S (\rho_1 x_1 + \ldots + \rho_n x_n)^{-n}ds. \tag{5}$$

Equations (4) and (5) are in affine invariant form.

4. Associated Functions[1]

4.1. Definition

Homogeneous functions are by definition eigenfunctions of the similarity transformation operator u acting according to

$$u\, f(x) = f(\alpha x).$$

Indeed, if $f(x)$ is a homogeneous function of degree λ, then

$$u\, f(x) = f(\alpha x) = \alpha^\lambda f(x).$$

In addition to an eigenfunction f_0 belonging to a given eigenvalue, a linear transformation will ordinarily also have so-called *associated functions* of various orders. The functions $f_1, f_2, \ldots, f_k, \ldots$ are said to be *associated with the eigenfunction f_0 of the transformation u* if

$$uf_0 = af_0,$$
$$uf_1 = af_1 + bf_0,$$
$$uf_2 = af_2 + bf_1,$$
$$\cdot \quad \cdot \quad \cdot \quad \cdot \quad \cdot \quad \cdot$$
$$uf_k = af_k + bf_{k-1},$$

[1] This section may be omitted on first reading.

or if, in other words, u reproduces the associated function of kth order except for some multiple of the associated function of $(k-1)$st order.

It is easily verified that the sum of an associated function of kth order with one of lower order is again an associated function of kth order. In a finite dimensional space, when the basis is chosen to consist of the eigenvectors and associated vectors of a given linear transformation, the matrix of this transformation is in Jordan normal form. (The associated vectors must then chosen so that $b = 1$.)

Let us return now to our functions and to the similarity transformation operator. We shall say that $f_1(x)$ is an *associated function of first order and of degree* λ if for any $\alpha > 0$

$$f_1(\alpha x) = \alpha^\lambda f_1(x) + h(\alpha) f_0(x),$$

where $f_0(x)$ is a homogeneous function of degree λ. The function $h(\alpha)$ is determined uniquely from the identity

$$h(\alpha\beta) = \alpha^\lambda h(\beta) + \beta^\lambda h(\alpha),$$

which can be obtained from consideration of $f_1(\alpha\beta x)$. Dividing both sides of this identity by $\alpha^\lambda \beta^\lambda$, we find that the function $h_1(\alpha) = h(\alpha)/\alpha^\lambda$ satisfies an equation of the form

$$h_1(\alpha\beta) = h_1(\alpha) + h_1(\beta).$$

It is well known that the only continuous solution of this equation is the logarithm. Noting further that $h(1) = 0$, we arrive at

$$h(\alpha) = \alpha^\lambda \ln \alpha.$$

Finally, we shall call $f_1(x)$ an *associated function of first order and of degree* λ if for any $\alpha > 0$ we have

$$f_1(\alpha x) = \alpha^\lambda f_1(x) + \alpha^\lambda \ln \alpha\, f_0(x), \tag{1}$$

where $f_0(x)$ is a homogeneous function of degree λ. For instance, $\ln |x|$ is an associated function of first order and of degree zero, since for any $\alpha > 0$,

$$\ln |\alpha x| = \ln |x| + \ln \alpha.$$

We shall now define generalized associated functions similarly to the way we defined generalized homogeneous functions. The generalized function f_1 is called an *associated homogeneous generalized function of first order and of degree* λ if for any $\alpha > 0$ we have

$$\left(f_1, \varphi\left(\frac{x}{\alpha}\right)\right) = \alpha^{\lambda+1}(f_1, \varphi) + \alpha^{\lambda+1} \ln \alpha\, (f_0, \varphi), \tag{2}$$

where f_0 is a generalized homogeneous function of degree λ. Generally speaking, for any k the generalized function f_k is called an *associated homogeneous generalized function of order k and of degree λ* if for any $\alpha > 0$ we have

$$\left(f_k, \varphi\left(\frac{x}{\alpha}\right)\right) = \alpha^{\lambda+1}(f_k, \varphi) + \alpha^{\lambda+1} \ln \alpha \, (f_{k-1}, \varphi), \tag{3}$$

where f_{k-1} is an associated generalized function of order $k - 1$.

Let us now attempt to clarify the nature of associated homogeneous functions of various orders and of arbitrary degree λ. For this purpose we note that if f_λ is a generalized homogeneous function of degree λ and is differentiable with respect to λ, its derivative with respect to λ will be associated of first order.

Indeed, differentiating the identity

$$\left(f_\lambda, \varphi\left(\frac{x}{\alpha}\right)\right) = \alpha^{\lambda+1}(f_\lambda, \varphi),$$

with respect to λ, we have

$$\left(\frac{df_\lambda}{d\lambda}, \varphi\left(\frac{x}{\alpha}\right)\right) = \alpha^{\lambda+1}\left(\frac{df_\lambda}{d\lambda}, \varphi\right) + \alpha^{\lambda+1} \ln \alpha \, (f_\lambda, \varphi)$$

so that $df_\lambda/d\lambda$ is associated of first order, as asserted. Similarly, the derivative with respect to λ of an associated generalized function of order k is associated of order $k + 1$.

4.2. Taylor's and Laurent Series for x_+^λ and x_-^λ

The Taylor's and Laurent series for x_+^λ and similar generalized functions will lead in a natural way to associated generalized functions.

The Taylor's series expansion of x_+^λ in the neighborhood of a regular point λ_0 is

$$x_+^\lambda = x_+^{\lambda_0} + (\lambda - \lambda_0) \frac{\partial}{\partial \lambda} x_+^{\lambda_0} + \frac{1}{2}(\lambda - \lambda_0)^2 \frac{\partial^2}{\partial \lambda^2} x_+^{\lambda_0} + \cdots$$

$$= x_+^{\lambda_0} + (\lambda - \lambda_0) x_+^{\lambda_0} \ln x_+ + \frac{1}{2}(\lambda - \lambda_0)^2 x_+^{\lambda_0} \ln^2 x_+ + \cdots . \tag{1}$$

This expansion contains the new generalized functions

$$x_+^\lambda \ln^m x_+ = \frac{\partial^m}{\partial \lambda^m} x_+^\lambda \qquad (m = 1, 2, \ldots).$$

Here the mth function is an associated function of order m. Explicit expressions for these functions can be obtained either in the same way as one defines x_+^λ, or by m-fold differentiation of Eqs. (3) and (4) of Section 3.2 with respect to λ. Thus for $\operatorname{Re}\lambda > -n - 1$ and $\lambda \neq -1, -2, \ldots, -n$ we have

$$
(x_+^\lambda \ln^m x_+, \varphi) = \int_0^1 x^\lambda \ln^m x \left[\varphi(x) - \varphi(0) - x\varphi'(0) - \ldots \right.
$$
$$
\left. \ldots - \frac{x^{n-1}}{(n-1)!} \varphi^{(n-1)}(0) \right] dx
$$
$$
+ \int_1^\infty x^\lambda \ln^m x\, \varphi(x)\, dx + \sum_{k=1}^\infty \frac{(-1)^m m!\; \varphi^{(k-1)}(0)}{(k-1)!\,(\lambda+k)^{m+1}}.
$$
(2)

Similarly, for $-n - 1 < \operatorname{Re}\lambda < -n$ we have

$$
\int_0^\infty x^\lambda \ln^m x\, \varphi(x)\, dx
$$
$$
= \int_0^\infty x^\lambda \ln^m x \left[\varphi(x) - \varphi(0) - \ldots - \frac{x^{n-1}}{(n-1)!} \varphi^{(n-1)}(0) \right] dx.
$$
(3)

In the neighborhood of the pole at $\lambda = -n$, we can expand x_+^λ in a Laurent series whose dominant term is of degree -1. To obtain this expansion explicitly, we turn to the regularized integral

$$
\int_0^\infty x^\lambda \varphi(x)\, dx
$$
$$
= \int_0^\infty x^\lambda \left[\varphi(x) - \varphi(0) - x\,\varphi'(0) - \ldots - \frac{x^{n-1}}{(n-1)!} \varphi^{(n-1)}(0) \right] dx,
$$

and isolate the term that fails to converge at $\lambda = -n$. In other words, we write this integral in the form[2]

$$
\int_0^\infty x^\lambda \varphi(x)\, dx = \int_0^1 x^\lambda \left[\varphi(x) - \ldots - \frac{x^{n-1}}{(n-1)!} \varphi^{(n-1)}(0) \right] dx
$$
$$
+ \int_1^\infty x^\lambda \left[\varphi(x) - \ldots - \frac{x^{n-2}}{(n-2)!} \varphi^{(n-2)}(0) \right] dx + \frac{\varphi^{(n-1)}(0)}{(n-1)!(\lambda+n)}.
$$

The sum of the integrals on the right-hand side of this equation is the

---

[2] Instead of writing $\int_1^\infty x^{\lambda+n-1}\, dx$ we could have considered the integral $\int_a^\infty x^{\lambda+n-1}\, dx$ for any $a > 0$. But we would then have had to expand $a^{\lambda+n}$ in a Taylor's series, and the resulting expressions would end up being quite formidable.

regular part of the Laurent expansion. It is an analytic function of λ in the strip $|\operatorname{Re} \lambda + n| < 1$. We shall denote this functional by $F_{-n}(x_+, \lambda)$. We thus have

$$(F_{-n}(x_+, \lambda), \varphi) = \int_0^\infty x^\lambda \left[\varphi(x) - \varphi(0) - x\varphi'(0) - \ldots \right.$$

$$\ldots - \frac{x^{n-2}}{(n-2)!} \varphi^{(n-2)}(0) - \frac{x^{n-1}}{(n-1)!} \varphi^{(n-1)}(0) \left. \theta(1-x) \right] dx,$$

where $\theta(x)$ is equal to zero for $x < 0$ and to one for $x > 0$. Another way of putting this is to say that the last term in the integrand is included only for $0 < x < 1$, while for $x > 1$ it is replaced by zero. The resulting integral thus converges both at $x = 0$ and at $x = \infty$.

Of particular interest is the value of this functional at $\lambda = -n$, which we shall call x_+^{-n}; this is the value at $\lambda = -n$ of the regular part of the Laurent expansion of x_+^λ about $\lambda = -n$, namely

$$x_+^{-n} = \lim_{\lambda \to -n} \frac{\partial}{\partial \lambda} \left[(\lambda + n) x_+^\lambda \right].$$

We now see that the generalized function x_+^{-n} is an associated generalized function of first order and of degree $-n$. Operating on a φ in K this functional yields

$$(x_+^{-n}, \varphi) = \int_0^\infty x^{-n} \left[\varphi(x) - \varphi(0) - \ldots - \frac{x^{n-1}}{(n-1)!} \varphi^{(n-1)}(0) \theta(1-x) \right] dx. \quad (4)$$

It should be emphasized that the generalized function x_+^{-n} is not the value of x_+^λ at $\lambda = -n$, for x_+^λ has a pole and therefore does not exist at this point. Nevertheless the functional x_+^{-n} is in a certain sense a regularization of the ordinary function x_+^{-n}.

The formula for the derivative of x_+^{-n} with respect to x is of some interest. We have

$$\left(\frac{d}{dx} x_+^{-n}, \varphi(x) \right) = - (x_+^{-n}, \varphi'(x))$$

$$= - \int_0^\infty x^{-n} \left[\varphi'(x) - \varphi'(0) - \ldots - \frac{x^{n-1}}{(h-1)!} \varphi^{(n)}(0) \theta(1-x) \right] dx$$

$$= - \int_0^1 x^{-n} \left[\varphi'(x) - \varphi'(0) - \ldots - \frac{x^{n-1}}{(n-1)!} \varphi^{(n)}(0) \right] dx$$

$$- \int_1^\infty x^{-n} \left[\varphi'(x) - \varphi'(0) - \ldots - \frac{x^{n-2}}{(n-2)!} \varphi^{(n-1)}(0) \right] dx.$$

We now integrate by parts in each of the terms, obtaining

$$\left(\frac{d}{dx} x_+^{-n}, \varphi\right)$$

$$= -\int_0^\infty n x^{-n-1} \left[\varphi(x) - \varphi(0) - \dots - \frac{x^n}{n!} \varphi^{(n)}(0)\, \theta(1-x)\right] dx + \frac{\varphi^{(n)}(0)}{n!}.$$

Hence

$$\frac{d}{dx} x_+^{-n} = -n x_+^{-n-1} + \frac{(-1)^n}{n!}\, \delta^{(n)}(x). \tag{5}$$

Thus although we have succeeded in establishing a correspondence between the ordinary function x_+^{-n} and a generalized function, we have had to sacrifice the ordinary formula for the derivative. We see, therefore, that Eq. (4) is not a canonical regularization of the ordinary function x_+^{-n}.

Of some interest also are the derivatives of $F_{-n}(x_+, \lambda)$ with respect to λ at $\lambda = -n$. We shall write these in the form

$$\frac{\partial}{\partial \lambda} F_{-n}(x_+, \lambda)\bigg|_{\lambda=-n} = x_+^{-n} \ln x_+,$$

$$\frac{\partial^2}{\partial \lambda^2} F_{-n}(x_+, \lambda)\bigg|_{\lambda=-n} = x_+^{-n} \ln^2 x_+,$$

. . . .

and then it is a simple matter to verify that the explicit definitions of these generalized functions are obtained from Eq. (4) by multiplying x^{-n} on the right-hand side of Eq. (4) by the power of the logarithm of x that appears on the right-hand side of the appropriate equation above.

It is clear that these generalized functions are associated functions of degree $-n$ and of order 2, 3, 4, ..., respectively.

Having defined the generalized functions x_+^{-n}, $x_+^{-n} \ln x_+$, ..., we can now obtain the Laurent expansion of the generalized function x_+^λ about $\lambda = -n$. To do this, we expand $F_{-n}(x_+, \lambda)$ in a Taylor's series, which yields

$$x_+^\lambda = \frac{(-1)^{n-1}\delta^{(n-1)}(x)}{(n-1)!\,(\lambda+n)} + x_+^{-n}$$

$$+ (\lambda+n) x_+^{-n} \ln x_+ + \dots + \frac{(\lambda+n)^k}{k!} x_+^{-n} \ln^k x_+ + \dots.$$

The Taylor's series for x_-^λ in the neighborhood of a regular point λ_0 is

$$x_-^\lambda = x_-^{\lambda_0} + (\lambda - \lambda_0) x_-^{\lambda_0} \ln x_- + \tfrac{1}{2}(\lambda-\lambda_0)^2 x_-^{\lambda_0} \ln^2 x_- + \dots.$$

This expansion contains the new generalized functions $x_-^\lambda \ln^k x_-$ (with $k = 1, 2, ...$), of which the kth is an associated function of degree λ and of order k. All these are analytic in λ for $\lambda \neq -1, -2, ...$, and are defined in the strip $-n - 1 < \operatorname{Re} \lambda < -n$ by

$$(x_-^\lambda \ln^k x_-, \varphi) = \int_0^\infty x^\lambda \ln^k x \left[\varphi(-x) - \varphi(0) + x\varphi'(0) - ... \right.$$

$$\left. ... - \frac{(-1)^{n-1} x^{n-1}}{(n-1)!} \varphi^{(n-1)}(0) \right] dx. \tag{6}$$

As before, in order to obtain the Laurent series for x_-^λ about pole at $\lambda = -n$, we separate out the term which fails to converge at this point. We have

$$\int_{-\infty}^0 x^\lambda \varphi(x) \, dx = \int_0^1 x^\lambda \left[\varphi(--x) - \varphi(0) + x\varphi'(0) - ... \right.$$

$$\left. ... - (-1)^{n-1} \frac{x^{n-1}}{(n-1)!} \varphi^{(n-1)}(0) \right] dx$$

$$+ \int_1^\infty x^\lambda \left[\varphi(-x) - \varphi(0) + x\varphi'(0) - ... \right.$$

$$\left. ... - (-1)^{n-2} \frac{x^{n-2}}{(n-2)!} \varphi^{(n-2)}(0) \right] dx + \frac{(-1)^{n-1}}{(n-1)!} \frac{\varphi^{(n-1)}(0)}{\lambda + n}.$$

The sum of the integrals on the right-hand side gives the regular part of the desired Laurent expansion. It is analytic for $| \operatorname{Re} \lambda + n | < 1$. We shall denote this functional by $F_{-n}(x_-, \lambda)$. Thus

$$(F_{-n}(x_-, \lambda), \varphi) = \int_0^\infty x^\lambda \left[\varphi(-x) - \varphi(0) + x\varphi'(0) - ... \right.$$

$$\left. ... - (-1)^{n-1} \frac{x^{n-1}}{(n-1)!} \varphi^{(n-1)}(0) \, \theta(1-x) \right] dx.$$

The value of this generalized function at $\lambda = -n$ shall be denoted by x_-^{-n}; this is the value at $\lambda = -n$ of the regular part of the Laurent expansion for x_-^λ, namely

$$x_-^{-n} = \lim_{\lambda \to -n} \frac{\partial}{\partial \lambda} \left[(\lambda + n) x_-^\lambda \right].$$

This generalized function acts on a φ in K according to

$$(x_-^{-n}, \varphi) = \int_0^\infty x^{-n} \Big[\varphi(-x) - \varphi(0) + x\varphi'(0) - \dots$$

$$\dots - (-1)^{n-1} \frac{x^{n-1}}{(n-1)!} \varphi^{(n-1)}(0)\, \theta(1-x) \Big]\, dx \qquad (7)$$

and is a regularization of the ordinary function x_-^{-n}. We emphasize again that the generalized function x_-^{-n} is not the value at $\lambda = -n$ of the analytic generalized function x_-^λ.

We shall denote the derivatives with respect to λ of $F_{-n}(x_-, \lambda)$ at $\lambda = -n$ by

$$\frac{\partial}{\partial \lambda} F_{-n}(x_-, \lambda) \Big|_{\lambda=-n} = x_-^{-n} \ln x_-,$$

$$\frac{\partial^2}{\partial \lambda^2} F_{-n}(x_-, \lambda) \Big|_{\lambda=-n} = x_-^{-n} \ln^2 x_-,$$

. . . .

The explicit definitions of these generalized functions are obtained from Eq. (7) by multiplying x^{-n} on the right-hand side of Eq. (7) by the power of the logarithm of x that appears on the right-hand side of the appropriate equation above.

By expanding $F_{-n}(x_-, \lambda)$ in a power series in $\lambda + n$, we obtain the Laurent expansion for x_-^λ, namely

$$x_-^\lambda = \frac{\delta^{(n-1)}(x)}{(n-1)!\,(\lambda+n)} + x_-^{-n} + (\lambda+n)\, x_-^{-n} \ln x_- + \dots . \qquad (8)$$

In this series the coefficient of $(\lambda + n)^{-1}$ is a homogeneous function of degree $-n$, while the coefficient of $(\lambda + n)^m$, with $m = 0, 1, 2, \dots$, is an associated function of degree $-n$ and of order $m + 1$.

4.3. Expansion of $|x|^\lambda$ and $|x|^\lambda \operatorname{sgn} x$

We now wish to develop the Taylor's and Laurent series for $|x|^\lambda$ and $|x|^\lambda \operatorname{sgn} x$. We shall obtain them simply by combining suitable expansions of x_+^λ and x_-^λ. In doing this we shall again introduce new generalized functions.

If λ_0 is a regular point of the generalized function $|x|^\lambda$, then we shall write

$$|x|^\lambda = |x|^{\lambda_0} + (\lambda - \lambda_0)\,|x|^{\lambda_0} \ln|x| + \tfrac{1}{2}(\lambda - \lambda_0)^2\,|x|^{\lambda_0} \ln^2 |x| + \dots . \qquad (1)$$

We have here introduced the new generalized functions $|x|^\lambda \ln^k |x|$, the kth of which is an associated function of degree λ and order k. These functions are all analytic in λ for $\lambda \neq 1, 3, \ldots$ and are defined in the region $-2m - 1 < \text{Re } \lambda < -2m + 1$ by

$$(|x|^\lambda \ln^k |x|, \varphi) = \int_0^\infty x^\lambda \ln^k x \left\{ \varphi(x) + \varphi(-x) \right.$$

$$- 2 \left[\varphi(0) + \frac{\varphi''(0)}{2!} x^2 + \ldots + \frac{x^{2m-2}}{(2m-2)!} \varphi^{(2m-2)}(0) \right] \right\} dx. \tag{2}$$

Similarly, if λ_0 is a regular point of the generalized function $|x|^\lambda \text{ sgn } x$ the Taylor's series expansion of this function about λ_0 is

$$|x|^\lambda \text{ sgn } x = |x|^{\lambda_0} \text{ sgn } x + (\lambda - \lambda_0) |x|^{\lambda_0} \ln |x| \text{ sgn } x$$

$$+ \tfrac{1}{2} (\lambda - \lambda_0)^2 |x|^{\lambda_0} \ln^2 |x| \text{ sgn } x + \ldots . \tag{3}$$

Here the $|x|^\lambda \ln^k |x| \text{ sgn } x$ are new generalized functions of which the kth is an associated function of degree λ and of order k. They are all analytic in λ for $\lambda \neq -2, -4, \ldots$, and in the region $-2m - 2 < \text{Re } \lambda < -2m$ they are defined by

$$(|x|^\lambda \ln^k |x| \text{ sgn } x, \varphi) = \int_0^\infty x^\lambda \ln^k x \left\{ \varphi(x) - \varphi(-x) \right.$$

$$- 2 \left[x\varphi'(0) + \frac{x^3}{3!} \varphi'''(0) + \ldots + \frac{x^{2m-1}}{(2m-1)!} \varphi^{(2m-1)}(0) \right] \right\} dx. \tag{4}$$

Now $\lambda_0 = -2m - 1$ is a simple pole of the generalized function $|x|^\lambda$; the Laurent expansion about λ_0 is

$$|x|^\lambda = 2 \frac{\delta^{(2m)}(x)}{(2m)!} \frac{1}{\lambda + 2m + 1} + x_+^{-2m-1} + x_-^{-2m-1}$$

$$+ (\lambda + 2m + 1) [x_+^{-2m-1} \ln x_+ + x_-^{-2m-1} \ln x_-] + \ldots \tag{5}$$

Similarly, $\lambda_0 = -2m$ is a simple pole of $|x|^\lambda \text{ sgn } x$; the Laurent expansion about λ_0 is

$$|x|^\lambda \text{ sgn } x = -2 \frac{\delta^{(2m-1)}(x)}{(2m-1)!} \frac{1}{\lambda + 2m} + x_+^{-2m} - x_-^{-2m}$$

$$+ (\lambda + 2m) [x_+^{-2m} \ln x_+ - x_-^{-2m} \ln x_-] + \ldots . \tag{6}$$

We shall denote the generalized function $x_+^{-2m-1} + x_-^{-2m-1}$ by $|x|^{-2m-1}$. When operating on some $\varphi(x)$ in K, it acts according to

$$(|x|^{-2m-1}, \varphi) = \int_0^\infty x^{-2m-1} \left\{ \varphi(x) + \varphi(-x) - 2\left[\varphi(0) + \frac{x^2}{2!}\varphi''(0) + \ldots \right.\right.$$

$$\left.\left. \ldots + \frac{x^{2m-2}}{(2m-2)!}\varphi^{(2m-2)}(0) + \frac{x^{2m}}{(2m)!}\varphi^{(2m)}(0)\,\theta(1-x)\right]\right\}\,dx \tag{7}$$

where $\theta(x)$ has its usual meaning.

Similarly, we shall denote the generalized function $x_+^{-2m} - x_-^{-2m}$ by $|x|^{-2m}\,\mathrm{sgn}\,x$. It acts on a $\varphi(x)$ in K according to

$$(|x|^{-2m}\,\mathrm{sgn}\,x, \varphi)$$

$$= \int_0^\infty x^{-2m} \left\{ \varphi(x) - \varphi(-x) - 2\left[x\varphi'(0) + \frac{x^3}{3!}\varphi'''(0) + \ldots \right.\right.$$

$$\left.\left. \ldots + \frac{x^{2m-3}}{(2m-3)!}\varphi^{(2m-3)}(0) + \frac{x^{2m-1}}{(2m-1)!}\varphi^{(2m-1)}(0)\,\theta(1-x)\right]\right\}\,dx. \tag{8}$$

Finally, our required Laurent expansions can be written

$$|x|^\lambda = 2\frac{\delta^{(2m)}(x)}{(2m)!}\frac{1}{\lambda+2m+1} + |x|^{-2m-1}$$

$$+ (\lambda+2m+1)\,|x|^{-2m-1}\ln|x| + \ldots; \tag{9}$$

$$|x|^\lambda\,\mathrm{sgn}\,x = -2\frac{\delta^{(2m-1)}(x)}{(2m-1)!}\frac{1}{\lambda+2m} + |x|^{-2m}\,\mathrm{sgn}\,x$$

$$+ (\lambda+2m)\,|x|^{-2m}\ln|x|\,\mathrm{sgn}\,x + \ldots. \tag{10}$$

The first coefficients $\delta^{(2m)}(x)$ and $\delta^{(2m-1)}(x)$ in these expansions are homogeneous functions of degrees $-2m-1$ and $-2m-2$, respectively; the other coefficients are associated functions of the same degrees and or order 1, 2, 3, ...

In particular, setting $m = 0$ in Eq. (9), we obtain the expansion of $|x|^\lambda$ about the pole at $\lambda = -1$. This is

$$|x|^\lambda = 2\frac{\delta(x)}{\lambda+1} + |x|^{-1} + (\lambda+1)\,|x|^{-1}\ln|x| + \ldots. \tag{11}$$

Here the functional $| x |^{-1}$ is defined by

$$(| x |^{-1}, \varphi) = \int_0^\infty \frac{\varphi(x) + \varphi(-x) - 2\varphi(0)\, \theta(1-x)}{x}\, dx$$

$$= \int_{-\infty}^\infty \frac{\varphi(x) - \varphi(0)\, \theta(1-x^2)}{x}\, dx, \qquad (12)$$

where $\theta(1 - x^2)$ causes the $\varphi(0)$ term to contribute only for $| x | \leqslant 1$.
The generalized functions

$$| x |^{-2m-1} = x_+^{-2m-1} + x_-^{-2m-1},$$

$$| x |^{-2m} \operatorname{sgn} x = x_+^{-2m} - x_-^{-2m}$$

that we have introduced above are not the analytic functionals $| x |^\lambda$ and $| x |^\lambda \operatorname{sgn} x$ at the corresponding values of λ (namely $-2m - 1$ and $-2m$), for at these values of λ the latter have simple poles. They are the values at the poles of the regular parts of the Laurent expansions about these poles.

We could have considered also the generalized functions

$$x_+^{-2m} + x_-^{-2m},$$

$$x_+^{-2m-1} - x_-^{-2m-1}.$$

It is easily shown that the first of these coincides with x^{-2m} (i.e., $| x |^\lambda$ at the regular point $\lambda = -2m$), and the second with x^{-2m-1}.

In Section 3.5 we defined

$$\frac{x_+^\lambda}{\Gamma(\lambda+1)}, \qquad \frac{x_-^\lambda}{\Gamma(\lambda+1)}, \qquad \frac{| x |^\lambda}{\Gamma\left(\dfrac{\lambda+1}{2}\right)}, \qquad \frac{| x |^\lambda \operatorname{sgn} x}{\Gamma\left(\dfrac{\lambda+2}{2}\right)}. \qquad (13)$$

These generalized functions have no singularities. By expanding the numerators and denominators in Taylor's or Laurent series and performing the indicated divisions in the ordinary way, one can obtain the Taylor's series for these functions about any point in the λ plane.

The power series expansion

$$\Gamma(\lambda + 1) = 1 + c_1\lambda + c_2\lambda^2 + c_3\lambda^3 + \dots,$$

for the gamma function is well known.[3] Here $c_1 = c = 0.505\dots$ is

---

[3] I. M. Ryzhik and I. S. Gradshtein, "Tables of Integrals, Sums, Series, and Products" (in Russian), p. 331. Gostekhizdat, 1951. German translation: I. M. Ryshik und I. S. Gradstein, "Summen-, Produkt- und Integraltafeln," p. 299. Berlin, 1957. Henceforth we shall refer to the German translation as Ryshik and Gradstein, "Tables." See also N. Nielsen, "Handbuch der Theorie der Gammafunktion," p. 40. Treubner, Leipzig, 1906.

Euler's constant, and the rest of the c_i are obtained from the recursion formula

$$c_{n+1} = \frac{1}{n+1} \sum_{k=0}^{n} (-1)^{k+1} c_{n-k} s_{k+1}, \qquad s_1 = c, \qquad s_n = \zeta(n).$$

In particular, expanding about $\lambda = 0$ one obtains

$$\frac{x_+^\lambda}{\Gamma(\lambda+1)} = \frac{\theta(x) + \lambda \ln x_+ + \cdots}{1 + c\lambda + \cdots} = \theta(x) + \lambda(\ln x_+ - c\,\theta(x)) + \cdots \tag{14}$$

4.4. The Generalized Functions $(x+i0)^\lambda$ and $(x-i0)^\lambda$

These generalized functions were defined in Section 3.6. We shall study them here in more detail. For $\mathrm{Re}\,\lambda > -1$ they were defined as the limits of $(x+iy)^\lambda$ and $(x-iy)^\lambda$ as $y \to +0$. From this we obtained the formulas

$$(x+i0)^\lambda = x_+^\lambda + e^{i\lambda\pi} x_-^\lambda, \tag{1}$$

$$(x-i0)^\lambda = x_+^\lambda + e^{-i\lambda\pi} x_-^\lambda. \tag{2}$$

The right-hand sides of these equations can be extended analytically to the rest of the λ plane, which we then take as the *definitions* of their left-hand sides for $\mathrm{Re}\,\lambda \leqslant -1$. We have already remarked that in the analytic continuations of the right-hand sides, the singularities at $\lambda = -1, -2, \ldots$ cancel out. We shall now find the values of these functions at these points.

In Section 4.2 we obtained the Laurent expansions for the generalized functions x_+^λ and x_-^λ about $\lambda = -n$, namely

$$x_+^\lambda = \frac{(-1)^{n-1}\delta^{(n-1)}(x)}{(n-1)!\,(\lambda+n)} + F_{-n}(x_+, \lambda), \tag{3}$$

$$x_-^\lambda = \frac{\delta^{(n-1)}(x)}{(n-1)!\,(\lambda+n)} + F_{-n}(x_-, \lambda). \tag{4}$$

Here $F_{-n}(x_+, \lambda)$ and $F_{-n}(x_-, \lambda)$ are the regular parts of the Laurent expansions; we have denoted their values at $\lambda = -n$ by x_+^{-n} and x_-^{-n} respectively.

Let us also write

$$e^{\pm i\lambda\pi} = (-1)^n e^{\pm i(\lambda+n)\pi} = (-1)^n[1 \pm i(\lambda+n)\pi + \cdots]. \tag{5}$$

Inserting Eqs. (3)–(5) into the right-hand sides of (1) and (2), canceling out the singularities, and going to the limit as $\lambda \to -n$, we arrive at

$$(x \pm i0)^{-n} = x_+^{-n} + (-1)^n x_-^{-n} \mp \frac{(-1)^{n-1} i\pi}{(n-1)!} \delta^{(n-1)}(x).$$

In Section 4.3 we noted that for even $n = 2m$,

$$x_+^{-2m} + x_-^{-2m} = x^{-2m} = [\,|\,x\,|^\lambda\,]_{\lambda=-2m}$$

while for odd $n = 2m + 1$,

$$x_+^{-2m-1} - x_-^{-2m-1} = x^{-2m-1} = [\,|\,x\,|^\lambda \operatorname{sgn} x\,]_{\lambda=-2m-1}.$$

Thus we finally obtain

$$(x + i0)^{-n} = x^{-n} - \frac{i\pi\,(-1)^{n-1}}{(n-1)!} \delta^{(n-1)}(x), \tag{6}$$

$$(x - i0)^{-n} = x^{-n} + \frac{i\pi(-1)^{n-1}}{(n-1)!} \delta^{(n-1)}(x). \tag{7}$$

In particular, for $n = 2$ we have

$$\left(\frac{1}{(x + i0)^2}, \varphi(x) \right) = \int_0^\infty \frac{\varphi(x) + \varphi(-x) - 2\varphi(0)}{x^2}\,dx - i\pi\,\varphi'(0); \tag{8}$$

$$\left(\frac{1}{(x - i0)^2}, \varphi(x) \right) = \int_0^\infty \frac{\varphi(x) + \varphi(-x) - 2\varphi(0)}{x^2}\,dx + i\pi\,\varphi'(0). \tag{9}$$

Note that it follows from (1) and (2) for $\lambda \neq -1, -2, \dots$ and from (6) and (7) for $\lambda = -1, -2, \dots$ that

$$\frac{d}{dx}(x + i0)^\lambda = \lambda(x + i0)^{\lambda-1}, \tag{10}$$

$$\frac{d}{dx}(x - i0)^\lambda = \lambda(x - i0)^{\lambda-1}. \tag{11}$$

Since differentiation decreases the exponent by one unit, Eqs. (10) and (11) could have been used as *defining relations* for the generalized functions $(x + i0)^\lambda$ and $(x - i0)^\lambda$ for $\lambda \neq -1, -2, \dots$. For instance, $(x + i0)^{-\frac{3}{2}}$ could have been defined as $-2(d/dx)(x + i0)^{-\frac{1}{2}}$, where d/dx is differentiation in the sense of generalized functions, and $(x + i0)^{-\frac{1}{2}}$ is the locally summable function defined by Eq. (1) with $\lambda = -\frac{1}{2}$.

Equations (10) and (11) have also another important implication; for $\lambda \neq -1, -2, ...,$

$$\lim_{y \to +0} (x + iy)^{\lambda} = (x + i0)^{\lambda}, \tag{12}$$

$$\lim_{y \to -0} (x + iy)^{\lambda} = (x - i0)^{\lambda} \tag{13}$$

in the sense of generalized functions. Indeed, for Re λ sufficiently large, this follows by definition, and the derivative in the sense of generalized functions of a convergent sequence is also convergent sequence.

For negative integral exponents, $(x + i0)^{-n}$ and $(x - i0)^{-n}$ can also be defined by differentiation. Recall (Section 2.2, Example 6) the (locally summable) function

$$\ln (x + i0) = \lim_{y \to +0} \ln (x + iy) = \ln |x| + i\pi \, \theta(-x), \tag{14}$$

whose derivative is given by

$$\frac{d}{dx} \ln (x + i0) = \frac{1}{x} - i\pi \, \delta(x). \tag{15}$$

Together with (6) (in which we set $n = 1$) this implies that

$$\frac{d}{dx} \ln (x + i0) = \frac{1}{x + i0}. \tag{16}$$

Similarly,

$$\ln (x - i0) = \lim_{y \to -0} \ln (x + iy) = \ln |x| - i\pi \, \theta(-x). \tag{17}$$

The derivative of this function is given by

$$\frac{d}{dx} \ln (x - i0) = \frac{1}{x} + i\pi \, \delta(x). \tag{18}$$

which, together with (7) (in which we set $n = 1$) leads to

$$\frac{d}{dx} \ln (x - i0) = \frac{1}{x - i0}. \tag{19}$$

Thus the generalized functions $(x + i0)^{-1}$ could have been defined by Eqs. (16) and (19), and $(x \pm i0)^{-n}$ for other negative integral exponents could be obtained by using (10) and (11).

Now the limits in (14) and (17) are taken in the sense of generalized functions. This implies that the limits in (12) and (13) converge in the sense of generalized functions *for all* λ.

This can be used, for instance, to rewrite (8) and (9) in the form

$$\lim_{y \to +0} \int \frac{\varphi(x)}{(x + iy)^2}\, dx = \int_0^\infty \frac{\varphi(x) + \varphi(-x) - 2\varphi(0)}{x^2}\, dx - i\pi\, \varphi'(0),$$

$$\lim_{y \to +0} \int \frac{\varphi(x)}{(x - iy)^2}\, dx = \int_0^\infty \frac{\varphi(x) + \varphi(-x) - 2\varphi(0)}{x^2}\, dx + i\pi\, \varphi'(0).$$

4.5. Taylor's Series for $(x + i0)^\lambda$ and $(x - i0)^\lambda$

Since $(x + i0)^\lambda$ is an entire function of λ, it can be expanded in a Taylor's series about any point λ_0, and this series will converge for all λ. We thus write

$$(x + i0)^\lambda = (x + i0)^{\lambda_0} + (\lambda - \lambda_0)\, (x + i0)^{\lambda_0} \ln (x + i0)$$
$$+ \tfrac{1}{2} (\lambda - \lambda_0)^2\, (x + i0)^{\lambda_0} \ln^2 (x + i0) + \ldots . \qquad (1)$$

The new generalized functions $(x + i0)^{\lambda_0} \ln (x + i0)$, ... thus introduced are the derivatives of $(x + i0)^\lambda$ with respect to λ at $\lambda = \lambda_0$. Explicit expressions for these can be written by comparing Eq. (1) with the expansion in powers of $\lambda - \lambda_0$ of the relation

$$(x + i0)^\lambda = x_+^\lambda + e^{i\lambda \pi} x_-^\lambda . \qquad (2)$$

Recall that for $\lambda_0 \neq -1, -2, \ldots$ we had

$$x_+^\lambda = x_+^{\lambda_0} + (\lambda - \lambda_0)\, x_+^{\lambda_0} \ln x_+ + \tfrac{1}{2} (\lambda - \lambda_0)^2\, x_+^{\lambda_0} \ln^2 x_+ + \ldots \qquad (3)$$

$$e^{i\lambda \pi} x_-^\lambda = e^{i\lambda_0 \pi} \left(1 + i\pi (\lambda - \lambda_0) - \frac{\pi^2}{2} (\lambda - \lambda_0)^2 + \ldots \right)$$
$$\times \left(x_-^{\lambda_0} + (\lambda - \lambda_0)\, x_-^{\lambda_0} \ln x_- + \tfrac{1}{2} (\lambda - \lambda_0)\, x_-^{\lambda_0} \ln^2 x_- + \ldots \right) . \qquad (4)$$

By comparing coefficients and replacing λ_0 by λ we thus obtain

$$(x + i0)^\lambda \ln (x + i0) = x_+^\lambda \ln x_+ + i\pi e^{i\lambda \pi} x_-^\lambda + e^{i\lambda \pi} x_-^\lambda \ln x_- , \qquad (5)$$

$$(x + i0)^\lambda \ln^2 (x + i0) = x_+^\lambda \ln^2 x_+ + e^{i\lambda \pi} [x_-^\lambda \ln^2 x_- + 2i\pi x_-^\lambda \ln x_- - \pi^2 x_-^\lambda] \qquad (6)$$

and similar expressions for the higher terms.

These relations are valid for all $\lambda \neq -1, -2, \ldots$.

The Taylor's series expansion for $(x - i0)^\lambda$ is

$$(x - i0)^\lambda = (x - i0)^{\lambda_0} + (\lambda - \lambda_0)\,(x - i0)^{\lambda_0} \ln (x - i0) \tag{7}$$

$$+ \tfrac{1}{2}(\lambda - \lambda_0)^2\,(x - i0)^\lambda \ln^2 (x - i0) + \dots$$

Here

$$(x - i0)^\lambda \ln (x - i0),\ (x - i0)^\lambda \ln^2 (x - i0),\ \dots$$

are the derivatives with respect to λ of the entire function $(x - i0)^\lambda$. Proceeding as above, we will arrive at

$$(x - i0)^\lambda \ln (x - i0) = x_+^\lambda \ln x_+ - i\pi e^{-i\lambda\pi} x_-^\lambda + e^{-i\lambda\pi} x_-^\lambda \ln x_-, \tag{8}$$

$$(x - i0)^\lambda \ln^2 (x - i0) \tag{9}$$

$$= x_+^\lambda \ln^2 x_+ + e^{-i\lambda\pi}[x_-^\lambda \ln^2 x_- - 2i\pi x_-^\lambda \ln x_- - \pi^2 x_-^\lambda]$$

$$(\lambda \neq -1, -2, \dots)$$

and similar expressions for the higher terms.

To obtain the corresponding formulas for negative integral λ, we turn instead of to (3) and (4) to the appropriate Laurent expansions, namely to

$$x_+^\lambda = \frac{(-1)^{n-1}\delta^{(n-1)}\,(x)}{(n-1)!\,(\lambda + n)} + x_+^{-n} + (\lambda + n)\,x_+^{-n} \ln x_+$$

$$+ \tfrac{1}{2}(\lambda + n)^2\,x_+^{-n} \ln^2 x_+ + \dots, \tag{10}$$

$$x_-^\lambda = \frac{\delta^{(n-1)}\,(x)}{(n-1)!\,(\lambda + n)} + x_-^{-n} + (\lambda + n)\,x_-^{-n} \ln x_-$$

$$+ \tfrac{1}{2}(\lambda + n)^2\,x_-^{-n} \ln^2 x_- + \dots, \tag{11}$$

$$e^{i\lambda\pi} = (-1)^n \left[1 + i\pi(\lambda + n) - \frac{\pi^2}{2}(\lambda + n)^2 - i\frac{\pi^3}{6}(\lambda + n)^3 + \dots \right]. \tag{12}$$

Multiplying (11) by (12) and adding to (10), we obtain

$$x_+^\lambda + e^{i\lambda\pi} x_-^\lambda = \left[x^{-n} + \frac{i\pi\,(-1)^n}{(n-1)!}\,\delta^{(n-1)}\,(x) \right] + (\lambda + n)\left[i\pi(-1)^n x_-^{-n} \right.$$

$$+ (-1)^{n-1}\frac{\pi^2}{2}\frac{\delta^{(n-1)}(x)}{(n-1)!} + x^{-n} \ln |\,x\,| \left] + \frac{(\lambda + n)^2}{2}\left[(-1)^{n-1}i\frac{\pi^3}{3}\frac{\delta^{(n-1)}\,(x)}{(n-1)!} \right.$$

$$+ (-1)^{n-1}\pi^2 x_-^{-n} + 2i\pi(-1)^n x_-^{-n} \ln x_- + x^{-n} \ln^2 |\,x\,| \left.\right]. \tag{13}$$

Thus we arrive at

$$(x + i0)^{-n} \ln (x + i0)$$

$$= (-1)^n i\pi x_-^{-n} + (-1)^{n-1} \frac{\pi^2}{2} \frac{\delta^{(n-1)}(x)}{(n-1)!} + x^{-n} \ln |x|, \qquad (14)$$

$$(x + i0)^{-n} \ln^2 (x + i0) = (-1)^{n-1} i \frac{\pi^3}{3} \frac{\delta^{(n-1)}(x)}{(n-1)!}$$

$$+ (-1)^{n-1}\pi^2 x_-^{-n} + 2i\pi (-1)^n x_-^{-n} \ln x_- + x^{-n} \ln^2 |x| \qquad (15)$$

and similar expressions for the higher terms.

General considerations imply that Eqs. (14) and (15) are the limits as $\lambda_0 \to -n$ of Eqs. (5) and (6). Similarly,

$$(x - i0)^{-n} \ln (x - i0) = (-1)^{n-1} i\pi x_-^{-n}$$

$$+ (-1)^{n-1} \frac{\pi^2}{2} \frac{\delta^{(n-1)}(x)}{(n-1)!} + x^{-n} \ln |x|, \qquad (16)$$

$$(x - i0)^{-n} \ln^2 (x - i0) = (-1)^n i \frac{\pi^3}{3} \frac{\delta^{(n-1)}(x)}{(n-1)!} + (-1)^{n-1}\pi^2 x_-^{-n} \qquad (17)$$

$$+ 2i\pi(-1)^{n-1} x_-^{-n} \ln x_- + x^{-n} \ln^2 |x|.$$

It is left to the reader to determine which of the generalized functions here introduced are homogeneous or associated, and of what degree and order.

4.6. Expansion of r^λ

The generalized function r^λ was defined in Section 3.9 by

$$(r^\lambda, \varphi) = \int r^\lambda \varphi(x) dx, \qquad (1)$$

an expression which can be used for Re $\lambda > -n$. It is obvious that r^λ is, for these values of λ, a homogeneous function of degree λ. In Section 3.9 we rewrote (1) in the form

$$(r^\lambda, \varphi) = \Omega_n \int_0^\infty r^{\lambda+n-1} S_\varphi(r) dr,$$

where

$$S_\varphi(r) = \frac{1}{\Omega_n} \int_\Omega \varphi(x) d\omega$$

is the mean value of the test function $\varphi(x) = \varphi(x_1, ..., x_n)$ on a sphere (hypersphere) of radius r. Thus (r^λ, φ) is equivalent to the result obtained when the functional $\Omega_n x_+^\mu$ (with $\mu = \lambda + - 1$) operates on the even test function $S_\varphi(r)$, and it can therefore be analytically continued into the λ plane except for the points $\lambda = -n, -n - 2, ...$. It is clear that the process of analytic continuation leaves r^λ a homogeneous function of degree λ.

The Taylor's or Laurent series for r^λ can be obtained directly from the corresponding expansion of $\Omega_n x_+^\mu$. For instance, the Taylor's series about the regular point λ_0 is

$$r^\lambda = r^{\lambda_0} + (\lambda - \lambda_0)r^{\lambda_0} \ln r + \tfrac{1}{2}(\lambda - \lambda_0)^2 r^{\lambda_0} \ln^2 r + ... ,$$

where $r^{\lambda_0} \ln^k r$ is the functional defined by

$$(r^{\lambda_0} \ln^k r, \varphi) = \Omega_n \int_0^\infty r^{\lambda_0 + n - 1} \ln^k r \left[S_\varphi(r) - \varphi(0) - ... - \frac{r^{m-1}}{(m-1)!} S_\varphi^{(m-1)}(0) \right] dr$$

in the region $-m - n < \mathrm{Re}\, \lambda_0 < -m - n + 1$ [recall that $S_\varphi(0) = \varphi(0)$].

Let us turn to the Laurent expansion of r^λ about $\lambda = -n - 2k$. This expansion will, of course, coincide with the Laurent expansion of $\Omega_n x_+^\mu$ about $\mu = -2k - 1$ (see Section 4.2), so that we have

$$r^\lambda = \Omega_n \frac{\delta^{(2k)}(r)}{(2k)!} \frac{1}{\lambda + n + 2k} + \Omega_n r^{-2k-n}$$

$$+ \Omega_n (\lambda + n + 2k)r^{-2k-n} \ln r + ... \qquad (2)$$

In this equation the left-hand side operates on $\varphi(x)$, the right-hand side on $S_\varphi(r)$. Further, $r^{-2k-n} \ln^m r$ is understood as the functional defined by

$$(r^{-2k-n} \ln^m r, S_\varphi(r)) = \int_0^\infty r^{-n-2k} \ln^m r \left[S_\varphi(r) - \varphi(0) - ... \right.$$

$$\left. ... - \frac{r^{2k-2}}{(2k-2)!} S_\varphi^{(2k-2)}(0) - \frac{r^{2k}}{(2k)!} S_\varphi^{(2k)}(0)\, \theta(1-r) \right] dr. \qquad (3)$$

Note that r^{-2k-n} is not the value of the generalized function r^λ at $\lambda = -2k - n$ (in fact it has a pole at this point), but the value at this point of the regular part of the Laurent expansion of r^λ.

In the expansions we have been discussing, the generalized function $r^{\lambda_0} \ln^k r$ is an associated function of degree λ_0 and order k, while $r^{-2k-n} \ln^m r$ is an associated function of degree $-2k - n$ and order m.

5. Convolutions of Generalized Functions

In classical analysis one often deals with the *convolution* of two functions $f(x)$ and $g(x)$, defined by

$$f(x) * g(x) = \int f(\xi)g(x - \xi)d\xi. \tag{1}$$

In generalized functional analysis the analogous operation is perhaps even of greater importance.

For generalized functions the definition of the convolution depends on the concept of the direct product. We shall therefore start with a discussion of this concept (Section 5.1). The remaining sections will be concerned with the definition and properties of the convolution, and will present some examples of its application.

5.1. Direct Product of Generalized Functions

Let $f(x)$ be a generalized function defined on X_k, the space of test functions (i.e., infinitely differentiable functions with bounded support) of the k independent variables x_1, x_2, ..., x_k, and let $g(y)$ be a generalized function defined on Y_m, the space of test functions of the m independent variables y_1, y_2, ..., y_m. We may use $f(x)$ and $g(y)$ to define a new generalized function $h(z)$ on Z_n, the space of test functions of the $n = k + m$ independent variables z_1, z_2, ..., z_n, where $z_i = x_i$ for $i \leqslant k$, and $z_i = y_{i-k}$ for $k < i \leqslant n$. This we do as follows. Instead of writing $\varphi(z)$, we shall write $\varphi(x, y)$. By fixing x we treat $\varphi(x, y)$ as a function only of y, which is then clearly in Y_m. We apply $g(y)$ to this function, obtaining some new function $\psi(x)$. This one is infinitely differentiable since

$$\frac{\psi(x + \Delta x_j) - \psi(x)}{\Delta x_j} = \left(g(y), \frac{\varphi(x + \Delta x_j, y) - \varphi(x, y)}{\Delta x_j}\right) \to \left(g(y), \frac{\partial\varphi(x, y)}{\partial x_j}\right),$$

which follows from the fact that $[\varphi(x + \Delta x_j, y) - \varphi(x, y)]/\Delta x_j$ converges in Y_m to $\partial\varphi(x, y)/\partial x_j$ and that $g(y)$ is a continuous functional. It is obvious also that $\psi(x)$ has bounded support. Thus $\psi(x)$ is in X_k, and we may now apply $f(x)$ to it. Hence the expression

$$(f(x), (g(y), \varphi(x, y))) \tag{2}$$

is a meaningful one. This defines a functional on Z_n, and it is implied by the continuity of $g(y)$ and $f(x)$ that this functional is continuous.[1] We shall denote it by $h(z) = f(x) \times g(y)$, and call it the *direct product* of $f(x)$ and $g(y)$.

The direct product is of a particularly simple form when $\varphi(x, y)$ is the product of two functions $\varphi_1(x)$ and $\varphi_2(y)$. Then by definition we obtain

$$(f(x) \times g(y), \varphi_1(x)\varphi_2(y)) = (f(x), (g(y), \varphi_1(x)\varphi_2(y)))$$

$$= (f(x), \varphi_1(x)(g(y), \varphi_2(y))) = (f, \varphi_1)(g, \varphi_2). \tag{3}$$

Examples. The direct product of two delta functions is given by $\delta(x) \times \delta(y) = \delta(x, y)$. Further, $\delta(x) \times 1(y)$ is defined by

$$(\delta(x) \times 1(y), \varphi(x, y)) = \int_Y \varphi(0, y)dy.$$

If $f(x)$ and $g(y)$ are regular functionals, their direct product is the regular functional corresponding to the function $f(x) g(y)$.

THE SUPPORT OF THE DIRECT PRODUCT

Recall that the set of essential points of f (points in every neighborhood of which the generalized function f fails to vanish) is called the *support* of f. Let F and G be the supports of f and g; we wish to determine H, the support of $h = f \times g$. This support will be found to be the direct product $F \times G$, in other words H is the set of ordered pairs (x, y) such that x is in F and y is in G.

---

[1] Let $\varphi_\nu(x, y)$ be a sequence converging to zero in Z_n. We wish to show that the number sequence

$$(f(x), (g(y), \varphi(x, y)))$$

converges to zero. For this purpose it is sufficient to show that all the $\psi_\nu(x) = (g(y), \varphi_\nu(x, y))$ vanish outside a fixed bounded region and that they converge uniformly to zero together with all their derivatives. The first of these assertions is obvious, since the $\varphi_\nu(x, y)$ vanish outside a fixed region in Z_n. Now let us assume that the $\psi_\nu(x)$ do not converge uniformly to zero. This means that for some $\epsilon > 0$ there exists a sequence of points $x^{(1)}, ..., x^{(\nu)}, ...$ such that

$$|\psi_\nu(x_\nu)| = |(g(y), \varphi_\nu(x_\nu, y))| \geqslant \epsilon.$$

But since the $\varphi_\nu(x, y)$ converge uniformly to zero with all their derivatives, the $\varphi_\nu^*(y) = \varphi_\nu(x_\nu, y)$ converge to zero in Y_m. The continuity of $g(y)$ then implies that

$$(g(y), \varphi_\nu^*(y)) = (g(y), \varphi_\nu(x_\nu, y)) \to 0,$$

in contradiction to our assumption. Thus the $\psi_\nu(x)$ converge uniformly to zero. The uniform convergence of all the derivatives of this sequence can be proved similarly, and this then completes the proof.

Indeed, let (x_0, y_0) be a point one of whose coordinates, say x_0, does not lie in the corresponding support. This means that f vanishes when applied to any $\varphi(x)$ in X_k which vanishes outside some fixed neighborhood U of x_0. Consider any function $\varphi(x, y)$ in Z_n whose support is in U. We shall show that $(f \times g, \varphi(x, y)) = 0$. In fact, by definition

$$(f \times g, \varphi(x, y)) = (f(y), (g(x), \varphi(x, y))) = (f(y), 0) = 0$$

so that $f \times g$ vanishes in the neighborhood of (x_0, y_0).

If, on the other hand, x_0 and y_0 are in F and G, there exists a function $\varphi(x) \psi(y)$ vanishing outside an arbitrary neighborhood of (x_0, y_0) and such that the application of $f \times g$ to this function does not give zero. This establishes the truth of the assertion.

Commutativity and Associativity

We have

$$f(x) \times g(y) = g(y) \times f(x),$$

$$f(x) \times \{g(y) \times h(z)\} = \{f(x) \times g(y)\} \times h(z). \tag{4}$$

To prove the first of these, note first that only continuous functionals appear in these equations, which therefore need be proved only on a dense set of functions in Z_n. We shall choose a dense set of the form $\sum_{j+1}^{\nu} \varphi_j(x)\psi_j(y)$, where $\varphi_j(x)$ and $\psi_j(y)$ $(j = 1, 2, ..., \nu = 1, 2, ...)$ are in X_k and Y_m, respectively.[2] For such functions we have

$$(f(x) \times g(y), \sum \varphi_j(x)\psi_j(y))$$

$$= \sum (f(x) \times g(y), \varphi_j(x)\psi_j(y)) = \sum (f(x), \varphi_j(x))(g(y), \psi_j(y))$$

and similarly

$$(g(y) \times f(x), \sum \varphi_j(x)\psi_j(y)) = \sum (g(y), \psi_j(y))(f(x), \varphi_j(x)),$$

as asserted.

The second equation is proved similarly on test functions of the form $\sum_j \varphi_j(x) \psi_j(y) \chi_j(z)$.

---

[2] This set is indeed dense in Z_n. For instance, let $\varphi(x, y)$ have support in $Q = \{|x| \leqslant a, |y| \leqslant a\}$. Then by Weierstrass' approximation theorem, for $\epsilon = 1/\nu$ there exists a polynomial $P_\nu(x, y)$ such that in $Q' = \{|x| \leqslant 2a, |y| \leqslant 2a\}$ it differs from $\varphi(x, y)$ by less than ϵ, together with its derivatives up to order ν. Further, let $b(x)$ be a fixed test function equal to unity for $|x| \leqslant a$ and to zero for $|x| \geqslant 2a$. Then as $\epsilon \to 0$, the $P_\nu(x, y) b(x) b(y)$ converge to $\varphi(x, y)$ in Z_n.

5.2. Convolutions of Generalized Functions

Let $f(x)$ and $g(x)$ be two absolutely integrable functions on the line, and let $h(x) = f(x) * g(x)$ be their convolution. Then the functional defined by the (also absolutely integrable) function $h(x)$ can be written in the form

$$(h(x), \varphi(x)) = \int h(x)\varphi(x)dx$$

$$\int \left\{ \int f(\xi)g(x)g(x - \xi)d\xi \right\}\varphi(x)dx$$

$$\int \int f(\xi)g(\eta)\varphi(\xi + \eta)d\xi d\eta.$$

In other words, the desired result is equivalent to applying the functional $f(x)\, g(y)$, which may be considered the direct product of $f(x)$ and $g(y)$, to the function $\varphi(x + y)$.

It is now natural to define the convolution of any two generalized functions f and g in general by

$$(f * g, \varphi) = (f(x) \times g(y), \varphi(x + y)). \tag{1}$$

One must bear in mind, however, that $\varphi(x + y)$ is not a function of bounded support in (x, y)-space, so that Eq. (1) is not in general meaningful.

However, under rather general assumptions which we shall soon make explicit, the definition is still workable.

We have seen below that the support of the direct product of $f(x)$ and $g(y)$ is the direct product of the supports of these functionals. Now Eq. (1) will be meaningful if the strip $|\, x + y\,| \leqslant a$ that contains the support of $\varphi(x + y)$ has a bounded intersection with the support of $f \times g$. Then in $|\, x + y \,| \leqslant a$ we can replace $\varphi(x + y)$ by a function $\varphi(x, y)$ with bounded support having the same values in the intersection of the strip and the support of $f \times g$. Now we can calculate the value of

$$(f \times g, \varphi(x, y)),$$

and it is easily shown that it is independent of the behavior of $\varphi(x, y)$ outside the intersection.

In particular, Eq. (1) is meaningful under either of the following conditions.

(a) Either f or g has bounded support.

(b) In one dimension the supports of both f and g are bounded on the same side (for instance, $f = 0$ for $x < a$, and $g = 0$ for $y < b$).

Consider first case (a). Let, for instance, $f(x)$ have bounded support and $g(y)$ be any functional. Then

$$(f * g, \varphi) = (f(x) \times g(y), \varphi(x + y)) = (g(y), (f(x), \varphi(x + y))). \tag{2}$$

Now $(f(x), \varphi(x + y))$ is an infinitely differentiable function which vanishes for $|y|$ so large that the supports of $\varphi(x + y)$ and $f(x)$ do not intersect. It is therefore in Y_m, and it is therefore meaningful to apply $g(y)$ to it.

If, on the other hand, $g(y)$ has bounded support while $f(x)$ may not, $(f(x), \varphi(x + y))$ is infinitely differentiable but does not in general have bounded support. Equation (2) remains valid, however, since $g(y)$ has bounded support and can therefore be applied meaningfully to this function.

Now consider case (b). Let us assume that the supports of $f(x)$ and $g(y)$ are bounded, say, on the left. Consider $(f(x), \varphi(x + y))$. As before, this is an infinitely differentiable function of y. For sufficiently large positive values of y, the support of $\varphi(x + y)$ does not intersect that of $f(x)$. Thus for such y the function $(f(x), \varphi(x + y))$ vanishes, and its support is therefore bounded on the right. But the support of $g(y)$ is bounded on the left by assumption, so that the intersection of the supports is bounded, and the right-hand side of Eq. (2) is meaningful.

Thus for cases (a) and (b) the convolution of f and g is well defined. In particular, the convolutions $\delta * f$ and $D\delta * f$ are well defined, where D is any differential operator; this follows from the fact that δ and $D\delta$ are concentrated at a point.

Let us obtain an explicit expression for $\delta * f$. By definition

$$(\delta * f, \varphi) = (\delta(x) \times f(y), \varphi(x + y))$$

$$= (f(y), (\delta(x), \varphi(x + y))) = (f(y), \varphi(y)) = (f, \varphi)$$

Thus for any f

$$\delta * f = f. \tag{3}$$

Similarly

$$D\delta * f = Df. \tag{4}$$

What we know about the direct product of functionals shows that their convolution is commutative, at least in cases (a) and (b); in other words,

$$f * g = g * f.$$

A similar statement may be made about associativity; we may write

$$(f * g) * h = f * (g * h),$$

assuming that the supports of two of the three functionals involved are bounded on both sides or that all three are bounded on the same side.

We now wish to prove the following formula for the derivative of the convolution:

$$D(f * g) = Df * g = f * Dg. \tag{5}$$

This is a formula which will be found quite useful in the future.

We have

$$(D(f * g), \varphi) = (f * g, D^*\varphi),$$

where, for instance, $D^* = (-1)^\nu D$ if D is a homogeneous differential operator of order ν. Further, by definition of the convolution we have

$$(f * g, D^*\varphi) = (g(y), (f(x), D^*\varphi(x + y)))$$

$$= (g(y), (Df(x), \varphi(x + y))) = (Df * g, \varphi),$$

from which it follows that $D(f * g) = Df * g$. The rest follows from commutativity:

$$D(f * g) = D(g * f) = Dg * f = f * Dg,$$

which establishes the desired result.

We wish further to prove a useful lemma on the continuity of the convolution.

Lemma. If $f_\nu \to f$, then $f_\nu * g \to f * g$ if any one of the following conditions is fulfilled:

(a) All of the f_ν are concentrated on a single bounded region.

(b) g is concentrated on a bounded region.

(c) The supports of the f_ν and g are bounded on the same side by a constant independent of ν.

Proof. By definition of the convolution, for any φ in K we have

$$(f_\nu * g, \varphi) = (f_\nu(y), (g(x), \varphi(x + y))). \tag{6}$$

In case (a) we may replace $(g(x), \varphi(x + y))$ by some $\psi(y)$ in K which vanishes outside the region in which all the $f_\nu(y)$ are concentrated. Thus

$$(f_\nu * g, \varphi) = (f_\nu(y), \psi(y)) \to (f, \psi) = (f * g, \varphi),$$

so that

$$f_\nu * g \to f * g,$$

as asserted.

In case (b) the function $\psi(y) = (g(x), \varphi(x + y))$ is in K; proceeding then as in case (a) we obtain the desired result.

For case (c), let us assume to be specific that the supports of the f_ν and g are bounded on the left. Then $\psi(y) = (g(x), \varphi(x + y))$ has support bounded on the right. This function can be replaced by another in K which vanishes outside the region on which all the $f_\nu(y)$ are concentrated, and then proceeding as above we obtain the desired result. This proves the lemma.

Corollary. If $f = f_t$ is a functional depending on a parameter t so that $\partial f_t/\partial t$ exists,[3] then

$$\frac{\partial}{\partial t}(f_t * g) = \frac{\partial f_t}{\partial t} * g$$

will hold under each of the following conditions:

(a) All the f_t are concentrated on a given bounded region.

(b) g is concentrated on a bounded region.

(c) The supports of the f_t and g are bounded on the same side by a constant independent of t.

To prove this corollary reduce it to the lemma just proven by using the fact that $\partial f_t/\partial t$ is the limit as $\Delta t \to 0$ of

$$\frac{f_{t+\Delta t} - f_t}{\Delta t}.$$

5.3. Newtonian Gravitational Potential and Elementary Solutions of Differential Equations

The classical (Newtonian) gravitational potential for a piecewise continuous mass density distribution $\mu(x_1, x_2, x_3)$ in a bounded region is given by

$$u(x_1, x_2, x_3) = -\frac{1}{4\pi} \int\int\int \frac{\mu(\xi_1, \xi_2, \xi_3)d\xi_1\, d\xi_2\, d\xi_3}{\sqrt{(\xi_1 - x_1)^2 + (\xi_2 - x_2)^2 + (\xi_3 - x_3)^2}}. \quad (1)$$

It is known that $u(x_1, x_2, x_3)$ is a function having derivatives through the second order and that it satisfies Poisson's equation

$$\Delta u(x_1, x_2, x_3) = \mu(x_1, x_2, x_3).$$

---

[3] Differentiation with respect to a parameter is discussed in more detail in Appendix 2.

Now Eq. (1) can also be written as the convolution

$$u = \mu * \left(-\frac{1}{4\pi r}\right) \qquad (r^2 = x_1^2 + x_2^2 + x_3^2).$$

Here μ may be any generalized function with bounded support. Then in general u will also be a generalized function. It turns out, in fact, that Poisson's formula holds in general. Indeed, recalling that $\Delta(1/r) = -4\pi\delta$ (Section 2.3) we may write

$$\Delta u = \Delta\left[\mu * \left(-\frac{1}{4\pi r}\right)\right] = \mu * \Delta\left(-\frac{1}{4\pi r}\right) = \mu * \delta = \mu.$$

Now let

$$P(D)u = \mu \tag{2}$$

be any linear differential equation with constant coefficients, and let μ be any generalized function. We shall call a functional $E(x)$ an *elementary solution* of Eq. (2) if

$$P(D)E = \delta$$

(to E may be added any solution of the homogeneous equation). When an elementary solution E is known, a solution of (2) can be written as the convolution

$$u = \mu * E$$

if, for instance, μ is a generalized function with bounded support. Indeed,

$$P(D)u = \mu * P(D)E = \mu * \delta = \mu.$$

For example, the regular functional $E = -1/4\pi r$ is an elementary solution for the Laplacian in three dimensions. In $n > 2$ dimensions an elementary solution for the Laplacian is the regular functional $-[(n-2)\,\Omega_n]^{-1}r^{2-n}$, where Ω_n is the surface (hypersurface) area of the unit sphere; for $n = 2$ the regular functional $-(2\pi)^{-1}\ln r^{-1}$ will do.

As an example, let us see how to construct an elementary solution of an ordinary differential equation with constant coefficients,[4] namely

$$a_0 E^{(n)} + a_1 E^{(n-1)} + \dots + a_n E = \delta(x). \tag{3}$$

Let $u_1(x), \dots, u_n(x)$ be a fundamental set of solutions of the homogeneous equation

$$a_0 u^{(n)}(x) + a_1 u^{(n-1)}(x) + \dots + a_n u(x) = 0. \tag{4}$$

[4] Actually the proof depends only on the existence of a fundamental set of solutions, which is true also for equations with variable coefficients so long as a_0 fails to vanish.

We write

$$E(x) = \begin{cases} A(x) \equiv \alpha_1 u_1(x) + ... + \alpha_n u_n(x) & \text{for } x > 0 \\ B(x) \equiv \beta_1 u_1(x) + ... + \beta_n u_n(x) & \text{for } x < 0 \end{cases}$$

and shall choose the α_i and β_j to satisfy Eq. (3). Since the delta function is the derivative of $\theta(x)$, it is sufficient to require that at $x = 0$ we have

$$A(0) = B(0), \qquad A'(0) = B'(0), ..., A^{(n-2)}(0) = B^{(n-2)}(0),$$
$$a_0[A^{(n-1)}(0) - B^{(n-1)}(0)] = 1. \tag{5}$$

Setting $\alpha_i - \beta_i = \gamma_i$ $(i = 1, ..., n)$, we obtain the following set of equations for the γ_i:

$$\gamma_1 u_1(0) + ... + \gamma_n u_n(0) = 0,$$
$$\gamma_1 u_1'(0) + ... + \gamma_n u_n'(0) = 0,$$
$$\cdots\cdots\cdots\cdots\cdots\cdots \tag{6}$$
$$\gamma_1 u_1^{(n-2)}(0) + ... + \gamma_n u_n^{(n-2)}(0) = 0,$$
$$\gamma_1 u_1^{(n-1)}(0) + ... + \gamma_n u_n^{(n-1)}(0) = \frac{1}{a_0}.$$

This set always has a solution, for its determinant is the Wronskian of the fundamental set of solutions $u_i(x)$ and thus never vanishes.

We see consequently that the elementary solution is quite arbitrary: only the γ_i, the differences of the coefficients of the fundamental set, are uniquely determined. This arbitrariness is very easily explained when we recall that the fundamental solution is determined only up to an additive solution of the homogeneous equation. This fact is used to construct Green's functions, which are elementary solutions satisfying particular boundary conditions.[5]

Example. Consider the equation

$$E'' = \delta(x). \tag{7}$$

For this case we may take $u_1 = 1$, $u_2 = x$ so that

$$A(x) = \alpha_1 + \alpha_2 x, \qquad B(x) = \beta_1 + \beta_2 x.$$

---

[5] See, for instance, M. A. Naimark, "Linear differential operators" (in Russian) Gostekhizdat, 1954. German translation: M. A. Neumark, "Lineare Differential-operatoren." Berlin, 1960.

Then the $\gamma_i = \alpha_i - \beta_i$ are

$$\gamma_1 = 0, \qquad \gamma_2 = 1.$$

Hence

$$E = \beta_1 + \beta_2 x + x_+.$$

This could, of course, have been seen immediately by inspection of Eq. (7).

In Section 6 and in Chapter III, Section 2, we shall construct elementary solutions for a large class of partial differential equations.

5.4. Poisson's Integral and Elementary Solutions of Cauchy's Problem

Poisson's classical formula in the theory of heat conduction is

$$u(x, t) = \frac{1}{2\sqrt{\pi t}} \int_{-\infty}^{\infty} \exp\left(-\frac{(x - \xi)^2}{4t}\right) \mu(\xi)d\xi, \qquad (1)$$

where we shall take $\mu(\xi)$ to be a summable function with bounded support. It is well known[6] that $u(x, t)$ has a first derivative with respect to t and the first two derivatives with respect to x, and that it satisfies the heat equation

$$\frac{\partial u}{\partial t} = \frac{\partial^2 u}{\partial x^2} \qquad (2)$$

with the initial condition

$$u(x, 0) = \mu(x). \qquad (3)$$

The function $u(x, t)$ is the temperature at any point x of an infinite conducting rod at any time t if the initial temperature $\mu(x)$ is known at $t = 0$. Equation (1) can be written in the convolution form

$$u(x, t) = \mu(x) * \frac{1}{2\sqrt{\pi t}} \exp\left(-\frac{x^2}{4t}\right). \qquad (4)$$

In this equation we may now take μ to be any generalized function with bounded support. Then in general $u(x, t)$ as given by (4) will, of course, be a generalized function depending on the parameter t. We shall prove that this generalized function is a solution of the heat equation (2) with initial condition (3).

---

[6] Smirnov, "Higher Math," Vol. 2, p. 548. See also R. Courant and D. Hilbert, "Methods of Mathematical Physics," Vol. II, p. 198. Interscience, New York, 1962.

Note first that for $t > 0$ differentiation of

$$v(x, t) = \frac{1}{2\sqrt{\pi t}} \exp\left(-\frac{x^2}{4t}\right),$$

in the sense of generalized functions, both with respect to x and with respect to the parameter t, is equivalent to ordinary differentiation with respect to x or t. From the lemma we have proven on the continuity of the convolution (Section 5.2) we have

$$\frac{\partial}{\partial t}\left(\mu(x) * v(x, t)\right) = \mu(x) * \frac{\partial}{\partial t} v(x, t)$$

and from the formula for the derivative of the convolution in the same section we have

$$\frac{\partial^2}{\partial x^2}(\mu * v) = \mu * \frac{\partial^2}{\partial x^2} v.$$

Thus

$$\left(\frac{\partial}{\partial t} - \frac{\partial^2}{\partial x^2}\right)(\mu * v) = \mu * \left(\frac{\partial^2}{\partial t} - \frac{\partial^2}{\partial x^2}\right) v.$$

But

$$\left(\frac{\partial}{\partial t} - \frac{\partial^2}{\partial x^2}\right) \frac{1}{2\sqrt{\pi t}} \exp\left(-\frac{x^2}{4t}\right) \equiv 0,$$

so that $\mu * v$ is indeed a solution of the heat equation.

On the other hand, since $(2\sqrt{\pi t})^{-1} \exp(-x^2/4t)$ converges in the sense of generalized functions to $\delta(x)$ as $t \to 0$ (see Section 2.5, Example 2), it follows from the lemma on the continuity of the convolution that

$$u(x, t) = \mu * \frac{1}{2\sqrt{\pi t}} \exp\left(-\frac{x^2}{4t}\right) \to \mu * \delta = \mu,$$

as asserted.

Now let

$$\frac{\partial u}{\partial t} = P\left(\frac{\partial}{\partial x}\right) u$$

be any differential equation with constant coefficients. We may now state Cauchy's problem: To find the solution to this equation (i.e., a generalized function depending on the parameter t) which at $t = 0$ is equal to a given generalized function $u_0(x)$. The particular solution of Cauchy's problem which at $t = 0$ is equal to $\delta(x)$ will be called the *elementary solution of this problem* and will be denoted by $E(x, t)$. For the case of the heat equation $(2\sqrt{\pi t})^{-1} \exp(-x^2/4t)$ is such a solution.

If the elementary solution is known, Cauchy's problem with initial condition $u_0(x)$ may be written, assuming that $u_0(x)$ has bounded support, in the form of the convolution

$$u(x, t) = E(x, t) * u_0(x).$$

Indeed, for $t > 0$ we have, on the one hand,

$$\left[\frac{\partial}{\partial t} - P\left(\frac{\partial}{\partial x}\right)\right] u(x, t)$$

$$= \frac{\partial}{\partial t}\left[E\left(x, t\right) * u_0(x)\right] - P\left(\frac{\partial}{\partial x}\right)\left[E(x, t) * u_0(x)\right]$$

$$= \frac{\partial E\left(x, t\right)}{\partial t} * u_0(x) - P\left(\frac{\partial}{\partial x}\right) E(x, t) * u_0(x)$$

$$= \left[\frac{\partial E}{\partial t} - P\left(\frac{\partial}{\partial x}\right)E\right] * u_0(x) = 0 * u_0(x) = 0,$$

and, on the other hand,

$$\lim_{t \to 0} u(x, t) = \left[\lim_{t \to 0} E(x, t)\right] * u_0(x) = \delta(x) * u_0(x) = u_0(x),$$

as asserted.

We may consider also the more general linear equation

$$P\left(\frac{\partial}{\partial t}, \frac{\partial}{\partial x}\right) u(x, t) = 0 \tag{5}$$

of, say, mth order with respect to t, still with constant coefficients. For this case Cauchy's problem is to find the solution $u(x, t)$ of (5) which satisfies the initial conditions

$$u(x, 0) = u_0(x), \quad \frac{\partial u(x, 0)}{\partial t} = u_1(x), \ldots, \frac{\partial^{m-1} u(x, 0)}{\partial t^{m-1}} = u_{m-1}(x). \tag{6}$$

We shall call an *elementary solution* of (5) a solution $E(x, t)$ which satisfies the initial conditions

$$E(x, 0) = 0, \quad \frac{\partial E\left(x, 0\right)}{\partial t} = 0, \ldots, \frac{\partial^{m-2} E\left(x, 0\right)}{\partial t^{m-2}} = 0,$$

$$\frac{\partial^{m-1} E(x, 0)}{\partial t^{m-1}} = \delta(x). \tag{7}$$

If the generalized function $u_{m-1}(x)$ has bounded support, the solution to Cauchy's problem for Eq. (5) with initial conditions

$$u(x, 0) = 0, \quad \frac{\partial u(x, 0)}{\partial t} = 0, \ ..., \ \frac{\partial^{m-2} u(x, 0)}{\partial t^{m-2}} = 0, \quad \frac{\partial^{m-1} u(x, 0)}{\partial t^{m-1}} = u_{m-1}(x) \quad (8)$$

can be written in the form

$$u(x, t) = E(x, t) * u_{m-1}(x). \tag{9}$$

This form is also valid for any $u_{m-1}(x)$ if $E(x, t)$ has bounded support.

Indeed, our function $u(x, t)$ is, on the one hand, a solution of (5), since the differential operator $P(\partial/\partial t, \partial/\partial x)$ can be applied to $E(x, t)$. Further, as $t \to 0$, we have

$$u(x, t) \to E(x, 0) * u_{m-1}(x) = 0,$$

$$\frac{du(x, t)}{\partial t} \to \frac{\partial E(x, 0)}{\partial t} * u_{m-1}(x) = 0,$$

$$\cdot \quad \cdot \quad \cdot \quad \cdot \quad \cdot \quad \cdot \quad \cdot \quad \cdot \quad \cdot \quad \cdot$$

$$\frac{\partial^{m-1} u(x, t)}{\partial t^{m-1}} \to \frac{\partial^{m-1} E(x, 0)}{\partial t^{m-1}} * u_{m-1}(x) = u_{m-1}(x).$$

Example.   Consider the vibrating-string equation

$$\frac{\partial^2 u}{\partial t^2} = \frac{\partial^2 u}{\partial x^2} \qquad (-\infty < x < \infty). \tag{10}$$

We wish to show that the elementary solution of Cauchy's problem for this equation is

$$E(x, t) = \begin{cases} \frac{1}{2} & \text{for } |x| < t, \\ 0 & \text{for } |x| > t. \end{cases}$$

Indeed, for fixed $t > 0$ we have

$$\frac{\partial E(x, t)}{\partial x} = \frac{1}{2} \delta(x + t) - \frac{1}{2} \delta(x - t),$$

$$\frac{\partial^2 E(x, t)}{\partial x^2} = \frac{1}{2} \delta'(x + t) - \frac{1}{2} \delta'(x - t),$$

and for fixed x we may differentiate $E(x, t)$ with respect to the parameter t, obtaining

$$\frac{\partial E(x, t)}{\partial t} = \frac{1}{2} \delta(x + t) + \frac{1}{2} \delta(x - t),$$

$$\frac{\partial^2 E(x, t)}{\partial t^2} = \frac{1}{2} \delta'(x + t) - \frac{1}{2} \delta'(x - t),$$

$$\tag{11}$$

which then shows that (10) is satisfied.

Further, allowing t to approach zero in (11), we obtain

$$\frac{\partial E(x,\, t)}{\partial t}\bigg|_{t=0} = \delta\,(x),$$

so that the initial condition defining the elementary solution of Cauchy's problem is also satisfied (it is obvious that as $t \to 0$ the solution $E(x,\, t)$ itself converges to zero).

This leads us to a general formula for solving Cauchy's problem for the vibrating string equation (10) with initial conditions

$$u(x,\, 0) = 0, \qquad \frac{\partial u(x,\, 0)}{\partial t} = u_1(x).$$

According to (9) we have [assuming $u_1(x)$ to be a locally summable function]

$$u(x,\, t) = E(x,\, t) * u_1(x) = \int_{-\infty}^{\infty} E(\xi,\, t) u_1(x - \xi)\, d\xi$$

$$= \tfrac{1}{2} \int_{-t}^{t} u_1(x - \xi)\, d\xi = \tfrac{1}{2} \int_{x-t}^{x+t} u_1(\eta)\, d\eta.$$

The elementary solution $E(x,\, t)$ we have obtained in this case has bounded support, so that $E(x,\, t) * u_1(x)$ exists for any generalized function $u_1(x)$.

We may now go over to the most general case in which we are given the functions

$$u(x,\, 0) = u_0(x), \qquad \frac{\partial u(x,\, 0)}{\partial t} = u_1(x),\, ..., \qquad \frac{\partial^{m-1} u(x,\, 0)}{\partial t^{m-1}} = u_{m-1}(x),$$

each of which has bounded support (although if $E(x,\, t)$ has bounded support we may drop this last condition). First let v be a solution satisfying the initial conditions

$$\frac{\partial^k v(x,\, 0)}{\partial t^k} = 0 \qquad (k = 0,\, 1,\, 2,\, ...,\, m - 2), \qquad \frac{\partial^{m-1} v(x,\, 0)}{\partial t^{m-1}} = u_{m-2}(x),$$

then $u_1(x,\, t) = \partial v/\partial t$ satisfies the same equation but with initial conditions in which only $\partial^{m-2} u_1(x,\, 0)/\partial t^{m-2}$ and $\partial^{m-1} u_1(x,\, 0)/\partial t^{m-1}$ are nonzero. If we subtract from $u_1(x,\, t)$ the solution $u_2(x,\, t)$ of the same problem with the initial conditions

$$\frac{\partial^k u_2(x,\, 0)}{\partial t^k} = 0 \qquad (k = 0,\, 1,\, 2,\, ...,\, m - 2), \qquad \frac{\partial^{m-1} u_2(x,\, 0)}{\partial t^{m-1}} = \frac{\partial^{m-1} u_1(x,\, 0)}{\partial t^{m-1}},$$

we will obtain a third solution u_3 which at $t = 0$ will have only the $(m - 2)$nd partial derivative with respect to time not equal to zero; moreover, this derivative will be equal to the given function $u_{m-2}(x)$. Similarly, for any $k \leqslant m - 1$ there exists a solution such that at $t = 0$ of the first $m - 1$ partial derivatives with respect to t only the kth will be nonzero, and moreover this one will be equal to the given function $u_k(x)$. The sum of such particular solutions will then give the solution to the general problem.

Let us use this method to solve Cauchy's problem for the vibrating-string equation (10) with the general initial conditions

$$u(x, 0) = u_0(x), \qquad \frac{\partial u(x, 0)}{\partial t} = u_1(x). \tag{12}$$

We proceed as follows. First we find a solution $u_1(x, t)$ for the initial conditions

$$v(x, 0) = 0, \qquad \frac{\partial v(x, 0)}{\partial t} = u_0(x).$$

We have already seen that this solution is

$$v(x, t) = \tfrac{1}{2} \int_{x-t}^{x+1} u_0(\xi)d\xi.$$

By differentiating $v(x, t)$ with respect to t, we arrive at the new solution

$$\frac{\partial v(x, t)}{\partial t} = u_1(x, t) = \tfrac{1}{2}\left[u_0(x + t) + u_0(x - t)\right],$$

which satisfies the initial conditions

$$u_1(x, 0) = \frac{\partial v(x, 0)}{\partial t} = u_0(x),$$

$$\frac{\partial u_1(x, 0)}{\partial t} = \tfrac{1}{2}\left[\frac{\partial u_0(x)}{\partial t} - \frac{\partial u_0(x)}{\partial t}\right] = 0.$$

From this we see that the solution corresponding to the initial conditions of Eq. (12) is

$$u(x, t) = \frac{u_0(x + t) + u_0(x - t)}{2} + \tfrac{1}{2} \int_{x-t}^{x+t} u_1(\xi)d\xi.$$

This is the celebrated formula of D'Alembert.

Later (Section 6) we shall use similar methods to obtain elementary solutions of Cauchy's problem for a wide class of partial differential equations.

5.5. Integrals and Derivatives of Higher Orders

Cauchy's well-known formula

$$g_n(x) = \int_0^x \int_0^{\xi_{n-1}} \cdots \int_0^{\xi_2} \int_0^{\xi_1} g(\xi) d\xi d\xi_1 \cdots d\xi_{n-1}$$

$$= \frac{1}{(n-1)!} \int_0^x g(\xi)(x-\xi)^{n-1} d\xi$$

reduces the calculation of the n-fold primitive of a function $g(x)$ defined for $x \geqslant 0$ to a single integral. This formula may also be written in the form

$$g_n(x) = g(x) * \frac{x^{n-1}}{(n-1)!} = g(x) * \frac{x^{n-1}}{\Gamma(n)},$$

where for $x < 0$ both $g(x)$ and x^{n-1} are replaced by zero.

It would seem quite natural to generalize this formula to the case of arbitrary index λ and arbitrary generalized function g concentrated on the half-line $x \geqslant 0$. We thus define the *primitive of order* λ of g as the convolution

$$g_\lambda(x) = g(x) * \frac{x_+^{\lambda-1}}{\Gamma(\lambda)}. \tag{1}$$

This formula holds quite simply for Re $\lambda > 0$. For other values of λ we must understand $x_+^{\lambda-1}/\Gamma(\lambda)$ as the generalized function constructed in Section 3.5. Since it remains concentrated on the half-line $x \geqslant 0$, the definition in terms of the convolution remains consistent. We shall find it convenient to call this generalized function Φ_λ, writing

$$g_\lambda(x) = g(x) * \Phi_\lambda. \tag{2}$$

Recalling that

$$\Phi_{-k} = \delta^{(k)}(x) \qquad (k = 0, 1, 2, \ldots)$$

(see Section 3.5), we may write

$$g_0(x) = g(x) * \Phi_0 = g(x) * \delta(x) = g(x),$$

$$g_{-1}(x) = g(x) * \Phi_{-1} = g(x) * \delta'(x) = g'(x),$$

and similar expressions for other orders. Thus Eq. (2) with various λ will give not only the integrals but also the derivatives of $g(x)$. We shall thus henceforth say that the convolution

$$g_{-\lambda} = g * \Phi_{-\lambda}$$

is *the derivative of order λ of the generalized function g*, writing

$$g_{-\lambda} = \frac{d^\lambda}{dx^\lambda} g.$$

The Φ_λ have the property that

$$\Phi_\lambda * \Phi_\mu = \Phi_{\lambda+\mu}, \tag{3}$$

or

$$\frac{x_+^{\lambda-1}}{\Gamma(\lambda)} * \frac{x_+^{\mu-1}}{\Gamma(\mu)} = \frac{x_+^{\lambda+\mu-1}}{\Gamma(\lambda+\mu)}. \tag{3'}$$

Let us first prove this formula for Re $\lambda > 0$ and Re $\mu > 0$. Since

$$\Phi_\lambda * \Phi_\mu = \frac{x_+^{\lambda-1}}{\Gamma(\lambda)} * \frac{x_+^{\mu-1}}{\Gamma(\mu)} = \int_0^x \frac{\xi^{\lambda-1}}{\Gamma(\lambda)} \cdot \frac{(x-\xi)^{\mu-1}}{\Gamma(\mu)} \, d\xi,$$

and

$$\Phi_{\lambda+\mu} = \frac{x_+^{\lambda+\mu-1}}{\Gamma(\lambda+\mu)},$$

we need only prove that

$$\int_0^x \xi^{\lambda-1}(x-\xi)^{\mu-1} d\xi = \frac{\Gamma(\lambda)\Gamma(\mu)}{\Gamma(\lambda+\mu)} x^{\lambda+\mu-1}.$$

We set $\xi = xt$ on the left-hand side, so that the integral becomes

$$x^{\lambda+\mu-1} \int_0^1 t^{\lambda-1}(1-t)^{\mu-1} dt,$$

and what we wish to prove is seen to follow from the well-known relation

$$B(\lambda, \mu) = \int_0^1 t^{\lambda-1}(1-t)^{\mu-1} dt = \frac{\Gamma(\lambda)\Gamma(\mu)}{\Gamma(\lambda+\mu)}.$$

Equation (3) can now be proven for other values of λ, μ by analytic continuation.

A further implication of (3) is that if g is any generalized function concentrated on the half-line $x \geqslant 0$, then

$$(g * \Phi_\lambda) * \Phi_\mu = g * (\Phi_\lambda * \Phi_\mu) = g * \Phi_{\lambda+\mu}. \tag{4}$$

Setting $\mu = -\lambda$ we see that differentiation and integration of the same order are mutually inverse processes. It follows further from (4) that

$$\frac{d^\beta}{dx^\beta}\left(\frac{d^\gamma g}{dx^\gamma}\right) = \frac{d^{\beta+\gamma}g}{dx^{\beta+\gamma}} \tag{5}$$

for any β and γ.

Let us point out some other implications of (3). Replacing λ by $-\lambda$, we may write

$$\frac{d^\lambda}{dx^\lambda}\left(\frac{x_+^{\mu-1}}{\Gamma(\mu)}\right) = \frac{x_+^{\mu-\lambda-1}}{\Gamma(\mu-\lambda)}. \tag{6}$$

In particular, for $\mu = 1$ we have

$$\frac{d^\lambda}{dx^\lambda}\,\theta(x) = \frac{x_+^{-\lambda}}{\Gamma(-\lambda+1)}. \tag{6'}$$

Writing $\mu = -k$ in (6), where k is a nonnegative integer, we find that

$$\frac{d^\lambda}{dx^\lambda}\,(\delta^{(k)}(x)) = \frac{x_+^{-k-\lambda-1}}{\Gamma(-k-\lambda)}. \tag{7}$$

If, on the other hand, $\mu - \lambda = -k$, where k is a nonnegative integer, Eq. (6) implies that

$$\frac{d^\lambda}{dx^\lambda}\left(\frac{x_+^{\lambda-k-1}}{\Gamma(\lambda-k)}\right) = \delta^{(k)}(x). \tag{8}$$

Example 1.   Consider *Abel's integral equation*

$$g(x) = \frac{1}{\Gamma(1-\alpha)}\int_0^x \frac{f(\xi)d\xi}{(x-\xi)^\alpha}. \tag{9}$$

Here g is given and f is the unknown function.

In the classical theory[7] α is assumed less than 1, which ensures the convergence of the integral on the right. We, however, need not make this assumption, since the right-hand side can be understood for any α as the integral of order $\lambda = -\alpha + 1$ of a generalized function f, that is, as the convolution

$$g(x) = f(x) * \Phi_\lambda. \tag{9'}$$

In order to obtain an expression for f in terms of g, we must obviously apply the differential operator of order λ to the latter. We thus have

---

[7] See Fikhtengolts, "Calculus," Vol. III, p. 290 ($\alpha = \frac{1}{2}$). See also E. T. Whittaker and G. N. Watson, "A Course of Modern Analysis," p. 229. Cambridge, Univ. Press, London and New York, 1943. Hereafter referred to as Whittaker and Watson, "Modern Analysis."

$f(x) = (d^\lambda/dx^\lambda) g(x)$. The solution is obtained by taking the convolution with $\Phi_{-\lambda}$, so that

$$g(x) * \Phi_{-\lambda} = (f(x) * \Phi_\lambda) * \Phi_{-\lambda} = f(x) * (\Phi_\lambda * \Phi_{-\lambda})$$
$$= f(x) * \delta = f(x). \tag{10}$$

Let us assume, in particular, that $0 < \alpha < 1$, so that Φ_λ becomes the ordinary function $[\Gamma(1 - \alpha)]^{-1}x_+^{-\alpha}$ and Eq. (9') can be written in the form of (9). In this case $-\lambda = \alpha - 1$ and Φ_λ is a singular functional so that in general the solution of (9) cannot be written in the classical form. If we assume further, however, that $g(x)$ is differentiable, then it becomes possible to use the classical form. Explicitly, we have

$$g(x) * \Phi_{\alpha-1} = g(x) * \frac{d}{dx} \Phi_\alpha = \frac{d}{dx} g(x) * \Phi_\alpha = \frac{1}{\Gamma(\alpha)} \int_0^x \frac{g'(\xi)d\xi}{(x - \xi)^{1-\alpha}}$$

or in other words

$$f(x) = \frac{1}{\Gamma(\alpha)} \int_0^x \frac{g'(\xi)d\xi}{(x - \xi)^{1-\alpha}}. \tag{11}$$

This expression was in fact obtained by Abel in solving Eq. (9) with $\alpha = \frac{1}{2}$.

Example 2. Many special functions can be written as derivatives of nonintegral order of elementary functions. There exist two such expressions for the *hypergeometric function*.[8] This function, written $F(\alpha, \beta, \gamma; x)$, is defined for $\text{Re}\,\gamma > \text{Re}\,\beta > 0$ and for $|x| < 1$ by the integral

$$F(\alpha, \beta, \gamma; x) = \frac{\Gamma(\gamma)}{\Gamma(\beta)\Gamma(\gamma - \beta)} \int_0^1 t^{\beta-1}(1 - t)^{\gamma-\beta-1}(1 - tx)^{-\alpha}\, dt. \tag{12}$$

For the remaining values of β, γ ($\gamma \neq 0, -1, -2, \dots$) and $|x| < 1$ it is defined by analytic continuation or by regularizing the above integral (see Section 3.8). Let us write $w = tx$ in this expression. We then obtain

$$F(\alpha, \beta, \gamma; x) = \frac{\Gamma(\gamma)}{\Gamma(\beta)\Gamma(\gamma - \beta)x^{\gamma-1}} \int_0^x w^{\beta-1}(1 - w)^{-\alpha}(x - w)^{\gamma-\beta-1}\, dw, \tag{13}$$

which we may write in the form

$$\frac{x^{\gamma-1}}{\Gamma(\gamma)} F(\alpha, \beta, \gamma; x) = \frac{d^{\beta-\gamma}}{dx^{\beta-\gamma}} \left(\frac{x_+^{\beta-1}(1 - x)_+^{-\alpha}}{\Gamma(\beta)} \right). \tag{14}$$

---

[8] See Fikhtengol'ts, "Calculus," Vol. II, p. 793. See also Whittaker and Watson, "Modern Analysis," p. 293.

Thus $[x^{\gamma-1}/\Gamma(\gamma)]\, F(\alpha, \beta, \gamma; x)$ is the derivative of order $\beta - \gamma$ of the function $x_{+}^{\beta-1}(1 - x)_{+}^{-\alpha}/\Gamma(\beta)$. This may also be written in the convolution form

$$\frac{x^{\gamma-1}}{\Gamma(\gamma)}\, F(\alpha, \beta, \gamma; x) = \frac{x_{+}^{\gamma-\beta-1}}{\Gamma(\gamma - \beta)} * \frac{x_{+}^{\beta-1}(1 - x)_{+}^{-\alpha}}{\Gamma(\beta)}. \tag{15}$$

Another expression for the hypergeometric function as a derivative of nonintegral order can be obtained by writing $w = x(1 - t)/(1 - tx)$ in the integrand of (12). This gives

$$F(\alpha, \beta, \gamma; x) = \frac{\Gamma(\gamma)(1 - x)^{\gamma-\alpha-\beta}}{\Gamma(\beta)\Gamma(\gamma - \beta)x^{\gamma-1}} \int_{0}^{x} w^{\gamma-\beta-1}(1 - w)^{\alpha-\gamma}(x - w)^{\beta-1}\, dw. \tag{16}$$

In other words,

$$\frac{x^{\gamma-1}(1 - x)^{\alpha+\beta-\gamma}}{\Gamma(\gamma)}\, F(\alpha, \beta, \gamma; x) = \frac{d^{-\beta}}{dx^{-\beta}} \left[\frac{x_{+}^{\gamma-\beta-1}(1 - x)_{+}^{\alpha-\gamma}}{\Gamma(\gamma - \beta)} \right]. \tag{17}$$

By comparing (14) and (17) we obtain the well-known relation

$$F(\alpha, \beta, \gamma; x) = (1 - x)^{\gamma-\alpha-\beta}F(\gamma - \alpha, \gamma - \beta, \gamma; x). \tag{18}$$

An interesting relation for the hypergeometric function is implied by the previously derived Eq. (5)

$$\frac{d^{\beta}}{dx^{\beta}} \left(\frac{d^{\gamma}g}{dx^{\gamma}} \right) = \frac{d^{\beta+\gamma}g}{dx^{\beta+\gamma}}. \tag{19}$$

Let us take the derivative or order $-\delta$ of both sides of (14). This gives

$$\frac{d^{-\delta}}{dx^{-\delta}} \left[\frac{x^{\gamma-1}}{\Gamma(\gamma)}\, F(\alpha, \beta, \gamma; x) \right] = \frac{d^{\beta-\gamma-\delta}}{dx^{\beta-\gamma-\delta}} \left[\frac{x_{+}^{\beta-1}(1 - x)_{+}^{-\alpha}}{\Gamma(\beta)} \right]$$

$$= \frac{x^{\gamma+\delta-1}}{\Gamma(\gamma + \delta)}\, F(\alpha, \beta, \gamma + \delta; x). \tag{20}$$

In integral form this is

$$F(\alpha, \beta, \gamma + \delta; x) = \frac{\Gamma(\gamma + \delta)}{\Gamma(\gamma)\Gamma(\delta)x^{\gamma+\delta-1}} \int_{0}^{x} w^{\gamma-1}F(\alpha, \beta, \gamma; w)\,(x - w)^{\delta-1}dw$$

$$= \frac{\Gamma(\gamma + \delta)}{\Gamma(\gamma)\Gamma(\delta)} \int_{0}^{1} w^{\gamma-1}(1 - w)^{\delta-1}F(\alpha, \beta, \gamma; xw)\, dw. \tag{21}$$

This formula will hold for all values of $\alpha, \beta, \gamma, \delta$, if we understand the integral on the right in the sense of its regularization (see Section 3.8).

The differential representations of the hypergeometric function in (14) and (17) may be used to establish the conditions under which the function is a polynomial or a polynomial multiplied by $(1 - x)^p$.

Let $\beta = -k$, where k is a nonnegative integer. Then $x^{\beta-1}/\Gamma(\beta) = \delta^{(k)}(x)$. Thus (14) becomes

$$\frac{x^{\gamma-1}F(\alpha, -k, \gamma; x)}{\Gamma(\gamma)} = \frac{d^{-k-\gamma}}{dx^{-k-\gamma}} [(1 - x)^{-\alpha}\delta^{(k)}(x)]. \tag{22}$$

Now the expression[9] $(1 - x)^{-\alpha}\delta^{(k)}(x)$ is a linear combination of delta functions and its derivatives:

$$(1 - x)^{-\alpha}\delta^{(k)}(x) = \sum_{r=0}^{k} (-1)^r C_k^r \frac{\Gamma(\alpha + r)}{\Gamma(\alpha)} \delta^{(k-r)}(x). \tag{23}$$

According to (7) we have

$$\frac{d^{-k-\gamma}}{dx^{-k-\gamma}} [\delta^{(k-r)}(x)] = \frac{x^{r+\gamma-1}}{\Gamma(r + \gamma)}.$$

Inserting (23) into (22), performing the required differentiation, and cancelling out $x^{\gamma-1}/\Gamma(\gamma)$, we arrive at

$$F(\alpha, -k, \gamma; x) = \sum_{r=0}^{k} (-1)^r C_k^r \frac{\Gamma(\alpha + r)}{\Gamma(\alpha)} \frac{\Gamma(\gamma)}{\Gamma(\gamma + r)} x^r \tag{24}$$

(where γ need not be a negative integer or zero). The function $F(\alpha, -n, \gamma; x)$ is called a *Jacobi polynomial* and is denoted by $G(\alpha, -n, \gamma; x)$. It follows from (18) and (24) that if $\beta - \gamma = n$ is a nonnegative integer, the hypergeometric function becomes

$$F(\alpha, \gamma + n, \gamma; x) = (1 - x)^{-\alpha-n}G(\gamma - \alpha, -n, \gamma; x).$$

From the symmetry relation $F(\alpha, \beta, \gamma; x) = F(\beta, \alpha, \gamma; x)$ we may obtain a similar relation for the cases in which α and $\gamma - \alpha$ are negative integers or zero.

Let us now treat the hypergeometric function as a function of γ. From (14) we see that it may have a singularity if γ is a negative integer or zero. At $\gamma = -n$ the $\Gamma(\gamma)$ function has a pole with residue $(-1)^n/n!$. If α or β is also a negative integer or zero, for instance if $\beta = -m$ but with $m < n$, according to what we have said the hypergeometric function

---

[9] This product is well defined, for since $\delta^{(k)}(x)$ is concentrated on the origin, we may replace $(1 - x)^{-\alpha}$ by an infinitely differentiable function equal to $(1 - x)^{-\alpha}$ in a neighborhood of the origin.

degenerates into a polynomial. If, however, this condition is not fulfilled, it has a pole (as a function of γ) with residue

$$\frac{(-1)^n x^{n+1}}{n!} \frac{d^{\beta+n}}{dx^{\beta+n}} \left(\frac{x_+^{\beta-1}(1-x)_+^{-\alpha}}{\Gamma(\beta)} \right) = \frac{(-1)^n x^{n+1}}{n!} \frac{d^{n+1}F(\alpha, \beta, 1; x)}{dx^{n+1}}. \quad (25)$$

Let us now consider the *Bessel functions*[10] and show that $u^{p/2} J_p(\sqrt{u})$ can also be represented as the derivative of nonintegral order of an elementary function. Recall that for $\mathrm{Re}\, p > -\frac{1}{2}$ we may write

$$J_p(z) = \frac{2(\frac{1}{2}z)^p}{\Gamma(p + \frac{1}{2})\sqrt{\pi}} \int_0^1 (1 - t^2)^{p-\frac{1}{2}} \cos zt \, dt. \quad (26)$$

For other values of p we may understand this integral in the sense of its regularization. We now set $zt = w$, which gives

$$J_p(z) = \frac{2}{(2z)^p \Gamma(p + \frac{1}{2})\sqrt{\pi}} \int_0^z (z^2 - w^2)^{p-\frac{1}{2}} \cos w \, dw, \quad (27)$$

or

$$2^p \sqrt{\pi} u^{p/2} J_p(\sqrt{u}) = \int_0^u \frac{(u - v)^{p-\frac{1}{2}}}{\Gamma(p + \frac{1}{2})} \frac{\cos \sqrt{v}}{\sqrt{u}} \, dv. \quad (28)$$

This may be written in the form

$$2^p \sqrt{\pi}\, u^{p/2} J_p(\sqrt{u}) = \frac{d^{-p-\frac{1}{2}}}{du^{-p-\frac{1}{2}}} \left[\frac{\cos \sqrt{u}}{\sqrt{u}} \right]. \quad (29)$$

Let us now take the derivative of order $-q - 1$ of both sides of this equation. Then

$$\frac{d^{-q-1}}{du^{-q-1}} [2^p \sqrt{\pi}\, u^{p/2} J_p(\sqrt{u})] = \frac{d^{-p-q-\frac{3}{2}}}{du^{-p-q-\frac{3}{2}}} \left[\frac{\cos \sqrt{u}}{\sqrt{u}} \right]$$

$$= 2^{p+q+1} \sqrt{\pi}\, u^{\frac{1}{2}(p+q+1)} J_{p+q+1}(\sqrt{u}). \quad (30)$$

In integral form this equation reads

$$2^{q+1} u^{\frac{1}{2}(p+q+1)} J_{p+q+1}(\sqrt{u}) = \int_0^u v^{\frac{1}{2}p} J_p(\sqrt{v}) \frac{(u - v)^q}{\Gamma(q + 1)} \, dv. \quad (31)$$

---

[10] See Smirnov, "Higher Math," Vol. 3, Part 2, Chapter VI, Section 2. See also G. N. Watson, "Treatise on the Theory of Bessel Functions." Cambridge Univ. Press, London and New York, 1922.

The substitution $u = x^2$, $v = x^2 \sin \theta$ transforms (31) to

$$J_{p+q+1}(x) = \frac{x^{q+1}}{\Gamma(q+1)2^q} \int_0^{\frac{1}{2}\pi} J_p(x \sin \theta) \sin^{p+1} \theta \cos^{2q+1} \theta \, d\theta. \qquad (32)$$

This is called *Sonin's integral*, since the expression was obtained by N. J. Sonin (Sonine) in 1880 for nonnegative integral p and q.

6. Elementary Solutions of Differential Equations with Constant Coefficients

6.1. Elementary Solutions of Elliptic Equations

Consider a linear differential operator $L(\partial/\partial x_1, ..., \partial/\partial x_n)$ of order $2m$ and with constant coefficients. Let L_0 be its principal part, that is, the sum of terms which contain only derivatives of order $2m$. We shall call L an *elliptic operator* if, when each partial derivative $\partial/\partial x_i$ in L_0 is replaced by a number ω_i, the polynomial $L_0(\omega_1, ..., \omega_n)$ obtained is nonzero for $\omega = (\omega_1, ..., \omega_n) \neq 0$.

The equation to be solved is

$$L\left(\frac{\partial}{\partial x_1}, ..., \frac{\partial}{\partial x_n}\right) u(x_1, ..., x_n) = \delta(x_1, ..., x_n). \qquad (1)$$

We shall do this in the following way.

1. Replace the δ function on the right-hand side by

$$\frac{2r^\lambda}{\Omega_n \Gamma(\frac{1}{2}\lambda + \frac{1}{2}n)},$$

which according to Eq. (9) of Section 3.9 is equal to the delta function for $\lambda = -n$.

2. Expand

$$\frac{2r^\lambda}{\Omega_n \Gamma(\frac{1}{2}\lambda + \frac{1}{2}n)}$$

in plane waves [i.e., write it according to Eq. (4) of Section 3.10]. It is then represented as the average over all directions $\omega = (\omega_1, ..., \omega_n)$ of a function of the form $C \mid \omega_1 x_1 + ... + \omega_n x_n \mid^\lambda$. We then solve the much simpler differential equation which has $\mid \omega_1 x_1 + ... + \omega_n x_n \mid^\lambda$ on the right-hand side instead of the δ function.

Thus, proceeding, we write (1) in the form

$$L\left(\frac{\partial}{\partial x_1}, \ldots, \frac{\partial}{\partial x_n}\right)u = \frac{2r^\lambda}{\Omega_n \Gamma(\frac{1}{2}\lambda + \frac{1}{2}n)}, \tag{2}$$

which yields (1) when $\lambda = -n$.

Using Eq. (4) of Section 3.10 to represent the right-hand side of this, namely

$$\frac{2r^\lambda}{\Omega_n \Gamma(\frac{1}{2}\lambda + \frac{1}{2}n)} = \frac{1}{\Omega_n \pi^{\frac{1}{2}(n-1)} \Gamma(\frac{1}{2}\lambda + \frac{1}{2})} \int_\Omega |\omega_1 x_1 + \ldots + \omega_n x_n|^\lambda \, d\omega. \tag{3}$$

If now we solve

$$L\left(\frac{\partial}{\partial x_1}, \ldots, \frac{\partial}{\partial x_n}\right)v = \frac{|\omega_1 x_1 + \ldots + \omega_n x_n|^\lambda}{\Omega_n \pi^{\frac{1}{2}(n-1)} \Gamma(\frac{1}{2}\lambda + \frac{1}{2})} \tag{4}$$

for v, a function depending only on $\xi = \omega_1 x_1 + \ldots + \omega_n x_n$, we can write the solution of (2) in the form

$$u(x_1, \ldots, x_n) = \int_\Omega v(\omega_1 x_1 + \ldots + \omega_n x_n) d\omega.$$

Now the solution of (4) reduces to solving an ordinary differential equation, for when applied to $v(\xi) = v(\omega_1 x_1 + \ldots + \omega_n x_n)$ the partial derivative operator may be written

$$\frac{\partial}{\partial x_k} = \omega_k \frac{d}{d\xi}.$$

When this is done, the ordinary differential equation of order $2m$ for $v(\xi)$ obtained from (4) is

$$L\left(\omega_1 \frac{d}{d\xi}, \ldots, \omega_n \frac{d}{d\xi}\right)v = \frac{|\xi|^\lambda}{\Omega_n \pi^{\frac{1}{2}(n-1)} \Gamma(\frac{1}{2}\lambda + \frac{1}{2})}. \tag{5}$$

The right-hand side of this equation depends on λ in addition to ξ, while the coefficients on the left depend on the ω_i. Consequently v depends on ξ, λ, and ω. We shall thus write

$$v = v_\omega(\xi, \lambda).$$

Let $G(\xi, \omega)$ be an elementary solution of (5), so that

$$L\left(\omega_1 \frac{d}{d\xi}, \ldots, \omega_n \frac{d}{d\xi}\right)G(\xi, \omega) = \delta(\xi).$$

Then

$$v_\omega(\xi, \lambda) = \frac{1}{\Omega_n \pi^{\frac{1}{2}(n-1)} \Gamma(\frac{1}{2}\lambda + \frac{1}{2})} \int_{-\infty}^{\infty} G(\xi - \eta, \omega) \, | \, \eta \, |^\lambda \, d\eta \qquad (6)$$

and

$$u(x_1, ..., x_n) = \int_\Omega v_\omega(\omega_1 x_1 + ... + \omega_n x_n, \lambda) d\omega. \qquad (7)$$

To obtain the desired elementary solution of (1) we set $\lambda = -n$ in (6) and (7).

A particularly simple result is obtained when the dimension is odd. For this case we have

$$\frac{| \, \eta \, |^\lambda}{\Gamma(\frac{1}{2}\lambda + \frac{1}{2})} \bigg|_{\lambda = -n} = C\delta^{(n-1)}(\eta),$$

so that

$$v_\omega(\xi, -n) = C \frac{\partial^{n-1} G(\xi, \omega)}{\partial \xi^{n-1}}. \qquad (8)$$

Hence in the case of odd dimension the elementary solution of the elliptical equation (1) can be written in the form

$$u(x_1, ..., x_n) = C_1 \int_\Omega \frac{\partial^{n-1}}{\partial \xi^{n-1}} G(\xi, \omega) d\omega. \qquad (9)$$

Here $\xi = \omega_1 x_1 + ... + \omega_n x_n$, and $G(\xi, \omega)$ is an elementary solution for the operator $L(\omega_1 \, d/d\xi, ..., \omega_n \, d/d\xi)$. It can be shown that

$$C_1 = \frac{(-1)^{\frac{1}{2}(n-1)}}{\Omega_n (2\pi)^{\frac{1}{2}(n-1)} \cdot 1 \cdot 3 ... (n-2)}.$$

Speaking more rigorously, the existence of the convolution of Eq. (6) and the ability to integrate (7) has yet to be established. The simplest and most direct way to do this is to obtain an explicit expression for $G(\xi, \omega)$. We shall, however, proceed differently. We shall avoid the existence problem and present a different method for obtaining $v_\omega(\xi)$, a method that leaves no doubt as to the possibility of integrating over the unit sphere.

Note that if the generalized function $v(\xi)$ depends continuously on several variables, so does its derivative of any order with respect to ξ (see Section 2). If, moreover, $v(\xi)$ is a solution of $L_\xi v(\xi) = f(\xi)$ (where L_ξ is a differential operator with constant coefficients), its derivative $w(\xi) = dv(\xi)/d\xi$ is a solution of $L_\xi w(\xi) = df(\xi)/d\xi$. We note also that if two generalized functions $v_\lambda(\xi)$ and $f_\lambda(\xi)$ depend continuously on a parameter λ and if for $\lambda \neq \lambda_0$ (in a neighborhood of λ_0) $v_\lambda(\xi)$ is a solution of $L_\xi v_\lambda(\xi) = f_\lambda(\xi)$, then $v_{\lambda_0}(\xi)$ is a solution of $L_\xi v_{\lambda_0}(\xi) = f_{\lambda_0}(\xi)$.

Let us assume first that Re $\lambda > 0$. Then the right-hand side of (5) is a continuous function of ξ, and that a solution exists is known from classical existence theorems. Further, for Re $\lambda > 0$, the right-hand side depends continuously on λ, so that v also depends continuously on λ. Further, the coefficients on the left-hand side of (5) depend continuously on ω (with $\Sigma \omega_i^2 = 1$), and the leading coefficient $L_0(\omega_1, ..., \omega_n)$ has positive minimum modulus, for the operator is elliptic. Therefore v also depends continuously on ω. In particular v depends on λ and ω continuously in the sense of generalized functions.

We may now note that when $\lambda \neq 0$ double differentiation with respect to ξ of the right-hand side of Eq. (5) reproduces it (up to a factor of 2λ) with index reduced by two. Then according to the preceding remarks we can obtain a solution of (5) depending continuously on ω for all λ by multiple differentiation of a solution $v_\omega(\xi, \lambda)$ corresponding to Re $\lambda > 0$ and, if necessary, by passing to the limit.

For the case of odd n we start by differentiating $v_\omega(\xi, \lambda)$ for $\lambda = 1$. Its second derivative is $G(\xi, \omega)$ up to a factor, since the right-hand side of (5), up to a factor, becomes $\delta(\xi)$ when $\lambda = -1$. Further differentiation yields Eq. (8).

For even n the solution $v_\omega(\xi, \lambda)$ is found for small positive λ and then doubly differentiated with respect to ξ and divided by λ; then λ is allowed to approach zero. The result yields a solution of (5) corresponding to $\lambda = -2$. Further differentiation will give $v_\omega(\xi, \lambda)$ for $\lambda = -4, -6, ...$.

Let us consider in more detail the case of a *homogeneous* elliptic differential operator $L = L_0$. The ordinary differential equation (5) in this case becomes

$$L(\omega_1, ..., \omega_n) v^{(2m)}(\xi) = \frac{|\xi|^\lambda}{\Omega_n \pi^{\frac{1}{2}(n-1)} \Gamma(\frac{1}{2}\lambda + \frac{1}{2})}$$

so that

$$v^{(2m)}(\xi) = \frac{|\xi|^\lambda}{\Omega_n \pi^{\frac{1}{2}(n-1)} \Gamma(\frac{1}{2}\lambda + \frac{1}{2}) L(\omega_1, ..., \omega_n)}.$$

To find $v(\xi)$, we need only integrate the right-hand side of this equation $2m$ times. As we have shown in Section 3.4, the result is

$$v(\xi) = \frac{1}{\Omega_n \pi^{\frac{1}{2}(n-1)} \Gamma(\frac{1}{2}\lambda + \frac{1}{2}) L(\omega_1, ..., \omega_n)} \left\{ \frac{|\xi|^{\lambda+2m}}{(\lambda+1)...(\lambda+2m)} + Q_\lambda(\xi) \right\},$$

where

$$Q_\lambda(\xi) = \sum_{k=1}^{m} \frac{\xi^{2m-2k}}{(2k-1)!\,(2m-2k)!\,(\lambda+2k)}.$$

Thus the solution of Eq. (2) when L is a *homogeneous elliptic* operator is given by

$$u(x_1, ..., x_n) = \frac{1}{\Omega_n \pi^{\frac{1}{2}(n-1)} \Gamma(\frac{1}{2}\lambda + \frac{1}{2})}$$

$$\times \int_\Omega \left\{ \frac{|\,\omega_1 x_1 + ... + \omega_n x_n\,|^{\lambda+2m}}{(\lambda+1) ... (\lambda+2m)} + Q_\lambda\left(\sum \omega_k x_k\right) \right\} \frac{d\omega}{L(\omega_1 ... \omega_n)}. \qquad (10)$$

In particular, for $\lambda = -n$ this formula will give an elementary solution of a homogeneous elliptical differential equation of order $2m$.

Note that the integral over the unit sphere Ω of each term in $Q_\lambda(\xi)$ is a polynomial of degree less than $2m$ and therefore satisfies the equation $L(\partial/\partial x_1, ..., \partial/\partial x_n)\, u = 0$, so that we need include in the elementary solution only that term in this expression which is required for the total function obtained, not to have a pole at $\lambda = -n$; all other terms may be dropped.

Let us analyze the form of the elementary solution in more detail. We shall consider separately the cases in which the order of the equation is no less than and less than the dimension, i.e., the case $2m \geqslant n$ and $2m < n$. Let us first treat the former. We shall need further to consider even and odd n separately. For odd n and $2m > n$ we may pass to the limit in (10), obtaining the expression

$$u(x_1, ..., x_n) = \frac{(-1)^{\frac{1}{2}(n-1)}}{4(2\pi)^n\,(2m-n)!} \int_\Omega \frac{|\,\omega_1 x_1 + ... + \omega_n x_n\,|^{2m-n}}{L(\omega_1, ..., \omega_n)}\, d\omega \qquad (11)$$

for the elementary solution. We have dropped the polynomial $Q_\lambda(\xi)$, since in the limit it serves no purpose.

For even n and $2m \geqslant n$, to obtain a finite expression for $\lambda = -n$ we must add to

$$\frac{\left|\sum \omega_k x_k\right|^{\lambda+2m}}{(\lambda+1) ... (\lambda+2m)}$$

that term of $Q_\lambda(\xi)$ which contains the factor $1/(\lambda+n)$, namely

$$\frac{\xi^{2m-n}}{(n-1)!\,(2m-n)!\,(\lambda+n)}.$$

Then as $\lambda \to -n$, the sum

$$\frac{|\xi|^{\lambda+2m}}{(\lambda+1)\dots(\lambda+2m)} + \frac{\xi^{-n+2m}}{(n-1)!\,(2m-n)!\,(\lambda+n)}$$

approaches

$$-\frac{\xi^{2m-n}\ln|\xi|}{(n-1)!\,(2m-n)!},$$

and the elementary solution is given by

$$u(x_1, \dots, x_n) \tag{12}$$
$$= \frac{(-1)^{\frac{1}{2}n-1}}{(2\pi)^n\,(2m-n)!} \int_\Omega \frac{(\omega_1 x_1 + \dots + \omega_n x_n)^{2m-n}\ln|\omega_1 x_1 + \dots + \omega_n x_n|}{L(\omega_1, \dots, \omega_n)}\,d\omega.$$

Note that when the order of the equation is no less than the dimension of the space, the integral over the sphere converges and the Green's function becomes an ordinary (not a generalized) function of the x_i, continuous at $x_i = 0$ $(i = 1, 2, \dots, n)$.

Let us now consider the case in which the order of the equation is less than the dimension of the space (that is, $2m < n$).

In this case we have

$$\frac{|\xi|^{\lambda+2m}}{\Gamma(\frac{1}{2}\lambda+\frac{1}{2})}\bigg|_{\lambda=-n} = \frac{(-1)^{\frac{1}{2}(n-1)}(\frac{1}{2}n-\frac{1}{2})!}{(n-2m-1)!}\,\delta^{(n-2m-1)}(\xi)$$

for odd n, and

$$\frac{|\xi|^{\lambda+2m}}{\Gamma(\frac{1}{2}\lambda+\frac{1}{2})}\bigg|_{\lambda=-n} = \frac{|\xi|^{-n+2m}}{\Gamma(\frac{1}{2}-\frac{1}{2}n)}$$

for even n.

Consequently for the elementary solution for odd n and $2m < n$, we obtain

$$u(x_1, \dots, x_n) = \frac{(-1)^{\frac{1}{2}(n-1)}}{2(2\pi)^{n-1}} \int_\Omega \delta^{(n-2m-1)}(\omega_1 x_1 + \dots + \omega_n x_n) \frac{d\omega}{L(\omega_1, \dots, \omega_n)},$$
$$\tag{13}$$

and for even n and $2m < n$, the expression

$$u(x_1, \dots, x_n)$$
$$= \frac{(-1)^{\frac{1}{2}n}(n-2m-1)!}{(2\pi)^n} \int_\Omega |\omega_1 x_1 + \dots + \omega_n x_n|^{-n+2m} \frac{d\omega}{L(\omega_1, \dots, \omega_n)}. \tag{14}$$

It can be shown that the elementary solution $u(x_1, ..., x_n)$ of Eqs. (11) to (14) is always an ordinary (not generalized) function with the following properties: at $x \neq 0$ it is analytic, and in the neighborhood of the origin

$$u(x_1, ..., x_n) = \begin{cases} O(r^{2m-n}), & \text{if } n \text{ odd or if } n \text{ even and } 2m < n, \\ O(r^{2m-n} \ln r), & \text{if } n \text{ even and } 2m \geqslant n, \end{cases}$$

where $r = \sqrt{\sum x_i^2}$; further, if $2m > n$, then $u(x_1, ..., x_n)$ has continuous derivatives of all orders up to $2m - n - 1$ at the origin.

Example. Let $L(\partial/\partial x_1, ..., \partial/\partial x_n) = (\partial^2/\partial x_1^2 + ... + \partial^2/\partial x_n^2)^m$ be the Laplacian operator raised to the mth power or iterated m times. Then on Ω the polynomial $L_0(\omega_1, ..., \omega_n) = (\sum \omega_i^2)^m$ is equal to one. The integrals in (10) and (11) can now be calculated by using Eq. (3), and for both $2m < n$ and odd n with $2m \geqslant n$ we obtain, up to a proportionality factor,

$$u(x) = C_{mn} r^{2m-n}.$$

If $2m \geqslant n$ and n is even, we transform to spherical coordinates in (12), obtaining

$$u(x) = C'_{mn} r^{2m-n} \ln r + C''_{mn} r^{2m-n}.$$

The second term may be dropped, as it is a solution of the homogeneous equation $\Delta^m u = 0$.

6.2. Elementary Solutions of Regular Homogeneous Equations

Consider the equation

$$L\left(\frac{\partial}{\partial x_1}, \frac{\partial}{\partial x_2}, ..., \frac{\partial}{\partial x_n}\right) u(x_1, ..., x_n) = \delta(x_1, ..., x_n), \qquad (1)$$

in which L is a differential operator such that if each $\partial/\partial x_i$ is replaced by ω_i, we obtain a homogeneous polynomial of degree m.[1] We shall assume that the cone $L(\omega_1, ..., \omega_n) = 0$ contains no singular points (except the origin), i.e., that for $L(\omega_1, ..., \omega_n) = 0$ and $\sum \omega_j^2 \neq 0$ the gradient of $L(\omega_1, ..., \omega_n)$ does not vanish. Then Eq. (1) and the operator $L(\partial/\partial x)$ itself will be called *regular*.

In Section 6.1 we treated the elliptic case, in which m is even and,

[1] Recall that in Section 6.1 we denoted the order of L by $2m$.

most important, $L(\omega_1, ..., \omega_n) \neq 0$ for $\sum \omega_j^2 \neq 0$. Formally, we may use the same procedure as in that section to arrive at

$$u(x_1, ..., x_n) = \int_\Omega \frac{f_{mn}\left(\sum x_i\omega_i\right) d\omega}{L(\omega_1, ..., \omega_n)}, \qquad (2)$$

where Ω is the unit sphere (hypersphere) $\sum_{i=1}^n \omega_i^2 = 1$, $d\omega$ is the element of area on this sphere, and the f_{mn} are given as follows: for n even and $m \geqslant n$,

$$f_{mn}(x) = \frac{(-1)^{\frac{1}{2}n-1}}{(2\pi)^n (m-n)!} x^{m-n} \ln |x|; \qquad (3)$$

for n even and $m < n$,

$$f_{mn}(x) = \frac{(-1)^{\frac{1}{2}n+m} (n-m-1)!}{(2\pi)^n} x^{m-n}; \qquad (4)$$

for n odd and $m \geqslant n$,

$$f_{mn}(x) = \frac{(-1)^{\frac{1}{2}(n-1)}}{4(2\pi)^{n-1} (m-n)!} x^{m-n} \operatorname{sgn} x; \qquad (5)$$

for n odd and $m < n$,

$$f_{mn}(x) = \frac{(-1)^{\frac{1}{2}(n-1)}}{2(2\pi)^{n-1}} \delta^{(n-m-1)}(x). \qquad (6)$$

But now the integral of (2) will in general diverge, since in general on Ω there exists a manifold P on which $L(\omega_1, ..., \omega_n) = 0$.

If L is a regular operator, however, so that P contains no singular points, the integral of (2) can be regularized in the following way:

$$u(x_1, ..., x_n) = \int_\Omega \frac{f_{mn}\left(\sum x_i\omega_i\right)}{L(\omega_1, ..., \omega_n)} d\omega = \lim_{\epsilon \to 0} u_\epsilon(x_1, ..., x_n),$$

where

$$u_\epsilon(x_1, ..., x_n) = \int_{\Omega_\epsilon} \frac{f_{mn}\left(\sum x_i\omega_i\right) d\omega}{L(\omega_1, ..., \omega_n)}. \qquad (7)$$

Here Ω_ϵ denotes the set of points on the unit sphere such that $|L(\omega_1, ..., \omega_n)| > \epsilon$.

We must first show that the generalized function u_ϵ converges, that is that for any φ in K the limit

$$\lim_{\epsilon \to 0} (u_\epsilon, \varphi(x_1, ..., x_n)) = (u, \varphi(x_1, ..., x_n))$$

exists.

Note that each of the definitions (3)–(6) of the $f_{mn}(\Sigma x_i\omega_i)$ implies that the (f_{mn}, φ) are infinitely differentiable functions of the ω_i. This can be verified by a simple change of variables in the integral expression for the (f_{mn}, φ) which transfers the ω_i-dependence to the argument of φ.

Let us denote $(f_{mn}(\Sigma x_i\omega_i), \varphi)$ by $r(\omega_1, ..., \omega_n)$; we wish to show that the limit

$$\lim_{\epsilon \to 0} \int_{\Omega_\epsilon} \frac{r(\omega_1, ..., \omega_n)\,d\omega}{L(\omega_1, ..., \omega_n)} = (u, \varphi)$$

exists. To do this we write $r = \Sigma_{i=1}^n r_i$, where each of the $r_i(\omega_i, ..., \omega_n)$ is an infinitely differentiable function and vanishes outside a sufficiently small region S_i. We need consider the integral only for those r_i such that the closure of S_i has nonempty intersection with P. We may thus assume that each r_i for which we wish to establish the existence of

$$\lim_{\epsilon \to 0} \int_{\Omega_\epsilon} \frac{r_i d\omega}{L}$$

will vanish outside a sufficiently small neighborhood of some point A belonging to P. Now on P, and therefore also at A, the gradient of L is nonzero. Consequently at least one of the derivatives $\partial L/\partial\omega_i$ fails to vanish at A. Further, since $\Sigma \omega_i^2 = 1$ at this point, at least one of the ω_i fails to vanish. To be specific let us write $\partial L/\partial\omega_1 \neq 0$ and $\omega_n \neq 0$. Then at we may introduce the coordinate system $L, \omega_2, ..., \omega_{n-1}$. In terms of these coordinates $d\omega$ may be written

$$d\omega = I(L, \omega_2, ..., \omega_{n-1})\,dL d\omega_2 ... d\omega_{n-1},$$

where $I(L, \omega_2, ..., \omega_{n-1})$ is a smooth function. We may now write

$$\int_{\Omega_\epsilon} \frac{r_i(\omega_1, ..., \omega_n)}{L(\omega_1, ..., \omega_n)}\,d\omega$$

$$= \int \underset{\Omega_\epsilon}{...} \int \frac{r_i(L, \omega_2, ..., \omega_{n-1})\,I(L, \omega_2, ..., \omega_{n-1})}{L}\,dL\,d\omega_2 ... d\omega_{n-1}$$

$$= \int_{-\infty}^{\infty} \underset{}{...} \int \left[\int_{|L|>\epsilon} \frac{r_i(L, \omega_2, ..., \omega_{n-1})\,I(L, \omega_2, ..., \omega_{n-1})\,dL}{L} \right] d\omega_2 ... d\omega_{n-1}.$$

Since as $\epsilon \to 0$ the integral $\int_{|L|>\epsilon} r_i I dL/L$ approaches its Cauchy principal value, this expression has a limit. Then by taking the sum over i we finally prove the existence of

$$\lim_{\epsilon \to 0} \int_{\Omega_\epsilon} \frac{r(\omega_1, ..., \omega_n)\,d\omega}{L(\omega_1, ..., \omega_n)} = \lim_{\epsilon \to 0} (u_\epsilon, \varphi).$$

This proves the existence of the Cauchy principal value regularization of

$$\int_\Omega \frac{r(\omega_1, ..., \omega_n)\, d\omega}{L(\omega_1, ..., \omega_n)}. \tag{8}$$

But every regularization of (8) will in general define a different generalization of Eq. (2) by

$$(u, \varphi(x_1, ..., x_n)) = \int_\Omega \frac{(f_{mn}, \varphi)\, d\omega}{L(\omega_1, ..., \omega_n)},$$

where the integral is understood as the particular regularization being used.

For our purposes, any regularization will do, so long as it has the property that if r and r_1 are infinitely differentiable functions such that

$$r(\omega_1, ..., \omega_n) = L(\omega_1, ..., \omega_n) r_1(\omega_1, ..., \omega_n),$$

then

$$\int_\Omega \frac{r(\omega_1, ..., \omega_n)\, d\omega}{L(\omega_1, ..., \omega_n)} = \int_\Omega r_1(\omega_1, ..., \omega_n)\, d\omega. \tag{9}$$

Two regularizations which satisfy this condition will differ from each other by

$$\int_P r(\omega_1, ..., \omega_n)\, d\mu, \tag{10}$$

where P is, as before, the set on the unit sphere Ω on which $L(\omega_1, ..., \omega_n) = 0$, and μ is a measure on this set.

Indeed, all regularizations of (8) that satisfy (9) will give the same result for any r which vanishes on P (since any such r can be written $r = r_1 L$). The difference between two regularizations cannot depend on the values of r off P or on the derivatives of r on P, so that it must be of the form of (10).

Further, two different regularizations of (2) which satisfy condition (9) can be shown to differ only by a solution of the homogeneous equation

$$L\left(\frac{\partial}{\partial x_1}, ..., \frac{\partial}{\partial x_n}\right) u = 0.$$

Indeed, the difference between two such generalized functions can be written, according to (10), in the form of the functional

$$s(x_1, ..., x_n) = \int_P f_{mn}\left(\sum x_i \omega_i\right) d\mu.$$

Applying L to this expression, we have

$$L\left(\frac{\partial}{\partial x_1}, \,...,\, \frac{\partial}{\partial x_n}\right) s(x_1,\,...,\,x_n) = \int_P L(\omega_1,\,...,\,\omega_n) f_{mn}^{(m)}\left(\sum x_i \omega_i\right) d\mu,$$

and since the integral is taken over P (where $L = 0$) it vanishes.

We have yet to show that the integral of (2) is indeed a solution of (1). We proceed by applying the differential operator $L(\partial/\partial x_1,\,...,\,\partial/\partial x_n)$ to Eq. (2). This gives

$$L\left(\frac{\partial}{\partial x_1}, \,...,\, \frac{\partial}{\partial x_n}\right) u(x_1,\,...,\,x_n) = \int_\Omega f_{mn}^{(m)}\left(\sum x_i \omega_i\right) d\omega.$$

We now make use of Eqs. (3)–(6) for the f_{mn} and the plane-wave expansion formulas for the δ function (Section 3). Then for n even we have

$$L\left(\frac{\partial}{\partial x_1}, \,...,\, \frac{\partial}{\partial x_n}\right) u(x_1,\,...,\,x_n)$$

$$= \frac{(-1)^{\frac{1}{2}n}\,(n-1)!}{(2\pi)^n} \int_\Omega \left(\sum x_i \omega_i\right)^{-n} d\omega = \delta(x_1,\,...,\,x_n);$$

for n odd we have

$$L\left(\frac{\partial}{\partial x_1}, \,...,\, \frac{\partial}{\partial x_n}\right) u(x_1,\,...,\,x_n)$$

$$= \frac{(-1)^{\frac{1}{2}(n-1)}}{2(2\pi)^{n-1}} \int_\Omega \delta^{(n-1)}\left(\sum x_i \omega_i\right) d\omega = \delta(x_1,\,...,\,x_n).$$

Thus if the cone $L(\omega_1,\,...,\,\omega_n) = 0$ has no singular points, Eqs. (2)–(6) will in fact give a solution of $L(\partial/\partial x)\,u = \delta(x)$.

6.3. Elementary Solutions of Cauchy's Problem

Consider the differential equation

$$P\left(\frac{\partial}{\partial t},\, \frac{\partial}{\partial x_1},\, ...,\, \frac{\partial}{\partial x_n}\right) u = 0 \qquad (1)$$

of order m in the variable t.

Somewhat later we shall formulate the restrictions we wish to place upon this equation. Here we wish to find the elementary solution of

Cauchy's problem for it (see Section 5.4). In other words, we wish to obtain a solution $u(t, x)$ satisfying the initial conditions

$$u(0, x) = 0, \qquad \frac{\partial u(0, x)}{\partial t} = 0, \ ..., \ \frac{\partial^{m-1} u(0, x)}{\partial t^{m-1}} = \delta(x).$$

We shall solve Cauchy's problem for Eq. (1) in an arbitrary number of independent variables by reducing it to Cauchy's problem in two independent variables. We shall assume further that the differential operator

$$P_\omega\left(\frac{\partial}{\partial t}, \frac{\partial}{\partial \xi}\right) = P\left(\frac{\partial}{\partial t}, \omega_1 \frac{\partial}{\partial \xi}, \ ..., \ \omega_n \frac{\partial}{\partial \xi}\right) \tag{2}$$

is such that Cauchy's problem for the equation

$$P_\omega\left(\frac{\partial}{\partial t}, \frac{\partial}{\partial \xi}\right) u_\omega = 0 \tag{3}$$

is *well posed*.

We now proceed to obtain the elementary solution to the problem stated.

Recall (Section 3.9) that

$$\delta(x) = \frac{2r^\lambda}{\Omega_n \Gamma(\frac{1}{2}\lambda + \frac{1}{2}n)} \bigg|_{\lambda=-n},$$

and consider Eq. (1) with the initial conditions

$$\frac{\partial^k u}{\partial t^k}\bigg|_{t=0} = 0 \qquad (k = 0, 1, \ ..., \ m - 2),$$

$$\frac{\partial^{m-1} u}{\partial t^{m-1}}\bigg|_{t=0} = \frac{2r^\lambda}{\Omega_n \Gamma(\frac{1}{2}\lambda + \frac{1}{2}n)}. \tag{4}$$

As before, we write

$$\frac{2r^\lambda}{\Omega_n \Gamma(\frac{1}{2}\lambda + \frac{1}{2}n)} = \frac{1}{\Omega_n \pi^{\frac{1}{2}(n-1)} \Gamma(\frac{1}{2}\lambda + \frac{1}{2})} \int_\Omega |\, \omega_1 x_1 + \ ... \ + \omega_n x_n \,|^\lambda \, d\omega.$$

We shall solve (1) with the initial conditions of (4), except that the last condition will be replaced by

$$\frac{1}{\Omega_n \pi^{\frac{1}{2}(n-1)} \Gamma(\frac{1}{2}\lambda + \frac{1}{2})} |\, \omega_1 x_1 + \ ... \ + \omega_n x_n \,|^\lambda.$$

Let us write the solution to this problem in the form

$$u_\omega(t, \xi, \lambda) = u_\omega(t, \omega_1 x_1 + \ ... \ + \omega_n x_n, \lambda),$$

which leads us to the following two-dimensional problem: to obtain the solution $u_\omega(t, \xi, \lambda)$ of (3) with the initial conditions

$$\frac{\partial^k u_\omega(t, \xi, \lambda)}{\partial t^k}\bigg|_{t=0} = 0 \qquad (k = 0, 1, ..., m - 2),$$

$$\frac{\partial^{m-1} u_\omega(t, \xi, \lambda)}{\partial t^{m-1}}\bigg|_{t=0} = \frac{1}{\Omega_n \pi^{\frac{1}{2}(n-1)} \Gamma(\frac{1}{2}\lambda + \frac{1}{2})} \, |\, \xi \,|^\lambda. \tag{5}$$

Then the elementary solution of our initial Cauchy problem for Eq. (1) will be written in the form

$$u(t, x) = \int_\Omega u_\omega(t, x_1\omega_1 + ... + x_n\omega_n, - n) \, d\omega. \tag{6}$$

We have set $\lambda = - n$, since it is for this value of λ that

$$\frac{2r^\lambda}{\Omega_n \Gamma(\frac{1}{2}\lambda + \frac{1}{2}n)}$$

becomes the delta function.

Now $u_\omega(t, \xi, \lambda)$ can in turn be expressed in terms of the elementary solution $G_\omega(t, \xi)$ of Cauchy's problem for Eq. (3), namely

$$u_\omega(t, \xi, \lambda) = \frac{1}{\Omega_n \pi^{\frac{1}{2}(n-1)} \Gamma(\frac{1}{2}\lambda + \frac{1}{2})} \int_{-\infty}^{\infty} G_\omega(t, \xi - \eta) \, |\, \eta \,|^\lambda \, d\eta. \tag{7}$$

A particularly simple formula is obtained when the dimension is odd. For this case we have

$$u_\omega(t, \xi, - n) = C \frac{d^{n-1}}{d\xi^{n-1}} G_\omega(t, \xi). \tag{8}$$

In general the elementary solution $u(t, x)$ of Cauchy's problem for Eq. (1) is given in terms of $G_\omega(t, \xi)$ by

$$u(t, x) = \frac{1}{\Omega_n \pi^{\frac{1}{2}(n-1)} \Gamma(\frac{1}{2}\lambda + \frac{1}{2})} \int_\Omega \left\{ \int_{-\infty}^{\infty} G_\omega(t, \xi - \eta) \, |\, \eta \,|^\lambda \, d\eta \right\} d\omega \bigg|_{\lambda = -n} \tag{9}$$

For the case of odd dimension we obtain the simpler formula

$$u(t, x) = \frac{(- 1)^{\frac{1}{2}(n-1)} (\frac{1}{2}n - \frac{1}{2})!}{\Omega_n \pi^{\frac{1}{2}(n-1)} (n - 1)!} \int_\Omega \frac{d^{n-1}}{d\xi^{n-1}} G_\omega(t, \xi) \, d\omega. \tag{10}$$

The differentiation here is to be understood in the sense of generalized functions, and the integration as the integral of a generalized function which depends continuously on a parameter.

We emphasize that the above formulas will hold for all equations for which Cauchy's problem is well posed, for instance, for any parabolic or hyperbolic equations.

If one wishes to avoid the question of the existence of the convolution in Eq. (7) and to obtain $u_\omega(t, \xi, -n)$ in a form which makes explicit the possibility of integrating it over the unit sphere, one may proceed by considerations similar to those brought forth in Section 6.1. We shall, however, not go into this here.

The Solution of Cauchy's Problem for Homogeneous Hyperbolic Equations. The Herglotz-Petrovskii Formulas

Let us consider in more detail the hyperbolic case of an equation containing no derivatives of order less than m.

A homogeneous operator $P(\partial/\partial t, \partial/\partial x) = P(\partial/\partial t, \partial/\partial x_1, ..., \partial/\partial x_n)$ is called *hyperbolic* if for any values of $\omega_1, ..., \omega_n$ such that $\Sigma \omega_j^2 = 1$, the mth degree equation in v

$$P(v, \omega_1, ..., \omega_n) = 0 \tag{11}$$

has m real and distinct roots.

For such an operator we again consider the problem

$$P\left(\frac{\partial}{\partial t}, \frac{\partial}{\partial x}\right) u = 0 \tag{12}$$

with the conditions

$$\frac{\partial^k u}{\partial t^k}\bigg|_{t=0} = 0 \qquad (k = 0, 1, ..., m-2),$$

$$\frac{\partial^{m-1} u}{\partial t^{m-1}}\bigg|_{t=0} = \frac{2r^\lambda}{\Omega_n \Gamma(\frac{1}{2}\lambda + \frac{1}{2}n)}. \tag{13}$$

As we have said before, this problem reduces to solving the one-dimensional Cauchy problem

$$P_\omega\left(\frac{\partial}{\partial t}, \frac{\partial}{\partial \xi}\right) u_\omega = P\left(\frac{\partial}{\partial t}, \omega_1 \frac{\partial}{\partial \xi}, ..., \omega_n \frac{\partial}{\partial \xi}\right) u_\omega = 0 \tag{14}$$

with the conditions

$$\frac{\partial^k u_\omega}{\partial t^k}\bigg|_{t=0} = 0 \qquad (k = 0, 1, ..., m-2),$$

$$\frac{\partial^{m-1} u_\omega}{\partial t^{m-1}}\bigg|_{t=0} = \frac{|\xi|^\lambda}{\Omega_n \pi^{\frac{1}{2}(n-1)} \Gamma(\frac{1}{2}\lambda + \frac{1}{2})}. \tag{15}$$

It is easily shown that any function of the form $f(\sum x_k \omega_k + v_j t)$ will satisfy (12), where v_j is any root of the algebraic equation (11). Since (12) is hyperbolic, all these roots are real and distinct. It is seen that this solution represents a plane wave propagating at velocity $-v_j$.

We shall attempt to find a solution of Eqs. (14) and (15) in the form of the sum of similar plane waves

$$u_\omega\left(t, \sum \omega_k x_k\right) = \sum_{j=1}^{m} c_j f\left(\sum_{k=1}^{n} x_k \omega_k + v_j t\right) \tag{16}$$

propagating at different speeds v_j. The sum in (16) is taken over all the roots of (11). According to the initial conditions, the c_j and f are given by

$$\sum c_j = 0, \quad \sum c_j v_j = 0, \dots, \sum c_j v_j^{m-2} = 0, \quad \sum c_j v_j^{m-1} = 1, \tag{17}$$

$$f^{(m-1)}(\xi) = \frac{1}{\Omega_n \pi^{\frac{1}{2}(n-1)} \Gamma(\frac{1}{2}\lambda + \frac{1}{2})} \, |\,\xi\,|^{\lambda}.$$

This gives

$$c_j = \frac{1}{(v_j - v_1) \dots (v_j - v_{j-1})(v_j - v_{j+1}) \dots (v_j - v_m)} . \tag{18}$$

To obtain $f(\xi)$ we must integrate $|\,\xi\,|^{\lambda}$ a total of $m - 1$ times. This gives

$$f(\xi) = \frac{1}{\Omega_n \pi^{\frac{1}{2}(n-1)} \Gamma(\frac{1}{2}\lambda + \frac{1}{2})} \left\{ \frac{|\,\xi\,|^{\lambda+m-1} (\operatorname{sgn} \xi)^{m-1}}{(\lambda + 1) \dots (\lambda + m - 1)} + Q_\lambda(\xi) \right\}, \tag{19}$$

where

$$Q_\lambda(\xi) = \sum_{k=1}^{[\frac{1}{2}(m-1)]} \frac{\xi^{m-2k-1}}{(2k - 1)!(m - 2k - 1)!(\lambda - 2k)} \tag{20}$$

(cf. Section 3.4).

We thus obtain the explicit expression

$$u_\omega\left(t, \sum \omega^k x_k\right) = \frac{1}{\Omega_n \pi^{\frac{1}{2}(n-1)} \Gamma(\frac{1}{2}\lambda + \frac{1}{2})}$$

$$\times \sum_{j=1}^{m} c_j \left\{ \frac{\left|\sum \omega_k x_k + v_j t\right|^{\lambda+m-1} \left[\operatorname{sgn}\left(\sum \omega_k x_k + v_j t\right)\right]^{m-1}}{(\lambda + 1)(\lambda + 2) \dots (\lambda + m - 1)} \right.$$

$$\left. + Q_\lambda\left(\sum \omega_i x_i + v_j t\right) \right\} \tag{21}$$

for the solution of Cauchy's problem stated in (14) and (15), where the c_j are given by (18), and the $Q_\lambda(\xi)$ are given by (20).

The solution of the original Cauchy's problem stated in (12) and (13) is obtained from (21) by integrating over the unit sphere $\sum \omega_i^2 = 1$:

$$u(t, x_1, ..., x_n; \lambda) = \int_\Omega u_\omega\Big(t, \sum \omega_k x_k\Big)\, d\omega. \qquad (22)$$

By setting $\lambda = -n$ in this equation we obtain, as before, the elementary solution of Cauchy's problem.

Now it is just by chance that the integral in (22) is over the sphere. Some other surface would do as well. For this reason we shall change this expression to a form which is more closely related to the essential aspects of our problem.

Equation (21) shows that the u_ω in (22) is the sum of m terms. Consider the jth of these, namely

$$c_j \left\{ \frac{\Big|\sum x_k \omega_k + v_j t\Big|^{\lambda+m-1} \Big[\operatorname{sgn}\Big(\sum x_k \omega_k + v_j t\Big)\Big]^{m-1}}{(\lambda + 1)(\lambda + 2)\,...\,(\lambda + m - 1)} + Q_\lambda\Big(\sum x_k \omega_k + v_j t\Big)\right\}$$

In taking the integral over Ω of this term, let us make the change of variables $\omega_i/v_j = \xi_i$. Our aim is to replace the integral over Ω by an integral over the hypersurface $P(1, \xi_1, ..., \xi_n) \equiv H(\xi_1, ..., \xi_n) = 0$. Now for each fixed set of ω_i the equation $P(v, \omega_1, ..., \omega_n) = 0$ has m distinct roots v_j. Since these roots depend continuously on the ω_i, we may follow the motion of one of the roots as we vary the ω_i. As these are varied, the point they describe runs over Ω, and the point $(1, \xi_1, ..., \xi_n)$

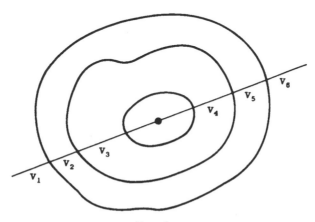

Fig. 5.

runs over a component on the surface $P(1, \xi_1, ..., \xi_n) = 0$. Every such component is of one of two kinds: when the ω_i are changed into their negatives, the corresponding root v_j either (1) is transformed into another root v_k or (2) returns to its initial value. We shall call components of the first kind "ovals," and components of the second kind "unpaired pieces." It can be shown that if the equation is of even order $P(1, \xi_1, ..., \xi_n) = 0$ consists of $m/2$ ovals, while if m is odd it consists of $(m - 1)/2$ ovals and one unpaired piece. We shall restrict our considerations to the first case (Fig. 5); the second can be treated similarly.

Thus we assume that when the ω_i run over Ω, the point $(1, \xi_1, ..., \xi_n)$ runs over an oval belonging to two different roots v_j and v_k. Consequently when we add up the integrals over Ω corresponding to different terms in (21), we obtain twice the integral over $H(\xi_1, ..., \xi_n) = 0$. Let the element of (hyper-) surface area on $H = 0$ subtended by the same angle as $d\omega$ be denoted by $d\sigma$, and let φ be the angle between the normals to $d\sigma$ and $d\omega$. Then clearly

$$d\omega = \frac{|\cos \varphi| \, d\sigma}{|\xi|^{n-1}} = \frac{\left|\sum \xi_i H_{\xi_i}\right| \, d\sigma}{|\xi|^n \, |\operatorname{grad} H|}.$$

The c_j are given by

$$c_j = \frac{1}{(v_j - v_1) \, ... \, (v_j - v_{j-1}) \, (v_j - v_{j+1}) \, ... \, (v_j - v_m)} = \frac{1}{\dfrac{\partial P}{\partial v}\Big|_{v=v_j}}.$$

Since $P(v, \omega_1, ..., \omega_n)$ is a homogeneous polynomial, we may write

$$v_j \frac{\partial P}{\partial v}\Big|_{v=v_j} + \sum \omega_k \frac{\partial P}{\partial \omega_k} = mP(v_j, \omega_1, ..., \omega_n) = 0$$

[for v_j is a root of Eq. (11)]. Thus

$$c_j = \frac{1}{\dfrac{dP}{dv}\Big|_{v=v_j}} = -\frac{v_j}{\sum \omega_k \dfrac{\partial P}{\partial \omega_k}\Big|_{v=v_j}}.$$

Now replacing ω_k by $v_j \xi_k$ and $P(1, \xi_1, ..., \xi_n)$ by $H(\xi_1, ..., \xi_n)$, we have

$$c_j = -\frac{v_j^{1-m}}{\sum_{k=1}^{m} \xi_k H_{\xi_k}} = -\frac{(\operatorname{sgn} v_j)^{m-1} |\xi|^{m-1}}{\sum \xi_k H_{\xi_k}}.$$

Further,

$$\left| \sum x_k \omega_k + v_j t \right|^{\lambda+m-1} \left[\mathrm{sgn} \left(\sum x_k \omega_k + v_j t \right) \right]^{m-1}$$

$$= | v_j |^{\lambda+m-1} (\mathrm{sgn}\ v_j)^{m-1} \left| \sum x_k \xi_k + t \right|^{\lambda+m-1} \left[\mathrm{sgn} \left(\sum x_k \xi_k + t \right) \right]^{m-1},$$

and for any j we have

$$\left(\sum x_k \omega_k + t v_j \right)^{m-2k-1} = v_j^{m-2k-1} \left(\sum x_k \xi_k + t \right)^{m-2k-1}$$

$$= \left(\frac{\sum \xi_k x_k + t}{|\xi|} \right)^{m-2k-1} (\mathrm{sgn}\ v_j)^{m-1}.$$

Now inserting the expressions for c_j, $d\omega$, and the functions of $\sum x_k \omega_k + v_j t$ into (22), we arrive at

$$u(t, x_1, ..., x_n; \lambda) = - \frac{2}{\Omega_n \pi^{\frac{1}{2}(n-1)} \Gamma(\frac{1}{2}\lambda + \frac{1}{2})}$$

$$\times \int_{H=0} \left\{ \frac{|\xi|^{-\lambda-n} \left| \sum x_k \xi_k + t \right|^{\lambda+m-1} \left[\mathrm{sgn} \left(\sum x_k \xi_k + t \right) \right]^{m-1}}{(\lambda+1)(\lambda+2)...(\lambda+m-1)} \right.$$

$$\left. + Q_\lambda \left(\frac{\sum x_k \xi_k + t}{|\xi|} \right) \right\} \frac{d\sigma}{|\ \mathrm{grad}\ H\ |\ \mathrm{sgn} \left(\sum \xi_k H_{\xi_k} \right)}. \qquad (23)$$

By passing to the limit $\lambda = -n$, we obtain the Herglotz-Petrovskii formula for the elementary solution of Cauchy's problem. For odd n, the formula we obtain is

$$u(x_1, ..., x_n) = - \frac{(-1)^{\frac{1}{2}(n-1)}}{2(2\pi)^{n-1}(m-n-1)!}$$

$$\times \int_{H=0} \left(\sum x_k \xi_k + t \right)^{m-n-1} \left[\mathrm{sgn} \left(\sum x_k \xi_k + t \right) \right]^{m-1} \omega, \qquad (24)$$

where ω denotes

$$\frac{d\sigma}{|\ \mathrm{grad}\ H\ |\ \mathrm{sgn} \left(\sum_{k=1}^{n} \xi_k H_{\xi_k} \right)}.$$

We shall analyze similar formulas in detail in Chapter III.

In the limit we may (and do) drop the polynomial Q_λ, since its integral vanishes. For even n we obtain

$$u(x_1, ..., x_n) = \frac{2(-1)^{\frac{1}{2}n}}{(2\pi)^n (m - n - 1)!}$$

$$\times \int_{H=0} \left(\sum x_k \xi_k + t \right)^{m-n-1} \ln \left| \frac{\sum x_k \xi_k + t}{\sum x_k \xi_k} \right| \omega. \qquad (25)$$

When the order m of the equation is less than $n - 1$, these formulas for the elementary solution of Cauchy's problem simplify considerably: for odd n we have

$$u(x_1, ..., x_n) = \frac{(-1)^{\frac{1}{2}(n+1)}}{(2\pi)^{n-1}} \int_{H=0} \delta^{(n-m)} \left(\sum x_k \xi_k + t \right) \omega \qquad (26)$$

and for even n

$$u(x_1, ..., x_n) = \frac{(-1)^{\frac{1}{2}(n+1)} (n - m)!}{(2\pi)^n} \int_{H=0} \frac{\omega}{\left(\sum x_k \xi_k + t \right)^{n-m+1}}. \qquad (27)$$

The affine invariant notation for the plane-wave expansion of the delta function [Section 3.11, Eqs. (4) and (5)] with the choice of the $H = 0$ surface as the region of integration could have been used to obtain Eqs. (24)–(27) more directly, without referring to the integration over the unit sphere.

Appendix 1

Local Properties of Generalized Functions

We mentioned in Section 1.4 that generalized functions can be locally defined, that is, that they can be defined in terms of their operation on test functions with support in arbitrarily small given neighborhoods of every point.

In this appendix we shall prove this as well as some other assertions concerning the local properties of generalized functions, results which are of great value for the theory.

We shall first return to a study of the space K of test functions, and shall show that in K there exist functions which take on given constant values on given nonintersecting bounded closed sets. These functions can then be used to study the local properties of generalized functions in the simplest way possible.

A1.1. Test Functions as Averages of Continuous Functions

We shall show that given any continuous function $f(x)$ (not necessarily with bounded support), there exists a sequence of infinitely differentiable functions $f_\delta(x)$ such that as $\delta \to 0$,

$$f_\delta(x) \to f(x)$$

uniformly in any bounded region.

We shall choose the $f_\delta(x)$ to be the averages

$$f_\delta(x) = C_\delta \int_{|x-\xi| \leqslant \delta} f(\xi)\varphi(x - \xi, \delta)d\xi \tag{1}$$

of $f(x)$, where $\varphi(x, \delta)$ is the function defined in Section 1.2, namely

$$\varphi(x, \delta) = \begin{cases} \exp\left(-\dfrac{\delta^2}{\delta^2 - |x|^2}\right) & \text{for } |x| < \delta, \\ 0 & \text{for } |x| > \delta; \end{cases} \qquad \frac{1}{C_\delta} = \int_{|x| \leqslant \delta} \varphi(x, \delta)dx.$$

From. Eq. (1) we see that since $\varphi(x, \delta)$ is infinitely differentiable, so is $f_\delta(x)$. We have, further,

$$f(x) - f_\delta(x)$$

$$= C_\delta \int_{|x-\xi| \leqslant \delta} f(x)\varphi(x - \xi, \delta)\, d\xi - C_\delta \int_{|x-\xi| \leqslant \delta} f(\xi)\varphi(x - \xi, \delta)\, d\xi$$

$$= C_\delta \int_{|x-\xi| \leqslant \delta} [f(x) - f(\xi)]\varphi(x - \xi, \delta)d\xi.$$

Now $f(x)$ is a continuous function of x, so that for sufficiently small δ and for $|x - \xi| \leqslant \delta$, the value of $|f(x) - f(\xi)|$ is less than some given ϵ. Thus by suitably choosing δ we arrive at

$$|f(x) - f_\delta(x)| \leqslant \epsilon \cdot C_\delta \int_{|x-\xi| \leqslant \delta} \varphi(x - \xi, \delta)d\xi = \epsilon,$$

as asserted.

In particular, if $f(x)$ has bounded support, then so does $f_\delta(x)$; in fact $f_\delta(x) = 0$ outside a δ-neighborhood of the support of $f(x)$.

If $f(x)$ is constant in a ball of radius δ with center at x_0, i.e., if

$$f(x) \equiv f(x_0) \qquad \text{for} \qquad |x - x_0| \leqslant \delta,$$

then

$$f_\delta(x_0) = f(x_0)\, C_\delta \int_R \varphi(x - \xi, \delta)d\xi = f(x_0).$$

Consequently if $f(x)$ is constant in a region G, then $f_\delta(x)$ is constant [and equal to $f(x)$] in G_δ, the region consisting of those points $x \in G$ such that a ball of radius δ with center at x lies entirely in G.

We remark also that if $f(x)$ is bounded everywhere by one above and zero below, then so is $f_\delta(x)$.

We now show that if F is a bounded closed set and U is an open region containing it, there exists a $\varphi(x)$ in K equal to unity on F and zero outside of U, and bounded by one above and zero below at other points.

Indeed, since F is bounded, it is contained in U together with its ϵ-neighborhood for some $\epsilon > 0$. Let F_1 be the closure of the $(\epsilon/3)$-neighborhood of F, let U_1 be the (open) $(2\epsilon/3)$-neighborhood of F, and let W be its (closed) complement in the whole of R_n. The continuous function of x defined by

$$\rho(x, W) = \min_{y \in W} \rho(x, y)$$

[where $\rho(x, y)$ is the Euclidean distance between x and y] is positive in U_1 and has a positive minimum μ on F_1. The new function

$$f(x) = \min \left\{ \frac{1}{\mu} \rho(x, W), 1 \right\}$$

is also continuous, vanishes outside U_1, is equal to one on F_1, and is bounded by one above and zero below at other points. Now the desired function $\varphi(x)$ may be chosen as $f_\delta(x)$ with $\delta = \epsilon/3$.

A1.2. Partition of Unity

Consider a given countable covering of R_n by open bounded regions or neighborhoods $U_1, U_2, ..., U_m, ...$, and let this covering be locally finite in the sense that every point is covered by only a finite number of the U_i. We wish to construct infinitely differentiable functions $e_1(x), e_2(x), ..., e_m(x), ...$ such that

(a) $0 \leqslant e_k(x) \leqslant 1$;
(b) $e_k(x) = 0$ outside of U_k, $k = 1, 2, ...$;
(c) $e_1(x) + e_2(x) + ... + e_m(x) + ... \equiv 1$.

Because of (b), once x is chosen there is only a finite number of terms on the left-hand side of (c). The set $\{e_i(x)\}$ is called a *partition of unity*, or more accurately a *partition of unity subordinated to the covering* $\{U_i\}$.

Such a partition can be constructed in the following way. First we note that there exist open neighborhoods V_1, V_2, ..., V_m, ... which form a covering of R_n such that V_k and its closure $\bar{V}_k$ are contained in U_k for each k. Indeed, assume that V_1, ..., V_{k-1} have been found such that $\bar{V}_j \subset U_j$, $j = 1, ..., k - 1$, and such that V_1, ..., V_{k-1}, U_k, U_{k+1}, ... is a locally finite covering of R_n. Then the complement to

$$V_1 \cup \cdots \cup V_{k-1} \cup U_{k+1} \cup \cdots$$

is a closed set F_k which is entirely covered by U_k. We may choose V_k to be any open neighborhood containing F_k and contained in U_k together with its closure; the rest follows by induction.

Since $\bar{V}_k$ is bounded, it follows from A1.1 that there exists an infinitely differentiable function $h_k(x)$ whose values everywhere lie between 0 and 1 and which is equal to 1 on V_k and to 0 outside of U_k. Let us write

$$h(x) = \sum_{k=1}^{\infty} h_k(x) \, ,$$

which is a function that exists for all x and whose values are obviously no less than 1.

We now need only write

$$e_k(x) = \frac{h_k(x)}{h(x)} \qquad (k = 1, 2, ...)$$

to obtain functions $e_k(x)$ satisfying the requirements we have set up.

Remark.   Let $\{e_i(x)\}$ be a partition of unity subordinated to the (locally finite) covering $\{U_i\}$ of R_n. Then for any $\varphi(x)$ in K we have

$$\varphi(x) = \sum_1^{\infty} \varphi_i(x), \qquad (1)$$

where

$$\varphi_i(x) = \varphi(x)e_i(x)$$

is in K and vanishes outside of U_i. Moreover, the number of terms on the right-hand side of (1) will be finite if we assume in addition that each ball $| x | \leqslant n$ intersects only a finite number of the U_i. We note further that if a sequence of functions $\varphi_\nu(x)$ converges to zero in K as $\nu \to \infty$, so does every sequence $\varphi_{\nu i}(x) = \varphi_\nu(x) \, e_i(x)$.

This result has already been used and will find much application in the sequel.

A1.3. Local Properties of Generalized Functions

In Section 1.4 we made the following definition: a generalized function f is equal to zero in the neighborhood of a given point if for every φ in K with support within this neighborhood, we have $(f, \varphi) = 0$.

In addition we said that the generalized function f vanishes on some open region G if it vanishes in a neighborhood of every point in this region. This is a typically local definition.

It is possible to give also another, nonlocal definition of the same property as follows. The generalized function f vanishes in a region G if $(f, \varphi) = 0$ for any φ in K with support in a set Q which, together with its closure, is contained in G.

We indicate the proof of the equivalence of these two definitions. It is sufficient to verify that the nonlocal definition follows from the local one (the converse is obvious). Let us therefore assume that f vanishes in G in the sense of the local definition, and let $\varphi(x)$ in K fail to vanish only on some set Q whose closure $\bar{Q}$ is contained in G. For every point $x \in \bar{Q}$ there exists, by assumption, a neighborhood U in which f vanishes. We may assume without loss of generality that this neighborhood is bounded. The set of these neighborhoods forms a covering of $\bar{Q}$, and then according to the Heine-Borel theorem there exists a countable covering $U_1, U_2, ..., U_m, ...$, such that every closed ball $|x| \leqslant n$ intersects only a finite number of the neighborhoods U_i. According to the remark at the end of A1.2, we may write $\varphi(x)$ as a (finite) sum of the form

$$\varphi(x) = \varphi_1(x) + ... + \varphi_m(x) + ...,$$

where each $\varphi_k(x)$ is in K and vanishes outside of U_k. Then by assumption $(f, \varphi_k) = 0$, so that $(f, \varphi) = \Sigma (f, \varphi_k) = 0$. This proves the equivalence.

An obvious but very important consequence of this proposition is that a generalized function f vanishing in the neighborhood of every point is the null generalized function, which means that for any φ in K we have

$$(f, \varphi) = 0.$$

The following consequence may also be noted. If $\varphi(x)$ in K vanishes in a neighborhood U of the support F of the generalized function f, then $(f, \varphi) = 0$. Indeed, according to the local definition $f = 0$ outside of F, and the rest follows from the equivalent nonlocal definition.

We have said that two generalized functions f and g coincide in a neighborhood of x_0 if their difference $f - g$ vanishes in such a neighborhood. The proposition we have proven above shows that generalized functions which coincide in a neighborhood of every point are equal to each other. Thus every generalized function is uniquely determined by its local properties.

This fact can be used to construct a generalized function in its entirety when it is defined everywhere only locally. In fact let us assume that for *every* point x_0 there exists a neighborhood $U(x_0)$ such that for every $\varphi(x)$ in K with support in $U(x_0)$ the numbers (f, φ) exist and are known, and that these numbers depend linearly and continuously on φ. Let us assume further that (f, φ) depends only on φ itself, but not on the particular choice of the point x_0 outside of whose neighborhood $U(x_0)$ the function $\varphi(x)$ vanishes. Then we may assert that there exists a unique functional on K which coincides with f for those $\varphi(x)$ on which the latter is defined.

We proceed to the proof in the following way.

Neighborhoods $U(x_0)$ of the type described exist, by assumption, at every point, and therefore form a covering of R_n. Without loss of generality we may consider these neighborhoods bounded. As above, we make use of the Heine-Borel theorem to choose from this covering a countable one $U_1, ..., U_m, ...$ which has the property that every closed ball $|x| \leqslant n$ intersects only a finite number of the U_i. According to the remark at the end of A1.2, every $\varphi(x)$ in K can be written in the form

$$\varphi(x) = \sum_1^\infty \varphi_i(x), \tag{1}$$

where each $\varphi_i(x)$ is in K and vanishes outside of U_i, such that there actually appear only a finite number of terms in the sum. The functional f is defined for each term of the sum. We now define in general

$$(f, \varphi) = \sum_1^\infty (f, \varphi_i). \tag{2}$$

We have clearly obtained a linear functional defined on all of K. It is moreover a continuous functional. Indeed, if a sequence of functions φ_ν converges to zero in K, so does each $\varphi_{\nu m}(x)$ for each fixed m (as was remarked in A1.2). The sum on the right-hand side of (1) contains a fixed number of terms, since the supports of the φ_ν are contained in a fixed ball. Therefore it follows that $(f, \varphi_\nu) \to 0$.

Obviously, when acting on some φ in K which vanishes outside of U_m, the functional we have constructed coincides with the one defined originally, for in this case all the φ_i on the right-hand side of (1) vanish outside of U_m.

This implies that the definition of f does not depend on the choice of covering $\{U_i\}$ with the required properties. In fact let $\{V_i\}$ be some other covering with these properties, and let g be the functional obtained from this new covering; then $f = g$ locally, and therefore also nonlocally.

A1.4. Differentiation as a Local Operation

We have said that the generalized function f vanishes in a neighborhood of x_0 if it vanishes when applied to any function in K whose support is in this neighborhood. We shall now show that all the derivatives of f also vanish in this neighborhood. Indeed, if φ in K fails to vanish only within the neighborhood $U(x_0)$, all its derivatives also fail to vanish only in this neighborhood. Thus for any such function we have

$$\left(\frac{\partial f}{\partial x_j}, \varphi\right) = \left(f, -\frac{\partial \varphi}{\partial x_j}\right) = 0,$$

as asserted.

This implies that two functionals which coincide in a region G have derivatives of all orders which also coincide in this region.

For instance let f coincide in a region G with a regular functional corresponding to a differentiable function $f(x)$. Then $\partial f/\partial x_j$ coincides in G with the regular functional corresponding to $\partial f(x)/\partial x_j$. It thus follows even for singular generalized functions that when it is possible to take the derivative in the ordinary way, the ordinary derivative is what we obtain.

Another consequence is that any functional f concentrated on a set F has derivatives $\partial f/\partial x_j$, $\partial^2 f/\partial x_j \partial x_k$, etc., also concentrated on F.

For instance, the derivatives of $\delta(x - x_0)$ are concentrated on x_0 (as is quite obvious anyway). Given a continuous function which vanishes outside a closed set F, its derivatives of any order are concentrated on F.

It is remarkable that the above propositions can be inverted:

1. Every generalized function concentrated at a single point x_0 can be represented as a (finite) linear combination of $\delta(x - x_0)$ and its derivatives.

2. Every generalized function concentrated on a bounded closed set F can be represented for any $\epsilon > 0$ as a (finite) linear combination of derivatives of continuous functions which vanish outside the ϵ-neighborhood of F.

These theorems will be proven in Volume II (Chapter II, Section 4).

Appendix 2

Generalized Functions Depending on a Parameter

In this appendix we shall study the properties of generalized functions depending on a parameter, particularly when this dependence is analytic. These properties are in many ways analogous to those of ordinary functions, and many of the proofs are obtained essentially by reducing the statements to statements about ordinary functions. Of special importance is the theorem on the *completeness* of the space of generalized functions: given a sequence $f_1, f_2, ..., f_\nu, ...$ of generalized functions such that every number sequence (f_ν, φ) converges, then the equation $\lim_{\nu \to \infty} (f_\nu, \varphi) = (f, \varphi)$ defines a generalized function. This theorem is proven in the Appendix at the end of the present volume.

A2.1. Continuous Functions

Let λ be a real or complex parameter taking on values in a certain region Λ of the complex plane, and assume that for each λ there exists a generalized function f_λ. In accordance with the definition of Section 1.8, we shall call f the limit of f_λ as $\lambda \to \lambda_0$ if the ordinary function (f_λ, φ) approaches (f, φ) for every φ in K. Then f_λ is called *continuous in λ throughout* Λ if for every $\lambda_0 \in \Lambda$ we have $f_{\lambda_0} = \lim_{\lambda \to \lambda_0} f_\lambda$.

Let us now turn to the very important question of *using continuity in λ to extend the definition* of a generalized function f_λ. Consider such a generalized function f_λ continuous on a set Λ and let λ_0 be a limit point of Λ at which f_λ is not initially defined. We wish to establish the possibility of extending f_λ to λ_0 so as to obtain a generalized function continuous on $\Lambda + \lambda_0$.

An obvious necessary condition is that it must be possible to extend the definition of each of the numerical functions (f_λ, φ) by continuity to λ_0. This is also a sufficient condition. Indeed, if for any φ in K and for any sequence $\lambda_\nu \to \lambda_0$, $\lambda_\nu \in \Lambda$, the sequence $(f_{\lambda_\nu}, \varphi)$ converges, then the completeness of K' implies that there exists a generalized function $f = f_{\lambda_0}$ which is the limit of the sequence f_{λ_ν}. It can be proven in the usual way that this newly defined generalized function is independent of the choice of the sequence $\lambda_\nu \to \lambda_0$.

Note further that if f_λ is a generalized function that depends continuously on λ, then its derivatives (with respect to x) are also continuous in λ. Indeed, as we have seen in Section 2.4, $f_{\lambda_\nu} \to f_{\lambda_0}$ implies that

$$\frac{\partial}{\partial x_j} f_{\lambda_\nu} \to \frac{\partial}{\partial x_j} f_{\lambda_0}.$$

If a generalized function is continuous in a parameter, then it can be *integrated with respect to this parameter* (cf. Section 3.10). For instance, let f_λ be continuous in λ on the rectifiable curve Γ. We construct the sum

$$s_n = \sum_{j=1}^{n} f_{\lambda_j'} \Delta\lambda_j,$$

where we have cut Γ into n parts at the points $\lambda_0, \lambda_1, ..., \lambda_n$ and have chosen arbitrary points λ_i' in the intervals $[\lambda_{j-1}, \lambda_j]$. Now let φ be any function in K and let $\max |\Delta\lambda_j| \to 0$; since (f_λ, φ) is continuous, the expression

$$(s_n, \varphi) = \sum_{j=1}^{n} (f_{\lambda_j'}, \varphi)\Delta\lambda_j$$

converges to the integral of (f_λ, φ) independent of the choice of the λ_j' or the λ_j. This limit defines a continuous linear functional on K, and this functional is called the integral of f_λ along Γ.

It is clearly possible to integrate not only along a curve but also over a region of any dimension.

A2.2. Differentiable Functions

In Section 3.1 we make the following definition: the generalized function g is called *the derivative of the generalized function f_λ with respect to λ at $\lambda = \lambda_0$*, if

$$g = \lim_{\lambda \to \lambda_0} \frac{f_\lambda - f_{\lambda_0}}{\lambda - \lambda_0}.$$

For the existence of $\partial f_\lambda/\partial\lambda$ at $\lambda = \lambda_0$ it is necessary and sufficient that all the ordinary functions (f_λ, φ) be differentiable in λ at $\lambda = \lambda_0$. The necessity is obvious. The proof of the sufficiency follows. By assumption, for every φ in K and for any sequences $\lambda_\nu \to \lambda_0$ the limit of

$$\frac{(f_\lambda, \varphi) - (f_{\lambda_0}, \varphi)}{\lambda - \lambda_0} = \left(\frac{f_\lambda - f_{\lambda_0}}{\lambda - \lambda_0}, \varphi\right)$$

exists. But then, as was pointed out above, the generalized function $(f_\lambda - f_{\lambda_0})/(\lambda - \lambda_0)$, defined for $\lambda \neq \lambda_0$, can be defined by continuity also at $\lambda = \lambda_0$. In other words, there exists a generalized function which is the limit of $(f_\lambda - f_{\lambda_0})/(\lambda - \lambda_0)$ as $\lambda \to \lambda_0$, as asserted.

If the derivative of f_λ with respect to λ exists for all $\lambda \in \Lambda$, we say that f_λ is differentiable with respect to λ in Λ.

The higher derivatives and higher order differentiability are defined similarly.

It is easily seen that if f_λ is differentiable with respect to λ throughout Λ, so are all the derivatives of f_λ with respect to x, and further that

$$\frac{\partial}{\partial \lambda} \frac{\partial}{\partial x_j} f_\lambda = \frac{\partial}{\partial x_j} \frac{\partial}{\partial \lambda} f_\lambda. \tag{1}$$

Indeed, for any φ in K the ordinary function

$$\left(\frac{\partial}{\partial x_j} f_\lambda, \varphi\right) = \left(f_\lambda, -\frac{\partial \varphi}{\partial x_j}\right)$$

is differentiable with respect to λ, and its derivative is

$$\frac{\partial}{\partial \lambda} \left(f_\lambda, -\frac{\partial \varphi}{\partial x_j}\right) = \left(\frac{\partial f_\lambda}{\partial \lambda}, -\frac{\partial \varphi}{\partial x_j}\right) = \left(\frac{\partial}{\partial x_j} \frac{\partial}{\partial \lambda} f_\lambda, \varphi\right).$$

This means that the functional $\partial f_\lambda / \partial x_j$ is differentiable with respect to λ and that Eq. (1) is satisfied, as asserted.

A2.3. Analytic Functions

If λ is a complex parameter taking on values in the open region Λ, a generalized function f_λ differentiable in λ is called *an analytic function of* λ. Then all the (f_λ, φ) are ordinary analytic functions of λ throughout Λ. Conversely, if for a generalized function f_λ all the ordinary functions (f_λ, φ) are analytic functions of λ in some region Λ, then f_λ is also an analytic function of λ (cf. Section 3.1). Then all the derivatives $\partial f / \partial \lambda$, $\partial^2 f / \partial \lambda^2$, ... exist at every point λ of Λ, and in the neighborhood of $\lambda_0 \in \Lambda$ we have the *Taylor's series* expansion

$$f_\lambda = f_{\lambda_0} + (\lambda - \lambda_0) \frac{\partial f_{\lambda_0}}{\partial \lambda} + \tfrac{1}{2} (\lambda - \lambda_0)^2 \frac{\partial^2 f_{\lambda_0}}{\partial \lambda^2} + \dots. \tag{2}$$

Indeed, $\partial f_{\lambda_0} / \partial \lambda$ exists, since by assumption all the derivatives of the ordinary functions (f_λ, φ) exist at $\lambda = \lambda_0$. Similar reasoning demonstrates the existence of the higher derivatives $\partial^2 f_{\lambda_0} / \partial \lambda^2$, ... Further, for every $\varphi(x)$ in K we may form the Taylor's series of the ordinary analytic function (f_λ, φ):

$$(f_\lambda, \varphi) = (f_{\lambda_0}, \varphi) + (\lambda - \lambda_0) \frac{\partial}{\partial \lambda} (f_\lambda, \varphi) \Big|_{\lambda=\lambda_0} + \tfrac{1}{2} (\lambda - \lambda_0)^2 \frac{\partial^2}{\partial \lambda^2} (f_\lambda, \varphi) \Big|_{\lambda=\lambda_0} + \cdots$$

$$= (f_{\lambda_0}, \varphi) + (\lambda - \lambda_0) \left(\frac{\partial f_{\lambda_0}}{\partial \lambda}, \varphi \right) + \tfrac{1}{2} (\lambda - \lambda_0)^2 \left(\frac{\partial^2 f_{\lambda_0}}{\partial \lambda^2}, \varphi \right) + \cdots$$

$$= \left(f_{\lambda_0} + [\lambda - \lambda_0] \frac{\partial f_{\lambda 0}}{\partial \lambda} + \tfrac{1}{2} [\lambda - \lambda_0]^2 \frac{\partial^2 f_{\lambda 0}}{\partial \lambda^2} + \cdots, \varphi \right),$$

and this implies the validity of (2).

Consider two analytic functions f_λ and g_λ defined on a region $\varLambda$ and assume that they coincide on some set in $\varLambda$ that has a limit point also in $\varLambda$. Then they will coincide for all $\lambda \in \varLambda$. This is because for any $\varphi(x)$ in K the functions (f_λ, φ) and (g_λ, φ) coincide in $\varLambda$ by the uniqueness of analytic continuation.

This property is the one on which we base the important *method of analytic continuation* in λ of a functional f_λ. Assume that f_λ is analytic in some region $\varLambda$ and further that all the ordinary functions (f_λ, φ) can be analytically continued to a larger region $\varLambda_1$. Then for any $\lambda_1 \subset \varLambda_1$ the function (f_{λ_1}, φ) also defines a continuous linear functional on K. For the proof, recall that the analytic continuation to any point of $\varLambda_1$ can always be obtained by a finite number of Taylor's series expansions. But every Taylor's series

$$(f_\lambda, \varphi) = (f_{\lambda_0}, \varphi) + (\lambda - \lambda_0) \left(\frac{\partial f_{\lambda_0}}{\partial \lambda}, \varphi \right) + \tfrac{1}{2} (\lambda - \lambda_0)^2 \left(\frac{\partial^2 f_{\lambda_0}}{\partial \lambda^2}, \varphi \right) + \cdots$$

converges for every φ in K with a radius of convergence which depends only on the configuration of $\varLambda$ and $\varLambda_1$, but not on φ. This means that for each λ within the radius of convergence this series represents a continuous linear functional, which is what we wish to prove.

Obviously the derivatives with respect to x of an analytic generalized function f_λ are also generalized functions analytic in λ. We may note also that analytic continuation preserves many properties of f_λ. For instance, if f_λ is invariant in $\varLambda$ under some operation u, so that $f_\lambda(ux) = f_\lambda(x)$, so is its analytic continuation into $\varLambda_1$. In fact, if the equation

$$(f_\lambda(ux), \varphi(x)) = (f_\lambda(x), \varphi(x)) = (f_\lambda(x), \varphi(u^{-1}x))$$

holds in $\varLambda$, the uniqueness of the analytic continuation implies that it holds also in $\varLambda_1$.

Thus, for instance, spherically symmetric analytic functionals have spherically symmetric analytic continuations.

In conclusion, we remark that analytic continuation of a generalized function depending on a parameter may lead, as is the case for analytic continuation of ordinary functions, to a functions with isolated singularities (poles or essential singularities) or to multiple-valued functions.

In the neighborhood of an isolated singularity λ_0 an analytic generalized function f can be expanded in a *Laurent series* in the classical form

$$f_\lambda = \sum_{-\infty}^{\infty} c_n(\lambda - \lambda_0)^n,$$

where the c_n $(n = 0, \pm 1, \pm 2, ...)$ are fixed functionals independent of λ. This is because the Laurent expansion can be formed for every φ in K according to

$$(f_\lambda, \varphi) = \sum_{-\infty}^{\infty} c_n(\varphi)(\lambda - \lambda_0)^n,$$

where the $c_n(\varphi)$ are given by the Cauchy integral formulas

$$c_n(\varphi) = \frac{1}{2\pi i} \int_\Gamma \frac{(f_\lambda, \varphi)\, d\lambda}{(\lambda - \lambda_0)^{n+1}} \qquad (n = 0, \pm 1, \pm 2, ...), \tag{3}$$

where Γ is a contour lying entirely within the region of analyticity of f_λ with the singularity λ_0 lying in its interior. The nth integral in (3) can be written in the form

$$\frac{1}{2\pi i} \int_\Gamma \left(\frac{f_\lambda}{(\lambda - \lambda_0)^{n+1}}, \varphi \right) d\lambda = \frac{1}{2\pi i} \int_\Gamma (g_\lambda, \varphi)\, d\lambda, \qquad g_\lambda = \frac{f_\lambda}{(\lambda - \lambda_0)^{n+1}},$$

which is seen from earlier considerations to be a continuous linear functional of λ. We can thus write $c_n(\varphi) = (c_n, \varphi)$, where c_n is a continuous linear functional. Then $f_\lambda = \sum_{-\infty}^{\infty} c_n(\lambda - \lambda_0)^n$, as asserted.

CHAPTER II

FOURIER TRANSFORMS
OF GENERALIZED FUNCTIONS

1. Fourier Transforms of Test Functions

1.1. Fourier Transforms of Functions in K

Henceforth we shall deal with the space K of *complex* test functions (Chapter I, Section 1.9) and the corresponding *complex* generalized-function space K'.

Let us first consider the case of a single variable.

Let $\varphi(x)$ be in K. We construct its *Fourier transform* according to

$$\psi(\sigma) = \int_{-\infty}^{\infty} \varphi(x)\, e^{i\sigma x}\, dx. \tag{1}$$

On occasion we shall denote $\psi(\sigma)$ by $\widetilde{\varphi(x)}$ or by $F[\varphi(x)]$.

Since $\varphi(x)$ has bounded support, the integral in (1) is in actual fact taken only over a finite region, say $-a \leqslant x \leqslant a$. Therefore $\psi(\sigma)$ can be defined also for *complex* values of its argument $s = \sigma + i\tau$ by

$$\psi(\sigma + i\tau) = \int_{-a}^{a} \varphi(x)\, e^{isx}\, dx = \int_{-a}^{a} \varphi(x)\, e^{i\sigma x}\, e^{-\tau x}\, dx. \tag{2}$$

Now the integral in (2) can be differentiated with respect to the complex parameter s, so that $\psi(\sigma + i\tau)$ is an entire analytic function. If we take the derivative of $\varphi(x)$, we multiply $\psi(s)$ by $-is$:

$$\int_{-a}^{a} \varphi'(x)\, e^{ixs}\, dx = \varphi(x)\, e^{ixs}\Big|_{-a}^{a} - \int_{-a}^{a} is\varphi(x)\, e^{ixs}\, dx = -is\psi(s).$$

Continuing to differentiate, we find that for any $q = 0, 1, 2, \dots$,

$$F[\varphi^{(q)}(x)] = (-is)^q\, F[\varphi(x)] \tag{3}$$

153

and more generally

$$F\left[P\left(\frac{d}{dx}\right)\varphi(x)\right] = P(-is)\,F[\varphi(x)], \tag{4}$$

where $P(t)$ is any polynomial with constant coefficients. Moreover, we obtain the inequality

$$\left|\,s\,|^q\,|\psi(s)\,\right| = \left|\int_{-a}^{a}\varphi^{(q)}(x)\,e^{izs}\,dx\right| \leqslant C_q\,e^{a\,|\tau|}.$$

Thus the *Fourier transform* $\psi(s)$ of any $\varphi(x)$ in K which vanishes for $|\,x\,| \geqslant a$ is an *entire analytic function of its argument* $s = \sigma + i\tau$, and for every $q = 0, 1, 2, \ldots$ it satisfies the inequality

$$|\,s^q\psi(s)\,| \leqslant C_q\,e^{a\,|\tau|}. \tag{5}$$

We maintain that the converse also is true: every entire function $\psi(s)$ that satisfies (5) for every q is the Fourier transform of some infinitely differentiable function $\varphi(x)$ which vanishes for $|\,x\,| \geqslant a$.

Proof. We start the proof with the usual expression for $\varphi(x)$, namely

$$\varphi(x) = \frac{1}{2\pi}\int_{-\infty}^{\infty}\psi(\sigma)\,e^{-i\sigma x}\,d\sigma.$$

The Cauchy integral theorem can be used to replace this integral along the real axis by an integral along a line parallel to it, so that

$$\varphi(x) = \frac{1}{2\pi}\int_{-\infty}^{\infty}\psi(\sigma + i\tau)\,e^{-i(\sigma+i\tau)x}\,d\sigma$$

$$= \frac{1}{2\pi}e^{\tau x}\int_{-\infty}^{\infty}\psi(\sigma + i\tau)\,e^{-i\sigma x}\,d\sigma,$$

and since (5) is assumed fulfilled this integral converges absolutely. In fact it remains absolutely convergent even after formal differentiation of the integrand with respect to x. Thus $\varphi(x)$ is infinitely differentiable, and

$$\varphi^{(q)}(x) = \frac{1}{2\pi}\int_{-\infty}^{\infty}(-is)^q\,\psi(s)\,e^{-isx}\,d\sigma.$$

Let $|\,x\,| > a$; given some $t > 0$ we choose τ such that $x\tau = -\,t\,|\,x\,|$. We now make use of (5) with $q = 0$ and $q = 2$, obtaining

$$|\,\psi(s)\,| \leqslant e^{a\,|\tau|}\min\left\{C_0,\,\frac{C_2}{|\,s\,|^2}\right\} \leqslant C\,\frac{e^{a\,|\tau|}}{1+|\,s\,|^2} \leqslant C\,\frac{e^{a\,|\tau|}}{1+|\,\sigma\,|^2}\,,$$

so that

$$| \varphi(x) | \leqslant \frac{1}{2\pi} e^{\tau x} \int_{-\infty}^{\infty} C \, \frac{e^{a \, |\tau|}}{1 + | \sigma |^2} \, d\sigma = C'e^{-t|x|+at} = C'e^{t(a-|x|)}.$$

Now C' is independent of t, so that by letting t approach infinity we find that $\varphi(x) = 0$. Thus $\varphi(x)$ vanishes for $| x | > a$. Hence $\varphi(x)$ has all the asserted properties. The well-known theorem on the Fourier integral[1] then implies that its Fourier transform coincides with $\psi(\sigma)$, which completes the proof.

Before proceeding we make note of one more useful formula: for any $\varphi(x)$ in K we have

$$FF[\varphi(x)] = 2\pi\varphi(-x). \tag{6}$$

Indeed, if $F[\varphi(x)] = \psi(\sigma)$, then

$$F^{-1}[\psi(\sigma)] = \frac{1}{2\pi} \int \psi(\sigma) \, e^{-ix\sigma} \, d\sigma = \varphi(x),$$

which implies

$$\int \psi(\sigma) \, e^{ix\sigma} \, d\sigma \equiv F[\psi(\sigma)] = 2\pi\varphi(-x),$$

as asserted.

1.2. The Space Z

The study of the Fourier transforms of functions in K, on which we have entered above, leads us naturally to define of the new *space Z of slowly increasing functions*, namely, *of all entire functions* $\psi(s)$ *satisfying the inequalities*

$$| s |^q | \psi(s) | \leqslant C_q e^{a|\tau|} \qquad (q = 0, 1, 2, ...) \tag{1}$$

(where the constants a and C_q may depend on ψ), with the obvious definition of the fundamental linear operations of addition and multiplication by a number.

As has been shown in Section 1.1, the *Fourier transform establishes a one-to-one mapping* between K and Z. This mapping obviously preserves these linear operations.

---

[1] Smirnov, "Higher Math," Vol. 2, p. 424. See also E. C. Titchmarsh, "Introduction to the Theory of Fourier Integrals," p. 48. Oxford Univ. Press (Clarendon), London and New York, 1948.

This implies that the every linear operator defined on K there corresponds a *dual* operator defined on Z. For example, it is seen from Eq. (3) of Section 1.1 that to differentiation on K corresponds multiplication by $-is$ on Z. Similarly, the formula

$$\frac{d\dot{\psi}(s)}{ds} = \int_{-\infty}^{\infty} ixe^{isx}\,\varphi(x)\,dx$$

shows that to multiplication by ix on K corresponds differentiation on Z. Repeating this operation, we find that for any $q = 0, 1, 2, ...,$

$$\frac{d^q}{ds^q}\,F[\varphi] = F[(ix)^q\,\varphi(x)]. \tag{2}$$

The existence of the right-hand side implies the existence of the left, so that a function $\psi(s)$ in Z can be differentiated an arbitrary number of times to yield a function in Z. We may, in fact, write the more general formula

$$P\!\left(\frac{d}{ds}\right)F[\varphi] = F[P(ix)\,\varphi(x)], \tag{3}$$

where $P(t)$ is any polynomial with constant coefficients.

The translation operation $\varphi(x) \to \varphi(x - h)$ in K corresponds to multiplication by e^{ish} in Z. This can be seen immediately by writing

$$F[\varphi(x - h)] = \int_{-\infty}^{\infty} e^{isx}\,\varphi(x - h)\,dx$$

$$= \int_{-\infty}^{\infty} e^{is(y+h)}\,\varphi(y)\,dy = e^{ish}F[\varphi(x)].$$

Conversely, multiplication by e^{ixh} in K (for arbitrary, even complex, h) corresponds to translation in Z:

$$F[e^{ixh}\,\varphi(x)] = \int_{-\infty}^{\infty} e^{ixh}e^{isx}\,\varphi(x)\,dx$$

$$= \int_{-\infty}^{\infty} e^{ix(h+s)}\,\varphi(x)\,dx = \psi(s + h).$$

We see consequently that all possible translations are allowed on the functions in Z.

The definition of *convergence in Z* can also be carried over from K. That is, the sequence of functions $\psi_\nu(s)$ converges to zero in Z if the sequence of their inverse images (inverse Fourier transforms) $\varphi_\nu(x)$

converges to zero in K. Convergence in Z can also be defined entirely intrinsically. We shall say that a sequence $\psi_\nu(s)$ converges to zero in Z if for each function in this sequence we have

$$| s^q \psi_\nu(s) | \leqslant C_q e^{a|\tau|}$$

with C_q and a independent of ν, and if the functions converge to zero uniformly on every interval of the (real) σ axis.

Note that the Taylor's series expansion

$$\sum_{q=0}^{\infty} \psi^{(q)}(s) \frac{h^q}{q!} = \psi(s + h)$$

for any fixed (complex) h is valid in the sense of convergence in Z. This follows from the dual formula

$$\sum_{q=0}^{\infty} (ix)^q \frac{h^q}{q!} \varphi(x) = e^{ixh} \varphi(x)$$

in the sense of convergence in K.

1.3. The Case of Several Variables

The above considerations can be carried over almost without change to the case of n independent variables. The Fourier transform of a function $\varphi(x) \equiv \varphi(x_1, ..., x_n)$ in K is defined by

$$\psi(\sigma) = \psi(\sigma_1, ..., \sigma_n)$$

$$= \int_{-\infty}^{\infty} ... \int_{-\infty}^{\infty} \varphi(x_1, ..., x_n) \exp\left[i(x_1\sigma_1 + ... + x_n\sigma_n)\right] dx_1 ... dx_n$$

or, more briefly, by

$$\psi(\sigma) = \int_{R_n} \varphi(x)\, e^{i(x,\sigma)}\, dx, \tag{1}$$

where (x, σ) denotes $x_1\sigma_1 + ... + x_n\sigma_n$.

The bounded support of $\varphi(x)$ makes it possible for ψ to be continued to complex values of its argument $s = (s_1, ..., s_n) = (\sigma_1 + i\tau_1, ..., \sigma_n + i\tau_n)$:

$$\psi(s) = \int_{R_n} \varphi(x)\, e^{i(x,s)}\, dx. \tag{2}$$

Our new function $\psi(s)$, defined in C_n, the space of n complex dimensions, is continuous and analytic in each of its variables s_k. If $\varphi(x)$ vanishes for $|x_k| > a_k$, $k = 1, 2, ..., n$, then $\psi(s)$ satisfies the inequality

$$|s_1^{q_1} ... s_n^{q_n} \psi(\sigma_1 + i\tau_1, ..., \sigma_n + i\tau_n)| \leqslant C_q \exp(a_1|\tau_1| + ... + a_n|\tau_n|). \quad (3)$$

Conversely, every entire function $\psi(s_1, ..., s_n)$ satisfying (3) is the Fourier transform of some $\varphi(x_1, ..., x_n)$ in K which vanishes for $|x_k| > a_k$, $k = 1, 2, ..., n$. (The proof is the same as for a single variable.) Further, we may obtain equations similar to Eq. (4) of Section 1.1 and Eq. (3) of Section 1.2:

$$P\left(\frac{\partial}{\partial s_1}, ..., \frac{\partial}{\partial s_n}\right) F[\varphi] = F[P(ix_1, ..., ix_n)\, \varphi(x)], \quad (4)$$

$$F\left[P\left(\frac{\partial}{\partial x_1}, ..., \frac{\partial}{\partial x_n}\right) \varphi(x)\right] = P(-is_1, ..., -is_n)\, F[\varphi], \quad (5)$$

where P is any polynomial with constant coefficients.

The space of all slowly increasing entire functions $\psi(s)$ satisfying inequalities such as (3) with the natural definitions of the linear operations (addition and multiplication by a number) will be called Z as before. The Fourier transform establishes a one-to-one mapping between K and Z which conserves the linear operations. We define convergence in Z in the following way: a sequence $\psi_\nu(s)$, $\nu = 1, 2, ...,$ is said to converge to zero in Z if the sequence of inverse Fourier transforms converges to zero in K. This convergence can also be defined intrinsically: we require that the inequalities

$$|s^q \psi_\nu(s)| \leqslant C_q \exp(a_1|\tau_1| + ... + a_n|\tau_n|)$$

be fulfilled with C_q and a independent of ν, and that the $\psi_\nu(\sigma)$ converge uniformly to zero on every bounded set in the real space R_n.

1.4. Functionals on Z

We may construct *generalized functions*, i.e., continuous linear functionals, on Z as well as on K. Consider again the case of a single independent variable. We shall call a *regular functional* any functional which is given by an expression of the form

$$(g, \psi) = \int_{-\infty}^{\infty} \bar{g}(\sigma)\, \psi(\sigma)\, d\sigma. \quad (1)$$

We shall call functionals of the form

$$(g, \psi) = \int_\Gamma \bar{g}(s)\, \psi(s)\, ds, \quad (2)$$

where Γ is some contour, *analytic functionals*. Thus, the delta function given by $(\delta(s - s_0), \psi) = \psi(s_0)$ (where s_0 is any complex number) is not a regular, but an analytic functional, since according to the Cauchy integral formula we have

$$\psi(s_0) = \frac{1}{2\pi i} \int_\Gamma \frac{\psi(s)\, ds}{s - s_0},$$

where Γ is any contour enclosing s_0. Thus $\delta(s - s_0)$ is an analytic functional corresponding to the function $(2\pi i)^{-1}(s - s_0)^{-1}$.

We shall denote by Z' *the set of all generalized functions on* Z. In Z' we may introduce operations similar to those in K'. The linear operations are defined in the obvious way, for neither addition, multiplication by a number, nor convergence present anything new. Multiplication by a function $\bar{h}(s)$, formally defined by

$$(\bar{h}(s)\, g, \psi) = (g, h\psi), \tag{3}$$

is, however, now possible only for a much smaller class of functions $h(s)$. In fact the consistency of this definition requires that the product of $h(s)$ with some $\psi(s)$ in Z should again give a function in Z. Those functions $h(s)$ which have this property will be called *multipliers* in Z. A function $h(s)$ is a multiplier in Z if it is an entire analytic function and satisfies an inequality of the form $| h(s) | \leqslant C e^{b|\tau|} (1 + | s |)^q$ for some b, q, and C.

The derivative of a functional $g \in Z'$ will be defined by

$$\left(\frac{dg}{ds}, \psi\right) = -\left(g, \frac{d\psi}{ds}\right);$$

it could also have been defined as the limit of $(1/h)\,[g(s + h) - g(s)]$.

As is true for K', the generalized functions in Z' have derivatives of all orders. There is a difference, however, in that the generalized functions of Z' are not only infinitely differentiable, but also expandable or *analyzable* in the sense that for every $g \in Z'$

$$\sum_{q=0}^{\infty} g^{(q)}(s)\, \frac{h^q}{q!} = g(s + h), \tag{4}$$

where the series on the left converges in Z', and $g(s + h)$ is the generalized function obtained from $g(s)$ by translation through h. Indeed, for any $\psi(s) \in Z$ we have

$$\left(g^{(q)}(s)\, \frac{h^q}{q!}, \psi(s)\right) = \left(g(s), \frac{(-1)^q\, h^q}{q!}\, \psi^{(q)}(s)\right)$$

and, as we have already noted, $\Sigma\,[(-1)^q\,h^q/q!]\psi^{(q)}\,(s)$ converges in Z to $\psi(s-h)$. Thus

$$\left(\sum g^{(q)}(s)\,\frac{h^q}{q!},\,\psi(s)\right) = (g(s),\psi(s-h)) = (g(s+h),\psi(s)),$$

as asserted.

We note in particular that the series expansion

$$\delta(s+h) = \sum_{q=0}^{\infty}\delta^{(q)}(s)\,\frac{h^q}{q!} \tag{5}$$

will hold for every (complex) h, where the translated delta function $\delta(s+h)$ is defined by

$$(\delta(s+h),\psi(s)) = (\delta(s),\psi(s-h)) = \psi(-h). \tag{6}$$

When we are dealing with several independent variables, Z' is constructed similarly. Addition, multiplication by a number, convergence, and multiplication by a function $\bar{h}(s) = \bar{h}(s_1,\,...,\,s_n)$ are defined in complete analogy with the above. A function $h(s)$ will be a multiplier in Z if it is continuous, analytic (that is, analytic in each s_i for fixed s_k, $i \neq k$) and if

$$|\,h(s)\,| \leqslant C\exp\,(b_1\,|\,\tau_1\,| + ... + b_n\,|\,\tau_n\,|)\,(1 + |\,s_1\,|)^{q_1}\,...\,(1 + |\,s_n\,|)^{q_n}.$$

The partial derivatives of the functionals $g \in Z'$ are also defined in analogy with the above. Every functional in Z' is then not only infinitely differentiable, but also analyzable, which means that it can be expanded in a Taylor's series which converges in Z'.

1.5. Analytic Functionals[2]

We have called a functional g on Z analytic if it can be represented in the form

$$(g,\psi) = \int_{\Gamma}\bar{g}(s)\,\psi(s)\,ds,$$

where $g(s)$ is a function and Γ is some contour in the complex plane (we shall start by considering the case of a single independent variable).

---

[2] These analytic functionals associated with integration along a contour should not be confused with analytic generalized functions of a complex variable, which are discussed in Appendix B at the end of this volume. The latter are associated with integration over the entire complex space.

From the theory of analytic functions we know that if $g(s)$ is an analytic function, Γ can be deformed continuously without changing (g, ψ) so long as in this deformation the end points of Γ remain fixed and the contour does not pass through any singular points of $g(s)$. For instance, the unit functional

$$(1, \psi) = \int_{-\infty}^{\infty} 1 \cdot \psi(s) \, ds$$

can be defined not only by integration along the real axis, but equivalently by integration on any line going from $-\infty$ to $+\infty$ and lying within some strip $|\operatorname{Im} s| \leqslant C$. An allowable contour is, for example, any straight line parallel to the real axis. We shall call two lines *equivalent* if for any ψ in Z the integrals along these lines are equal. The integral of 1 along some other, nonequivalent contour Γ will give a different result. For instance, if Γ is a closed contour lying within the strip $|\operatorname{Im} s| \leqslant C$ and starting and ending at $-\infty$ (or at $+\infty$), the functional we obtain is clearly zero. Such contours, i.e., contours such that the integral of any ψ in Z along them vanishes, will be called *null or generalized closed contours*.

Consider the function $g(s) = 1/s$. We may use it to construct two different analytic functionals, namely,

$$(g_+, \psi) = \int_{-\infty+ai}^{+\infty+ai} \frac{\psi(s)}{s} \, ds \qquad a > 0),$$

$$(g_-, \psi) = \int_{-\infty-ai}^{+\infty-ai} \frac{\psi(s)}{s} \, ds \qquad (a > 0).$$

In both cases we take the integral along a straight line parallel to the real axis and lying either above it (in the first case) or below it (in the second). These functionals both satisfy the equation

$$\bar{s}g = 1.$$

The difference between them can be reduced to the form

$$(g_+ - g_-, \psi) = \int_{|s|=1} \frac{\psi(s)}{s} \, ds$$

in which the integral is taken, for instance, clockwise along the boundary of the unit circle. Cauchy's residue theorem gives

$$(g_+ - g_-, \psi) = -2\pi i \psi(0),$$

so that

$$g_+ - g_- = -2\pi i \delta(s).$$

The new functional $g_0 = g_+ - g_-$ obviously satisfies the equation

$$\bar{s} g_0 = 0.$$

Consider a general rational fractional function

$$g(s) = \frac{P(s)}{Q(s)}.$$

Let the distinct roots of $Q(s)$ be denoted by $s_1, ..., s_n$. Integrating along any contour Γ equivalent to the real axis, we obtain the functional

$$(\bar{g}, \psi) = \int_{\Gamma} g(s)\, \psi(s)\, ds,$$

which depends in general on Γ. Every such functional satisfies the equation

$$Qg = P.$$

Integrating along any null contour Γ_0, we obtain the functional

$$(\bar{g}_0, \psi) = \int_{\Gamma_0} g(s)\, \psi(s)\, ds,$$

which also depends on Γ_0. All of these new functionals satisfy the equation

$$Qg_0 = 0.$$

For instance, if Γ_0 is a contour which encircles the simple root s_1 in the positive direction, we have

$$(\bar{g}_0, \psi) = \int_{\Gamma_0} g(s)\, \psi(s)\, ds = 2\pi i \operatorname*{res}_{s=s_1} [g(s)\, \psi(s)] = C\psi(s_1),$$

so that $g_0 = C\delta(s - s_1)$. If Γ_0 encircles a k-fold root s_1, we have

$$(g_0, \psi) = 2\pi i \operatorname*{res}_{s=s_1} [g(s)\, \psi(s)]$$

$$= \frac{2\pi i}{(k-1)!} \frac{d^{k-1}}{ds^{k-1}} [(s-s_1)^k\, g(s)\, \psi(s)]_{s=s_1}.$$

Consider the analytic functionals associated with

$$g(s) = \exp s^n.$$

The fact that

$$|\exp s^n| = \exp(\mathrm{Re}\, s^n) = \exp(|s|^n \cos n\theta), \qquad \text{where} \quad \theta = \arg s,$$

implies that the s plane can be divided into $2n$ equal sectors subtending an angle of π/n each such that $|\exp s^n|$ alternately increases or decreases exponentially in each of these sectors. We shall call the sectors in which it increases "ridges" and those in which it decreases "valleys." Consider an arbitrary path Γ_k starting at ∞ in the first valley and ending at ∞ in the kth (with $k = 2, ..., n$). We form the obviously convergent integral

$$(F_k, \psi) = \int_{\Gamma_k} \exp(s^n)\,\psi(s)\,ds$$

along this path. [That this integral converges follows from the exponential decrease of $|\exp s^n|$ in both valleys and from the slowly increasing nature of $\psi(s)$.]

In this way we obtain $n - 1$ different functionals which we shall denote by $\exp_1(s^n), ..., \exp_{n-1}(s^n)$. They all satisfy a certain first-order differential equation. This equation is easily obtained by differentiating $\exp_k(s^n)$, obtaining

$$\left(\frac{d}{ds}\exp_k(s^n), \psi\right) = \left(\exp_k(s^n), -\frac{d\psi}{ds}\right) = -\int_{\Gamma_k} \exp(s^n)\frac{d\psi}{ds}\,ds.$$

Integrating by parts (the resulting integrated term vanishes due to the exponential damping of $\exp s^n$ in the valleys), we obtain

$$\left(\frac{d}{ds}\exp_k(s^n), \psi\right) = \int_{\Gamma_k} n s^{n-1}\exp(s^n)\,\psi(s)\,ds = (\overline{ns^{n-1}}\exp(s^n), \psi).$$

Thus the $\exp_k(s^n)$ satisfy the equation

$$\frac{d}{ds}\exp_k s^n = \overline{ns^{n-1}}\exp_k s^n.$$

We see thus that a first-order homogeneous differential equation for a functional on Z may have any number of linearly independent solutions.

Of great interest also are analytic functionals defined on functions of *several* complex variables $s_1, ..., s_n$. For this case they are defined by

$$(g, \psi) = \int_\Gamma \bar{g}(s)\,\psi(s)\,ds,$$

where Γ is some $(2n - 1)$-dimensional hypersurface in the $2n$-dimensional real space of the n complex s_i.

Poincaré's generalization[3] of Cauchy's integral theorem to analytic functions of several variables implies that Γ can also be deformed arbitrarily (without changing the result) so long as its boundary is fixed and so long as it is not allowed to pass through any singularities of $g(s)$.

We shall call a hypersurface Γ *equivalent to the real hyperplane* if for any ψ in Z we have

$$\int_\Gamma \psi(s)\, ds = \int_{-\infty}^{\infty} \psi(\sigma)\, d\sigma$$

(where the integral on the right-hand side is over all the real variables). An example of such a surface is what we shall call "Hörmander's staircase," the locus of all the s_i such that $s_2, \ldots, s_n$ are real, and for each choice of the s_i, $i \neq 1$, the remaining variable s_1 traces out a contour equivalent (in the s_1 plane) to the real axis. A functional $\bar{g}$ defined by

$$(\bar{g}, \psi) = \int_\Gamma \frac{1}{P(s)} \psi(s)\, ds,$$

where Γ is a Hörmander's staircase on which the polynomial (or entire function) $P(s)$ fails to vanish, satisfies the equation

$$Pg = 1. \tag{1}$$

In Volume II (Chapter II, Section 3.3) we shall show that such a staircase can be constructed for every polynomial $P(s)$ and therefore that Eq. (1) always has a solution in Z'. We shall call Γ a *null surface* or *generalized closed* if for every $\psi(s)$ in Z we have

$$\int_\Gamma \psi(s)\, ds = 0.$$

An example of such a null surface is the product of a null contour in the s_1 plane by any surface in the space of the remaining variables.

Let $P(s)$ be any polynomial whose inverse $1/P(s)$ is bounded on the null surface Γ. Then the functional

$$(\bar{g}, \psi) = \int_\Gamma \frac{1}{P(s)} \psi(s)\, ds$$

[3] B. A. Fuks, "Theory of Analytic Functions of Several Complex Variables" (in Russian), p. 299. Moscow-Leningrad, 1948. See also H. Behnke and P. Thullen, "Theorie der Funktionen mehrerer komplexer Veränderlichen." Springer, Berlin, 1934. In this connection see also M. E. Martinelli, "Colloque sur les Fonctions de Plusieurs Variables," p. 109. George Thone, Liège, 1953.

is a solution of

$$Pg = 0. \tag{2}$$

Remark. By taking the Fourier transforms of Eqs. (1) and (2) (see Section 2) we obtain solutions of the equations

$$P\left(i\frac{\partial}{\partial x}\right)f = \delta(x) \tag{3}$$

and

$$P\left(i\frac{\partial}{\partial x}\right)f = 0, \tag{4}$$

where $i\,\partial/\partial x$ denotes $(i\,\partial/\partial x_1, ..., i\,\partial/\partial x_n)$. In this way we shall be able to prove in Volume II that every partial differential equation with constant coefficients possesses an elementary solution.

Later we shall use these simple considerations to obtain explicit solutions for such equations.

1.6. Fourier Transforms of Functions in S

Toward the end of Chapter I, Section 1, we defined the space S of infinitely differentiable functions $\varphi(x)$ satisfying inequalities of the form

$$|\,x^k D^q \varphi(x)\,| \leqslant C_{kq}. \tag{1}$$

Let us obtain the class of functions which are the Fourier transforms of these.[4] Every $\varphi(x) \in S$ has, of course, a classical Fourier transform given by

$$\psi(\sigma) = \int \varphi(x)\,e^{i(x,\sigma)}\,dx,$$

where $\psi(\sigma)$ is infinitely differentiable, for the integral

$$D^q\psi(\sigma) = \int (ix)^q\,\varphi(x)\,e^{i(x,\sigma)}\,dx$$

is absolutely convergent. Now $(ix)^q\varphi(x)$, as well as $\varphi(x)$, is infinitely differentiable, and all its derivatives are absolutely integrable. Thus $D^q\psi(\sigma)$ converges to zero more rapidly than any power of $1/|\,\sigma\,|$. Con-

---

[4] The present material will be treated in more detail in Volume II, Chapter III, Section 1.

sequently $\psi(\sigma)$ possesses all the same properties as a function of $\sigma = (\sigma_1,$ ..., $\sigma_n)$ as $\varphi(x)$ possesses as a function of x. We may thus deduce that the Fourier transform maps S into itself. Moreover, since the same considerations hold for the inverse Fourier transform, this must be a mapping of S *onto* itself, which means that every $\psi(\sigma) \in S$ has an inverse image under the mapping. It can be shown that every sequence $\varphi_\nu(x)$ that converges in S is mapped by the Fourier transform into a sequence $\psi_\nu(\sigma)$ which also converges in S.

All this implies, in particular, that every $\psi(\sigma)$ in Z is an element of S. (This result can also be obtained more directly: the definition of Z implies that every $\psi(\sigma) \in Z$ is infinitely differentiable and that as $|\sigma| \to \infty$ it converges to zero more rapidly than any power of $1/|\sigma|$. The Cauchy integral formula can be used to deduce this same property also for any of its derivatives.) It implies in addition that any sequence $\psi_\nu(\sigma)$ which converges in Z converges also in S. Further, since K is dense in S, its image, namely Z, is also dense in S, so that any $\psi(\sigma) \in S$ can be obtained as the limit (in the sense of convergence in S) of a sequence $\psi_\nu(\sigma) \in Z$.

2. Fourier Transforms of Generalized Functions. A Single Variable

2.1. Definition

We have seen that there exists a one-to-one mapping between K and Z which preserves the linear operations and convergence. A similar mapping can be established now between the continuous linear functionals on these spaces. We shall set up this mapping so that when applied to a regular functional corresponding to an absolutely integrable function it induces a mapping of this function into its classical Fourier transform.

We shall start again with the case of a single independent variable. Let $f(x)$ be an absolutely integrable function whose Fourier transform is $g(\sigma)$. Then for any $\varphi(x)$ in K and its Fourier transform $\psi(\sigma)$ we have

$$(f, \varphi) = \int_{-\infty}^{\infty} \overline{f(x)}\, \varphi(x)\, dx = \frac{1}{2\pi} \int_{-\infty}^{\infty} \overline{f(x)} \left\{ \int_{-\infty}^{\infty} \psi(\sigma)\, e^{-ix\sigma}\, d\sigma \right\} dx$$

$$= \frac{1}{2\pi} \int_{-\infty}^{\infty} \psi(\sigma) \left\{ \int_{-\infty}^{\infty} \overline{f(x)\, e^{ix\sigma}}\, dx \right\} d\sigma = \frac{1}{2\pi} \int_{-\infty}^{\infty} \overline{g(\sigma)}\, \psi(\sigma)\, d\sigma = \frac{1}{2\pi} (g, \psi),$$

a relation which is sometimes called Parseval's theorem and remains valid also if $f(x)$ and $\varphi(x)$, and therefore also their Fourier transforms

$g(\sigma)$ and $\psi(\sigma)$, are square-integrable on the real line. Parseval's theorem shows that $g(\sigma)$ is the generalized function defined by

$$(g, \psi) = 2\pi(f, \varphi). \tag{1}$$

We may use this equation to define a generalized function g in Z' for every generalized function f in K'. Then we shall call g as defined by (1) the *Fourier transform of* f and shall denote it by $F[f]$ or by $\hat{f}$.

We emphasize that $F[f]$ is a functional not on K, but on Z.

The formulas for the derivatives of the Fourier transforms are preserved also for generalized functions:

$$P\left(\frac{d}{ds}\right)F[f] = F[P(ix)f], \tag{2}$$

$$F\left[P\left(\frac{d}{dx}\right)f\right] = P(-is)F[f]. \tag{3}$$

In particular, multiplication by ix in K' corresponds to differentiation in Z', and differentiation in K' corresponds to multiplication by $-is$ in Z.

To prove these statements we need only set $P(d/dx) = d/dx$. For this case we have

$$(F[ixf], F[\varphi]) = 2\pi(ixf, \varphi) = 2\pi(f, -ix\varphi)$$

$$= (F[f], F[-ix\varphi]) = \left(F[f], -\frac{d}{ds}F[\varphi]\right) = \left(\frac{d}{ds}F[f], F[\varphi]\right),$$

which will then give Eq. (2). Equation (3) can be verified similarly.

The inverse operator F^{-1} defined on Z' maps g into f again according to Eq. (1) (which we may now read from right to left) so that

$$F^{-1}[F[f]] = f, \qquad F[F^{-1}[g]] = g,$$

$$(F^{-1}[g], \varphi) = \frac{1}{2\pi}(g, F[\varphi]). \tag{4}$$

The formula $FF[\varphi(x)] = 2\pi\,\varphi(-x)$ [Section 1.1, Eq. (6)] is also easily translated into one for generalized functions. Indeed, for φ in S we have

$$(FF[f], FF[\varphi(x)]) = 2\pi(F[f], F[\varphi(x)]) = (2\pi)^2(f, \varphi(x)).$$

But we may replace $FF[\varphi(x)]$ on the left by $2\pi\varphi(-x)$. Dividing by 2π and replacing x by $-x$, we arrive at

$$(FF[f], \varphi(x)) = (2\pi f, \varphi(-x)),$$

that is,

$$FF[f(x)] = 2\pi f(-x),\qquad(5)$$

as asserted.

2.2. Examples

Example 1. *To find $F[\delta]$.* By definition we have

$$(\tilde{\delta}, \tilde{\varphi}) = 2\pi(\delta, \varphi) = 2\pi\varphi(0) = \int_{-\infty}^{\infty} \psi(\sigma)\, d\sigma = (1, \psi),$$

whence

$$F[\delta] \equiv \tilde{\delta} = 1, \qquad F^{-1}[1] = \delta.\qquad(1)$$

Example 2. *To find $F[1]$.* This is not very different; in fact,

$$(\tilde{1}, \tilde{\varphi}) = 2\pi(1, \varphi) = 2\pi \int_{-\infty}^{\infty} \varphi(x)\, dx = 2\pi \int_{-\infty}^{\infty} \varphi(x)\, e^{-ix,0}\, dx$$

$$= 2\pi\psi(0) = 2\pi(\delta, \psi),$$

whence

$$F[1] = \tilde{1} = 2\pi\delta, \qquad F^{-1}[\delta] = \frac{1}{2\pi}.\qquad(2)$$

Example 3. *Fourier transform of a polynomial.* From Eqs. (2) and (3) of Section 2.1, we have

$$F[P(x)] = F[P(x) \cdot 1]$$

$$= P\left(-i\frac{d}{ds}\right)\tilde{1} = 2\pi P\left(-i\frac{d}{ds}\right)\delta(s),\qquad(3)$$

$$F\left[P\left(\frac{d}{dx}\right)\delta(x)\right] = P(-is)\tilde{\delta} = P(-is) \cdot 1 = P(-is).\qquad(4)$$

In particular,

$$F[\delta^{(2m)}(x)] = (-1)^m s^{2m},$$
$$F[\delta^{(2m+1)}(x)] = (-1)^{m+1} i s^{2m+1}.\qquad(5)$$

Example 4. *Differential equation of order $n - 1$.* Consider the differential equation

$$n y^{(n-1)}(x) = xy(x),\qquad(6)$$

which, after Fourier transformation, becomes the first-order equation

$$n(-i\sigma)^{n-1} u(\sigma) = -i\frac{du(\sigma)}{d\sigma} \qquad (u = \tilde{y}).\qquad(7)$$

Since Eq. (6), of order $n - 1$, has $n - 1$ linearly independent (ordinary) solutions, so has the first-order equation (7), but its solutions are generalized functions in Z'. We have already treated this problem in Section 1.5.

Example 5. *Fourier transform of the exponential e^{bx}.* We may use the fact that

$$e^{bx} = \sum_{n=0}^{\infty} \frac{b^n x^n}{n!}$$

converges in K' to calculate the Fourier transform of the exponential by applying F to this series term by term. This leads to [see Eqs. (5) and (6) of Section 1.4]

$$\widetilde{e^{bx}} = \sum_{n=0}^{\infty} \frac{b^n}{n!} \widetilde{x^n} = 2\pi \sum_{0}^{\infty} \frac{b^n}{n!} \left(-i \frac{d}{ds}\right)^n \delta(s) = 2\pi\delta(s - ib). \tag{8}$$

With this equation we can easily obtain the Fourier transforms of the sine, the cosine, the hyperbolic sine, and the hyperbolic cosine:

$$F[\sin bx] \ = F\left[\frac{e^{ibx} - e^{-ibx}}{2i}\right] = i\pi[\delta(s - b) - \delta(s + b)], \tag{9}$$

$$F[\cos bx] \ = F\left[\frac{e^{ibx} + e^{-ibx}}{2}\right] = \pi[\delta(s - b) + \delta(s + b)], \tag{10}$$

$$F[\sinh bx] = F\left[\frac{e^{bx} - e^{-bx}}{2}\right] = \pi[\delta(s - ib) - \delta(s + ib)], \tag{11}$$

$$F[\cosh bx] = F\left[\frac{e^{bx} + e^{-bx}}{2}\right] = \pi[\delta(s - ib) + \delta(s + ib)]. \tag{12}$$

Example 6. *Fourier transforms of generalized functions extended to S.* Consider a functional f defined on K which can be extended by continuity to S (see Chapter I, Section 1.10). We shall show that its Fourier transform $g = \tilde{f}$ can also be extended (from Z) to a functional on S.

Indeed, the equation

$$(g, \psi) = 2\pi(f, \varphi), \qquad \psi = \tilde{\varphi}, \tag{13}$$

defines the functional g immediately for any ψ which is the Fourier transform of some φ, and since φ may be any function in S, so may ψ. Now convergence of a sequence $\psi_\nu(\sigma)$ to zero in S implies convergence of the inverse image sequence $\varphi_\nu(x)$ also to zero in S. This means that g

as defined by (13) is continuous on S. Indeed, if $\psi_\nu(\sigma) \to 0$, then $(g, \psi_\nu) = 2\pi(f, \varphi_\nu) \to 0$.

Thus (13) defines a continuous functional g on S. It is clear, however, that for $\psi \in Z$ this functional coincides with $\tilde{f}$, a functional defined on K. This demonstrates that $\tilde{f}$ can be extended from Z to S, as asserted.

In particular, the Fourier transforms of periodic functions, generalized functions with bounded support, and all ordinary functions increasing at infinity no faster than some power of their argument, as well as the Fourier transforms of their derivatives, can be considered functionals on S. Such, for instance, are the generalized functions $F[x_+^\lambda]$, $F[x_-^\lambda]$, $F[\,|\,x\,|^\lambda]$, and $F[\,|\,x\,|^\lambda \operatorname{sgn} x]$, which we shall study in detail in Section 2.3.

Example 7. *Fourier transforms of periodic functions.* Every periodic locally summable function $f(x)$ (with period 2π), as we have seen in Chapter I, Section 2.4, can be written in the form of the Fourier series

$$f(x) = \sum_{-\infty}^{\infty} c_n e^{inx}, \tag{14}$$

which always converges in the sense of generalized functions (even in S').

Taking the Fourier transform of (14) term by term (which we are permitted to do by continuity) and using the result of example 5, we arrive at

$$\widetilde{f(x)} = \sum_{-\infty}^{\infty} c_n \delta(s + n). \tag{15}$$

Thus $\widetilde{f(x)}$ is a generalized function in S' concentrated on the countable set of points $s = 0, \pm 1, \pm 2, \ldots$.

2.3. The Fourier Transforms of x_+^λ, x_-^λ, $|\,x\,|^\lambda$, $|\,x\,|^\lambda \operatorname{sgn} x$

Let us calculate the Fourier transform of x_+^λ. We shall first restrict our considerations to values of λ such that $-1 < \operatorname{Re} \lambda < 0$.

Consider the expression

$$F[x_+^\lambda e^{-\tau x}] = \int_0^\infty x^\lambda e^{i\sigma x} e^{-\tau x}\, dx = \int_0^\infty x^\lambda e^{isx}\, dx, \tag{1}$$

where $\tau = \operatorname{Im} s > 0$, so that $0 < \arg s < \pi$. The integral obviously converges. As $\tau \to +0$, it is seen that $x_+^\lambda e^{-\tau x}$ converges to x_+^λ in the sense of generalized functions, so that its Fourier transform converges to the desired Fourier transform of x_+^λ.

We shall calculate the integral in (1) by making the change of variables

$$isx = -\xi, \quad is\,dx = -d\xi, \quad x = -\frac{\xi}{is}, \quad dx = -\frac{d\xi}{is}.$$

Then we obtain

$$F[x_+^\lambda e^{-\tau x}] = \left(\frac{e^{i(\pi/2)}}{s}\right)^{\lambda+1} \int_L \xi^\lambda e^{-\xi}\,d\xi,$$

where the contour L of integration is a ray from the origin to infinity whose angle with respect to the real axis is given by $\arg\xi = \arg s - \pi/2$. It follows that

$$-\frac{\pi}{2} < \arg\xi < \frac{\pi}{2}.$$

This represents the right half of the ξ plane, where the $e^{-\xi}$ is exponentially damped. Then by Cauchy's theorem the integral is equal to the integral along the positive real axis, namely,

$$\int_0^\infty \xi^\lambda e^{-\xi}\,d\xi = \Gamma(\lambda+1),$$

so that we obtain

$$F[x_+^\lambda e^{-\tau x}] = ie^{i\lambda(\pi/2)}(\sigma + i\tau)^{-\lambda-1}\,\Gamma(\lambda+1).$$

Note that because $-1 < \mathrm{Re}\,\lambda < 0$, we have

$$-1 < \mathrm{Re}\,(-\lambda - 1) < 0.$$

We now pass to the limit $\tau \to +0$, arriving at (see Chapter I, Section 3.6)

$$F[x_+^\lambda] = ie^{i\lambda(\pi/2)}\,\Gamma(\lambda+1)(\sigma + i0)^{-\lambda-1}. \tag{2}$$

Now by analytic continuation[1] this formula will be valid for all $\lambda \neq -1$,

---

[1] Here (and often in the future) we make use of the following simple considerations. Let f_λ and g_λ be generalized functions depending analytically on λ in a region G, and assume that in some smaller region $G_1 \subset G$ we know that g_λ is the Fourier transform of f_λ, i.e., that $g_\lambda = \tilde{f}_\lambda$. Analyticity means that the ordinary functions of λ

$$(f_\lambda, \varphi) \quad \text{and} \quad (g_\lambda, \psi), \tag{*}$$

where φ and ψ are test functions (for instance, in K and Z, respectively), are analytic in G. The Fourier transform relation means that in G_1 and for $\psi = \tilde{\varphi}$ the functions of (*) are related by

$$(g_\lambda, \psi) = 2\pi(f_\lambda, \varphi). \tag{**}$$

Uniqueness of analytic continuation then leads to the result that (**) is true also in G, and therefore that in G the generalized function g_λ is again the Fourier transform of f_λ.

— 2, ... By dividing both sides of the equation by $\Gamma(\lambda + 1)$, we obtain entire functions of λ on both sides of the equation, so that for all λ we may write

$$F\left[\frac{x_+^\lambda}{\Gamma(\lambda + 1)}\right] = ie^{i\lambda(\pi/2)}(\sigma + i0)^{-\lambda-1}. \tag{3}$$

We may insert the explicit expression for $(\sigma + i0)^{-\lambda-1}$ from Chapter I, Section 3. Then for $\lambda \neq 0, \pm 1, \pm 2, ...$ we obtain

$$F[x_+^\lambda] = i\Gamma(\lambda + 1)\,[e^{i\lambda(\pi/2)}\sigma_+^{-\lambda-1} - e^{-i\lambda(\pi/2)}\sigma_-^{-\lambda-1}]\;; \tag{4}$$

for $\lambda = n$ (where n is a nonnegative integer) we obtain

$$F[x_+^n] = i^{n+1}[n!\sigma^{-n-1} + (-1)^{n+1}\,i\pi\delta^{(n)}(\sigma)]. \tag{5}$$

In particular,

$$F[x_+^0] \equiv F[\theta(x)] = i\sigma^{-1} + \pi\delta(\sigma), \tag{6}$$

$$F[x_+^1] \equiv F[x_+] = -\sigma^{-2} - i\pi\delta'(\sigma), \tag{7}$$

and similar expressions for higher powers.

The Fourier transform of x_-^λ is calculated similarly. We start by assuming that $-1 < \operatorname{Re}\lambda < 0$, and calculate $F[x_-^\lambda e^{-\tau x}]$ for $\tau < 0$. We then go to the limit as $\tau \to -0$. In this way we arrive at

$$F[x_-^\lambda e^{-\tau x}] = \int_\infty^0 |x^\lambda|\,e^{i\sigma x}e^{-\tau x}\,dx$$

$$= \int_0^\infty t^\lambda e^{-ist}\,dt = \left(-\frac{i}{s}\right)^{\lambda+1}\Gamma(\lambda + 1).$$

This means that for $\lambda \neq -1, -2, ...$ we have

$$F[x_-^\lambda] = -ie^{-i\lambda(\pi/2)}\,\Gamma(\lambda + 1)\,(\sigma - i0)^{-\lambda-1} \tag{8}$$

and that for all λ we have

$$F\left[\frac{x_-^\lambda}{\Gamma(\lambda + 1)}\right] = -ie^{-i\lambda(\pi/2)}\,(\sigma - i0)^{-\lambda-1}. \tag{9}$$

Again inserting the explicit expression for $(\sigma - i0)^{-\lambda-1}$, we find that for $\lambda \neq 0, \pm 1, \pm 2, ...$

$$F[x_-^\lambda] = i\Gamma(\lambda + 1)\,[e^{i\lambda(\pi/2)}\,\sigma_-^{-\lambda-1} - e^{-i\lambda(\pi/2)}\,\sigma_+^{-\lambda-1}]. \tag{10}$$

For $\lambda = n$ (with n a nonnegative integer) we have

$$F[x_-^n] = i^{n+1}[(-1)^{n+1}n!\sigma^{-n-1} - i\pi\delta^{(n)}(\sigma)]. \tag{11}$$

Let us now proceed to $| x |^\lambda$ and $|x |^\lambda \operatorname{sgn} x$. Adding and subtracting Eqs. (4) and (10), we find that for $\lambda \neq 0, \pm 1, \pm 2, \ldots$

$$F[| x |^\lambda] \equiv F[x_+^\lambda] + F[x_-^\lambda] = -2\Gamma(\lambda + 1) \sin \frac{\lambda\pi}{2} | \sigma |^{-\lambda-1}, \qquad (12)$$

$$F[| x |^\lambda \operatorname{sgn} x] \equiv F[x_+^\lambda] - F[x_-^\lambda] = 2i\Gamma(\lambda + 1) \cos \frac{\lambda\pi}{2} | \sigma |^{-\lambda-1} \operatorname{sgn} \sigma. \qquad (13)$$

We now divide both sides of (12) by $\Gamma\left(\dfrac{\lambda + 1}{2}\right)$ and use the duplication formula for the Γ function.[2] We then obtain an elegant formula for the Fourier transform of the entire function

$$f_\lambda(x) = 2^{-\frac{1}{2}\lambda} \frac{| x |^\lambda}{\Gamma\left(\dfrac{\lambda + 1}{2}\right)},$$

namely,

$$F[f_\lambda(x)] = \sqrt{2\pi}\, \frac{2^{\frac{1}{2}(\lambda+1)} | \sigma |^{-\lambda-1}}{\Gamma\left(-\dfrac{\lambda}{2}\right)}$$
$$= \sqrt{2\pi} f_{-\lambda-1}(\sigma). \qquad (12')$$

Performing similar operations on (13) and setting

$$g_\lambda(x) = \frac{2^{-\frac{1}{2}\lambda} | x |^\lambda \operatorname{sgn} x}{\Gamma\left(\dfrac{\lambda + 2}{2}\right)},$$

we obtain

$$F[g_\lambda(x)] = \sqrt{2\pi}\, i\, \frac{2^{\frac{1}{2}(\lambda+1)} | \sigma |^{-\lambda-1} \operatorname{sgn} \sigma}{\Gamma\left(\dfrac{-\lambda + 1}{2}\right)} = \sqrt{2\pi}\, i g_{-\lambda-1}(\sigma). \qquad (13')$$

Similarly, for $\lambda = n$ (with n a nonnegative integer) Eqs. (5) and (11) give

$$F[| x |^n] \equiv F[x_+^n] + F[x_-^n] = i^{n+1}[\{1 + (-1)^{n+1}\} \sigma^{-n-1} n!$$
$$+ \{(-1)^{n+1} - 1\} i\pi\delta^{(n)}(\sigma)], \qquad (14)$$

$$F[| x |^n \operatorname{sgn} x] \equiv F[x_+^n] - F[x_-^n] = i^{n+1}[\{1 - (-1)^{n+1}\} \sigma^{-n-1} n!$$
$$+ \{(-1)^{n+1} + 1\} i\pi\delta^{(n)}(\sigma)]. \qquad (15)$$

[2] Fikhtengolts, "Calculus," Vol. 2, p. 784. See also Whittaker and Watson, "Modern Analysis," p. 240.

In particular, for $n = 2k$ even, we have

$$F[x^{2k}] = (-1)^k \, 2\pi \delta^{(2k)}(\sigma), \tag{16}$$

$$F[|\,x\,|^{2k} \, \text{sgn} \, x] = 2i(-1)^k \, (2k)! \, \sigma^{-2k-1}, \tag{17}$$

while for $n = 2k + 1$ odd, we have

$$F[x^{2k+1}] = 2\pi i(-1)^{k+1} \, \delta^{(2k+1)} \, (\sigma), \tag{18}$$

$$F[|\,x\,|^{2k+1}] = 2(-1)^{k+1}(2k + 1)! \, \sigma^{-2k-2}. \tag{19}$$

The generalized functions $|\,x\,|^\lambda$ and $|\,x\,|^\lambda \, \text{sgn} \, x$ remain continuous also for certain negative integral λ, the former for even $\lambda = -2k - 2$, and the latter for odd $\lambda = -2k - 1$, $k = 0, 1, 2, \dots$. Their Fourier transforms are obtained directly from (12) and (13) by reapplying the operator F and making use of the formula $FF[f(x)] = 2\pi f(-x)$ [Eq. (5) of Section 2.1]. Finally replacing x by σ and vice versa, we arrive in this way at

$$F[x^{-2k-2}] = (-1)^{k+1} \frac{\pi}{(2k + 1)!} \, |\, \sigma \,|^{2k+1}, \tag{20}$$

$$F[x^{-2k-1}] = \frac{(-1)^k \, i\pi}{(2k)!} \, |\, \sigma \,|^{2k} \, \text{sgn} \, \sigma. \tag{21}$$

It can be seen that these expressions are obtained from (12) and (13) simply by writing $\lambda = -2k - 2$ and $\lambda = -2k - 1$, respectively.

Proceeding in the same way, we apply F again to Eqs. (3) and (9) and replace $-\lambda - 1$ by λ. This gives the Fourier transforms of $(x + i0)^\lambda$ and $(x - i0)^\lambda$, namely,

$$F[(x + i0)^\lambda] = \frac{2\pi e^{i\lambda(\pi/2)}}{\Gamma(-\lambda)} \, \sigma_-^{-\lambda-1} \; ; \tag{22}$$

$$F[(x - i0)^\lambda] = \frac{2\pi e^{-i\lambda(\pi/2)}}{\Gamma(-\lambda)} \, \sigma_+^{-\lambda-1}. \tag{23}$$

The formulas we have here obtained will be found in the table at the end of the book.

2.4. Fourier Transforms of $x_+^\lambda \ln x_+$ and Similar Generalized Functions[3]

By differentiating the Fourier transforms of the preceding section *with respect to* λ, we obtain new Fourier transform formulas. For instance,

---

[3] On first reading this section may be omitted. The results we obtain here are presented in the table of Fourier transforms at the back of the book.

the derivative of Eq. (2) leads to the result that for $\lambda \neq -1, -2, \ldots$

$$F[x_+^\lambda \ln x_+] = ie^{i\lambda(\pi/2)} \left\{ \Gamma'(\lambda+1)(\sigma+i0)^{-\lambda-1} \right.$$

$$\left. - \Gamma(\lambda+1)(\sigma+i0)^{-\lambda-1} \ln(\sigma+i0) + i\frac{\pi}{2}\Gamma(\lambda+1)(\sigma+i0)^{-\lambda-1} \right\} \quad (1)$$

$$= ie^{i\lambda(\pi/2)} \left\{ \left[\Gamma'(\lambda+1) + i\frac{\pi}{2}\Gamma(\lambda+1) \right] (\sigma+i0)^{-\lambda-1} \right.$$

$$\left. - \Gamma(\lambda+1)(\sigma+i0)^{-\lambda-1} \ln(\sigma+i0) \right\}.$$

Similarly, differentiating Eq. (8) of Section 2.3 gives the result that for $\lambda \neq -1, -2, \ldots$

$$F[x_-^\lambda \ln x_-] = -ie^{-i\lambda(\pi/2)} \left\{ \left[\Gamma'(\lambda+1) - i\frac{\pi}{2}\Gamma(\lambda+1) \right] (\sigma-i0)^{-\lambda-1} \right.$$

$$\left. - \Gamma(\lambda+1)(\sigma-i0)^{-\lambda-1} \ln(\sigma-i0) \right\}. \quad (2)$$

In particular, setting $\lambda = 0$ gives

$$F[\ln x_+] = i\left\{ \left(\Gamma'(1) + i\frac{\pi}{2} \right)(\sigma+i0)^{-1} - (\sigma+i0)^{-1} \ln(\sigma+i0) \right\}; \quad (3)$$

$$F[\ln x_-] = -i\left\{ \left(\Gamma'(1) - i\frac{\pi}{2} \right)(\sigma-i0)^{-1} - (\sigma-i0)^{-1} \ln(\sigma-i0) \right\}. \quad (4)$$

Adding and subtracting Eqs. (1) and (2), we arrive at

$$F[|x|^\lambda \ln|x|] = ie^{i\lambda(\pi/2)} \left[\Gamma'(\lambda+1) + i\frac{\pi}{2}\Gamma(\lambda+1) \right] (\sigma+i0)^{-\lambda-1}$$

$$- ie^{-i\lambda(\pi/2)} \left[\Gamma'(\lambda+1) - i\frac{\pi}{2}\Gamma(\lambda+1) \right] (\sigma-i0)^{-\lambda-1}$$

$$- ie^{i\lambda(\pi/2)} \Gamma(\lambda+1)(\sigma+i0)^{-\lambda-1} \ln(\sigma+i0) \quad (5)$$

$$+ ie^{-i\lambda(\pi/2)} \Gamma(\lambda+1)(\sigma-i0)^{-\lambda-1} \ln(\sigma-i0),$$

$$F[|x|^\lambda \ln|x| \operatorname{sgn} x] = ie^{i\lambda(\pi/2)} \left[\Gamma'(\lambda+1) + i\frac{\pi}{2}\Gamma(\lambda+1) \right] (\sigma+i0)^{-\lambda-1}$$

$$+ ie^{-i\lambda(\pi/2)} \left[\Gamma'(\lambda+1) - i\frac{\pi}{2}\Gamma(\lambda+1) \right] (\sigma-i0)^{-\lambda-1}$$

$$- ie^{i\lambda(\pi/2)} \Gamma(\lambda+1)(\sigma+i0)^{-\lambda-1} \ln(\sigma+i0) \quad (6)$$

$$- ie^{-i\lambda(\pi/2)} \Gamma(\lambda+1)(\sigma-i0)^{-\lambda-1} \ln(\sigma-i0).$$

In particular, for $\lambda = 0$ we have

$$F[\ln \mid x \mid] = i\left[\Gamma'(1) + i\frac{\pi}{2}\right](\sigma + i0)^{-1} - i\left[\Gamma'(1) - i\frac{\pi}{2}\right](\sigma - i0)^{-1}$$

$$-i(\sigma + i0)^{-1}\ln(\sigma + i0) + i(\sigma - i0)^{-1}\ln(\sigma - i0), \qquad (7)$$

$$F[\ln \mid x \mid \operatorname{sgn} x] = i\left[\Gamma'(1) + i\frac{\pi}{2}\right](\sigma + i0)^{-1} + i\left[\Gamma'(1) - i\frac{\pi}{2}\right](\sigma - i0)^{-1}$$

$$- i(\sigma + i0)^{-1}\ln(\sigma + i0) - i(\sigma - i0)^{-1}\ln(\sigma - i0). \qquad (8)$$

If we take the limit as $\lambda \to -2k$ in Eq. (5) and as $\lambda \to -2k - 1$ in Eq. (6), we obtain expressions for $F[x^{-2k}\ln \mid x \mid]$ and $F[x^{-2k-1}\ln \mid x \mid]$. These expressions can also be obtained in a somewhat different fashion, as we shall show below.

The Fourier transforms of x_+^{-n}, x_-^{-n}, $x_+^{-n}\ln x_+$, and $x_-^{-n}\ln x_-$ can be obtained from the Laurent series for x_+^{λ} and x_-^{λ} about $\lambda = -n$. Recall that the Laurent series of x_+^{λ} about such a pole is

$$x_+^{\lambda} = \frac{(-1)^{n-1}\delta^{(n-1)}(x)}{(n-1)!(\lambda+n)} + x_+^{-n} + (\lambda + n)x_+^{-n}\ln x_+ + \dots.$$

Taking the Fourier transform term by term, we have

$$F[x_+^{\lambda}] = F\left[\frac{(-1)^{n-1}\delta^{(n-1)}(x)}{(n-1)!(\lambda+n)}\right] + F[x_+^{-n}] + (\lambda + n)F[x_+^{-n}\ln x_+] + \dots. \qquad (9)$$

If we now expand

$$F[x_+^{\lambda}] = ie^{i\lambda(\pi/2)}\Gamma(\lambda + 1)(\sigma + i0)^{-\lambda-1} \qquad (10)$$

in powers of $\lambda + n$ and compare coefficients with (9), we will obtain expressions for the Fourier transforms of the desired functions. Note that $e^{i\lambda(\pi/2)}$ and $(\sigma + i0)^{-\lambda-1}$ are entire functions, while $\Gamma(\lambda + 1)$ has a simple pole at $\lambda = -1$. Thus the desired expansion of (10) in powers of $\lambda + n$ can be obtained by multiplying the Taylor's series for $(\sigma + i0)^{-\lambda-1}$ by the Laurent series for

$$A(\lambda) = ie^{i\lambda(\pi/2)}\Gamma(\lambda + 1).$$

The first of these series is

$$(\sigma + i0)^{-\lambda-1} = \sigma^{n-1} - (\lambda + n)\sigma^{n-1}\ln(\sigma + i0)$$

$$+ \tfrac{1}{2}(\lambda + n)^2\sigma^{n-1}\ln^2(\sigma + i0) + \dots, \qquad (11)$$

and we write the second in the form

$$A(\lambda) = \frac{a_{-1}^{(n)}}{\lambda + n} + a_0^{(n)} + a_1^{(n)}(\lambda + n) + \ldots \tag{12}$$

The $a_{-1}^{(n)}$, $a_0^{(n)}$, ... can be calculated explicitly, and we shall present them without derivation somewhat later. The product of (11) and (12) yields

$$ie^{i\lambda(\pi/2)}\,\Gamma(\lambda + 1)\,(\sigma + i0)^{-\lambda - 1}$$

$$= \frac{a_{-1}^{(n)}}{\lambda + n}\,\sigma^{n-1} + [a_0^{(n)}\,\sigma^{n-1} - a_{-1}^{(n)}\,\sigma^{n-1}\ln(\sigma + i0)]$$

$$+ (\lambda + n)\,[a_1^{(n)}\,\sigma^{n-1} - a_0^{(n)}\,\sigma^{n-1}\ln(\sigma + i0)$$

$$+ \tfrac{1}{2}\,a_{-1}^{(n)}\,\sigma^{n-1}\ln^2(\sigma + i0)] + \ldots \tag{13}$$

We now compare coefficients of (9) and (13), arriving at

$$F[x_+^{-n}] = a_0^{(n)}\sigma^{n-1} - a_{-1}^{(n)}\sigma^{n-1}\ln(\sigma + i0), \tag{14}$$

$$F[x_+^{-n}\ln x_+] = a_1^{(n)}\sigma^{n-1} - a_0^{(n)}\sigma^{n-1}\ln(\sigma + i0) + \tfrac{1}{2}a_{-1}^{(n)}\sigma^{n-1}\ln^2(\sigma + i0). \tag{15}$$

Similar calculations will give the Fourier transforms of generalized functions such as x_-^{-n}, $x_-^{-n}\ln x_-$, etc. Again we start with the Laurent series

$$x_-^\lambda = \frac{\delta^{(n-1)}(x)}{(n-1)!(\lambda + n)} + x_-^{-n} + (\lambda + n)\,x_-^{-n}\ln x_- + \ldots,$$

and take the Fourier transform term by term:

$$F[x_-^\lambda] = F\left[\frac{\delta^{(n-1)}(x)}{(n-1)!(\lambda + n)}\right] + F[x_-^{-n}] + (\lambda + n)\,F[x_-^{-n}\ln x_-] + \ldots \tag{16}$$

Again we use the fact that

$$F[x_-^\lambda] = -ie^{-i\lambda(\pi/2)}\,\Gamma(\lambda + 1)\,(\sigma - i0)^{-\lambda - 1}$$

to obtain the desired formulas by expanding in powers of $\lambda + n$ and equating coefficients with (16). In this way we arrive at

$$(\sigma - i0)^{-\lambda - 1} = \sigma^{n-1} - (\lambda + n)\,\sigma^{n-1}\ln(\sigma - i0)$$

$$+ \tfrac{1}{2}(\lambda + n)^2\,\sigma^{n-1}\ln^2(\sigma - i0) + \ldots, \tag{17}$$

and

$$-ie^{-i\lambda(\pi/2)}\,\Gamma(\lambda + 1) \equiv B(\lambda) = \frac{b_{-1}^{(n)}}{\lambda + n} + b_0^{(n)} + b_1^{(n)}(\lambda + n) + \ldots, \tag{18}$$

which yields

$$F[x_-^\lambda] = \frac{b_{-1}^{(n)}}{\lambda + n} \sigma^{n-1} + [b_0^{(n)} \sigma^{n-1} - b_{-1}^{(n)} \sigma^{n-1} \ln(\sigma - i0)]$$

$$+ (\lambda + n) [b_1^{(n)} \sigma^{n-1} - b_0^{(n)} \sigma^{n-1} \ln(\sigma - i0)]$$

$$+ \tfrac{1}{2} b_{-1}^{(n)} \sigma^{n-1} \ln^2(\sigma - i0) + \dots. \qquad (19)$$

We now equate coefficients as planned, arriving at

$$F[x_-^{-n}] = b_0^{(n)} \sigma^{n-1} - b_{-1}^{(n)} \sigma^{n-1} \ln(\sigma - i0), \qquad (20)$$

$$F[x_-^{-n} \ln x_-] = b_1^{(n)} \sigma^{n-1} - b_0^{(n)} \sigma^{n-1} \ln(\sigma - i0)$$

$$+ \tfrac{1}{2} b_{-1}^{(n)} \sigma^{n-1} \ln^2(\sigma - i0). \qquad (21)$$

Let us now obtain the Fourier transforms of the generalized functions $|x|^{-n}$, $|x|^{-n} \operatorname{sgn} x$, $|x|^{-n} \ln |x|$, and $|x|^{-n} \ln |x| \operatorname{sgn} x$. We expand $|x|^\lambda$ in a Taylor's series about $\lambda = -2k$:

$$|x|^\lambda = x^{-2k} + (\lambda + 2k) x^{-2k} \ln |x| + \dots.$$

The term-by-term Fourier transform is

$$F[|x|^\lambda] = F[x^{-2k}] + (\lambda + 2k) F[x^{-2k} \ln |x|] + \dots. \qquad (22)$$

On the other hand, we have already found that

$$F[|x|^\lambda] = -2 \sin \frac{\lambda\pi}{2} \Gamma(\lambda + 1) |\sigma|^{-\lambda-1}.$$

Let us write

$$C(\lambda) = -2 \sin \frac{\lambda\pi}{2} \Gamma(\lambda + 1). \qquad (23)$$

At $\lambda = -2k$ this function is analytic, so that we may write

$$C(\lambda) = c_0^{(2k)} + (\lambda + 2k) c_1^{(2k)} + \dots. \qquad (24)$$

Moreover, in the neighborhood of $\lambda = -2k$ we have

$$|\sigma|^{-\lambda-1} = |\sigma|^{2k-1} - (\lambda + 2k) |\sigma|^{2k-1} \ln |\sigma| + \dots. \qquad (25)$$

We now multiply (24) by (25) and compare the results with (22), obtaining

$$F[x^{-2k}] = c_0^{(2k)} |\sigma|^{2k-1} \qquad (26)$$

[cf. Eq. (20) of Section 2.3] and

$$F[x^{-2k} \ln | x |] = c_1^{(2k)} | \sigma |^{2k-1} - c_0^{(2k)} | \sigma |^{2k-1} \ln | \sigma |. \qquad (27)$$

The higher order terms of (22), (24), and (25) can be used to find $F[x^{-2k} \ln^2 | x |]$, etc.

We now proceed similarly with $| x |^{\lambda} \operatorname{sgn} x$, expanding it about $\lambda = -2k - 1$ to obtain

$$| x |^{\lambda} \operatorname{sgn} x = x^{-2k-1} + (\lambda + 2k + 1) x^{-2k-1} \ln | x | + \dots.$$

Again we take the Fourier transform term by term:

$$F[| x |^{\lambda} \operatorname{sgn} x] = F[x^{-2k-1}] + (\lambda + 2k + 1) F[x^{-2k-1} \ln | x |] + \dots. \qquad (28)$$

On the other hand, we have previously calculated

$$F[| x |^{\lambda} \operatorname{sgn} x] = iD(\lambda)| \sigma |^{-\lambda-1} \operatorname{sgn} \sigma,$$

where

$$D(\lambda) = 2 \cos \frac{\lambda \pi}{2} \Gamma(\lambda + 1). \qquad (29)$$

Proceeding according to the same method, we expand $D(\lambda)$ and $| \sigma |^{-\lambda-1} \operatorname{sgn} \sigma$ in powers of $\lambda + 2k + 1$. We have

$$D(\lambda) = d_0^{(2k+1)} + (\lambda + 2k + 1) d_1^{(2k+1)} + \dots, \qquad (30)$$

$$| \sigma |^{-\lambda-1} \operatorname{sgn} \sigma = | \sigma |^{2k} \operatorname{sgn} \sigma - (\lambda + 2k + 1) | \sigma |^{2k} \ln | \sigma | \operatorname{sgn} \sigma + \dots \qquad (31)$$

Multiplication of (30) by (31) and comparison with (28) gives

$$F[x^{-2k-1}] = id_0^{(2k+1)} | \sigma |^{2k} \operatorname{sgn} \sigma \qquad (32)$$

[cf. Eq. (21) of Section 2.3] and

$$F[x^{-2k-1} \ln | x |] = id_1^{(2k+1)} | \sigma |^{2k} \operatorname{sgn} \sigma - id_0^{(2k+1)} | \sigma |^{2k} \ln | \sigma | \operatorname{sgn} \sigma. \qquad (33)$$

We wish now to calculate the Fourier transforms of $| x |^{-n}$ and $| x |^{-n} \ln | x |$ for odd n, and of $| x |^{-n} \operatorname{sgn} x$ and $| x |^{-n} \ln | x | \operatorname{sgn} x$ for even n. To do this we shall proceed as above, except that instead of Taylor's series we shall use Laurent series.

First we expand $| x |^{\lambda}$ in a Laurent series about $\lambda = -2k - 1$. This is

$$| x |^{\lambda} = 2 \frac{\delta^{(2k)}(x)}{(2k)!} \frac{1}{\lambda + 2k + 1} + | x |^{-2k-1}$$
$$+ (\lambda + 2k + 1) | x |^{-2k-1} \ln | x | + \dots.$$

Taking the Fourier transform term by term we have

$$F[|\,x\,|^{\lambda}] = 2\frac{F[\delta^{(2k)}(x)]}{(2k)!}\frac{1}{\lambda + 2k + 1} + F[|\,x\,|^{-2k-1}]$$

$$+ (\lambda + 2k + 1)F[|\,x\,|^{-2k-1}\ln|\,x\,|] + \dots. \qquad (34)$$

We now multiply the expansion

$$C(\lambda) = \frac{c_{-1}^{(2k+1)}}{\lambda + 2k + 1} + c_0^{(2k+1)} + (\lambda + 2k + 1)\,c_1^{(2k+1)} + \dots \qquad (35)$$

by

$$|\,\sigma\,|^{-\lambda-1} = \sigma^{2k} - (\lambda + 2k + 1)\,\sigma^{2k}\ln|\,\sigma\,|$$

$$+ \tfrac{1}{2}(\lambda + 2k + 1)^2\,\sigma^{2k}\ln^2|\,\sigma\,| + \dots \qquad (36)$$

to arrive at

$$F[|\,x\,|^{\lambda}] = \frac{c_{-1}^{(2k+1)}\,\sigma^{2k}}{\lambda + 2k + 1} + [c_0^{(2k+1)}\,\sigma^{2k} - c_{-1}^{(2k+1)}\,\sigma^{2k}\ln|\,\sigma\,|]$$

$$+ (\lambda + 2k + 1)\,[c_1^{(2k+1)}\,\sigma^{2k} - c_0^{(2k+1)}\,\sigma^{2k}\ln|\,\sigma\,|$$

$$+ \tfrac{1}{2}c_{-1}^{(2k+1)}\,\sigma^{2k}\ln^2|\,\sigma\,|] + \dots. \qquad (37)$$

Therefore

$$F[|\,x\,|^{-2k-1}] = c_0^{(2k+1)}\sigma^{2k} - c_{-1}^{(2k+1)}\sigma^{2k}\ln|\,\sigma\,|, \qquad (38)$$

$$F[|\,x\,|^{-2k-1}\ln|\,x\,|] = c_1^{(2k+1)}\sigma^{2k} - c_0^{(2k+1)}\sigma^{2k}\ln|\,\sigma\,| + \tfrac{1}{2}c_{-1}^{(2k+1)}\sigma^{2k}\ln^2|\,\sigma\,|. \qquad (39)$$

We now proceed similarly to expand the Fourier transform of $|\,x\,|^{\lambda}\,\mathrm{sgn}\,x$ about $\lambda = -\,2k$:

$$|\,x\,|^{\lambda}\,\mathrm{sgn}\,x = 2\frac{\delta^{(2k-1)}(x)}{(2k-1)!}\frac{1}{\lambda + 2k} + |\,x\,|^{-2k}\,\mathrm{sgn}\,x$$

$$+ (\lambda + 2k)\,|\,x\,|^{-2k}\ln|\,x\,|\,\mathrm{sgn}\,x + \dots,$$

$$F[|\,x\,|^{\lambda}\,\mathrm{sgn}\,x] = 2\frac{F[\delta^{(2k-1)}(x)]}{(2k-1)!}\frac{1}{\lambda + 2k} + F[|\,x\,|^{-2k}\,\mathrm{sgn}\,x]$$

$$+ (\lambda + 2k)\,F[|\,x\,|^{-2k}\ln|\,x\,|\,\mathrm{sgn}\,x] + \dots. \qquad (40)$$

In this case the series to be multiplied are

$$|\,\sigma\,|^{-\lambda-1}\,\mathrm{sgn}\,\sigma = \sigma^{2k-1} - (\lambda + 2k)\,\sigma^{2k-1}\ln|\,\sigma\,|$$

$$+ \tfrac{1}{2}(\lambda + 2k)^2\sigma^{2k-1}\ln^2|\,\sigma\,|, \qquad (41)$$

and

$$D(\lambda) = \frac{d_{-1}^{(2k)}}{\lambda + 2k} + d_0^{(2k)} + (\lambda + 2k)\, d_1^{(2k)} + \dots,\tag{42}$$

which gives

$$D(\lambda)|\,\sigma\,|^{-\lambda-1}\,\mathrm{sgn}\,\sigma = \frac{d_{-1}^{(2k)}}{\lambda + 2k}\,\sigma^{2k-1} + [d_0^{(2k)}\sigma^{2k-1} - d_{-1}^{(2k)}\sigma^{2k-1}\ln|\,\sigma\,|]$$

$$+ (\lambda + 2k)\,[d_1^{(2k)}\sigma^{2k-1} - d_0^{(2k)}\sigma^{2k-1}\ln|\,\sigma\,|$$

$$+ \tfrac{1}{2}\, d_{-1}^{(2k)}\sigma^{2k-1}\ln^2|\,\sigma\,|] + \dots.\tag{43}$$

Comparison of (40) and (43) leads to

$$F[|\,x\,|^{-2k}\,\mathrm{sgn}\,x] = id_0^{(2k)}\sigma^{2k-1} - id_{-1}^{(2k)}\sigma^{2k-1}\ln|\,\sigma\,|,$$

$$F[|\,x\,|^{-2k}\ln|\,x\,|\,\mathrm{sgn}\,x]$$

$$= id_1^{(2k)}\sigma^{2k-1} - id_0^{(2k)}\sigma^{2k-1}\ln|\,\sigma\,| + \frac{i}{2}\, d_{-1}^{(2k)}\sigma^{2k-1}\ln^2|\,\sigma\,|,$$

In the above calculations we have introduced the new functions

$$A(\lambda) = ie^{i\lambda(\pi/2)}\,\Gamma(\lambda + 1) = \frac{a_{-1}^{(n)}}{\lambda + n} + a_0^{(n)} + a_1^{(n)}\,(\lambda + n) + \dots,$$

$$B(\lambda) = -ie^{-i\lambda(\pi/2)}\,\Gamma(\lambda + 1) = \frac{b_{-1}^{(n)}}{\lambda + n} + b_0^{(n)} + b_1^{(n)}(\lambda + n) + \dots,$$

$$C(\lambda) = -2\sin\frac{\lambda\pi}{2}\,\Gamma(\lambda + 1) = \frac{c_{-1}^{(n)}}{\lambda + n} + c_0^{(n)} + c_1^{(n)}\,(\lambda + n) + \dots,$$

$$D(\lambda) = 2\cos\frac{\lambda\pi}{2}\,\Gamma(\lambda + 1) = \frac{d_{-1}^{(n)}}{\lambda + n} + d_0^{(n)} + d_1^{(n)}(\lambda + n).$$

Note that $B(\lambda) = \overline{A(\lambda)}$ for real λ, so that $B(\lambda)$ is obtained from $A(\lambda)$ by replacing i by $-i$. Further, $C(\lambda) = A(\lambda) + B(\lambda)$, which means that $C(\lambda) = 2\,\mathrm{Re}\,A(\lambda)$ for real λ. Similarly, $iD(\lambda) = A(\lambda) - B(\lambda)$, so that $D(\lambda) = 2\,\mathrm{Im}\,A(\lambda)$ for real λ. We here present $a_{-1}^{(n)}$, $a_0^{(n)}$, and $a_1^{(n)}$ without derivation:

$$a_{-1}^{(n)} = \frac{i^{n-1}}{(n-1)!}\,;$$

$$a_0^{(n)} = \frac{i^{n-1}}{(n-1)!}\left[1 + \frac{1}{2} + \dots + \frac{1}{n-1} + \Gamma'(1) + i\,\frac{\pi}{2}\right];$$

$$a_1^{(n)} = \frac{i^{n-1}}{(n-1)!} \left\{ \sum_{j,k=1}^{n-1} \frac{1}{jk} - \frac{\pi^2}{8} \right.$$

$$+ \left(1 + \frac{1}{2} + \dots + \frac{1}{n-1} \right) \Gamma'(1) + \Gamma''(1)$$

$$\left. + i\frac{\pi}{2} \left[1 + \frac{1}{2} + \dots + \frac{1}{n-1} + \Gamma'(1) \right] \right\};$$

$$b_i^{(n)} = \bar{a}_i^{(n)}; \quad c_i^{(n)} = 2 \operatorname{Re} a_i^{(n)}; \quad d_i^{(n)} = 2 \operatorname{Im} a_i^{(n)}.$$

In particular,

$$b_{-1}^{(n)} = \frac{(-i)^{n-1}}{(n-1)!}; \quad c_{-1}^{(n)} = \frac{2(-1)^{n-1}}{(n-1)!} \cos (n-1)\frac{\pi}{2};$$

$$d_{-1}^{(n)} = \frac{2(-1)^n}{(n-1)!} \sin (n-1)\frac{\pi}{2}.$$

2.5. Fourier Transform of the Generalized Function $(ax^2 + bx + c)_+^\lambda$

Consider the function $(ax^2 + bx + c)_+^\lambda$, which is equal to $(ax^2 + bx + c)^\lambda$ for those values of x for which $ax^2 + bx + c > 0$, and to zero for all others.

The corresponding generalized function is defined by

$$((ax^2 + bx + c)_+^\lambda, \varphi(x)) = \int_{ax^2+bx+c>0} (ax^2 + bx + c)^\lambda \varphi(x)\, dx \tag{1}$$

for those λ for which the integral exists and by its analytic continuation for other λ.

The region in which $ax^2 + bx + c > 0$, if it exists, may be of one of the following forms:

(1) An open interval $\alpha < x < \beta$;
(2) The entire line;
(3) Two half-lines $x < \alpha$ and $x > \beta$ (with $\alpha < \beta$);
(4) The entire line except for a point $x = \alpha$.

It is always possible to perform a linear transformation on x (a translation and a change of scale) such that $(ax^2 + bx + c)_+^\lambda$ is transformed to one of the following forms corresponding, respectively, to the four cases above:

(1) $(1 - x^2)^\lambda_+$;

(2) $(1 + x^2)^\lambda_+ = (1 + x^2)^\lambda$;

(3) $(x^2 - 1)^\lambda_+$;

(4) $x^{2\lambda}_+$.

We shall not be interested in this last case, since $x^{2\lambda}_+$ and its Fourier transform have been treated already.

In Case 1 the integral

$$\int_{-1}^{1} (1 - x^2)^\lambda \, \varphi(x) \, dx$$

converges for Re $\lambda > -1$, and for other values of λ it can be regularized (analytically continued) in accordance with the expressions given in Chapter I, Section 3. It is then clear that $(1 - x^2)^\lambda_+$ is a generalized function which is analytic everywhere except at $\lambda = -k$, where k is any positive integer, at which points it has poles with residues

$$\frac{(-1)^{k-1}\delta^{(k-1)}(1 - x^2)}{(k - 1)!}.$$

Here $\delta^{(k-1)}(1 - x^2)$ is the generalized function

$$\delta^{(k-1)}(1 - x^2) = \frac{(-1)^{k-1}}{2^k x^{k-1}} [\delta^{(k-1)}(x - 1) - \delta^{(k-1)}(x + 1)].$$

Similarly, $(1 + x^2)^\lambda$ is the generalized function defined by the integral

$$\int_{-\infty}^{\infty} (1 + x^2)^\lambda \, \varphi(x) \, dx,$$

which converges for Re $\lambda < -\frac{1}{2}$ and is analytic for all λ, while $(x^2 - 1)^\lambda_+$ is the generalized function defined by the sum of the integrals

$$\int_{-\infty}^{-1} (1 - x^2)^\lambda \, \varphi(x) \, dx + \int_{1}^{\infty} (1 - x^2)^\lambda \, \varphi(x) \, dx,$$

which converge for $-1 < $ Re $\lambda < -\frac{1}{2}$. This generalized function is analytic except at $\lambda = -k$, where k is any positive integer, where it has poles with residues

$$\frac{(-1)^{k-1}\delta^{(k-1)}(x^2 - 1)}{(k - 1)!}.$$

Here $\delta^{(k-1)}(x^2 - 1)$ is the generalized function

$$\delta^{(k-1)}(x^2 - 1) = (-1)^{k-1}\delta^{(k-1)}(1 - x^2)$$

$$= \frac{1}{2^k x^{k-1}} \left[\delta^{(k-1)}(x - 1) - \delta^{(k-1)}(x + 1) \right].$$

In connection with the above, it would perhaps be valuable as an aside to give the general definitions of $\delta(f(x))$ and $\delta^{(k)}(f(x))$.

Let us first assume that $f(x)$ is an infinitely differentiable function having the single simple root $x = x_0$, and that $f'(x_0) > 0$. Consider the integral

$$\int \delta(t)\,\varphi(t)\,dt = \varphi(0), \tag{I}$$

in which we make the substitution $t = f(x)$. Equation (I) then becomes

$$\int \delta(f(x))\,\varphi(f(x))\,f'(x)\,dx = \varphi(0)$$

in which we may write $\varphi(f(x))f'(x) = \psi(x)$ to obtain

$$\int \delta(f(x))\,\psi(x)\,dx = \frac{\psi(x_0)}{f'(x_0)}.$$

If $f'(x_0)$ were negative, we would have had to take the negative of the integral on the left-hand side if by the integral we mean integration in the direction of increasing x. Consequently, in general we must write

$$\int \delta(f(x))\,\psi(x)\,dx = \frac{\psi(x_0)}{|f'(x_0)|}. \tag{II}$$

These preliminary considerations lead in a natural way to the following definition. Let $f(x)$ be a differentiable function having the single simple root $x = x_0$. We define $\delta(f(x))$ by Eq. (II), where $\psi(x)$ is any function in K.

Assume now that $f(x)$ has any number of simple roots. Then it is natural to define $\delta(f(x))$ by

$$\int \delta(f(x))\,\psi(x)\,dx = \sum_n \frac{\psi(x_n)}{|f'(x_n)|}, \tag{III}$$

where the sum is taken over all the solutions of the $f(x) = 0$. This definition can be written in the form

$$\delta(f(x)) = \sum_n \frac{1}{|f'(x_n)|}\,\delta(x - x_n). \tag{IV}$$

For instance,

$$\delta(\sin x) = \sum_n \delta(x - \pi n).$$

It is quite simple to obtain a definition of $\delta^{(k)}(f(x))$ from (IV) by formal differentiation. Taking the first derivative, we obtain

$$\delta'(f(x)) f'(x) = \sum_n \frac{1}{|f'(x_n)|} \frac{d}{dx} \delta(x - x_n).$$

Since $f'(x)$ does not vanish at $x = x_n$, we may divide by it to obtain

$$\delta'(f(x)) = \sum_n \frac{1}{|f'(x_n)|} \frac{1}{f'(x)} \frac{d}{dx} \delta(x - x_n).$$

Proceeding in the same way, we arrive at

$$\delta^{(k)}(f(x)) = \sum_n \frac{1}{|f'(x_n)|} \left(\frac{1}{f'(x)} \frac{d}{dx}\right)^k \delta(x - x_n). \tag{V}$$

Let us now find the Fourier transforms of $(1 - x^2)_+^\lambda$, $(1 + x^2)^\lambda$, and $(x^2 - 1)_+^\lambda$. For those λ for which the corresponding integrals converge, these Fourier transforms can be given in terms of Bessel functions. It follows from the uniqueness of analytic continuation that these expressions will hold for all λ. It is interesting that for integral λ these Fourier transforms can be given in terms of elementary functions and derivatives of the δ function, which then establishes relations between the Bessel functions and the derivatives of the δ function.

Let us start with $(1 - x^2)_+^\lambda$. For Re $\lambda > -1$ we may write[4]

$$\int_{-1}^1 (1 - x^2)^\lambda e^{ixs} \, dx = \sqrt{\pi}\, \Gamma(\lambda + 1) \left(\frac{s}{2}\right)^{-\lambda - \frac{1}{2}} J_{\lambda + \frac{1}{2}}(s). \tag{2}$$

The function on the right-hand side, and therefore also that on the left, is analytic everywhere except at $\lambda = -n$, with n a positive integer, where it has simple poles.

Thus the Fourier transform of $(1 - x^2)_+^\lambda$ is given by

$$F[(1 - x^2)_+^\lambda] = \sqrt{\pi}\ \Gamma(\lambda + 1) \left(\frac{s}{2}\right)^{-\lambda - \frac{1}{2}} J_{\lambda + \frac{1}{2}}(s) \tag{2'}$$

and exists for all $\lambda \neq -1, -2, \dots$.

---

[4] Ryshik and Gradstein, "Tables," p. 312, Eq. (6.413,3). See also P. M. Morse and H. Feshbach, "Methods of Mathematical Physics," Part I, p. 619. McGraw-Hill, New York, 1953.

The residue at $\lambda = -n$ is

$$\frac{(-1)^{n-1}}{(n-1)!} \sqrt{\pi} \left(\frac{s}{2}\right)^{n-\frac{1}{2}} J_{-n+\frac{1}{2}}(s).\tag{3}$$

This residue can also be written in a different way. To do this we divide both sides of (2) by $\Gamma(\lambda + 1)$ and go to the limit as $\lambda \rightarrow -n$. Since

$$\lim_{\lambda \rightarrow -n} \frac{(1-x^2)_+^\lambda}{\Gamma(\lambda + 1)} = \delta^{(n-1)}(1-x^2),$$

we have

$$\sqrt{\pi} \left(\frac{s}{2}\right)^{n-\frac{1}{2}} J_{-n+\frac{1}{2}}(s) = \int_{-\infty}^{\infty} \delta^{(n-1)}(1-x^2) e^{ixs} dx.\tag{4}$$

The integral in this expression is easily shown to be

$$\frac{1}{2^{n-1}} s^{2n-1} \frac{d^{n-1}}{(s\,ds)^{n-1}} \left(\frac{\cos s}{s}\right).\tag{5}$$

This is the second expression we wanted to obtain for the residue at $\lambda = -n$.

For integral $\lambda = n > 0$, we may easily obtain the formula

$$\int_{-1}^{1} (1-x^2)^n e^{ixs} dx = 2^{n+1} n! (-1)^n \frac{d^n}{(s\,ds)^n} \left(\frac{\sin s}{s}\right).\tag{6}$$

To derive this, note first that it is true for $n = 0$. Now assume it to be true for some $n > 0$ and proceed by induction. In particular, we apply the differential operator $2s^{-1}\,d/ds$ to both sides and integrate by parts on the left, which then proves the assertion for $n + 1$.

From the above formulas, we may easily obtain an expression for the Bessel functions of half-odd integral order in terms of elementary functions. From Eqs. (3), (4), and (5) we have

$$J_{-n+\frac{1}{2}}(s) = \sqrt{\frac{2}{\pi}} (s)^{n-\frac{1}{2}} \frac{d^{n-1}}{(s\,ds)^{n-1}} \left(\frac{\cos s}{s}\right).\tag{7}$$

This expression is valid for negative orders. For positive orders we set $\lambda = n$ in Eq. (2), obtaining

$$\int_{-1}^{1} (1-x^2)^n e^{ixs} dx = n!\,\sqrt{\pi} \left(\frac{s}{2}\right)^{-n-\frac{1}{2}} J_{n+\frac{1}{2}}(s).\tag{8}$$

Comparing this with (6), we arrive at

$$J_{n+\frac{1}{2}}(x) = (-1)^n \, x^{n+\frac{1}{2}} \sqrt{\frac{2}{\pi}} \frac{d^n}{(x \, dx)^n} \left(\frac{\sin x}{x} \right). \tag{7'}$$

We have thus shown that for both negative and positive half odd integers r the Bessel function $J_r(x)$ can be expressed in terms of elementary functions.

We now turn to $(1 + x^2)^\lambda$. The Fourier transform of this function for Re $\lambda < 0$ is[5]

$$\int_{-\infty}^{\infty} (1 + x^2)^\lambda \, e^{ixs} \, dx = \frac{2\sqrt{\pi}}{\Gamma(-\lambda)} \left| \frac{s}{2} \right|^{-\lambda-\frac{1}{2}} K_{-\lambda-\frac{1}{2}}(|s|). \tag{9}$$

This equation holds for all λ. For positive integral $\lambda = n$ the Fourier transform is easily obtained more directly. For this case $(1 + x^2)^\lambda$ is the polynomial $(1 + x^2)^n$, while $|s|^{-\lambda-\frac{1}{2}} K_{-\lambda-\frac{1}{2}}(|s|)$ is singular (has a pole). Indeed, we already know that the Fourier transform of $(1 + x^2)^n$ is $(1 - d^2/ds^2)^n \delta(s)$.

This implies

$$\lim_{\lambda \to -n} \frac{1}{\Gamma(\lambda)} \left| \frac{s}{2} \right|^{\lambda-\frac{1}{2}} K_{\lambda-\frac{1}{2}}(|s|) = \sqrt{\pi} \left(1 - \frac{d^2}{ds^2} \right)^n \delta(s). \tag{10}$$

Finally, let us turn to the Fourier transform of $(x^2 - 1)^\lambda_+$. This is given by

$$\int_{-\infty}^{-1} (x^2 - 1)^\lambda \, e^{isx} \, dx + \int_{1}^{\infty} (x^2 - 1)^\lambda \, e^{isx} \, dx$$

$$= 2 \int_{1}^{\infty} (x^2 - 1)^\lambda \cos sx \, dx. \tag{11}$$

But for $-1 < $ Re $\lambda < 0$ we may write[6]

$$\int_{1}^{\infty} (x^2 - 1)^\lambda \cos sx \, dx = -\frac{\Gamma(\lambda + 1)\sqrt{\pi}}{2 \left| \dfrac{s}{2} \right|^{\lambda+\frac{1}{2}}} N_{-\lambda-\frac{1}{2}}(|s|). \tag{12}$$

---

[5] Ryshik and Gradstein, "Tables," p. 251, Eq. (5.116,1). See also W. Magnus and F. Oberhettinger, "Formulas and Theorems for the Functions of Mathematical Physics," p. 118. Chelsea, New York, 1954.

[6] Ryshik and Gradstein, "Tables," p. 314, Eq. (6.422). See also W. Magnus and F. Oberhettinger, *loc. cit.*, p. 27.

This means that for all $\lambda \neq -n$, where n is a positive integer, the Fourier transform of the generalized function $(x^2 - 1)_+^\lambda$ is

$$2 \int_1^\infty (x^2 - 1)^\lambda \cos sx \, dx$$

$$= -\Gamma(\lambda + 1) \sqrt{\pi} \left| \frac{s}{2} \right|^{-\lambda - \frac{1}{2}} N_{-\lambda - \frac{1}{2}}(|s|)$$

$$= \Gamma(\lambda + 1) \sqrt{\pi} \left| \frac{s}{2} \right|^{-\lambda - \frac{1}{2}}$$

$$\times \frac{\cos \pi(\lambda + \frac{1}{2}) \, J_{-\lambda - \frac{1}{2}}(|s|) - J_{\lambda + \frac{1}{2}}(|s|)}{\sin \pi(\lambda + \frac{1}{2})}. \qquad (13)$$

Particularly simple is the Fourier transform of $(x^2 - 1)_+^n$, where n is a nonnegative integer, for which the above expression becomes

$$2 \int_1^\infty (x^2 - 1)^n \cos sx \, dx$$

$$= \int_{-\infty}^\infty (x^2 - 1)^n e^{isx} \, dx + (-1)^{n+1} \int_{-1}^1 (1 - x^2)^n e^{isx} \, dx$$

$$= 2\pi(-1)^n \left(1 + \frac{d^2}{ds^2}\right)^n \delta(s) + (-1)^{n+1} \sqrt{\pi} \left(\frac{s}{2}\right)^{-n - \frac{1}{2}} J_{n + \frac{1}{2}}(s). \qquad (14)$$

2.6. Fourier Transforms of Analytic Functionals

Consider the analytic functional on Z defined by

$$(\bar{g}, \psi) = \int_\Gamma g(s) \, \psi(s) \, ds. \qquad (1)$$

We shall assume Γ to be either a finite or infinite contour, but in any case, one along which $g(s) e^{\tau b}$ is absolutely integrable for any real b. We then assert that the functional $\bar{g}$ is the Fourier transform of the regular functional on K corresponding to the function

$$f(x) = \frac{1}{2\pi} \overline{\int_\Gamma g(s) \, e^{isx} \, ds}. \qquad (2)$$

Before proceeding we point out that $f(x)$ is defined by (2) for all x by our assumption about Γ. Let $\varphi(x)$ be a function in K whose Fourier transform is $\psi(s)$. Then in (1) we replace $\psi(s)$ by its expression in terms of $\varphi(x)$, obtaining

$$(\bar{g}, \psi) = \int_\Gamma g(s) \left\{ \int_{-\infty}^{\infty} \varphi(x)\, e^{isx}\, dx \right\}$$

$$= \int_{-\infty}^{\infty} \varphi(x) \left\{ \int_\Gamma g(s)\, e^{isx}\, ds \right\} dx$$

$$= 2\pi \int_{-\infty}^{\infty} \overline{f(x)}\, \varphi(x)\, dx = 2\pi(f, \varphi),$$

where $f(x)$ is defined by (2). This completes the proof of the assertion.

Assume, for instance, that Γ is a finite closed contour in the interior of which $g(s)$ has a single isolated singular point s_0. Then according to the residue theorem, $\overline{f(x)}$ may be written

$$\overline{f(x)} = \int_\Gamma g(s)\, e^{isx}\, ds = 2\pi i \cdot \mathop{\mathrm{res}}_{s=s_0} [g(s)\, e^{isx}]$$

Example. Consider the analytic functional

$$(\bar{g}(s), \psi(s)) = \int_{-i\infty}^{i\infty} \exp\left(\frac{s^2}{2}\right) \psi(s)\, ds \tag{3}$$

in which the integral is taken along the imaginary axis (or any equivalent contour). By the above proof $\bar{g}$ is the Fourier transform of the function f defined by

$$\overline{f(x)} = \frac{1}{2\pi} \int_{-i\infty}^{i\infty} \exp\left(\frac{s^2}{2}\right) e^{isx}\, ds.$$

This integral is easily calculated. If we replace s by $i\tau$, it becomes

$$\frac{i}{2\pi} \int_{-\infty}^{\infty} \exp\left(-\frac{\tau^2}{2} - \tau x\right) d\tau = \frac{i}{2\pi} \exp\left(\frac{x^2}{2}\right) \int_{-\infty}^{\infty} \exp\left\{-\tfrac{1}{2}(\tau + x)^2\right\} d\tau$$

$$= \frac{i}{2\pi} \exp\left(\frac{x^2}{2}\right) \int_{-\infty}^{\infty} \exp\left(-\frac{\tau^2}{2}\right) d\tau = \frac{i}{\sqrt{2\pi}} \exp\left(\frac{x^2}{2}\right).$$

Thus the analytic functional of Eq. (3) is the Fourier transform of

$$f(x) = -\frac{i}{\sqrt{2\pi}} \exp\left(\frac{x^2}{2}\right).$$

Inverting this result, we find that the Fourier transform of $\exp(x^2/2)$ is the analytic functional corresponding to the function $i\sqrt{2\pi} \exp(s^2/2)$ and the contour $(-i\infty, i\infty)$.

3. Fourier Transforms of Generalized Functions. Several Variables

3.1. Definitions

The Fourier transform of a functional f acting on the space K of test functions $\varphi(x)$ of several independent variables $x = (x_1, ..., x_n)$ is defined as the functional g acting on the space Z of functions $\psi(s)$, with $s = (s_1, ..., s_n)$, according to

$$(g, \psi) = (2\pi)^n (f, \varphi),\tag{1}$$

where $\psi = \varphi$ is the Fourier transform of $\varphi(x)$. The functional g is both linear and continuous, and we shall denote it by $\tilde{f}$ or by $F[f]$. The results of Section 2 can be applied here with hardly any changes. Equations (2) and (3) of Section 2.1 now become the following: if P is a polynomial with constant coefficients in n variables, then

$$P\left(\frac{\partial}{\partial s_1}, ..., \frac{\partial}{\partial s_n}\right) F[f] = F[P(ix_1, ..., ix_n)f],\tag{2}$$

$$F\left[P\left(\frac{\partial}{\partial x_1}, ..., \frac{\partial}{\partial x_n}\right)f\right] = P(-is_1, ..., -is_n)\tilde{f}.\tag{3}$$

The inverse operator F^{-1} acts on Z' and maps it into K' according to

$$(F^{-1}g, \varphi) = \frac{1}{(2\pi)^n}(g, F\varphi).\tag{4}$$

If $f(x)$ is a function that has a Fourier transform in the classical sense, then $\tilde{f}$ is the regular functional corresponding to the Fourier transform of $f(x)$.

Of the singular functionals the simplest, as usual, is the delta function. It satisfies the two formulas

$$\widetilde{\delta(x)} = 1, \qquad \tilde{1} = (2\pi)^n \delta(s),\tag{5}$$

which are the analogs of Eqs. (2) of Section 2.2.

The Fourier transform of a polynomial is obtained by combining Eqs. (2) and (5) to give

$$F[P(x_1, ..., x_n)] = F[P(x_1, ..., x_n)\, 1]$$

$$= (2\pi)^n P\left(-i\frac{\partial}{\partial s_1}, ..., -i\frac{\partial}{\partial s_n}\right)\delta(s).\tag{6}$$

Let us study the behavior of the Fourier transform of some generalized function f when this latter is subjected to some linear operator u (see Chapter I, Section 1.6). For any $\varphi(x)$ in K we have, writing $u^{-1}x = y$, $x = uy$, and $dx = |u|\,dy$,

$$F[\varphi(u^{-1}x)] = \int \varphi(u^{-1}x)\,e^{i(\sigma,x)}\,dx = |u|\int \varphi(y)\,e^{i(\sigma,uy)}\,dy$$

$$= |u|\int \varphi(y)\,e^{i(u'\sigma,y)}\,dy = |u|\,\tilde{\varphi}(u'\sigma),$$

so that the u^{-1} operating on the x variables induces the transpose u' on the σ variables, and the result is multiplied by $|u|$. For a generalized function f we have, writing $\psi(\sigma) = F[\varphi(x)]$ as always,

$$(F[f(ux)], F[\varphi(x)]) = (2\pi)^n (f(ux), \varphi(x))$$

$$= (2\pi)^n (f(x), \varphi(u^{-1}x)) = (F[f(x)], F[\varphi(u^{-1}x)])$$

$$= (F[f(x)], |u|\,\psi(u'\sigma)) = |u|(F[f(x)], F[\varphi(u'\sigma)]).$$

Now putting $g(\sigma) = F[f(x)]$, $g_u(\sigma) = F[f(ux)]$, we obtain

$$(g_u(\sigma), \psi(\sigma)) = |u|(g(\sigma), \psi(u'\sigma)) = |u|(g(u'^{-1}\sigma), \psi(\sigma)),$$

whence

$$g_u(\sigma) = |u|\,g(u'^{-1}\sigma).$$

Thus the transformation $f(x) \to f(ux)$ in K' induces the transformation $g(\sigma) \to |u|\,g(u'^{-1}\sigma)$ in Z'.

If, in particular, f is invariant under u, so that $f(ux) = f(x)$, its Fourier transform is invariant under u'^{-1} followed by multiplication by $|u|$, so that $|u|\,g(u'^{-1}\sigma) = g(\sigma)$. For instance, if f is spherically symmetric, that is, if it is invariant under any rotation u, its Fourier transform is also spherically symmetric, since every rotation has the two properties $u'^{-1} = u$ and $|u| = 1$.

3.2. Fourier Transform of the Direct Product

Let $f(x)$ and $g(y)$ be given generalized functions of the variables $x = (x_1, ..., x_k)$ and $y = (y_1, ..., y_m)$, and let $\tilde{f}(\xi)$ and $\tilde{g}(\eta)$ be their Fourier transforms. Then the Fourier transform of the direct product $f(x) \times g(y)$ is given by

$$F[f \times g] = \tilde{f}(\xi) \times \tilde{g}(\eta).$$

In other words, the *Fourier transform of the direct product is the direct product of the Fourier transforms.* To prove this one need only verify the equation for functions $\varphi(x, y)$ of the form $\sum_{j=1}^{n} \varphi_j(x)\psi_j(y)$ (see Chapter I, Section 5.1). For such functions we have

$$\left(\widetilde{f \times g}, \overline{\sum \varphi_j \psi_j}\right) = (2\pi)^{m+k} \left(f \times g, \sum \varphi_j \psi_j\right)$$

$$= (2\pi)^{m+k} \sum (f, \varphi_j)(g, \psi_j) = \sum (\tilde{f}, \tilde{\varphi}_j)(\tilde{g}, \tilde{\psi}_j)$$

$$= \left(\tilde{f} \times \tilde{g}, \sum \tilde{\varphi}_j \tilde{\psi}_j\right) = \left(\tilde{f} \times \tilde{g}, \sum \overline{\varphi_j \psi_j}\right),$$

as asserted.

Example 1. For $f(x) = \delta(x)$, our formula becomes

$$F[\delta(x) \times g(y)] = 1(\xi) \times \tilde{g}(\eta).$$

In particular,

$$F[\delta(x) \times 1(y)] = 1(\xi) \times (2\pi)^m \delta(\eta).$$

In other words, the Fourier transform of the characteristic function of the subspace R_y is the characteristic function of the subspace R_ξ multiplied by $(2\pi)^m$.

Example 2. Consider the function $f(x, y)$ of two real variables which is equal to 1 for $x > 0$, $y > 0$, and to zero for other x and y; let us find its Fourier transform. Since the function we are dealing with is the product (and therefore also the direct product) of $\theta(x)$ and $\theta(y)$, we may use Eq. (6) of Section 2.3 to obtain

$$\widetilde{f(x, y)} = \widetilde{\theta(x)} \times \widetilde{\theta(y)} = [\pi\delta(\xi) + i\xi^{-1}] \times [\pi\delta(\eta) + i\eta^{-1}].$$

3.3. Fourier Transform of r^λ

In Chapter I, Section 3.9 the spherically symmetric generalized function r^λ is defined for $\lambda \neq -n, -n-2, \ldots$. Therefore its Fourier transform $g_\lambda(\sigma)$ is also a spherically symmetric generalized function. The Fourier integral

$$g_\lambda(\sigma) = \int r^\lambda e^{i(\sigma, x)} \, dx$$

converges for $-n < \operatorname{Re} \lambda < 0$, and is a spherically symmetric function, or a function of $\rho = \sqrt{\Sigma \sigma_j^2}$. Further, for any $t > 0$ we may write

$$g_\lambda(t\sigma) = \int r^\lambda e^{i(t\sigma, x)} \, dx = \int r^\lambda e^{i(\sigma, tx)} \, dx.$$

We now write

$$tx = y, \qquad x = t^{-1}y, \qquad dx = t^{-n} \, dy, \qquad r = |x| = t^{-1}|y|$$

which leads to the result

$$g_\lambda(t\sigma) = \int |y|^\lambda t^{-\lambda-n} e^{i(\sigma, y)} \, dy = t^{-\lambda-n} g_\lambda(\sigma).$$

This means that $g_\lambda(\sigma)$ is a homogeneous generalized function of degree $-\lambda - n$. It can therefore be written in the form $g_\lambda(\sigma) = C_\lambda \rho^{-\lambda-n}$. Let us calculate C_λ. For this purpose we use the general formula $(g_\lambda, \psi) = (2\pi)^n (r^\lambda, \varphi)$ in which we set[1] $\varphi(x) = \exp(-r^2/2) = \exp(-x_1^2/2) \ldots \exp(-x_n^2/2)$. Then

$$\psi(\sigma) = \left[\sqrt{2\pi} \exp\left(-\frac{\sigma_1^2}{2}\right)\right] \ldots \left[\sqrt{2\pi} \exp\left(-\frac{\sigma_n^2}{2}\right)\right] = (2\pi)^{\frac{1}{2}n} \exp\left(-\frac{\rho^2}{2}\right)$$

and we arrive at

$$C_\lambda(2\pi)^{\frac{1}{2}n} \int \exp\left(-\frac{\rho^2}{2}\right) \rho^{-n-\lambda} \, d\sigma = (2\pi)^n \int r^\lambda \exp\left(-\frac{r^2}{2}\right) \, dx.$$

Both integrals can be calculated by going over to spherical coordinates. We divide both sides by the area of the unit sphere and replace the multiple integral by a single one, substituting $\rho^{n-1} d\rho$ for $d\sigma$, and $r^{n-1} dr$ for dx. Expressed in terms of the gamma function, the results are

$$\int_0^\infty \exp\left(-\frac{\rho^2}{2}\right) \rho^{-\lambda-1} \, d\rho = 2^{-\frac{1}{2}\lambda-1} \, \Gamma\left(-\frac{\lambda}{2}\right),$$

$$\int_0^\infty \exp\left(-\frac{r^2}{2}\right) r^{\lambda+n-1} \, dr = 2^{\frac{1}{2}(\lambda+n-2)} \, \Gamma\left(\frac{\lambda+n}{2}\right).$$

Then we arrive at

$$C_\lambda = 2^{\frac{1}{2}n + \frac{1}{2}(\lambda+n-2) + \frac{1}{2}\lambda+1} \pi^{\frac{1}{2}n} \frac{\Gamma\left(\dfrac{\lambda+n}{2}\right)}{\Gamma\left(-\dfrac{\lambda}{2}\right)} = 2^{\lambda+n} \pi^{\frac{1}{2}n} \frac{\Gamma\left(\dfrac{\lambda+n}{2}\right)}{\Gamma\left(-\dfrac{\lambda}{2}\right)} \tag{1}$$

[1] We make use of the fact that $F[\exp(-x^2/2)] = \sqrt{2\pi} \exp(-\sigma^2/2)$ for $n = 1$. See Ryshik and Gradstein, "Tables," p. 251, Eq. 5.119. See also R. Courant, "Differential and Integral Calculus," Vol. II, p.320. Interscience, New York, 1950.

so that

$$F[r^\lambda] = 2^{\lambda+n}\pi^{\frac{1}{2}n}\frac{\Gamma\left(\dfrac{\lambda+n}{2}\right)}{\Gamma\left(-\dfrac{\lambda}{2}\right)}\rho^{-\lambda-n}. \tag{2}$$

This equation, which we have derived for $-n < \mathrm{Re}\,\lambda < 0$ remains valid throughout the region in which the analytic function r^λ exists, that is, for all values of λ except $\lambda = -n, -n-2, \ldots$, at which points it has poles.

If we divide Eq. (2) by $\Gamma(\frac{\lambda+n}{2})$, we obtain the following simple formula for the Fourier transform of the entire generalized function

$$f_\lambda(r) = 2^{-\frac{1}{2}\lambda}\frac{r^\lambda}{\Gamma\left(\dfrac{\lambda+n}{2}\right)}:$$

$$F[f_\lambda(r)] = (2\pi)^{\frac{1}{2}n}\,2^{\frac{1}{2}(\lambda+n)}\frac{r^{-\lambda-n}}{\Gamma\left(-\dfrac{\lambda}{2}\right)} = (2\pi)^{\frac{1}{2}n}f_{-\lambda-n}(\rho) \tag{2'}$$

[cf. Section 2.3, Eq. (12')].

The Taylor's and Laurent series expansions of r^λ give formulas for Fourier transforms of other generalized functions.

We know from Chapter I, Section 4.6 that in the neighborhood of the regular point λ_0 we may write

$$r^\lambda = r^{\lambda_0} + (\lambda - \lambda_0)\, r^{\lambda_0}\ln r + \tfrac{1}{2}(\lambda - \lambda_0)^2\, r^{\lambda_0}\ln^2 r + \ldots.$$

This implies that

$$\widetilde{r^\lambda} = \widetilde{r^{\lambda_0}} + (\lambda - \lambda_0)\,F[r^{\lambda_0}\ln r] + \tfrac{1}{2}(\lambda - \lambda_0)^2\,F[r^{\lambda_0}\ln^2 r] + \ldots. \tag{3}$$

On the other hand, we can also expand $\widetilde{r^\lambda} = C_\lambda\rho^{-\lambda-n}$ in a Taylor's series about of λ_0, obtaining

$$C_\lambda\rho^{-\lambda-n} = C_{\lambda_0}\rho^{-\lambda_0-n} + (\lambda - \lambda_0)\,[C'_{\lambda_0}\rho^{-\lambda_0-n} + C_{\lambda_0}\rho^{-\lambda_0-n}\ln\rho]$$

$$+\tfrac{1}{2}(\lambda - \lambda_0)^2\,[C''_{\lambda_0}\rho^{-\lambda_0-n} + 2C'_{\lambda_0}\rho^{-\lambda_0-n}\ln\rho + C_{\lambda_0}\rho^{-\lambda_0-n}\ln^2\rho] + \ldots. \tag{4}$$

By comparing coefficients in (4) and (3), we arrive at

$$F[r^{\lambda_0}\ln r] = C'_{\lambda_0}\rho^{-\lambda_0-n} + C_{\lambda_0}\rho^{-\lambda_0-n}\ln\rho, \tag{5}$$

$$F[r^{\lambda_0}\ln^2 r] = C''_{\lambda_0}\rho^{-\lambda_0-n} + 2C'_{\lambda_0}\rho^{-\lambda_0-n}\ln\rho + C_{\lambda_0}\rho^{-\lambda_0-n}\ln^2\rho \tag{6}$$

and similar expressions for the higher orders.

In the neighborhood of the singular point $\lambda = -n - 2m$ we may write

$$r^\lambda = \frac{a_{-1}}{\lambda + n + 2m} + a_0 + a_1(\lambda + n + 2m) + \dots, \qquad (7)$$

where a_{-1}, a_0, a_1, ... are given by [see Chapter I, Section 4.6 and the comment following Eq. (3) of that section]

$$a_{-1} = \Omega_n \frac{\delta^{(2m)}(r)}{(2m)!}, \qquad a_0 = \Omega_n r^{-2m-n}, \qquad a_1 = \Omega_n r^{-2m-n} \ln r, \dots. \qquad (8)$$

We take the Fourier transform of (7) term by term, which gives

$$\tilde{r}^\lambda = \frac{\Omega_n}{(2m)!} \frac{F[\delta^{(2m)}(r)]}{\lambda + n + 2m} + \Omega_n F[r^{-2m-n}]$$

$$+ \Omega_n(\lambda + n + 2m) F[r^{-2m-n} \ln r] + \dots. \qquad (9)$$

On the other hand, in the neighborhood of this same point we may write

$$C_\lambda \rho^{-\lambda-n} = \left[\frac{c_{-1}^{(n+2m)}}{\lambda + n + 2m} + c_0^{(n+2m)} + c_1^{(n+2m)}(\lambda + n + 2m) + \dots \right]$$

$$\times \left[\rho^{2m} + (\lambda + n + 2m) \rho^{2m} \ln \rho + \dots \right]$$

$$= \frac{c_{-1}^{(n+2m)} \rho^{2m}}{\lambda + n + 2m} + \left(c_{-1}^{(n+2m)} \rho^{2m} \ln \rho + c_0^{(n+2m)} \rho^{2m} \right)$$

$$+ (\lambda + n + 2m)\left(\tfrac{1}{2} c_{-1}^{(n+2m)} \rho^{2m} \ln^2 \rho + c_0^{(n+2m)} \rho^{2m} \ln \rho + c_1^{(n+2m)} \rho^{2m} \right) + \dots. \qquad (10)$$

As before, we compare coefficients of (10) and (8), to obtain

$$\frac{\Omega_n}{(2m)!} F[\delta^{(2m)}(r)] = c_{-1}^{(n+2m)} \rho^{2m},$$

$$\Omega_n F[r^{-2m-n}] = c_{-1}^{(n+2m)} \rho^{2m} \ln \rho + c_0^{(n+2m)} \rho^{2m},$$

$$\Omega_n F[r^{-2m-n} \ln r] = \tfrac{1}{2} c_{-1}^{(n+2m)} \rho^{2m} \ln^2 \rho + c_0^{(n+2m)} \rho^{2m} \ln \rho + c_1^{(n+2m)} \rho^{2m}. \qquad (11)$$

Here the $c_{-1}^{(n+2m)}$, $c_0^{(n+2m)}$, ... are the coefficients of the Laurent series for C_λ of Eq. (1) about $\lambda = -n - 2m$.

3.4. Fourier Transform of a Generalized Function with Bounded Support

We wish to prove that the Fourier transform of any generalized function f with bounded support is a regular functional that corresponds to a function of σ of the form

$$(\overline{f(x)}, e^{i(x,\sigma)}).$$

This expression is to be understood in the following way. By $e^{i(x,\sigma)}$ we mean any function $\varphi_0(x)$ in K which is equal to $e^{i(x,\sigma)}$ on the support of f and vanishes outside a sufficiently large region; then the functional $\bar{f}$ is applied to this function to give

$$(\bar{f}, \varphi_0(x)) = \overline{(f, \overline{\varphi_0(x)})}.$$

The number obtained in this way is independent of what particular function we choose for $\varphi_0(x)$ so long as it has the above properties (see Appendix 1.3 to Chapter I). In order to prove our assertion we must verify that

$$\int \overline{(\overline{f(x)}, e^{i(x,\sigma)})}\, \psi(\sigma)\, d\sigma = (2\pi)^n (f, \varphi)$$

for any $\varphi(x)$ in K with Fourier transform $\psi(\sigma)$. But we know that

$$(f, \varphi) = \left(f, \frac{1}{(2\pi)^n} \int \psi(\sigma)\, e^{-i(\sigma,x)}\, d\sigma \right).$$

If we can now move the functional f under the integral sign, our assertion will be proven, for we will have

$$(f, \varphi) = \frac{1}{(2\pi)^n} \int \psi(\sigma)\, (f, e^{-i(\sigma,x)})\, d\sigma = \frac{1}{(2\pi)^n} \int \overline{(\overline{f(x)}, e^{i(\sigma,x)})}\, \psi(\sigma)\, d\sigma. \qquad (1)$$

We must thus verify the validity of moving f into the integrand. We first show that this is a proper procedure if the integral is taken over a bounded region G, that is, that we may write

$$\left(f, \int_G \psi(\sigma)\, e^{-i(\sigma,x)}\, d\sigma \right) = \int_G \psi(\sigma)\, (f, e^{-i(\sigma,x)})\, d\sigma.$$

The integral over G is the limit of the Riemann sums

$$s_N(x) = \sum_{j=1}^{N} \psi(\sigma_j)\, e^{-i(\sigma_j,x)}\, \Delta\sigma_j.$$

The $s_N(x)$ converge to the integral over G uniformly for every bounded region of the variable x, as do all their derivatives with respect to x. We can therefore transform them by the usual trick to functions in K (that is, by multiplying them by a fixed function in K which is equal to one in a neighborhood of the support f), obtaining a sequence converging in K. This means that

$$(f, s_N) = \sum \psi(\sigma_j)\, (f, \sigma^{-i(\sigma_j, x)})\, \varDelta\sigma_j$$

converges to

$$\left(f, \int_G \psi(\sigma)\, e^{-i(\sigma, x)}\, d\sigma\right).$$

On the other hand, the (f, s_N) are seen to be the Riemann sums for the function (of σ)

$$\psi(\sigma)\, (f, e^{-i(\sigma, x)}),$$

which converge to the integral of this function. This establishes the validity of (1) for a bounded region G.

Now the integral over the entire space is the limit of integrals over bounded regions:

$$\int_{R_n} \psi(\sigma)\, e^{-i(\sigma, x)}\, d\sigma = \lim_{N \to \infty} \int_{|\sigma| \leqslant N} \psi(\sigma)\, e^{-i(\sigma, x)}\, d\sigma.$$

The sequence of functions of x whose limit we are taking converges (namely, to the integral on the left) again uniformly together with all derivatives in every bounded region of the variable x, which means that by the familiar trick we may convert this to convergence in K. Combining these results we have

$$\left(f, \int_{R_n} \psi(\sigma)\, e^{-i(\sigma, x)}\, d\sigma\right) = \lim_{N \to \infty} \left(f, \int_{|\sigma| \leqslant N} \psi(\sigma)\, e^{-i(\sigma, x)}\, d\sigma\right)$$

$$= \lim_{N \to \infty} \int_{|\sigma| \leqslant N} \psi(\sigma)\, (f, e^{-i(\sigma, x)})\, d\sigma = \int_{R_n} \psi(\sigma)\, (f, e^{-i(\sigma, x)})\, d\sigma,$$

which completes the proof.

As an example let us obtain the Fourier transform of the singular generalized function $\delta(r - a)$, which corresponds to a uniform mass distribution of unit density on the sphere U_a of radius a centered at the origin. In other words,

$$(\delta(r - a), \varphi) = \int_{U_a} \varphi(x)\, dx.$$

Now $\delta(r - a)$ has bounded support, so that according to the preceding considerations its Fourier transform is

$$F[\delta(r - a)] = (\delta(r - a), e^{i(x,\sigma)}) = \int_{U_a} e^{i(x,\sigma)} \, dx.$$

In spherical coordinates ($r = |x| = \alpha$, $\rho = |\sigma|$, and θ is the angle between the x and σ vectors) this becomes

$$F[\delta(r - a)] = \int e^{ia\rho\cos\theta} \, a^{n-1} \sin^{n-2}\theta \, d\theta \, d\omega$$

$$= a^{n-1} \Omega_{n-1} \int_0^\pi e^{ia\rho\cos\theta} \sin^{n-2}\theta \, d\theta,$$

where $d\omega$ is the element of area on the unit (hyper) sphere in the $(n - 1)$-dimensional subspace orthogonal to ρ.

It is known that the integral on the right-hand side can be expressed in terms of Bessel functions,[2] so that we obtain

$$N_n F[\delta(r - a)] = a^{n-1} \Omega_{n-1} (a\rho)^{1-\frac{1}{2}n} J_{\frac{1}{2}(n-2)}(a\rho)$$

$$= \Omega_{n-1} a^{\frac{1}{2}n} \rho^{1-\frac{1}{2}n} J_{\frac{1}{2}(n-2)}(a\rho). \tag{2}$$

Here N_n is given by

$$N_n = \frac{2^{1-\frac{1}{2}n}}{\Gamma\left(\dfrac{n-1}{2}\right) \Gamma\left(\dfrac{1}{2}\right)}.$$

For $n = 2m + 3$ odd, the Bessel functions can be written in terms of trigonometric functions [see Eq. (7') of Section 2.5] :

$$J_{\frac{1}{2}}(z) = \sqrt{\frac{2}{\pi z}} \sin z,$$

$$J_{m+\frac{1}{2}}(z) = (-1)^m z^{m+\frac{1}{2}} \left(\frac{1}{z}\frac{d}{dz}\right)^m \left(\frac{1}{\sqrt{z}} J_{\frac{1}{2}}(z)\right).$$

Thus for $n = 2m + 3$ we have

$$N_n F[\delta(r - a)]$$

$$= \Omega_{n-1} a^{\frac{1}{2}n} \rho^{1-\frac{1}{2}n} (-1)^m (a\rho)^{m+\frac{1}{2}} \left(\frac{1}{z}\frac{d}{dz}\right)^m \left(\frac{1}{\sqrt{z}} J_{\frac{1}{2}}(z)\right)\bigg|_{z=a\rho}. \tag{3}$$

[2] Ryshik and Gradstein, "Tables," p. 148, Eq. 3.227,3 (set $x = \cos\theta$). See also Whittaker and Watson, "Modern Analysis," p. 366.

In particular for $m = 0$, and therefore for $n = 3$, this becomes

$$N_n F[\delta(r - a)] = 2\pi a^{\frac{3}{2}} \rho^{-\frac{1}{2}} J_{\frac{1}{2}}(a\rho)$$

or

$$F[\delta(r - a)] = 4\pi a \frac{\sin \rho a}{\rho}. \tag{4}$$

It is known, on the other hand, that for any $n = 2m + 3$ the Bessel functions satisfy the formula[3]

$$\left(\frac{d}{z \, dz}\right)^m [z^{m+\frac{1}{2}} J_{m+\frac{1}{2}}(z)] = \sqrt{z} \, J_{\frac{1}{2}}(z) = \sqrt{\frac{2}{\pi}} \sin z. \tag{5}$$

We may replace $a\rho$ in Eq. (2) by z. Then it becomes

$$N_n \rho^{n-2} \frac{1}{a} F[\delta(r - a)] = \Omega_{n-1} J_{\frac{1}{2}n-1}(z) z^{\frac{1}{2}n-1},$$

and we may use (5) to write

$$N_n \rho^{n-2} \left(\frac{d}{z \, dz}\right)^m \frac{1}{a} F[\delta(r - a)] = \Omega_{n-1} \sqrt{\frac{2}{\pi}} \sin z.$$

Now returning to $z = a\rho$ (and $dz = \rho da$), we have

$$N_n \rho \left(\frac{d}{a \, da}\right)^m \frac{1}{a} F[\delta(r - a)] = \Omega_{n-1} \sqrt{\frac{2}{\pi}} \sin a\rho.$$

Now the Fourier transform operator acts on the x variables, while the differentiation acts on the a variables, so that we may write

$$\left(\frac{d}{a \, da}\right)^m \frac{1}{a} F[\delta(r - a)] = F\left[\left(\frac{d}{a \, da}\right)^m \frac{1}{a} \delta(r - a)\right].$$

We then arrive at the interesting result

$$N_n F\left[\left(\frac{d}{a \, da}\right)^m \frac{\delta(r - a)}{a}\right] = \Omega_{n-1} \sqrt{\frac{2}{\pi}} \frac{\sin a\rho}{\rho}. \tag{6}$$

---

[3] Ryshik and Gradstein, "Tables," p. 326, Eq. 6.482,3. See also Whittaker and Watson, "Modern Analysis," p. 360, 17.211 (use $J_{-n}(z) = (-)^n J_n(z)$).

3.5. The Fourier Transform as the Limit of a Sequence of Functions

The method described above also may be used to find the Fourier transforms of generalized functions without bounded support.

Let f be any generalized function. As we have seen (Chapter I, Section 1.8), f can be written as the limit of a sequence of generalized functions f_ν, each of which has bounded support. The Fourier transforms of the f_ν, as we have proven, are functions $g_\nu(s)$. Since convergent sequences are carried into convergent sequences under Fourier transformation, the desired Fourier transform of f can be obtained as the limit (in the sense of convergence in Z') of the sequence of the $g_\nu(s)$ as $\nu \to \infty$.

In particular, let $f(x)$ be any ordinary function (which may increase arbitrarily rapidly). Its Fourier transform can be defined as the limit (in the sense of convergence in Z') of the sequence of ordinary functions

$$g_\nu(\sigma) = \int_{|x| \leqslant \lambda} f(x)\, e^{i(x,\sigma)}\, dx. \tag{1}$$

If $f(x)$ increases no more rapidly than some power of its argument and therefore defines a functional on S, the integral of (1) converges to the Fourier transform of $f(x)$ also in the sense of convergence in S' [that is, when operating on any $\psi(\sigma) \in S$]. In particular, the $g_\nu(\sigma)$ converge to a limit $g(\sigma)$ also when operating on infinitely differentiable functions with bounded support, or in K'. We have already discussed such phenomena in the examples of Chapter I, Section 2.5.

4. Fourier Transforms and Differential Equations

4.1. Introductory Remarks

The Fourier transform in its classical form is one of the important tools used in solving differential equations and associated problems. But the applicability of Fourier transform methods is restricted essentially to functions which themselves, or whose powers, are integrable over the entire space (e.g., L^1 or L^2). The use of Fourier transforms in the complex domain has made it possible to include also exponentially increasing functions, but with the requirement that these functions vanish for negative values of their argument. We refer to the Laplace transform,[1] a modification of the Fourier transform. The two-sided

---

[1] See, for example, H. S. Carslaw and J. C. Jaeger, "Operational Methods in Applied Mathematics." Oxford Univ. Press, London and New York, 1943.

Laplace integral, on which van der Pol's celebrated book[2] is based, allows the inclusion of functions which increase exponentially as $x \to +\infty$ and remain nonzero but must be exponentially damped for $x < 0$, which ensures the existence of a strip in the complex plane in which the Laplace transform exists. Another elegant method for developing an operational calculus has been suggested by Mikusinski.[3] His method (as yet developed only for the case of a single variable) can be used to deal with functions which increase in an arbitrary way as $x \to +\infty$ and vanish for negative x.

The Fourier transform method for generalized functions as developed in the present book requires no assumptions concerning the growth of the functions treated, either as $x \to +\infty$ or as $x \to -\infty$, and can be used for functions of any number of variables. It is thus evident that this method can be used to solve, in particular, all types of problems to which the classical Fourier transform, Laplace transform, and Mikusinski's methods are applicable, as well as many other problems which will not yield to the earlier methods.

We shall restrict our considerations in the present volume to some rather simple examples.

4.2. The Iterated Laplace Equation $\Delta^m u = f$

This problem will be solved when we have obtained an elementary solution E and then form its convolution with f, namely $u = f * E$. An elementary solution, as we know, is a solution of the equation

$$\Delta^m E = \delta(x). \tag{1}$$

Let us attempt to find such a solution in K'. Taking the Fourier transform of (1), we have

$$(-1)^m \rho^{2m} V = 1, \qquad (\rho^2 = \sum \sigma_j^2), \tag{2}$$

where V denotes the Fourier transform of E. The problem is now to solve Eq. (2).

If $2m < n$, we may take the (locally summable) function $(-1)^m \rho^{-2m}$ as our solution. If, further, $\lambda = -2m < 0$ is not a pole of the analytic function ρ^λ (see Chapter I, Section 3.9, and recall that the poles

[2] B. van der Pol and H. Bremmer, "Operational Calculus Based on the Two-Sided Laplace Integral." Cambridge Univ. Press, London and New York, 1950.

[3] J. Mikusinski, "Operational Calculus." Pergamon, New York, 1959.

lie at $\lambda = -n, -m-2, ...$), the functional $(-1)^m \rho^{-2m}$ is a solution, as is seen by passing to the limit $\lambda \to -2m$ in the equation $\rho^{2m}\rho^{\lambda} = \rho^{2m+\lambda}$.

Finally, let $\lambda = -2m$ be a pole of ρ^{λ}, which we expand in a Laurent series about this point, obtaining

$$\rho^{\lambda} = \frac{a_{-1}}{\lambda + 2m} + a_0 + a_1(\lambda + 2m) + ..., \tag{3}$$

where the a_i are generalized functions (Chapter I, Section 4.6).

We multiply this equation term by term by ρ^{2m} and then go to the limit as $\lambda \to -2m$. As above, the left-hand side converges to unity. On the right-hand side all the terms higher than the second vanish in the limit, the second term $\rho^{2m}a_0$ remains constant, and if we assume that $\rho^{2m}a_{-1} \neq 0$, the first term increase without bound. But this would contradict the limit equation in which all the other terms are finite, so we conclude that $\rho^{2m}a_{-1} = 0$, and therefore that

$$\rho^{2m}a_0 = 1.$$

Thus for this case the solution of Eq. (2) is $(-1)^m a_0$, where a_0 is the regular part of the Laurent series at $\lambda = -2m$.[4] According to Eq. (2) of Section 3.3, the inverse Fourier transform and solution of (1) in the first and second cases is $C_{2m}r^{2m-n}$. In the third case, according to Eq. (5) of Section 3.3, the solution is

$$Ar^{2m-n} \ln r + Br^{2m-n}.$$

For our purposes, however, the second term may be dropped, since it is annihilated by the operator Δ^m. Therefore collecting the results we have

$$E(x) = \begin{cases} Cr^{2m-n} \ln r, & \text{if } 2m > n \text{ and } n \text{ is even,} \\ Cr^{2m-n} & \text{otherwise.} \end{cases}$$

4.3. The Wave Equation in Space of Odd Dimension

Particular solutions of the wave equation

$$\frac{\partial^2 u}{\partial t^2} = \frac{\partial^2 u}{\partial x_1^2} + ... + \frac{\partial^2 u}{\partial x_n^2} \tag{1}$$

---

[4] This coefficient is the generalized function $\Omega_n r^{-n-2m}$ when operating on $S\varphi(r)$ (see Chapter I, Section 4.6). In any case, its explicit expression is of no concern here.

are the running waves

$$e^{-i[(\sigma,x)\pm\rho t]}, \qquad \rho = |\sigma| = \sqrt{\sigma_1^2 + \dots + \sigma_n^2},$$

from which we can construct an arbitrary solution in the form

$$u(x, t) = \frac{1}{(2\pi)^n} \int \Psi_1(\sigma)\, e^{-i(\sigma,x)+i\rho t}\, d\sigma + \frac{1}{(2\pi)^n} \int \Psi_2(\sigma)\, e^{-i(\sigma,x)-i\rho t}\, d\sigma.$$

The particular solution $u(x, t)$ is determined by the initial conditions which we shall assume (cf. Chapter I, Section 5.4) to be

$$u(x, 0) = 0, \qquad \frac{\partial u(x, 0)}{\partial t} = \delta(x). \tag{2}$$

The first of these gives

$$\int [\Psi_1(\sigma) + \Psi_2(\sigma)]\, e^{-i(\sigma,x)}\, d\sigma = 0$$

which is satisfied if $\Psi_2(\sigma) = -\Psi_1(\sigma) \equiv (2i)^{-1}\, \Psi(\sigma)$, so that we may write

$$u(x, t) = \frac{1}{(2\pi)^n} \int \Psi(\sigma)\, e^{-i(\sigma,x)} \sin \rho t\, d\sigma = F^{-1}[\Psi(\sigma) \sin \rho t]. \tag{3}$$

The second of the initial conditions can now be written

$$\frac{\partial u(x, 0)}{\partial t} = \frac{1}{(2\pi)^n} \int \Psi(\sigma)\, e^{-i(\sigma,x)}\, \rho\, d\sigma = \delta(x)$$

or

$$F^{-1}[\rho\, \Psi(\sigma)] = \delta(x).$$

Taking the Fourier transform we have

$$\rho\Psi(\sigma) = \widetilde{\delta(x)} = 1$$

so that, according to (3),

$$u(x, t) = F^{-1}\left[\frac{\sin \rho t}{\rho}\right].$$

But we have already encountered the inverse Fourier transform of $\sin(\rho t)/\rho$. According to Eq. (6) of Section 3.4 for $n = 2m + 3$ we have

$$F^{-1}\left[\frac{\sin \rho t}{\rho}\right] = \sqrt{\frac{\pi}{2}}\frac{N_n}{\Omega_{n-1}}\left(\frac{d}{t\, dt}\right)^m \frac{\delta(r - t)}{t}.$$

From this we may obtain the solution of the Cauchy problem with the initial condition $\partial u\,(x, 0)/\partial t = f(x)$. This is

$$u(x, t) = \sqrt{\frac{\pi}{2}}\,\frac{N_n}{\Omega_{n-1}}\left(\frac{d}{t\,dt}\right)^m \frac{\delta(r - t)}{t} * f(x)$$

$$= \sqrt{\frac{\pi}{2}}\,\frac{N_n}{\Omega_{n-1}}\left(\frac{d}{t\,dt}\right)^m t^{n-2}\,\Omega^n M_t[f],$$

where $M_t[f]$ is the mean value of $f(x - \xi)$ in the closed ball $|\xi| \leqslant t$.

In particular, for $m = 0$ and consequently $n = 3$, we have

$$u(x, t) = tM_t[f].$$

The usual method of descent can be used to solve the wave equation in a space of even dimension.[5]

4.4. The Relation between the Elementary Solution of an Equation and the Corresponding Cauchy Problem

An elementary solution of the equation

$$\frac{\partial u}{\partial t} - P\left(i\frac{\partial}{\partial x}\right) u = f \tag{1}$$

is by definition a generalized function $E(x, t)$ in K' which satisfies the equation

$$\frac{\partial E(x, t)}{\partial t} - P\left(i\frac{\partial}{\partial x}\right) E(x, t) = \delta(x, t). \tag{2}$$

The elementary solution of Cauchy's problem for the equation

$$\frac{\partial u(x, t)}{\partial t} - P\left(i\frac{\partial}{\partial x}\right) u(x, t) = 0 \tag{3}$$

is a generalized function $u(x, t)$ in K', depending on t as a parameter and defined for $t \geqslant 0$, which satisfies Eq. (2) and is equal to $\delta(x)$ at $t = 0$. If $u(x, t)$ is known, $E(x, t)$ can be found. Indeed, we assert the following:

Theorem. Let $u(x, t)$ be the elementary solution of Cauchy's problem for Eq. (3). Let

$$E(x, t) = \begin{cases} 0 & \text{for } t < 0, \\ u(x, t) & \text{for } t \geqslant 0. \end{cases} \tag{4}$$

Then $E(x, t)$ is an elementary solution of Eq. (1).

[5] R. Courant and D. Hilbert, "Methods of Mathematical Physics," Vol. II, p. 686. Interscience, New York, 1962.

Proof. The definition in Eq. (4) requires some additional explanations for $t \geq 0$. We wish, in fact, to define a generalized function acting on test functions of x and t, while $u(x, t)$ is a generalized function acting on test functions of x and depends on t only as a parameter. To make the definition complete, let $\varphi(x, t)$ be a test function of x and t, and then $u(x, t)$ will be defined by

$$(u(x, t), \varphi(x, t)) = \int_0^\infty (u(x, t), \varphi(x, t))\, dt.$$

In the integrand, $u(x, t)$ is applied for fixed t to the function $\varphi(x, t)$ which is treated as a function of x only with t fixed. The result is then a function of t with bounded support, and the integration over t can be performed.

We now show that the generalized function $E(x, t)$ is indeed a solution of (2). It is sufficient to show that it satisfies the dual equation [obtained from (2) by taking the Fourier transform in x and t]. Let x_i (with $i = 1$, ..., n) be associated through the Fourier transform with σ_i, and let t be associated with σ_0. Then the dual equation may be written

$$[-i\sigma_0 - P(\sigma)]\, V(\sigma, \sigma_0) = 1. \qquad (5)$$

Now let us take the Fourier transform of $E(x, t)$ as defined by (4), but at first only in the x variables, with fixed t. We shall call this preliminary Fourier transform $v(\sigma, t)$. According to (3) we have

$$v(\sigma, t) = 0 \qquad (t < 0),$$
$$v(\sigma, 0) = 1 \qquad (t = 0), \qquad (6)$$
$$\frac{\partial}{\partial t}\, v(\sigma, t) - P(\sigma)\, v(\sigma, t) = 0 \qquad (t > 0).$$

The derivative with respect to t in the classical sense can be replaced by the derivative in the sense of generalized functions if we take account of the discontinuity in $v(\sigma, t)$ at $t = 0$ (Chapter I, Section 2.2, Example 2). In this way we can replace Eq. (6) by

$$\frac{\partial}{\partial t}\, v(\sigma, t) - P(\sigma)\, v(\sigma, t) - \delta(\sigma) = 0,$$

where $\partial/\partial t$ now operates in the sense of generalized functions on $v(\sigma, t)$. Proceeding to the Fourier transform in t, we arrive at

$$-i\sigma_0\, V(\sigma, \sigma_0) - P(\sigma)\, V(\sigma, \sigma_0) = 1,$$

which is what we wished to prove.

4.5. Classical Operational Calculus

In the classical operational calculus one considers differential equations and systems of differential equations of the form

$$\frac{\partial u(x, t)}{\partial t} = P\left(\frac{\partial}{\partial x}\right) u(x, t)$$

for $t > 0$ with some initial condition at $t = 0$ and some boundary conditions for certain values of x. The Laplace transform in t is applied to reduce the problem to a differential equation in x alone (or to an algebraic problem if the initial equation is an ordinary differential equation).

Let us consider such problems from the point of view of generalized functions. Let $U(x, t)$ be a generalized function in t which depends on x as a parameter, equal to $u(x, t)$ for $t > 0$ and to zero for $t < 0$. Now at $t = 0$ this function is discontinuous, undergoing a change from zero to the value $u(x, 0)$ given by the initial condition. Thus we may write

$$\frac{\partial U(x, t)}{\partial t} = \frac{\partial u(x, t)}{\partial t} + u(x, 0)\,\delta(t).$$

Consequently $U(x, t)$ satisfies the (system of) equations

$$\frac{\partial U(x, t)}{\partial t} = P\left(\frac{\partial}{\partial x}\right) U(x, t) + u(x, 0)\,\delta(t). \tag{1}$$

Taking the Fourier transform in t and calling the variable associated with it $p = p_1 + ip_2$, we obtain

$$-ipV(x, p) = P\left(\frac{\partial}{\partial x}\right) V(x, p) + u(x, 0) \cdot 1. \tag{2}$$

For a well-posed problem this equation with the given boundary conditions in x has a unique solution $V(x, p)$ which is an analytic function of p (in the case of an ordinary differential equation, it is even a rational function of p) which may have singularities (and perhaps be multiple-valued). Having found $V(x, p)$, we can construct a family of analytic functionals of the form

$$(V, \psi) = \int_\Gamma V(x, p)\,\psi(p)\,dp, \tag{3}$$

where Γ is any fixed contour that avoids the singularities of $V(x, p)$ and is equivalent to the real axis; this latter conditions means by definition that

$$\int_\Gamma \psi(s)\,ds = \int_{-\infty}^{\infty} \psi(\sigma)\,d\sigma$$

for any $\psi(s)$ in Z. In particular, Γ may be any straight line parallel to the real s axis and avoiding all the singularities of $V(x, p)$.

Each such functional is a solution of Eq. (2), and the inverse Fourier transform of such a functional is a solution of (1). But such a solution will not in general vanish for $t < 0$.

To find an analytic functional of the form of (3) whose Fourier transform will vanish for $t < 0$ is in general quite difficult. We may, however, assert the following:

Theorem. Assume that $V(x, p)$ has the following special properties:

(a) $V(x, p)$ has no singularities above some straight line whose equation is $\text{Im } p = p_2^0$;

(b) There exists an integrable majorant $W(p_1)$ for $V(x, p)$ in this region:

$$|V(x, p)| \leqslant W(p_1), \qquad \int_{-\infty}^{\infty} W(p_1)\, dp_1 < \infty.$$

Then the analytic functional

$$(V, \psi) = \int_{-\infty+ip_2}^{\infty+ip_2} V(x, p)\, \psi(p)\, dp \qquad (p_2 > p_2^0)$$

is the Fourier transform of a generalized (even ordinary) function $u(x, t)$ which vanishes for $t < 0$ and satisfies the (system of) Eqs. (1).

Proof. The conditions of this theorem are the usual conditions for the existence of the inverse Laplace transform of a function. The Fourier transform of V can be written, according to Eq. (2) of Section 2.6,

$$u(x, t) = \frac{1}{2\pi} \int_{-\infty+ip_2}^{\infty+ip_2} V(x, p)\, e^{-ipt}\, dp.$$

The fact that $u(x, t)$ vanishes for $t < 0$ is implied by the inequality

$$|u(x, t)| \leqslant \frac{1}{2\pi} e^{p_2 t} \int_{-\infty}^{\infty} W(p_1)\, dp_1,$$

in which we allow p_2 to approach $+\infty$.

Let us consider further, any function $V_1(x, p)$ that can be written in the form

$$V_1(x, p) = Q(p)\, V(x, p), \tag{4}$$

where $V(x, p)$ has the properties described above, and $Q(p)$ is some polynomial. The inverse Fourier transform of (4) is

$$u_1(x, t) = Q\left(i\frac{d}{dt}\right)u(x, t),$$

where $u_1(x, t)$ and $u(x, t)$ are the inverse Fourier transforms of $V_1(x, p)$ and $V(x, p)$. We see that for this case also $V_1(x, p)$ is the Fourier transform of a functional with support on the semiaxis $t \geqslant 0$.

CHAPTER III

PARTICULAR TYPES
OF GENERALIZED FUNCTIONS

1. Generalized Functions Concentrated on Smooth Manifolds of Lower Dimension

The simplest example of a generalized function concentrated on a manifold of dimension less than n is one defined by

$$(f, \varphi) - \int_S f(x)\, \varphi(x)\, d\sigma, \tag{1}$$

where S is the given manifold, $d\sigma$ is the induced measure, or Euclidean element of area on S, $f(x)$ is a fixed function, and $\varphi(x)$ is any function in K.[1] A somewhat more complicated example is obtained by replacing the integrand by some differential expression in $\varphi(x)$.

In this section we shall define and study important functionals concentrated on manifolds of dimension less than n imbedded in an n-dimensional space. We have seen for the case of $n = 1$ that functionals concentrated on a point are the delta function and its derivatives. Moreover, it shall be proven in the second volume that every functional concentrated on a point is a linear combination of the delta function and its derivatives. For $n > 1$, when S is $(n - 1)$-dimensional (we shall call such S hypersurfaces or simply surfaces) and is given by an equation of the form

$$P(x_1, ..., x_n) = 0, \tag{2}$$

we will find that an analogous role is played by generalized functions which we shall call $\delta(P)$, $\delta'(P)$, etc. It is with these that we shall be essentially concerned.

[1] In this chapter we shall assume all our test functions to be infinitely differentiable and of bounded support, i.e., in K.

If $P \equiv x_1$, that is, if S is the hyperplane $x_1 = 0$, we naturally put[2]

$$(\delta(x_1), \varphi(x)) \equiv \int \delta(x_1)\, \varphi(x)\, dx$$

$$= \int \left[\int \delta(x_1)\, \varphi(x_1, \ldots, x_n)\, dx_1 \right] dx_2 \ldots dx_n$$

$$= \int \varphi(0, x_2, \ldots, x_n)\, dx_2 \ldots dx_n.$$

In other words, it is natural to define the generalized function $\delta(x_1)$ by the equation[3]

$$\int \delta(x_1)\, \varphi(x)\, dx = \int \varphi(0, x_2, \ldots, x_n)\, dx_2 \ldots dx_n. \qquad (3)$$

For the same reason, we shall define $\delta^{(k)}(x_1)$ by

$$\int \delta^{(k)}(x_1)\, \varphi(x)\, dx = (-1)^k \int \varphi_{x_1}^{(k)}(0, x_2, \ldots, x_n)\, dx_2 \ldots dx_n. \qquad (4)$$

Now let $P(x_1, \ldots, x_n)$ be any sufficiently smooth function such that on $P = 0$ we have

$$\operatorname{grad} P \neq 0 \qquad (5)$$

(which means that there are no singular points on $P = 0$). Then the generalized function $\delta(P)$ can be defined in the following way.

In a sufficiently small neighborhood U of any point of the $P = 0$ hypersurface we can introduce a new coordinate system such that $P = 0$ becomes one of the coordinate hypersurfaces. For this purpose we write $P = u_1$ and choose the remaining u_i coordinates (with $i = 2$, ..., n) arbitrarily except that the Jacobian of the x_i with respect to the u_i, which we shall denote by $D(\genfrac{}{}{0pt}{}{x}{u})$, fail to vanish (which is always possible so long as $\operatorname{grad} P \neq 0$ on $P = 0$). We may assume without loss of generality that $\varphi(x)$ is nonzero only in U.[4] In the "integral" $\int \delta(P)\varphi\, dx$ whose definition we are trying to establish, we make a change of variables according to

$$\int \delta(P)\, \varphi(x)\, dx = \int \delta(P)\, \varphi_1(u)\, D\!\left(\genfrac{}{}{0pt}{}{x}{u}\right) du = \int \delta(u_1)\, \psi(u)\, du,$$

[2] We make use of the notational convention of Chapter I, Section 1.3.

[3] This definition means that $\delta(x_1)$ is the direct product $\delta(x_1) \times 1(x_2, \ldots, x_n)$, where $\delta(x_1)$ is the delta function on the x_1 line, and $1(x_2, \ldots, x_n)$ is the function identically equal to unity (see Chapter I, Section 5.1).

[4] See Appendix 1.2 to Chapter I.

where

$$\varphi_1(u_1, ..., u_n) = \varphi(x_1, ..., x_n) \qquad \text{and} \qquad \psi = \varphi_1(u) \, D\binom{x}{u}. \qquad (6)$$

We thus see that the generalized function $\delta(P)$ is to be defined by

$$(\delta(P), \varphi) = \int \delta(P) \, \varphi \, dx = \int \psi(0, u_2, ..., u_n) \, du_2 \, ... \, du_n. \qquad (7)$$

This last integral, as is seen, is taken over the $P = 0$ surface, which is why $\delta(P)$ is said to be concentrated on this surface.

Similarly, we shall put

$$(\delta^{(k)}(P), \varphi) = \int \delta^{(k)}(P) \, \varphi \, dx$$

$$= (-1)^k \int \psi_{u_1}^{(k)}(0, u_2, ..., u_n) \, du_2 \, ... \, du_n, \qquad (8)$$

where $\psi(u)$ is the function defined in (6), and the integral on the right-hand side is again taken over the $P = 0$ surface.

In order for this equation to be consistent, all we need require is that $P(x)$ have continuous derivatives up to and including the $(k + 1)$st.

At first it would seem that the definitions in (6) and (7) depend on the choice of the coordinate system. In fact this is not so, however, and we may show that the generalized functions $\delta(P)$, $\delta'(P)$, and derivatives of higher order are uniquely determined by P. It is, in fact, most convenient to define them in an invariant way, independent of the choice of u_i; we shall do this by introducing considerations from the theory of what are called differential forms. The reader wishing to avoid this theory may skip over to the resulting formulas we shall derive and list below. The techniques of differential forms are, however, extremely helpful, and we shall make systematic use of them in this section. The necessary minimum of information concerning differential forms will be dealt with in Section 1.1.

We may return to the initial coordinates in the integrand of (7), which will then contain linear combinations of $\varphi(x)$ and its derivatives with x-dependent coefficients. We shall show in Section 1.7 that any functional f of the form

$$(f, \varphi) = \sum_{i_1 + ... + i_n \leqslant k} \int_{P=0} a_{i_1 \cdots i_k}(x) \, \frac{\partial^{i_1 + ... + i_k} \varphi(x)}{\partial x_1^{i_1} ... \partial x_n^{i_n}} \, dx \qquad (9)$$

can be expressed in terms of $\delta(P)$ and its derivatives by

$$(f, \varphi) = \sum_{j=0}^{k} \int b_j(x) \, \delta^{(j)}(P) \, \varphi \, dx. \qquad (10)$$

Equation (10) contains only $k + 1$ derivatives, though (9) contains in general many more. Further, (10) has the additional advantage that it is unique: if $f = 0$, then $b_j(x) = 0$ for each $j = 0, ..., k$, which cannot be said of the $a_{i_1...i_n}(x)$.

Now we shall present a list of the formulas that will be proven later. In these formulas P is assumed smoother than above, but in order not to go into irrelevancies at this point, let us simply assume P to be infinitely differentiable.

Let $\theta(P)$ be the characteristic function of the region $P \geqslant 0$:

$$\theta(P) = \begin{cases} 0 & \text{for } P < 0, \\ 1 & \text{for } P \geqslant 0, \end{cases}$$

$$(\theta(P), \varphi) = \int_{P \geqslant 0} \varphi(x) \, dx. \tag{11}$$

Then it follows that

$$\theta'(P) = \delta(P), \tag{12}$$

which is understood in the sense that

$$\frac{\partial \theta(P)}{\partial x_j} = \frac{\partial P}{\partial x_j} \delta(P). \tag{12'}$$

Further, a similar "chain rule" applies to differentiation of $\delta^{(k)}(P)$

$$\frac{\partial}{\partial x_j} \delta^{(k)}(P) = \frac{\partial P}{\partial x_j} \delta^{(k+1)}(P), \qquad k = 0, 1, 2, ... \tag{13}$$

We shall also establish the following identities relating $\delta(P)$ to its derivatives:

$$P\delta(P) = 0, \tag{14}$$

$$P\delta'(P) + \delta(P) = 0, \tag{15}$$

$$P\delta''(P) + 2\delta'(P) = 0, \tag{16}$$

$$. \quad . \quad . \quad . \quad . \quad . \quad . \quad . \quad . \quad . \quad . \quad . \quad .$$

$$P\delta^{(k)}(P) + k\delta^{(k-1)}(P) = 0, \tag{17}$$

$$. \quad . \quad . \quad . \quad . \quad . \quad . \quad . \quad . \quad . \quad . \quad . \quad .$$

For two surfaces $P = 0$ and $Q = 0$ (where Q is a function with the same properties as P) *that do not intersect*, we have

$$\delta(PQ) = P^{-1}\delta(Q) + Q^{-1}\delta(P). \tag{18}$$

In particular, if $a(x)$ is a function nowhere equal to zero, we have

$$\delta(aP) = a^{-1}\delta(P). \tag{19}$$

From this we may derive

$$\delta^{(k)}(aP) = a^{-(k+1)}\delta^{(k)}(P). \tag{20}$$

The above applies to functions concentrated on a hypersurface, or an $(n-1)$-dimensional manifold in an n-dimensional space. Assume now that S is a manifold of lower dimension given by the equations

$$P_1(x_1, ..., x_n) = 0, ..., P_k(x_1, ..., x_n) = 0 , \tag{21}$$

where the P_i are sufficiently smooth functions. Let the $P_i(x) = 0$ surfaces for $i = 1, ..., k$ form a lattice such that in the neighborhood of every point of S we may take the $P_i(x)$ to be the first k coordinates $u_1, ..., u_k$; let the remaining coordinates $u_{k+1}, ..., u_n$ be chosen arbitrarily except that the Jacobian $D\left(\frac{x}{u}\right)$ fail to vanish. Then we can define a generalized function $\delta(P_1, ..., P_k)$ and its derivatives

$$\frac{\partial^m \delta(P_1, ..., P_k)}{\partial P_1^{\alpha_1} ... \partial P_k^{\alpha_k}}$$

in analogy with what we have done above. Specifically, if we wish to satisfy the identity

$$\int \delta(P_1, ..., P_k)\, \varphi\, dx = \int \delta(u_1, ..., u_k)\, \varphi_1(u)\, D\left(\frac{x}{u}\right) du$$

$$= \int \delta(u_1, ..., u_k)\, \psi(u_1, ..., u_n)\, du_1 ... du_n,$$

where

$$\varphi_1(u) = \varphi(x) \quad \text{and} \quad \psi(u) = \varphi_1(u)\, D\left(\frac{x}{u}\right), \tag{22}$$

we put

$$\int \delta(P_1, ..., P_k)\, \varphi\, dx = \int \psi(0, ..., 0, u_{k+1}, ..., u_n)\, du_{k+1} ... du_n. \tag{23}$$

Similar motivation leads to

$$\int \frac{\partial^{\alpha_1 + ... + \alpha_k}\, \delta(P_1, ..., P_k)}{\partial P_1^{\alpha_1} ... \partial P_k^{\alpha_k}}\, \varphi(x)\, dx = (-1)^{\alpha_1 + ... + \alpha_k}$$

$$\times \int \frac{\partial^{\alpha_1 + ... + \alpha_k}\, \psi(0, ..., 0, u_{k+1}, ..., u_n)}{\partial u_1^{\alpha_1} ... \partial u_k^{\alpha_k}}\, du_{k+1} ... du_n . \tag{24}$$

In Section 1.9 we shall show that these definitions are independent of the choice of coordinate system. We shall find the following identities to hold:

$$\frac{\partial}{\partial x_j}\,\delta(P_1, ..., P_k) = \sum_{i=1}^{k} \frac{\partial\delta(P_1, ..., P_k)}{\partial P_i}\,\frac{\partial P_i}{\partial x_j} \tag{25}$$

(the "chain rule");

$$P_i\delta(P_1, ..., P_k) = 0\,, \tag{26}$$

$$P_iP_j\delta(P_1, ..., P_k) = 0, \tag{27}$$

. . . .

$$P_1P_2 ... P_k\delta(P_1, ..., P_k) = 0\,, \tag{28}$$

and additional identities obtained by formal differentiation of the above.

1.1. Introductory Remarks on Differential Forms[5]

A differential form of kth degree on an *n*-dimensional manifold with coordinates x_1, ..., x_n is an expression of the form

$$\sum a_{i_1i_2...i_k}(x)\,dx_{i_1} ... dx_{i_k},$$

where the sum is taken over all possible combinations of k indices. The *coefficients* $a_{i_1 ... i_k}(x)$ are assumed to be infinitely differentiable functions of the coordinates. Two forms of degree k are considered equal if they are transformed into each other when products of differentials are transposed according to the *anticommutation rule*

$$dx_i\,dx_j = -dx_j\,dx_i \tag{1}$$

and all similar terms are collected.

This rule implies, among other things, that if a term in a differential form has two differentials with the same index, it must vanish. It can be used also to put any differential form into "canonical form," in which the indices in each term appear in increasing order.

[5] We shall have no need for the theory of differential forms in its greatest generality, and therefore in the interests of clarity we shall simplify wherever possible. The reader desiring a deeper understanding of the considerations touched on in this section may turn for instance, to P. K. Rashevskii, "Geometrical Theory of Partial Differential Equations" (in Russian). Gostekhizdat, 1947; or to J. de Rham, "Variétés Différentiables. Formes, Courants, Formes Harmoniques." Hermann, Paris, 1955.

We shall say that a differential form has bounded support if all of its coefficients are functions with bounded supports.

The *exterior product* of two forms $\Sigma a_{i_1 \ldots i_k}(x)\, dx_{i_1} \ldots dx_{i_k}$ and $\Sigma b_{j_1 \ldots j_m}(x)\, dx_{j_1} \ldots dx_{j_m}$ is the form of degree $k + m$ obtained by formal algebraic multiplication of the original two forms. The result of such multiplication can in general be further simplified by using Eq. (1) to put it in canonical form.[6]

This definition of multiplication implies that the anticommutation rule will hold for any differential forms of first degree. Indeed, let $\alpha = \Sigma a_j(x)dx_j$ and $\beta = \Sigma b_k(x)dx_k$; then

$$\alpha\beta = \sum_{j,k} a_j(x)\, b_k(x)\, dx_j\, dx_k = -\sum_{j,k} a_j(x)\, b_k(x)\, dx_k\, dx_j = -\beta\alpha.$$

Let us find how differential forms transform under an infinitely differentiable change of coordinates given by $x_i = x_i(x'_1, \ldots, x'_n)$. We have

$$dx_i = \sum_{j=1}^{n} \frac{\partial x_i}{\partial x'_j}\, dx'_j$$

and

$$\sum_{i_1 < \ldots < i_k} a_{i_1 \cdots i_k}\, dx_{i_1} \ldots dx_{i_k} = \sum_{i_1 < \ldots < i_k}\sum_{j} a_{i_1 \cdots i_k} \frac{\partial x_{i_1}}{\partial x'_{j_1}} \cdots \frac{\partial x_{i_k}}{\partial x'_{j_k}}\, dx'_{j_1} \ldots dx'_{j_k}.$$

In the sum we have obtained, terms in which the same differential occurs twice will vanish. Different terms containing the same combination of differentials can be combined using the anticommutation rule, which holds also for the dx'_i. Then it is easily establish that for $j_1 < j_2 \ldots < j_k$, the coefficient of $dx'_{j_1} \ldots dx'_{j_k}$ is multiplied by the Jacobian

$$D\begin{pmatrix} x_{i_1}\, x_{i_2} \ldots x_{i_k} \\ x'_{j_1}\, x'_{j_2} \ldots x'_{j_k} \end{pmatrix}.$$

We thus arrive at

$$\sum_{i_1 < \ldots < i_k} a_{i_1 \cdots i_k}\, dx_{i_1} \ldots dx_{i_k} = \sum_{j_1 < \ldots < j_k} a'_{j_1 \ldots j_k}\, dx'_{j_1} \ldots dx'_{j_k}, \tag{2}$$

where

$$a'_{j_1 \ldots j_k} = \sum_{i_1 < i_2 < \ldots < i_k} D\begin{pmatrix} x_{i_1} \ldots x_{i_k} \\ x'_{j_1} \ldots x'_{j_k} \end{pmatrix} a_{i_1 \ldots i_k}. \tag{3}$$

---

[6] We are, of course, assuming that the coefficients commute with each other and with the differentials.

In particular, a form of degree n, which can always be reduced to a single term, has the property that

$$a \, dx_1 \ldots dx_n = aD\!\left(\begin{matrix} x_1 \cdots x_n \\ x_1' \cdots x_n' \end{matrix}\right) dx_1' \ldots dx_n'. \tag{4}$$

Note now that this is the formula which describes the transformation of an *integrand* under a change of independent variables in a multiple integral. For this reason the techniques of differential forms can be used when performing such a change of variables. For instance, for the case of a double integral, writing $x = \varphi(u, v)$ and $y = \psi(u, v)$, we have

$$dx = \varphi_u \, du + \varphi_v \, dv; \qquad dy = \psi_u \, du + \psi_v \, dv,$$

$$dx \, dy = (\varphi_u \psi_v - \varphi_v \psi_u) \, du \, dv.$$

The *exterior derivative* of a differential form

$$\alpha = \sum a_{i_1 \cdots i_k} \, dx_{i_1} \ldots dx_{i_k}$$

is defined as the $(k + 1)$st degree differential form

$$d\alpha = \sum_{i_1, \ldots, i_k} \left(\sum_i \frac{\partial a_{i_1 \cdots i_k}}{\partial x_i} \, dx_i \right) dx_{i_1} \ldots dx_{i_k}, \tag{5}$$

which, of course, can be simplified by using the anticommutation rule (1).

For instance, the exterior derivative of a form of degree zero, which is understood as a *scalar* function $a(x)$, is essentially the ordinary derivative, in fact is the first degree form $\sum(\partial a/\partial x_i) \, dx_i$; the coefficients of this form represent the gradient of $a(x)$. The exterior derivative of the first degree form $\sum a_i(x) dx_i$ is the second degree form

$$\sum_{i<j} \left(\frac{\partial a_j}{\partial x_i} - \frac{\partial a_i}{\partial x_j} \right) dx_i \, dx_j$$

(if we associate with the coefficients of the first order form the vector $a = \{a_i\}$, those of the exterior derivative will be associated with the curl of a). The derivative of the $(n - 1)$st degree form

$$\sum a_j(x) \, dx_1 \ldots dx_{j-1} \, dx_{j+1} \ldots dx_n$$

is the nth degree form

$$\left\{ \sum (-1)^{j+1} \frac{\partial a_j(x)}{\partial x_j} \right\} dx_1 \ldots dx_n.$$

It is easily shown that according to the anticommutation rule (1), any differential form α satisfies the equation

$$d\,d\alpha = 0 \,. \tag{6}$$

Let us now turn to integration of differential forms over regions and manifolds.

Let us start by assuming that the nth degree form $\alpha = a(x)dx_1 \ldots dx_n$ is integrated over an n-dimensional region G of R_n in the usual way according to the usual rules of multiple integration.

We wish now to define the integral of a form of degree k over a k-dimensional manifold. Before proceeding we make some remarks about *orientations* of regions and manifolds.

Let V be a neighborhood (of arbitrary dimension m) in which we are given a local coordinate system $u_1, \ldots, u_m$. We shall say that this coordinate system defines the *orientation* of the neighborhood V. The same orientation is defined by any other local coordinate system $v_1, \ldots, v_m$ so long as the transformation from the u_1 to the v_i has positive Jacobian. It is possible, in addition, to give V the *opposite orientation*. This is defined by any local coordinate system $w_1, \ldots, w_m$, such that the transformation from the u_i to the w_i has a negative Jacobian.

A region or manifold Γ (of m dimensions) is called *orientable* if in the neighborhood of every point of Γ it is possible to define the orientation *consistently*, that is so that the coordinates in intersecting neighborhoods define in the intersection the same orientation. We shall not concern ourselves with nonorientable manifolds (for instance, the Möbius strip).

Let U be an m-dimensional neighborhood contained in an $(m + 1)$-dimensional neighborhood V, and let U divide V into two parts one of which we shall call the *interior*, and the other the *exterior*. Assume that the coordinate system $w_1, \ldots, w_m$ defined in U can be extended to a coordinate system $w_0, w_1, \ldots, w_m$ in V. Then we shall agree to say that the resulting orientation of V corresponds to the *positive direction of the normal* if w_0 increases in the exterior of U; in the opposite case we shall say that the orientation of V corresponds to the negative direction of the normal.

We shall now proceed to define the integral of a kth degree differential form α over a k-dimensional (for $k \leqslant n$) orientable manifold Γ. We shall assume this manifold closed, bounded, and sufficiently smooth. We break up Γ into the submanifolds $\Gamma^{(1)}, \ldots, \Gamma^{(m)}$, in each of which we can introduce a local coordinate system; we introduce such systems consistently, thereby choosing the orientation of Γ. We now express α in each of the $\Gamma^{(i)}$ in the coordinate system $u_1, \ldots, u_k$ chosen locally. If α was defined on an n-dimensional region containing Γ, the local coordinates $u_1, \ldots, u_n$ should be chosen so that $n - k$ of them, for instance $u_{k+1}, \ldots, u_n$, vanish on Γ. Then $du_{k+1} = 0, \ldots, du_n = 0$ on Γ so that α becomes

$$\alpha = a(u_1, \ldots, u_k)\, du_1 \ldots du_k.$$

We now integrate α in the usual way over each of the $\Gamma^{(i)}$ and add up the results. It is not difficult to see that the integral $\int_\Gamma \alpha$ so defined is independent of the way Γ is broken up into manifolds and of the choice of local coordinates in each of these, assuming the orientation to be fixed. One need only recall that on transforming to some new coordinates $u_1', \ldots, u_k'$ the differential form becomes

$$\alpha = a(u_1(u_1', \ldots, u_k'), \ldots, u_k(u_1', \ldots, u_k'))D\binom{u_1 \ldots u_k}{u_1' \ldots u_k'}\, du_1' \ldots du_k'.$$

If, however, we transform to local coordinates corresponding to the opposite orientation in each of the $\Gamma^{(i)}$, our integral $\int_\Gamma \alpha$ will change sign, since in this case the Jacobian $D\binom{u_1 \ldots u_k}{u_1' \ldots u_k'}$ becomes negative.

Thus $\int_\Gamma \alpha$ is defined uniquely up to sign, and the sign in turn is defined by the orientation of Γ.

The Gauss-Ostrogradskii Formula (Gauss' Theorem)

Let α be a differential form of degree $n - 1$ defined on some bounded n-dimensional region G with a piecewise smooth boundary Γ. We assume an orientation of G corresponding to the positive direction of the normal to Γ. By $d\alpha$ we shall denote, as above, the exterior derivative of α. Then

$$\int_G d\alpha = \int_\Gamma \alpha, \tag{7}$$

which is called the *Gauss-Ostrogradskii formula*. If the orientation of G corresponds to the negative direction of the normal, one of the integrals must be multiplied by minus one.

As an example, consider a second degree form α in three dimensions. In terms of x_1, x_2, x_3 this differential form may be written

$$\alpha = a_1\, dx_2\, dx_3 + a_2\, dx_3\, dx_1 + a_3\, dx_1\, dx_2,$$

and its exterior derivative is

$$d\alpha = \left(\frac{\partial a_1}{\partial x_1} + \frac{\partial a_2}{\partial x_2} + \frac{\partial a_3}{\partial x_3}\right) dx_1\, dx_2\, dx_3,$$

so that the Gauss-Ostrogradskii formula becomes

$$\int_{\Gamma} [a_1 \, dx_2 \, dx_3 + a_2 \, dx_3 \, dx_1 + a_3 \, dx_1 \, dx_2]$$

$$= \int_{G} \left(\frac{\partial a_1}{\partial x_1} + \frac{\partial a_2}{\partial x_2} + \frac{\partial a_3}{\partial x_3} \right) dx_1 \, dx_2 \, dx_3, \qquad (8)$$

in which form it is usually given in courses on analysis.

A generalization of this formula is Stokes' theorem.[7] This theorem can be written in the same form, namely

$$\int_{G} d\alpha = \int_{\Gamma} \alpha,$$

except that now G is a k-dimensional bounded region (with $k < n$), while α is a differential form of degree $k - 1$. As before we assume the orientation of G to correspond to the positive direction of the normal of its boundary Γ.

As an example, we present the classical form of Stokes' theorem. In three-dimensions and the coordinates x_1, x_2, x_3, a first degree differential form α may be written

$$\alpha = a_1 \, dx_1 + a_2 \, dx_2 + a_3 \, dx_3,$$

and its exterior derivative in the form

$$d\alpha = \left(\frac{\partial a_2}{\partial x_1} - \frac{\partial a_1}{\partial x_2} \right) dx_1 \, dx_2 + \left(\frac{\partial a_3}{\partial x_2} - \frac{\partial a_2}{\partial x_3} \right) dx_2 \, dx_3$$

$$+ \left(\frac{\partial a_1}{\partial x_3} - \frac{\partial a_3}{\partial x_1} \right) dx_3 \, dx_1.$$

Then Stokes' theorem becomes

$$\int_{G} \left[\left(\frac{\partial a_2}{\partial x_1} - \frac{\partial a_1}{\partial x_2} \right) dx_1 \, dx_2 + \left(\frac{\partial a_3}{\partial x_2} - \frac{\partial a_2}{\partial x_3} \right) dx_2 \, dx_3 \right.$$

$$\left. + \left(\frac{\partial a_1}{\partial x_3} - \frac{\partial a_3}{\partial x_1} \right) dx_3 \, dx_1 \right] = \int_{\Gamma} [a_1 \, dx_1 + a_2 \, dx_2 + a_3 \, dx_3], \qquad (9)$$

where G is a two-dimensional bounded region and Γ is the curve bounding it.

---

[7] See J. de Rahm, *loc. cit.*, who gives this formula in terms very similar to those used here.

1.2. The Form ω

Consider a hypersurface S given by $P(x_1, x_2, ..., x_n) = 0$, where P is an infinitely differentiable function such that grad $P = \{\partial P/\partial x_1, ..., \partial P/\partial x_n\}$ does not vanish on S (which therefore has no singular points). We wish to turn our attention in this section to a certain particular differential form ω of degree $n - 1$, which will be important in what follows. This form is intimately related to the $P = 0$ surface, or more accurately is uniquely determined by P on the $P = 0$ surface.

The form is *defined* by

$$dP \cdot \omega = dv, \tag{1}$$

where $dv = dx = dx_1 ... dx_n$, and dP is the differential of P.

Before proceeding let us convince ourselves that under the above assumptions ω does in fact exist in some n-dimensional region containing S.

By assumption, in the neighborhood of any point of the surface we can introduce a local coordinate system $u_1, ..., u_n$ such that one of the coordinates, say u_j, is $P(x)$, and such that the transformation from the x_i to the u_i (with $i = 1, ..., n$) is given by infinitely differentiable functions with positive Jacobian $D\binom{x}{u}$. In this coordinate system we may write $dv = D\binom{x}{u} du_1 ... du_{j-1} dP\, du_{j+1} ... du_n$, and therefore we may set

$$\omega = (-1)^{j-1} D\binom{x}{u} du_1 ... du_{j-1}\, du_{j+1} ... du_n. \tag{2}$$

Thus ω is seen to exist.

If, in particular, in the neighborhood of the given point $\partial P/\partial x_j \neq 0$, we may take the u_1 coordinates to be

$$u_1 = x_1, ..., u_j = P, ..., u_n = x_n.$$

Then

$$D\binom{x}{u} = \left[D\binom{u}{x}\right]^{-1} = \frac{1}{\partial P/\partial x_j},$$

and the differential form ω defined by (2) becomes

$$\omega = (-1)^{j-1} \frac{dx_1 ... dx_{j-1}\, dx_{j+1} ... dx_n}{\partial P/\partial x_j}. \tag{3}$$

Let us now study the *uniqueness* of ω. In general Eq. (1) does not specify ω uniquely, for it is clear that if ω satisfies (1), so does $\omega + \gamma$

where γ is any differential form "orthogonal" to dP, that is, any form such that $dP \cdot \gamma = 0$. Now it can be shown that any differential form γ of degree $n - 1$ orthogonal to dP can be written

$$\gamma = \alpha\, dP ,$$

where α is some differential form of degree $n - 2$. Indeed, in the coordinates $u_1 = P,\ u_2,\ ...,\ u_n$ we may write γ (or for that matter any other form of degree $n - 1$) in the form

$$\gamma = g_1\, du_2 \,...\, du_n + g_2\, dP\, du_3 \,...\, du_n + ... + g_n\, dP\, du_2 \,...\, du_{n-1},$$

where the g_i are functions of the coordinates. Then $dP \cdot \gamma = 0$ implies that $g_1 = 0$. Factoring out the dP from the remaining terms, we obtain the desired result.

Note (we shall make use of this somewhat later) that if γ has bounded support, so does α.

From the above result it follows, in particular, that on S itself γ vanishes identically (since $dP = 0$) so that on the $P = 0$ surface the differential form γ is uniquely determined by P.

Let us establish the geometrical meaning of ω. Consider the surfaces S and S_h, the first of which is defined, as before, by $P = 0$, and the second of which is defined by $P = h$ for some small h (Fig. 6). Con-

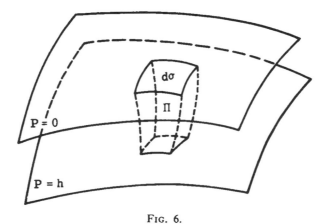

FIG. 6.

sider an element of area $d\sigma$ on S and transfer it to S_h along the coordinate lines $u_1,\ ...,\ u_{j-1},\ u_{j+1},\ ...,\ u_n$. We have then traced out the figure Π whose volume we shall call dv. Then by definition ω is the ratio of dv to $dP = h$. Thus, ω is the rate of change of the element of volume dv with respect to P.

This model serves to explain the invariance of this differential form ω on S. If, in fact, we replace the $u_1, \ldots, u_{j-1}, u_{j+1}, \ldots, u_n$ coordinates by new ones, we merely change the inclination of Π with respect to S. Nervertheless its base and its height do not change, since the first is compensated for by the Jacobian and the second is determined by the distance to S_h; therefore the volume does not change, and nor does $\omega = dv/dP$.

Consequently ω does not depend on the choice of the $u_1, \ldots, u_{j-1}, u_{j+1}, \ldots, u_n$ coordinates.

However, ω does of course depend on the function P by which we represent S through the equation $P = 0$. Let us see how ω changes if the equation $P = 0$ is replaced by $P_1 \equiv \alpha P = 0$ with some nonvanishing function $\alpha(x)$. At S we have the equation $dP_1 = \alpha dP$, from which it follows that the new differential form is given by

$$\omega_1 = \frac{dv}{dP_1} = \frac{1}{\alpha}\,\omega.$$

Note also that if $P(x)$ is the Euclidean distance of x from the $P = 0$ surface (or differs from it by a quantity of higher order), the differential form ω on S coincides with the Euclidean element of area $d\sigma$ on S.

1.3. The Generalized Function $\delta(P)$

We now proceed to the fundamental definition. With every $\varphi(x)$ in K we associate the number

$$\int_{P=0} \varphi(x)\,\omega.$$

This will clearly give a continuous linear functional on K. We shall denote this functional by $\delta(P)$, i.e., we shall write

$$(\delta(P), \varphi) = \int \delta(P)\,\varphi\,dx = \int_{P=0} \varphi(x)\,\omega. \tag{1}$$

From the above it follows that $\delta(P)$ is independent of the choice of the differential form ω, depending only on $P(x)$.

To check that this definition agrees with that at the beginning of the chapter, we need note only that according to Eq. (2) of Section 1.2, in the coordinate system $u_1 = P, u_2, \ldots, u_n$ we have

$$\omega = D\binom{x}{u}\,du_2 \ldots du_n.$$

Example 1. Let us return first to the generalized function $\delta(x_1)$ from which we started the general discussion of $\delta(P)$. The equation $x_1 = 0$ defines one of the coordinate hypersurfaces. For this case we may write

$$\omega = dx_2 \ldots dx_n,$$

so that

$$\int \delta(x_1)\,\varphi(x)\,dx = (\delta(x_1), \varphi) = \int \varphi(0, x_2, \ldots, x_n)\,dx_2 \ldots dx_n. \qquad (2)$$

Example 2. Consider the generalized function $\delta(\alpha_1 x_1 + \ldots + \alpha_n x_n)$, where $\Sigma \alpha_i^2 = 1$. The equation

$$\alpha_1 x_1 + \ldots + \alpha_n x_n = 0$$

determines a hypersurface which passes through the origin and is orthogonal to the unit vector α. In terms of the coordinates

$$u_1 = \alpha_1 x_1 + \ldots + \alpha_n x_n, \qquad u_2, \ldots, u_n$$

which we choose so as to make the transformation matrix orthogonal (so that we obtain a rotation), ω can be written $du_2 \ldots du_n$. We thus arrive at

$$(\delta(\alpha_1 x_1 + \ldots + \alpha_n x_n), \varphi) = \int_{\Sigma \alpha_i x_i = 0} \varphi\, du_2 \ldots du_n = \int_{\Sigma \alpha_i x_i = 0} \varphi\, d\sigma, \qquad (3)$$

where $d\sigma$ is the Euclidean element of area on the hypersurface $\Sigma \alpha_i x_i = 0$.

Example 3. Consider the generalized function $\delta(xy - c)$ in two dimensions. The equation $xy - c = 0$ defines a hyperbola.[8] Using the coordinates $u_1 = x$, $u_2 = xy - c$, we find from Eq. (3) of Section 1.2 that $\omega = -(1/x)dx$, so that

$$(\delta(xy - c), \varphi(x, y)) = \int \delta(xy - c)\,\varphi(x, y)\,dx\,dy = -\int \varphi\left(x, \frac{c}{x}\right)\frac{dx}{x}. \qquad (4)$$

Example 4. Consider the generalized function $\delta(r - c)$, where $r^2 = \Sigma x_i^2$, and $c > 0$. The equation $r - c = 0$ defines the sphere O_c of radius c. Since $P = r - c$ is the Euclidean distance from the surface of the sphere, at $r = c$ the differential form ω coincides with the Euclidean element of area dO_c on the sphere, so that we have[9]

$$(\delta(r - c), \varphi) = \int \delta(r - c)\,\varphi\,dx = \int_{O_c} \varphi\, dO_c, \qquad (5)$$

[8] We assume that $c \neq 0$. For a discussion of the case $c = 0$, see Section 4.5.
[9] See Example 4 in Section 1.5.

which means that $\delta(r - c)$ may be characterized as the functional corresponding to a uniform mass distribution of unit density over the surface of the sphere of radius $r = c$.

The same sphere of radius c is given by the equation $r^2 - c^2 = 0$. Let us find also $\delta(r^2 - c^2)$. We write $u_1 = r^2 - c^2$, $u_2 = \theta_1$, ..., $u_n = \theta_{n-1}$, where θ_1, ..., θ_{n-1} are the same angles as those used in ordinary spherical coordinates. We then have

$$\omega = \frac{1}{2r} dO_c = \frac{1}{2c} dO_c$$

so that

$$(\delta(r^2 - c^2), \varphi) = \int \delta(r^2 - c^2)\, \varphi \, dx = \frac{1}{2c} \int_{O_c} \varphi \, dO_c. \tag{6}$$

The generalized function $\delta(P)$ arises quite naturally on differentiating the characteristic function $\theta(P)$ for the region $P \geqslant 0$, namely, the function equal to unity for $P(x) \geqslant 0$, and to zero for $P(x) < 0$, so that

$$(\theta(P), \varphi) = \int_{P \geqslant 0} \varphi(x) \, dx.$$

We thus assert that

$$\theta'(P) = \delta(P), \tag{7}$$

in the sense that for each $j = 1, 2, ..., n$

$$\frac{\partial \theta(P)}{\partial x_j} = \frac{\partial P}{\partial x_j} \delta(P). \tag{7'}$$

To prove (7) let us verify first that for fixed j the functionals on the right and left-hand sides of (7') coincide in the neighborhood of any point of $P = 0$ at which $\partial P/\partial x_j \neq 0$ (off this surface both functionals vanish). For this purpose we operate with both sides on some $\varphi(x)$ in K with support in a small neighborhood of such a point.

The left-hand side becomes

$$\left(\frac{\partial \theta(P)}{\partial x_j}, \varphi \right) = -\left(\theta(P), \frac{\partial \varphi}{\partial x_j} \right) = -\int_{P \geqslant 1} \frac{\partial \varphi}{\partial x_j} \, dx,$$

and the right-hand side becomes

$$\left(\frac{\partial P}{\partial x_j} \delta(P), \varphi \right) = \left(\delta(P), \frac{\partial P}{\partial x_j} \varphi \right) = \int_{P=0} \frac{\partial P}{\partial x_j} \varphi \omega.$$

Let us first assume that $P \geqslant 0$ defines a bounded region. Then we may apply the Gauss-Ostrogradskii formula to the integral over this region and to the differential form of degree $n - 1$ in the integrand. We shall include P itself among the coordinates, and then using the fact that it increases into the interior of the region, we obtain

$$\int_{P=0} \frac{\partial P}{\partial x_j}\, \varphi\omega = -\int_{P \geqslant 0} d\left(\frac{\partial P}{\partial x_j}\, \varphi\omega\right).$$

Now we have assumed $\varphi(x)$ to be nonzero only in a small neighborhood of some point of the $P = 0$ surface, so we may use the local coordinates $u_i = x_i$ for $i \neq j$, and $u_j = P$. As before, we obtain

$$\omega = (-1)^{j-1} \frac{dx_1\, ...\, dx_{j-1}\, dx_{j+1}\, ...\, dx_n}{\partial P/\partial x_j}$$

so that

$$d\left(\varphi \frac{\partial P}{\partial x_j}\, \omega\right) = (-1)^{j-1} d(\varphi\, dx_1\, ...\, dx_{j-1}\, dx_{j+1}\, ...\, dx_n) = \frac{\partial \varphi}{\partial x_j}\, dx_1\, ...\, dx_n.$$

Of course outside the region of interest this equation

$$d\left(\varphi \frac{\partial P}{\partial x_j}\, \omega\right) = \frac{\partial \varphi}{\partial x_j}\, dx$$

remains valid, since there $\varphi(x) \equiv 0$. We thus arrive at

$$\int_{P=0} \frac{\partial P}{\partial x_j}\, \varphi\omega = -\int_{P \geqslant 0} \frac{\partial \varphi}{\partial x_j}\, dx.$$

The left-hand side of this equation is $(\delta(P)\, \partial P/\partial x_j, \varphi)$, while the right-hand side is $(\partial \theta(P)/\partial x_j, \varphi)$, so that (7') is established for the case we have been considering.

If $P \geqslant 0$ does not define a bounded region, we replace it by its intersection G_R with a sufficiently large ball $|x| \leqslant R$ outside of which $\varphi(x)$ is known to vanish. Let Γ_R be the boundary of G_R, so that the Gauss-Ostrogradskii formula gives

$$\int_{\Gamma_R} \frac{\partial P}{\partial x_j}\, \varphi\omega = -\int_{G_R} \frac{\partial \varphi}{\partial x_j}\, dx.$$

Now since $\varphi(x)$ vanishes outside of $|x| \leqslant R$, we arrive at

$$\int_{P=0} \frac{\partial P}{\partial x_j}\, \varphi\omega = -\int_{P \geqslant 0} \frac{\partial \varphi}{\partial x_j}\, dx,$$

which completes the first part of our proof.

Now let x be any point of the $P = 0$ surface. In its neighborhood we construct a new coordinate system $\xi_j = \Sigma \alpha_{jk} x_k$ $(j = 1, 2, ..., n)$ such that $\partial P / \partial \xi_j \neq 0$ for all j. Then according to the above theorem, in the neighborhood of this point and for each j we have

$$\frac{\partial \theta(P)}{\partial \xi_j} = \frac{\partial P}{\partial \xi_j} \delta(P). \tag{8}$$

It is true, on the other hand, that

$$\frac{\partial}{\partial x_k} = \sum_j \frac{\partial \xi_j}{\partial x_k} \frac{\partial}{\partial \xi_j} = \sum_j \alpha_{jk} \frac{\partial}{\partial \xi_j} .$$

If we now multiply (8) by α_{jk} and sum over j, we find that in the neighborhood of this point

$$\frac{\partial \theta(P)}{\partial x_k} = \sum \alpha_{jk} \frac{\partial \theta(P)}{\partial \xi_j} = \sum \alpha_{jk} \frac{\partial P}{\partial \xi_j} \delta(P) = \frac{\partial P}{\partial x_k} \delta(P).$$

Thus $\partial \theta(P) / \partial x_k$ and $\delta(P) \partial P / \partial x_k$ coincide in the neighborhood of any point of $P = 0$. But this means that the functionals are equal, which completes our proof.

It is sometimes convenient to deal with *generalized vector functions* (or *generalized vectors*), that is, vectors $f = (f_1, ..., f_n)$, whose components f_i are generalized functions. Two vector functions f and g are equal if $f_i = g_i, i = 1, ..., n$. If for any generalized function g we write

$$\operatorname{grad} g = \left(\frac{\partial g}{\partial x_1}, ..., \frac{\partial g}{\partial x_n} \right), \tag{9}$$

Eq. (7′) can be restated in the form

$$\operatorname{grad} \theta(P) = \delta(P) \operatorname{grad} P. \tag{7″}$$

We shall now use this formula for a very simple derivation of Green's theorem.

1.4. Example: Derivation of Green's Theorem

A generalized vector function can be considered a functional on vector test functions. Let f be a generalized vector with components f_i and let ψ be a vector test function with components ψ_i in K, $i = 1, ..., n$. We shall put

$$(f, \psi) = \sum_{j=1}^{n} (f_j, \psi_j). \tag{1}$$

For instance, if g is a generalized function and ψ (*not* a vector) is in K, we can write the functional Δg (where Δ is the Laplacian) as

$$(\Delta g, \psi) = -(\text{grad } g, \text{grad } \psi), \qquad (2)$$

where grad g is the generalized vector whose components are $\partial g/\partial x_i$, and grad ψ is the vector whose components are $\partial \psi/\partial x_i$.

In the preceding section we showed that

$$\text{grad } \theta(P) = \delta(P) \text{ grad } P. \qquad (3)$$

We shall now use this to give a simple derivation of Green's theorem.

We recall first that if g is a generalized function and $h(x)$ is an infinitely differentiable function, then in the usual way we have

$$\text{grad}(hg) = g \text{ grad } h + h \text{ grad } g.$$

Now let $u(x)$ be some infinitely differentiable function and let us consider the generalized function $\Delta[u\theta(P)]$. Using the vector notation described above, we may write this in the form

$$(\Delta[u\theta(P)], \varphi) = -(\text{grad}[u\theta(P)], \text{grad } \varphi)$$
$$= -(u \text{ grad } \theta(P), \text{grad } \varphi) - (\theta(P) \text{ grad } u, \text{grad } \varphi)$$
$$= -(u\delta(P) \text{ grad } P, \text{grad } \varphi) - (\theta(P) \text{ grad } u, \text{grad } \varphi).$$

Now we use the definitions of $\delta(P)$ and $\theta(P)$ to obtain

$$(\Delta[u\theta(P)], \varphi) = -\int_{P\geqslant 0} (\text{grad } u, \text{grad } \varphi) \, dx - \int_{P=0} u \cdot (\text{grad } P, \text{grad } \varphi) \, \omega. \qquad (4)$$

This, in fact, is Green's theorem in generalized-function notation. If $P(x)$ is (up to terms of higher order) the distance from x to the $P = 0$ surface, so that grad P is the unit normal, (grad P, grad φ) will be the normal derivative of φ, while ω will reduce to the Euclidean element of area $d\sigma$ on $P = 0$. Then Eq. (4) takes on the usual form

$$\int_{P\geqslant 0} \Delta u\varphi \, dv = -\int_{P\geqslant 0} (\text{grad } u, \text{grad } \varphi) \, dx - \int_{P=0} u \frac{\partial \varphi}{\partial n} \, d\sigma. \qquad (5)$$

We have given the proof for infinitely differentiable functions u and φ, but by passing to the appropriate limit we can extend it to functions having only those derivatives that appear in the final expression of the theorem.

1.5. The Differential Forms $\omega_k(\varphi)$ and the Generalized Functions $\delta^{(k)}(P)$

We shall find important also the derivatives of $\delta(P)$ with respect to the argument P. To define these in an invariant manner we shall define, in addition to the differential form ω, a set of differential forms of degree $n - 1$, which we shall call $\omega_0(\varphi)$, $\omega_1(\varphi)$, ... These forms depend both on P and on $\varphi(x)$, and are defined by

$$\omega_0(\varphi) = \varphi \cdot \omega, \tag{1}$$

$$d\omega_0(\varphi) = dP \cdot \omega_1(\varphi), \tag{2}$$

$$\cdots \cdots \cdots \cdots \cdots$$

$$d\omega_{k-1}(\varphi) = dP \cdot \omega_k(\varphi), \tag{3}$$

$$\cdots \cdots \cdots \cdots \cdots$$

where d denotes the exterior derivative.

The existence of these forms in a region containing the $P = 0$ surface is easily verified in the coordinate system $u_1 = P$, u_2, ..., u_n. Indeed, recalling that

$$\omega = D\binom{x}{u} du_2 \dots du_n$$

and setting $\varphi(x) = \varphi_1(u)$, we may write

$$d\omega_0 = d(\varphi_1\omega) = \frac{\partial}{\partial u_1} \left[\varphi_1 D\binom{x}{u} \right] du_1 \dots du_n = dP \cdot \omega_1(\varphi),$$

so that a possible solution is

$$\omega_1(\varphi) = \frac{\partial}{\partial u_1} \left[\varphi_1 D\binom{x}{u} \right] du_2 \dots du_n.$$

We may proceed in the same way, writing

$$d\omega_1(\varphi) = \frac{\partial^2}{\partial u_1^2} \left[\varphi_1 D\binom{x}{u} \right] dP \, du_2 \dots du_n = dP \cdot \omega_2(\varphi),$$

so that a possible solution is

$$\omega_2(\varphi) = \frac{\partial^2}{\partial u_1^2} \left[\varphi_1 D\binom{x}{u} \right] du_2 \dots du_n$$

and so on. In general

$$\omega_k(\varphi) = \frac{\partial^k}{\partial u_1^k} \left[\varphi_1 D\binom{x}{u} \right] du_2 \dots du_n. \tag{4}$$

Unlike ω, these new differential forms are not uniquely determined by $P(x)$ [and the choice of $\varphi(x)$ in K] even on the $P = 0$ surface. For our purposes, however, all that is important is that the integral of $\omega_k(\varphi)$ over this surface be unique. To prove that it is, we show that if $\tilde{\omega}_k(\varphi)$ satisfies the same equation as $\omega_k(\varphi)$, then

$$\omega_k - \tilde{\omega}_k = d\alpha + \beta \, dP, \tag{5}$$

where α and β are differential forms with bounded support of degrees $n - 1$ and $n - 2$, respectively. Before going into the proof, however, we must explain how uniqueness of the integral will follow from this. First, note that on the $P = 0$ surface the second term vanishes, so that on this surface

$$\omega_k - \tilde{\omega}_k = d\alpha.$$

Now by Stokes' theorem we have

$$\int_{P=0} d\alpha = \int_{\Gamma} \alpha,$$

where Γ is the boundary of the $P = 0$ surface. But this surface is either closed, so that it has no boundary, or extends to infinity. In this second case $\int_{\Gamma} \alpha = 0$ because α has bounded support.

We shall give an inductive proof of Eq. (5). Recall that we have shown in Section 1.2 that if γ is a differential form orthogonal to dP, then it can be written

$$\gamma = \gamma_1 \, dP,$$

where γ_1 has degree $n - 2$ and has bounded support if γ has bounded support. Thus for $k = 0$ we immediately arrive at

$$\omega_0 - \tilde{\omega}_0 = \beta \, dP,$$

where β has bounded support since ω_0 and $\tilde{\omega}_0$ contain the function φ as a factor. Let us now assume that Eq. (5) holds for suitable α and β, and then we take the inductive step from k to $k + 1$. For this purpose we differentiate (5), obtaining

$$d\omega_k - d\tilde{\omega}_k = d\beta \, dP,$$

where we have used the fact that the exterior derivative of an exterior derivative vanishes. Now by definition of ω_{k+1} and $\tilde{\omega}_{k+1}$ this equation can be written

$$dP \cdot (\omega_{k+1} - \tilde{\omega}_{k+1} - (-1)^{n-1} \, d\beta) = 0.$$

We therefore have

$$\omega_{k+1} - \tilde{\omega}_{k+1} - (-1)^{n-1} \, d\beta = \gamma \, dP,$$

and we have only to establish that γ has bounded support. But this follows from Eqs. (1)–(3), according to which both ω_{k+1} and $\tilde{\omega}_{k+1}$ have bounded support.

Thus the integral over the $P = 0$ surface of any of the $\omega_k(\varphi)$ is indeed uniquely determined by $P(x)$. We now make the following *definition*:

$$(\delta^{(k)}(P), \varphi) = \int \delta^{(k)}(P) \, \varphi \, dx = (-1)^k \int_{P=0} \omega_k(\varphi) \qquad (k = 0, 1, 2, ...). \qquad (6)$$

That this definition of the $\delta^{(k)}(P)$ coincides with the definition at the beginning of this chapter can be seen from Eq. (4).

Example 1. For $P(x) \equiv x_1 = 0$, Eq. (4) gives

$$\omega_k(\varphi) = \frac{\partial^k \varphi(x)}{\partial x_1^k} \, dx_2 ... dx_n$$

so that

$$\int \delta^{(k)}(x_1) \, \varphi \, dx = (-1)^k \int \frac{\partial^k \varphi(0, x_2, ..., x_n)}{\partial x_1^k} \, dx_2 ... dx_n. \qquad (7)$$

Example 2. Let us calculate $\delta^{(k)}(\alpha_1 x_1 + ... + \alpha_n x_n)$, where $\Sigma \alpha_i^2 = 1$. Performing the same rotation of coordinates as in Example 2 of Section 1.3, we arrive at

$$\omega_k(\varphi) = \frac{\partial^k \varphi}{\partial u_1^k} \, du_2 ... du_n$$

so that

$$\int \delta^{(k)} \left(\sum \alpha_i x_i \right) \varphi(x) \, dx = (-1)^k \int_{\Sigma \alpha_i x_i = 0} \frac{\partial^k \varphi}{\partial u_1^k} \, d\sigma, \qquad (8)$$

where the derivative on the right-hand side is taken normal to the hypersurface

$$\sum \alpha_i x_i = 0$$

(in the direction of increasing $\Sigma \alpha_i x_i$), and $d\sigma$ is the element of area on this hypersurface.

Example 3. Let us find $\delta^{(k)}(xy - c)$. As in Example 3 of Section 1.3, we transform to the coordinates $u_1 = x$, $u_2 = xy - c$. Since

$$D\begin{pmatrix} x & y \\ u_1 & u_2 \end{pmatrix} = \frac{1}{u_1},$$

we may write

$$\omega_k(\varphi) = -\frac{\partial^k}{\partial u_2^k}\left(\frac{\varphi\left(u_1, \dfrac{u_2 + c}{u_1}\right)}{u_1}\right) du_1$$

so that

$$\int \delta^{(k)}(xy - c)\,\varphi(x, y)\,dx\,dy = (-1)^{k+1} \int \frac{\partial^k \varphi(x, c/x)}{\partial y^k}\,\frac{dx}{x^{k+1}}. \tag{9}$$

Example 4. Let us now construct $\delta^{(k)}(r - c)$, where $r^2 = \Sigma x_i^2$. Again we use the spherical coordinates $u_1 = r$, $u_2 = \theta_1$, ..., $u_n = \theta_{n-1}$. In these coordinates it is a simple matter to calculate

$$\omega = r^{n-1}\,d\Omega,$$

where $d\Omega$ (previously $d\omega$) is the element of area on the unit sphere $r = 1$. This gives

$$\omega_0 = \varphi r^{n-1}\,d\Omega, \qquad \omega_1(\varphi) = \frac{\partial}{\partial r}(\varphi r^{n-1})\,d\Omega$$

and so on. In general,

$$\omega_k(\varphi) = \frac{\partial^k}{\partial r^k}(\varphi r^{n-1})\,d\Omega,$$

so that

$$\int \delta^{(k)}(r - c)\,\varphi\,dx = \frac{(-1)^k}{c^{n-1}} \int_{O_c} \frac{\partial^k}{\partial r^k}(\varphi r^{n-1})\,dO_c, \tag{10}$$

where O_c is the sphere $r - c = 0$, and dO_c is the Euclidean element of area of it.

Example 5. Finally, we calculate $\delta^{(k)}(r^2 - c^2)$ by transforming to the coordinates $u_1 = r^2 - c^2$, $u_2 = \theta_1$, ..., $u_n = \theta_{n-1}$. We obtain

$$\omega = \tfrac{1}{2}r^{n-2}\,d\Omega$$

and

$$\omega_k(\varphi) = \tfrac{1}{2}\left(\frac{\partial}{2r\,\partial r}\right)^k \varphi r^{n-2}\,d\Omega,$$

which means that

$$\int \delta^{(k)}(r^2 - c^2)\,\varphi\,dx = \frac{(-1)^k}{2c^{n-1}} \int_{O_c}\left(\frac{\partial}{2r\,\partial r}\right)^k (\varphi r^{n-2})\,dO_c \tag{11}$$

or that

$$(\delta^{(k)}(x_1^2 + \ldots + x_n^2 - c^2), \varphi) = \frac{(-1)^k}{2c^{n-1}} \int\left[\left(\frac{\partial}{2r\,\partial r}\right)^k \varphi r^{n-2}\right]_{r=c} dO_c. \tag{12}$$

1.6. Recurrence Relations for the $\delta^{(k)}(P)$

We shall show that the generalized function $\delta^{(k)}(P)$ can be differentiated by the chain rule, that is, that

$$\frac{\partial}{\partial x_j}\,\delta^{(k)}(P) = \frac{\partial P}{\partial x_j}\,\delta^{(k+1)}(P), \qquad k = 0, 1, 2, \ldots. \tag{1}$$

To prove this it is sufficient, as in Section 1.3, to verify that for fixed j the functionals on both sides of this equation coincide in a neighborhood of any point of the $P = 0$ surface such that $\partial P/\partial x_j \neq 0$.

As before, we apply both sides of the equation to a $\varphi(x)$ in K with support in a small neighborhood of such a point, and choose the coordinates $u_i = x_i$ for $i \neq j$, and $u_j = P$. In these coordinates

$$D\binom{u}{x} = \frac{\partial P}{\partial x_j}, \qquad D\binom{x}{u} = \frac{1}{\partial P/\partial x_j},$$

and ω_k becomes

$$\omega_k(\varphi) = \frac{\partial^k}{\partial P^k}\left(\frac{\varphi}{\partial P/\partial x_j}\right) dx_1 \ldots dx_{j-1}\, dx_{j+1} \ldots dx_n.$$

Now the functional on the right-hand side of Eq. (1) operates according to

$$\left(\frac{\partial P}{\partial x_j}\,\delta^{(k+1)}(P), \varphi\right) = \left(\delta^{(k+1)}(P), \frac{\partial P}{\partial x_j}\,\varphi\right)$$

$$= (-1)^{k+1} \int_{P=0} \omega_{k+1}\left(\varphi\,\frac{\partial P}{\partial x_j}\right)$$

$$= (-1)^{k+1} \int_{P=0} \frac{\partial^{k+1}\varphi}{\partial P^{k+1}}\, dx_1 \ldots dx_{j-1}\, dx_{j+1} \ldots dx_n.$$

The functional on the left-hand side operates according to

$$\left(\frac{\partial}{\partial x_j}\,\delta^{(k)}(P), \varphi\right) = -\left(\delta^{(k)}(P), \frac{\partial \varphi}{\partial x_j}\right) = (-1)^{k+1}\int_{P=0}\omega_k\left(\frac{\partial\varphi}{\partial x_j}\right)$$

$$= (-1)^{k+1}\int_{P=0}\frac{\partial^k}{\partial P^k}\left(\frac{\partial\varphi/\partial x_j}{\partial P/\partial x_j}\right) dx_1 \ldots dx_{j-1}\, dx_{j+1} \ldots dx_n.$$

Thus what we must prove is that

$$\frac{\partial^{k+1}\varphi}{\partial P^{k+1}} = \frac{\partial^k}{\partial P^k}\left(\frac{\partial\varphi/\partial x_j}{\partial P/\partial x_j}\right). \tag{2}$$

Now for $k = 0$ this is obviously true since $\varphi = \varphi(x_1, ..., x_j(P), ..., x_n)$ and therefore

$$\frac{\partial \varphi}{\partial P} = \frac{\partial \varphi / \partial x_j}{\partial P / \partial x_j} .$$

But then Eq. (2) follows without further ado for any k, since

$$\frac{\partial^{k+1} \varphi}{\partial P^{k+1}} = \frac{\partial^k}{\partial P^k} \left(\frac{\partial \varphi}{\partial P} \right) .$$

We have thus shown that Eq. (1) holds in the neighborhood of any point of the $P = 0$ surface when $\partial P / \partial x_j \neq 0$. As we have already seen, from this the result follows.

We shall now derive the following recurrence relations, identities between $\delta(P)$ and its derivatives:

$$P\delta(P) = 0, \tag{3}$$

$$P\delta'(P) + \delta(P) = 0, \tag{4}$$

$$P\delta''(P) + 2\delta'(P) = 0, \tag{5}$$

. . . .

$$P\delta^{(k)}(P) + k\delta^{(k-1)}(P) = 0, \tag{6}$$

. . . .

The first of these is obvious, since the integral of $P\varphi$ over the $P = 0$ surface clearly vanishes. We now take the derivative with respect to x_j and use Eq. (1), obtaining

$$\frac{\partial P}{\partial x_j} \delta(P) + P \frac{\partial P}{\partial x_j} \delta'(P) = 0.$$

But we know that for at least one j the derivative $\partial P / \partial x_j \neq 0$, so that the second of the identities follows immediately.[10]

The remaining recurrence relations are proven similarly.

---

[10] In general we may assert the following: if $a_j(x) f = 0$, $j = 1, 2, ..., m$ and if the $a_j(x)$ have no common roots [common, that is, to all the $a_j(x)$] in the support F of the functional f, then $f = 0$. The proof follows. Every point x_0 of F has a neighborhood U in which at least one of the $| a_j(x) |$, say for $j = 1$, is greater than some positive constant. Now consider a $\varphi(x)$ in K with support contained in U. For this function we may write $\varphi(x) = a_1(x) \psi(x)$, where $\psi(x)$ is some other function in K. Then we have $(f, \varphi) = (a_1 f, \psi) = 0$, which means that f vanishes in the neighborhood of x_0. Thus f vanishes in the neighborhood of every point, and consequently $f = 0$.

Example. *The generalized function $\delta^{(k)}(r^2 - t^2)$ as the solution of the wave equation in a space of odd dimension.* Let us apply operator $\varDelta - \partial^2/\partial t^2$ to the generalized function $\delta^{(k)}(r^2 - t^2)$. We have

$$\frac{\partial}{\partial x_j}\,\delta^{(k)}(r^2 - t^2) = 2x_j\delta^{(k+1)}(r^2 - t^2),$$

$$\frac{\partial^2}{\partial x_j^2}\,\delta^{(k)}(r^2 - t^2) = 2\delta^{(k+1)}(r^2 - t^2) + 4x_j^2\delta^{(k+2)}(r^2 - t^2),$$

$$\sum \frac{\partial^2}{\partial x_j^2}\,\delta^{(k)}(r^2 - t^2) = 2n\delta^{(k+1)}(r^2 - t^2)$$
$$+ 4(r^2 - t^2)\,\delta^{(k+2)}(r^2 - t^2) + 4t^2\delta^{(k+2)}(r^2 - t^2),$$

which, according to Eq. (6), can be written

$$\sum \frac{\partial^2}{\partial x_j^2}\,\delta^{(k)}(r^2 - t^2) = [2n - 4(k + 2)]\,\delta^{(k+1)}(r^2 - t^2) + 4t^2\delta^{(k+2)}(r^2 - t^2).$$

Similarly,

$$\frac{\partial^2}{\partial t^2}\,\delta^{(k)}(r^2 - t^2) = -2\delta^{(k+1)}(r^2 - t^2) + 4t^2\delta^{(k+2)}(r^2 - t^2),$$

so that

$$\sum \frac{\partial^2}{\partial x_j^2}\,\delta^{(k)}(r^2 - t^2) - \frac{\partial^2}{\partial t^2}\,\delta^{(k)}(r^2 - t^2) = (2n - 4k - 6)\,\delta^{(k+1)}(r^2 - t^2).$$

For $k = \frac{1}{2}(n - 3)$ this vanishes, so that if n is odd and $k = \frac{1}{2}(n - 3)$, the generalized function $\delta^{(k)}(r^2 - t^2)$ is a solution of the wave equation

$$\left(\varDelta - \frac{\partial^2}{\partial t^2}\right) u = 0.$$

In particular, for $n = 3$ such a solution is $\delta(r^2 - t^2)$.

Let us see to what initial conditions this solution belongs. From Eq. (11) of Section 1.5, we have

$$(\delta^{(k)}(r^2 - t^2), \varphi(x)) = \frac{(-1)^k}{2} \int_\Omega \left[\left(\frac{\partial}{2r\,\partial r}\right)^k \varphi r^{n-2}\right]_{r=t} d\Omega. \tag{7}$$

If we put

$$n = 2k + 3$$

the integrand becomes

$$\left(-\frac{\partial}{2r\,\partial r}\right)^{k}\varphi r^{2k+1}.$$

Each application of the operator $r^{-1}\partial/\partial r$ reduces the power of r by two. After k such operations on φr^{2k+1}, we obtain a sum of terms each of which contains r at most in the first power. We now set $r = t$ and allow t to approach zero, obtaining

$$\lim_{t\to 0}\delta^{(k)}(r^2 - t^2) = 0. \tag{8}$$

To obtain the second initial condition we study

$$\frac{\partial\delta^{(k)}(r^2 - t^2)}{\partial t} = -2t\delta^{(k+1)}(r^2 - t^2),$$

which, when applied to φ yields

$$2t\frac{(-1)^k}{2}\int_{\Omega}\left[\left(\frac{\partial}{2r\,\partial r}\right)^{k+1}(\varphi r^{2k+1})\right]_{r=t}d\Omega.$$

In this case the integrand contains a single term containing r to a negative power. A simple calculation shows that this term is

$$\frac{(2k + 1)!!}{2^{k+1}}\frac{\varphi(x)}{r}.$$

Thus on setting $r = t$ and allowing t to approach zero, we obtain

$$(-1)^k\frac{(2k + 1)!!}{2^{k+1}}\Omega_n\varphi(0) \tag{9}$$

where Ω_n is the area of the unit sphere in R_n. Now as is known,

$$\Omega_n = \frac{2\pi^{\frac{1}{2}n}}{\Gamma\left(\frac{n}{2}\right)} = \frac{2\pi^{\frac{1}{2}(2k+3)}}{\Gamma\left(\frac{2k+3}{2}\right)} = \frac{2^{k+2}}{(2k+1)!!}\pi^{k+1};$$

so that we arrive finally at

$$\frac{\partial\delta^{(k)}(r^2 - t^2)}{\partial t}\bigg|_{t=0} = (-1)^k\,2\pi^{k+1}\,\delta(x), \tag{10}$$

which means that the solution

$$u(x, t) = \delta^{(k)}(r^2 - t^2)$$

of the wave equation corresponds to the initial conditions

$$u(x, 0) = 0, \frac{\partial u(x, 0)}{\partial t} = (-1)^k 2\pi^{k+1} \delta(x). \tag{11}$$

We see consequently that to within a numerical factor this function is an elementary solution of Cauchy's problem for the wave equation (cf. Chapter I, Section 5.4).

1.7. Recurrence Relations for the $\delta^{(k)}(aP)$

Consider two functions $P(x)$ and $Q(x)$ such that the $P = 0$ and $Q = 0$ hypersurfaces have, as before, no singular points. Assume further that these surfaces fail to intersect and that the $PQ = 0$ surface also has no singular points. Then

$$\delta(PQ) = P^{-1} \delta(Q) + Q^{-1} \delta(P). \tag{1}$$

Proof. Let ω_P be the differential form corresponding to the $P = 0$ surface, ω_Q be the differential form corresponding to the $Q = 0$ surface, and ω be the differential form corresponding to the $PQ = 0$ surface. These differential forms have the following properties:

$$\omega_P \, dP = dv \qquad \text{on} \quad P = 0, \tag{2}$$

$$\omega_Q \, dQ = dv \qquad \text{on} \quad Q = 0, \tag{3}$$

$$\omega(P \, dQ + Q \, dP) = dv \qquad \text{on both} \quad P = 0 \quad \text{and} \quad Q = 0. \tag{4}$$

Now on the $P = 0$ surface Eq. (4) becomes

$$\omega Q \, dP = dv,$$

which, when we compare it with (2), shows that on this surface $\omega = Q^{-1}\omega_P$. Similarly, on the $Q = 0$ surface $\omega = P^{-1}\omega_Q$. This leads directly to the desired result:

$$(\delta(PQ), \varphi) = \int_{P=0} \varphi\omega + \int_{Q=0} \varphi\omega = \int_{P=0} \varphi Q^{-1}\omega_P + \int_{Q=0} \varphi P^{-1}\omega_Q$$

$$= (Q^{-1} \delta(P), \varphi) + (P^{-1} \delta(Q), \varphi).$$

In particular, if $a(x)$ is some nonvanishing function, this gives

$$\delta(aP) = a^{-1} \delta(P). \tag{5}$$

Further interesting formulas can be obtained by differentiating (5). In particular, the derivative with respect to x_j gives

$$\delta'(aP) \frac{\partial(aP)}{\partial x_j} = \delta'(aP) P \frac{\partial a}{\partial x_j} + \delta'(aP) a \frac{\partial P}{\partial x_j}$$

$$= a^{-1} \delta'(P) \frac{\partial P}{\partial x_j} - a^{-2} \delta(P) \frac{\partial a}{\partial x_j}.$$

Now Eq. (4) of Section 1.6 can be used to replace $\delta'(aP)P$ in the first term in the right-hand side by $- a^{-1}\delta(aP) = - a^{-2}\delta(P)$, to give the result

$$\delta'(aP) = a^{-2} \delta'(P).$$

In a similar way, for any k and any $a(x)$ we have

$$\delta^{(k)}(aP) = a^{-(k+1)} \delta^{(k)}(P). \tag{6}$$

Example.

$$\delta^{(k)}(r^2 - c^2) = \delta^{(k)}[(r + c)(r - c)] = (r + c)^{-k-1}\delta^{(k)}(r - c). \tag{7}$$

This formula means that for any φ in K

$$\int \delta^{(k)}(r^2 - c^2) \varphi(x)\, dx = \int (r + c)^{-k-1} \delta^{(k)}(r - c)\, \varphi(x)\, dx.$$

This formula would, of course, have been quite difficult to discover if we had had only Eqs. (10) and (11) of Section 1.5. For $k = 0$ it becomes

$$\int \delta(r^2 - c^2) \varphi(x)\, dx = \frac{1}{2c} \int \delta(r - c)\, \varphi(x)\, dx,$$

which, on the other hand, it was difficult to overlook in Section 1.3 [see Eqs. (5) and (6)].

1.8. Multiplet Layers

A functional of the form $\mu(x)\delta^{(k-1)}(P)$, or

$$\int \mu(x)\, \delta^{(k-1)}(P)\, \varphi(x)\, dx = \int_{P=0} \omega_{k-1}(\mu\varphi), \tag{1}$$

is called a *k-fold layer or distribution* on the $P = 0$ hypersurface. In particular, a *singlet or simple layer* ($k = 1$) is given by

$$(\mu\delta(P), \varphi) = \int_{P=0} \mu\varphi\omega = \int_{P=0} \omega_0(\mu\varphi) \tag{2}$$

while a *doublet or double layer* ($k = 2$) is given by

$$(\mu\delta'(P), \varphi) = \int_{P=0} \omega_1(\mu\varphi). \tag{3}$$

The function $\mu(x)$ in these expressions is called the *density* of the corresponding layer.

The definition we have given would not be consistent if it were to depend on the form in which the $P = 0$ equation is written. It is found, however, that the statement that some functional f is a k-fold layer is independent on the form of this equation, and that if it is transformed from $P = 0$ to $a(x)P = 0$, where $a(x)$ is some nonvanishing function, only the expression for $\mu(x)$ will change. This is seen by using Eq. (6) of Section 1.7, which gives

$$\mu(x)\,\delta^{(k-1)}(aP) = \mu(x)\,a^{-k}(x)\,\delta^{(k-1)}(P) = \mu_1(x)\,\delta^{(k-1)}(P).$$

We wish to show that every functional f of the form

$$(f, \varphi) = \int_{P=0} \sum_j a_j(x)\,D^j\,\varphi(x)\,d\sigma$$

$$\left(j = (j_1, \ldots, j_n),\ D^j = \frac{\partial^{j_1 + \ldots + j_n}}{\partial x_1^{j_1} \ldots \partial x_n^{j_n}}\right)$$

can be written as the sum of multiplet layers. From what we have just said is follows that we may use any convenient form to specify the $P = 0$ surface. Let us assume that we have written it in a way in which $P(x)$ is the distance from x to the surface, so that the associated differential form coincides with the Euclidean element of area $d\sigma$. Then we have

$$(f, \varphi) = \int_{P=0} \sum_j a_j(x)\,D^j\varphi(x)\,\omega = \int_{P=0} \omega_0\left(\sum_j a_j(x)\,D^j\varphi\right)$$

$$= \left(\delta(P), \sum_j a_j(x)\,D^j\varphi\right) = \sum_j (-1)^j (D^j a_j(x)\,\delta(P), \varphi)$$

$$= \sum_j (-1)^j \left(\sum_k a_{jk}(x)\,\delta^{(k)}(P), \varphi\right) = \left(\sum_k b_k(x)\,\delta^{(k)}(P), \varphi\right),$$

where

$$b_k(x) = \sum_j (-1)^j a_{jk}(x).$$

Hence

$$f = \sum_k b_k(x)\,\delta^{(k)}(P),$$

as asserted.

1.9. The Generalized Function $\delta(P_1, ..., P_k)$ and Its Derivatives

We have so far been dealing with generalized functions associated with hypersurfaces, or manifolds of dimension $n - 1$. We now wish to turn to new generalized functions associated with manifolds S of lower dimension defined by k equations of the form

$$P_1(x_1, ..., x_n) = 0, \qquad P_2(x_1, ..., x_n) = 0, ..., P_k(x_1, ..., x_n) = 0, \qquad (1)$$

where k is in general greater than one.

We shall make the following assumptions:

(1) The P_i are infinitely differentiable functions.

(2) The $P_i(x_1, ..., x_n) = \xi_i$ hypersurfaces ($i = 1, ..., k$) form a lattice such that in the neighborhood of every point of S there exist a local coordinate system in which $u_i = P_i(x_1, ..., x_n)$ for $i = 1, ..., k$ and the remaining $u_{k+1}, ..., u_n$ can be chosen so that the Jacobian $D(\frac{x}{u}) > 0$.

Consider the element of volume in R_n

$$dv = dx_1 ... dx_n,$$

a differential form of degree n, and let us write it as the product of the first-degree differential forms $dP_1, ..., dP_k$ with an additional differential form ω of degree $n - k$; in other words we write

$$dv = dP_1 ... dP_k \, \omega. \qquad (2)$$

Obviously if ω exists, it cannot be unique. Indeed, if ω is such a differential form, we may add to it any form such as $\sum_{i=1}^{k} \alpha_i dP_i$ where the α_i are any differential forms of degree $n - k - 1$, for then the sum $\omega + \sum_{i=1}^{k} \alpha_i dP_i$ will clearly satisfy Eq. (2) because $dP_i dP_i$ vanishes.

We shall prove the existence of ω by exhibiting it in the variables $x_1, ..., x_n$. In these variables $dv = dx_1 ... dx_n$. The dP_i are given by

$$dP_i = \frac{\partial P_i}{\partial x_1} dx_1 + ... + \frac{\partial P_i}{\partial x_n} dx_n$$

and it is easily shown [cf. Eqs. (2) and (3) of Section 1.1] that

$$dP_1 ... dP_k = \sum_{i_1 < i_2 < ... < i_k} D \begin{pmatrix} P_1 ... P_k \\ x_{i_1} ... x_{i_k} \end{pmatrix} dx_{i_1} ... dx_{i_k}. \qquad (3)$$

Now by the conditions we have imposed on the P_i functions, the matrix

$$\begin{pmatrix} \dfrac{\partial P_1}{\partial x_1} & \cdots & \dfrac{\partial P_1}{\partial x_n} \\ \cdots \cdots \cdots \\ \dfrac{\partial P_k}{\partial x_1} & \cdots & \dfrac{\partial P_k}{\partial x_n} \end{pmatrix}$$

has at least one minor of order k that fails to vanish identically. Let us assume it to be composed of the first k rows and columns, so that

$$D\begin{pmatrix} P_1 \cdots P_k \\ x_1 \cdots x_k \end{pmatrix} \neq 0.$$

Then we may write

$$\omega = \frac{dx_{k+1} \cdots dx_n}{D\begin{pmatrix} P_1 \cdots P_k \\ x_1 \cdots x_k \end{pmatrix}}, \tag{4}$$

since when we multiply this by (3) all the terms except

$$D\begin{pmatrix} P_1 \cdots P_k \\ x_1 \cdots x_k \end{pmatrix} dx_1 \cdots dx_k,$$

will be annihilated, since each will contain at least one differential twice. The sole remaining term gives

$$D\begin{pmatrix} P_1 \cdots P_k \\ x_1 \cdots x_k \end{pmatrix} dx_1 \cdots dx_k \frac{dx_{k+1} \cdots dx_n}{D\begin{pmatrix} P_1 \cdots P_k \\ x_1 \cdots x_k \end{pmatrix}} = dx_1 \cdots dx_n = dv.$$

Similarly, if for some collection $i_1 < i_2 < \ldots < i_k$ we have

$$D\begin{pmatrix} P_1 \cdots P_k \\ x_{i_1} \cdots x_{i_k} \end{pmatrix} \neq 0,$$

we replace Eq. (4) by

$$\omega = (-1)^{i_1 + \ldots + i_k - \frac{1}{2}k(k-1)} \frac{dx_{j_1} \cdots dx_{j_{n-k}}}{D\begin{pmatrix} P_1 \cdots P_k \\ x_{i_1} \cdots x_{i_k} \end{pmatrix}}, \tag{4'}$$

where $j_1 < \ldots < j_{n-k}$, and $j_r \neq j_s$ for $r \neq s$. We have thus exhibited the existence of ω.

A more symmetric, but not always more convenient expression for ω is obtained when the local coordinate system is chosen as $u_1 = P_1, \ldots,$

$u_k = P_k$, and the remaining $u_{k+1}, ..., u_n$ are sufficiently smooth and such that $D(\tfrac{x}{u}) > 0$. In this coordinate system the element of volume becomes

$$dv = D\binom{x}{u} du_1 \, ... \, du_n.$$

Then the first degree forms dP_i simply coincide with the differentials du_i of the variables for $i = 1, ..., k$ and for ω we can then write

$$\omega = D\binom{x}{u} du_{k+1} \, ... \, du_n. \tag{5}$$

We shall need a rather simple lemma from the theory of differential forms to define the generalized functions $\delta(P_1, ..., P_k)$.

Lemma. If γ is a differential form of degree $n - k$ such that its product with k *independent differentials* $dP_1 ... dP_k$ vanishes, there exist differential forms $\alpha_1, ..., \alpha_k$ such that

$$\gamma = \alpha_1 \, dP_1 + ... + \alpha_k \, dP_k.$$

Here by *differentials* we mean the exterior derivatives of zeroth-degree differential forms (or functions) P_i, $i = 1, ..., k$. Such differentials are *independent* if

$$dP_1 ... dP_k \neq 0.$$

In terms of the x_i, this differential form of degree k is

$$\sum_{j_1, ..., j_k} \frac{\partial P_1}{\partial x_{j_1}} \frac{\partial P_2}{\partial x_{j_2}} \cdots \frac{\partial P_k}{\partial x_{j_k}} dx_{j_1} \, ... \, dx_{j_k} = \sum_{j_1 < ... < j_k} D\binom{P_1 \, ... \, P_k}{x_{j_1} \, ... \, x_{j_k}} dx_{j_1} \, ... \, dx_{j_k}.$$

Thus the independence of the dP_i implies that the matrix of the $\partial P_i/\partial x_j$, $i = 1, ..., k; j = 1, ..., n$, is of rank k, or that the P_i may be chosen as the first k (local) coordinates in R_n.

Proof. To prove the lemma we transform to the variables $u_i = P_i$ $(i = 1, ..., k)$, $u_{k+1}, ..., u_n$. Then in these coordinates γ may in general consist of terms containing at least one of the first k of the du_i and a single term of the form $q du_{k+1} \, ... \, du_n$ containing none of the first k. Now by assumption we have

$$du_1 ... du_k \gamma = q \, du_1 \, ... \, du_n = 0$$

and since the du_i are all independent and their product cannot therefore vanish, it follows that $q \equiv 0$. Thus γ consists only of terms containing at least one of the first k of the du_i. This means that in general γ can be written

$$\gamma = \alpha_1 \, du_1 + ... + \alpha_k \, du_k,$$

or

$$\gamma = \alpha_1 \, dP_1 + ... + \alpha_k \, dP_k,$$

which completes the proof of the lemma.

We now define the generalized function $\delta(P_1, ..., P_k)$ by the equation

$$(\delta(P_1, ..., P_k), \varphi(x_1, ..., x_n)) = \int_S \varphi\omega, \tag{6}$$

where ω is the differential form we have already discussed [see Eq. (2)] and S is the manifold defined in Eq. (1).

Equation (5) shows that this definition coincides with that given at the beginning of the chapter.

It is easily shown that this definition is independent of the particular choice of ω. Let

$$dv = dP_1 ... dP_k\omega = dP_1 ... dP_k\tilde{\omega}.$$

We shall show that $\int \varphi(\omega - \tilde{\omega}) = 0$ for any $\varphi(x)$ in K. According to the lemma, we have $\omega - \tilde{\omega} = \alpha_1 \, dP_1 + ... + \alpha_k dP_k$, so that in the entire space ω and $\tilde{\omega}$ can differ at most by a form which can be written $\alpha_1 dP_1 + ... + \alpha_k dP_k$, where the α_i are differential forms of degree $n - k - 1$. Obviously, on the $P_i = 0$ manifold this differential form vanishes so that the generalized function $\delta(P_1, ..., P_k)$ defined by the P_i, $i = 1, ..., k$, is unique.

It should be emphasized that both the differential form ω and the functional $\delta(P_1, ..., P_k)$ change when we change the equations describing S. In order to study the way ω changes, let us transform from the equations $P_i = 0$ to $Q_i = 0$, $i = 1, ..., k$, where

$$Q_j(x) = \sum_{i=1}^{k} \alpha_{ij}(x) \, P_i(x).$$

Here the $\alpha_{ij}(x)$ are assumed to be infinitely differentiable functions and the matrix they form is assumed nonsingular. (Obviously these equations describes the same manifold.) The defining equations for the initial differential form ω and for the new one $\tilde{\omega}$ are

$$dP_1 \, dP_2 ... dP_k \, \omega = dv;$$

$$dQ_1 ... dQ_k \, \tilde{\omega} = \left(\sum \alpha_{i1} \, dP_i \right) ... \left(\sum \alpha_{ik} \, dP_i \right) \tilde{\omega} = dv.$$

By expanding the terms in parentheses and using the anticommutation rule $dP_i dP_j = - \, dP_j dP_i$, we may write

$$\det \| \alpha_{ij} \| \, dP_1 ... dP_k \, \tilde{\omega} = dv.$$

This means that

$$\tilde{\omega} = \frac{1}{\det \|\alpha_{ij}\|} \omega.$$

This is the formula which gives the transformation of ω. Then the generalized function $\delta(Q_1, ..., Q_k)$ is related to the original one by

$$(\delta(Q_1, ..., Q_k), \varphi) = \int \frac{\varphi \omega}{\det \|\alpha_{ij}\|} = \left(\delta(P_1, ..., P_k), \frac{\varphi}{\det \|\alpha_{ij}\|}\right).$$

We now wish to define the derivatives of these newly defined generalized functions with respect to their arguments P_i, or the new generalized functions

$$\frac{\partial^m \delta(P_1, ..., P_k)}{\partial P_1^{\alpha_1} ... \partial P_k^{\alpha_k}}.$$

These will be defined as the integrals over S (the same manifold as before) of certain differential forms $\omega_{\alpha_1 ... \alpha_k}$ which depend on the P_i and on the particular function $\varphi(x_1, ..., x_n)$ in K and its derivatives up to the mth order, inclusive.

Let us denote $\varphi \omega$ by $\omega_{00...0}(\varphi)$ [here ω is the differential form defined in Eq. (2)]. We then define, for instance, the differential form $\omega_{10...0}(\varphi)$ (whose integral over S will give $\partial \delta(P_1, ..., P_k)/\partial P_1$) as follows. We take the exterior derivative of the differential form of degree $n - 1$

$$dP_2 ... dP_k \omega_{00...0}(\varphi),$$

and write it in the form

$$d(dP_2 ... dP_k \omega_{00...0}(\varphi)) = dP_1 ... dP_k \omega_{10...0}(\varphi).$$

That it is possible to do this is easily shown in the previously chosen local coordinate system in which the $u_i = P_i$ for $i = 1, ..., k$. Let us denote $\varphi(x_1(u_1, ..., u_n), ..., x_n(u_1, ..., u_n))$ by $\tilde{\varphi}(u_1, ..., u_n) = \tilde{\varphi}(u)$; we then obtain

$$\omega_{00...0}(\varphi) = \tilde{\varphi} D\binom{x}{u} du_{k+1} ... du_n,$$

$$dP_2 ... dP_k \omega_{00...0}(\varphi) = \tilde{\varphi} D\binom{x}{u} du_2 ... du_n,$$

$$d(dP_2 ... dP_k \omega_{00...0}(\varphi)) = \frac{\partial}{\partial u_1}\left[\tilde{\varphi} D\binom{x}{u}\right] du_1 ... du_n,$$ \qquad (7)

$$\omega_{10...0}(\varphi) = \frac{\partial}{\partial u_1}\left[\tilde{\varphi}(u) D\binom{x}{u}\right] du_{k+1} ... du_n.$$

Any of the k indices of $\omega_{00\ldots0}(\varphi)$ can be changed from zero to one in the same way. In general, assuming that we know $\omega_{\alpha_1\alpha_2\ldots\alpha_k}(\varphi)$, we may raise its jth index by multiplying on the left by all the dP_i with $i \neq j$, taking the exterior derivative, and writing

$$d(dP_1 \ldots dP_{j-1}\, dP_{j+1} \ldots dP_k \omega_{\alpha_1\cdots\alpha_k}(\varphi))$$

$$= (-1)^j\, dP_1 \ldots dP_k \omega_{\alpha_1\ldots\alpha_{j-1},\alpha_j+1,\alpha_{j+1}\ldots\alpha_k}. \qquad (8)$$

This defines the $\omega_{\alpha_1\ldots\alpha_k}(\varphi)$ for any nonnegative integral indices.

Obviously, if $\omega_{00\ldots0}(\varphi)$ is not unique, neither are the $\omega_{\alpha_1\ldots\alpha_k}(\varphi)$.

We wish to show, however, that the $\omega_{\alpha_1\ldots\alpha_k}(\varphi)$ defined inductively by equations of the form of (8) can differ for any given $\alpha_1, \ldots, \alpha_k$ only by a differential form that can be written

$$d\tau + \beta_1\, dP_1 + \ldots + \beta_k\, dP_k,$$

where $d\tau$ is the exterior derivative of some differential form τ of degree $n - k - 1$, and the β_i are arbitrary differential forms of degree $n - k - 1$. We may assert further that both τ and the β_i have bounded support.

We shall proceed by induction. According to our lemma,

$$\omega_{00\ldots0} - \tilde{\omega}_{00\ldots0} = \beta_1\, dP_1 + \ldots + \beta_k\, dP_k$$

for any two forms defining $\delta(P_1, \ldots, P_k)$. Since each of the differential forms on the left-hand side has $\varphi(x)$ as a factor and since the dP_i are linearly independent, all the β_1 on the right-hand side of the equation must have bounded support. Let us now assume that given any two differential forms with indices $\alpha_1, \ldots, \alpha_k$ they can be written throughout our space in the form

$$\omega_{\alpha_1,\ldots,\alpha_k} - \tilde{\omega}_{\alpha_1,\ldots,\alpha_k} = d\tau + \beta_1\, dP_1 + \ldots + \beta_k\, dP_k, \qquad (9)$$

where τ and the β_i have bounded support. We shall show that in raising one of the indices (for simplicity we shall take the first) by one unit, this property is preserved. Proceeding, we multiply both sides of Eq. (9) by $dP_2 \ldots dP_k$ and differentiate. Since $d(dP_i) = d(d\tau) = 0$, we obtain

$$d[dP_2 \ldots dP_k(\omega_{\alpha_1\ldots\alpha_k} - \tilde{\omega}_{\alpha_1\ldots\alpha_k})]$$

$$= dP_1\, dP_2 \ldots dP_k \cdot (\omega_{\alpha_1+1,\alpha_2,\ldots,\alpha_k} - \tilde{\omega}_{\alpha_1+1,\alpha_2,\ldots,\alpha_k})$$

$$= (-1)^k\, dP_1 \ldots dP_k\, d\beta_1.$$

This leads to

$$dP_1 \, dP_2 \, ... \, dP_k \cdot (\omega_{\alpha_1+1, \alpha_2, ..., \alpha_k} - \tilde{\omega}_{\alpha_1+1, \alpha_2, ..., \alpha_k} - (-1)^k \, d\beta_1) = 0$$

and then by the lemma we may write

$$\omega_{\alpha_1+1, \alpha_2, ..., \alpha_k} - \tilde{\omega}_{\alpha_1+1, \alpha_2, ..., \alpha_k} = (-1)^k \, d\beta_1 + \gamma_1 \, dP_1 + ... + \gamma_k \, dP_k.$$

Now since the two differential forms on the left-hand side of this equation have bounded support, as does β_1, the γ_i must also have bounded support. This completes the proof.

Thus Eq. (9) is valid for any indices $\alpha_1, ..., \alpha_k$; let us study it on S. On this manifold $dP_i = 0$, and $d\tau$ is equal to some differential form $d\tau^*$ (the exterior derivative of the scalar function τ^* of bounded support which is obtained from τ by projecting it onto S). Thus on S we have

$$\omega_{\alpha_1, ..., \alpha_k} - \tilde{\omega}_{\alpha_1, ..., \alpha_k} = d\tau^*.$$

We now define the generalized function $\dfrac{\partial^m \delta(P_1, ..., P_k)}{\partial P_1^{\alpha_1} ... \partial P_k^{\alpha_k}}$ by

$$\left(\frac{\partial^m \delta(P_1, ..., P_k)}{\partial P_1^{\alpha_1} ... \partial P_k^{\alpha_k}} , \varphi \right) = (-1)^m \int_S \omega_{\alpha_1, ..., \alpha_k}(\varphi). \tag{10}$$

It is seen from equations such as (7) that this definition agrees with the one given at the beginning of the chapter. Further, it is independent of the particular choice of $\omega_{\alpha_1, ..., \alpha_k}$. Indeed, if $\omega_{\alpha_1, ..., \alpha_k}$ and $\tilde{\omega}_{\alpha_1, ..., \alpha_k}$ are two differential forms obtained from $\omega_{00...0}$ by successively raising the indices, then as we have seen

$$\int_S \omega_{\alpha_1, ..., \alpha_k}(\varphi) - \int_S \tilde{\omega}_{\alpha_1, ..., \alpha_k}(\varphi) = \int_S d\tau^*.$$

But according to Stokes' theorem,

$$\int_S d\tau^* = \int_\Gamma \tau^*,$$

where Γ is the boundary of S. From what we have assumed concerning the P_i, however, S is either closed or extends to infinity. In either case $\int_\Gamma \tau^* = 0$, for τ^* has bounded support. Therefore

$$\int_S \omega_{\alpha_1 ... \alpha_k} = \int_S \tilde{\omega}_{\alpha_1 ... \alpha_k}.$$

This proves that the generalized function $\dfrac{\partial^m \delta(P_1, ..., P_k)}{\partial P_1^{\alpha_1} ... \partial P_k^{\alpha_k}}$ is defined uniquely by the P_i functions on the $P_i = 0$ manifold ($i = 1, ..., k$).

For the generalized functions we have been discussing we have the following theorems.

Theorem 1. The generalized functions $\delta(P_1, ..., P_k)$ and $\dfrac{\partial^m \delta(P_1, ..., P_k)}{\partial P_1^{\alpha_1} ... \partial P_k^{\alpha_k}}$ can be differentiated according to the chain rule

$$\frac{\partial}{\partial x_j}\, \delta(P_1, ..., P_k) = \sum_{i=1}^{k} \frac{\partial \delta(P_1, ..., P_k)}{\partial P_i} \frac{\partial P_i}{\partial x_j}.$$

Here the partial derivatives with respect to the x_j are defined in the usual way, namely

$$\left(\frac{\partial}{\partial x_j}\, \delta(P_1, ..., P_k), \varphi\right) = -\left(\delta(P_1, ..., P_k), \frac{\partial \varphi}{\partial x_j}\right).$$

Theorem 2. The generalized functions $\delta(P_1, ..., P_k)$ and their derivatives satisfy the identities

$$P_i \delta(P_1, ..., P_k) = 0,$$

$$P_i P_j \delta(P_1, ..., P_k) = 0,$$

$$\cdots \cdots \cdots \cdots \cdots \cdots \cdots \cdots$$

$$P_1 P_2 ... P_k \delta(P_1, ..., P_k) = 0,$$

as well as other identities that can be obtained from these by formal differentiation with respect to the P_i. For instance, successive differentiation of the first identity leads to

$$\delta(P_1, ..., P_k) + P_i \delta'_{P_i}(P_1, ..., P_k) = 0,$$

$$\cdots \cdots \cdots \cdots \cdots \cdots \cdots \cdots \cdots \cdots \cdots$$

$$m\delta^{(m-1)}_{P_i...P_i}(P_1, ..., P_k) + P_i \delta^{(m)}_{P_i...P_i}(P_1, ..., P_k) = 0,$$

$$\cdots \cdots \cdots \cdots \cdots \cdots \cdots \cdots \cdots \cdots \cdots$$

We may differentiate with respect to any of the P_i. For instance, the derivative of the second identity with respect to P_i and P_j yields

$$\delta(P_1, ..., P_k) + P_i \delta'_{P_i}(P_1, ..., P_k) + P_j \delta'_{P_j}(P_1, ..., P_k)$$

$$+ P_i P_j \delta''_{P_i P_j}(P_1, ..., P_k) = 0.$$

These two theorems can be proven in the same way as their analogs for $\delta(P)$; we shall omit these proofs.

In conclusion we remark that the analogs of the m-fold multiplet layers can be defined as functionals of the form

$$\int \sum_{|\alpha|=m} \mu_\alpha(x)\, \delta^{(\alpha)}(P_1, ..., P_k)\, \varphi(x)\, dx,$$

where

$$\alpha = (\alpha_1, ..., \alpha_k), \qquad |\alpha| = \alpha_1 + ... + \alpha_k,$$

$$\delta^{(\alpha)}(P_1, ..., P_k) = \frac{\partial^{|\alpha|}\delta(P_1, ..., P_k)}{\partial P_1^{\alpha_1} ... \partial P_k^{\alpha_k}}.$$

In particular, a singlet or simple layer is given by

$$(\mu\delta(P_1, ..., P_k), \varphi) = \int \mu(x)\, \delta(P_1, ..., P_k)\, \varphi(x)\, dx = \int_S \omega_{00...0}(\mu\varphi).$$

Again it can be proven that every functional f of the form

$$(f, \varphi) = \int_S \sum a_j(x)\, D^j\varphi(x)\, d\sigma, \qquad D^j = \frac{\partial^{|j|}}{\partial x_1^{j_1} ... \partial x_n^{j_n}},$$

where $d\sigma$ is the element of the area on S, can be written as the sum of multiplet layers.

2. Generalized Functions Associated with a Quadratic Form

In this section we shall consider certain types of generalized functions associated with nondegenerate quadratic forms. Sections 2.1 and 2.2 are devoted to specific regularizations of certain integrals without extension to the complex domain. The rest of the section makes wide use of convergence from the complex domain.

This complex extension method is the more simple, and we suggest that on first reading the reader merely skim over Sections 2.1 and 2.2, which contain some useful information on delta functions associated with quadratic forms, and start the careful study with Section 2.3.

2.1. Definition of $\delta_1^{(k)}(P)$ and $\delta_2^{(k)}(P)$

In Section 1 we defined $\delta^{(k)}(P)$ for an infinitely differentiable function $P(x_1, ..., x_n)$ such that the $P = 0$ hypersurface has no singular points. But if

$$P(x_1, ..., x_n) = x_1^2 + ... + x_p^2 - x_{p+1}^2 - ... - x_{p+q}^2 \qquad (1)$$

(with $p + q = n$), the $P = 0$ hypersurface is a hypercone with a singular point (the vertex) at the origin.

For this case $\delta^{(k)}(P)$ can be defined as follows. Since the only singularity of the $P = 0$ surface is at the origin, we have in fact already defined $(\delta^{(k)}(P), \varphi)$ for all φ in K which vanish at the origin. This definition according to Section 1 is

$$(\delta^{(k)}(P), \varphi) = \int \delta^{(k)}(P)\, \varphi \, dx = (-1)^k \int_{P=0} \omega_k(\varphi). \tag{2}$$

But if φ is to be an arbitrary function in K, the integral in Eq. (2) may diverge. We shall then define $(\delta^{(k)}(P), \varphi)$ by regularizing the integral.

Let us proceed according to this plan, but in a relatively elementary way which does not involve the definition of $\int_{P=0} \omega_k(\varphi)$ from Section 1.

We thus start by assuming that $\varphi(x)$ vanishes in a neighborhood of the origin. Away from the origin, the $P = 0$ surface has no singularities, and in a small enough neighborhood of any such nonsingular point we may set up the coordinate system $u_1 = P, u_2, ..., u_n$ with Jacobian $D(\frac{x}{u}) > 0$. Then according to the definition of $\delta^{(k)}(P)$ of Section 1, we have

$$(\delta^{(k)}(P), \varphi) = (-1)^k \int \left[\frac{\partial^k}{\partial u_1^k} \left\{ \varphi D \binom{x}{u} \right\} \right]_{u_1=0} du_2 \, ... \, du_n, \tag{3}$$

where the integral is taken over $P = 0$. This definition is independent of the particular choice of u_i coordinates.

Indeed, as was shown in Section 1, the definition in Eq. (3) coincides with $(-1)^k \int_{P=0} \omega_k(\varphi)$ and is therefore coordinate-system independent.

By choosing the u_i in a special way, we can obtain an explicit expression for $(\delta^{(k)}(P), \varphi)$.

Assume that both $p > 1$ and $q > 1$; we shall deal later with the case in which either q or p is 1.

Let us transform to bipolar coordinates defined by

$$x_1 = r\omega_1, ..., x_p = r\omega_p, \qquad x_{p+1} = s\omega_{p+1}, ..., x_{p+q} = s\omega_{p+q}, \tag{4}$$

where

$$r = \sqrt{x_1^2 + \cdots + x_p^2}, \qquad s = \sqrt{x_{p+1}^2 + ... + x_{p+q}^2}. \tag{5}$$

In these coordinates the element of volume is given by

$$dx = r^{p-1}\, s^{q-1}\, dr\, ds\, d\Omega^{(p)}\, d\Omega^{(q)}, \tag{6}$$

where $d\Omega^{(p)}$ and $d\Omega^{(q)}$ are the elements of surface area on the unit sphere in R_p and R_q, respectively. Then Eq. (1) becomes

$$P = r^2 - s^2.$$

Now let us choose the coordinates to be P, r, and the ω_i. In these coordinates Eq. (6) becomes

$$dx = \tfrac{1}{2}(r^2 - P)^{\frac{1}{2}(q-2)} r^{p-1} \, dP \, dr \, d\Omega^{(p)} \, d\Omega^{(q)}.$$

Then we may rewrite the defining equation (3) in the form

$$(\delta^{(k)}(P), \varphi) = (-1)^k \int \left[\frac{\partial^k}{\partial P^k} \left\{ \tfrac{1}{2} \varphi (r^2 - P)^{\frac{1}{2}(q-2)} \right\} \right]_{P=0} r^{p-1} \, dr \, d\Omega^{(p)} \, d\Omega^{(q)}. \qquad (7)$$

Further, if we transform from P to $s = \sqrt{r^2 - P}$ and note that $\partial/\partial P = -(2s)^{-1}\partial/\partial s$, we may write this in the form

$$(\delta^{(k)}(P), \varphi) = \int \left[\left(\frac{\partial}{2s\,\partial s} \right)^k \left\{ s^{q-2} \frac{\varphi}{2} \right\} \right]_{s=r} r^{p-1} \, dr \, d\Omega^{(p)} \, d\Omega^{(q)}. \qquad (8)$$

Let us now write

$$\psi(r, s) = \int \varphi \, d\Omega^{(p)} \, d\Omega^{(q)}, \qquad (9)$$

which transforms (8) to the form

$$(\delta^{(k)}(P), \varphi) = \int_0^\infty \left[\left(\frac{\partial}{2s\,\partial s} \right)^k \left\{ s^{q-2} \frac{\psi(r, s)}{2} \right\} \right]_{s=r} r^{p-1} \, dr. \qquad (10)$$

We could equally well have interchanged the rules of r and s in these considerations. We would then have arrived at the equivalent expression

$$(\delta^{(k)}(P), \varphi) = (-1)^k \int_0^\infty \left[\left(\frac{\partial}{2r\,\partial r} \right)^k \left\{ r^{p-2} \frac{\psi(r, s)}{2} \right\} \right]_{r=s} s^{q-1} \, ds. \qquad (10')$$

Now we are still assuming that φ vanishes in a neighborhood of the origin, so that these integrals will converge for any k. If, further, $(p-1) + (q-2) \geqslant 2k$, that is, if $k < \tfrac{1}{2}(p + q - 2)$, these integrals will converge for *any* $\varphi(x)$ in K. Thus we may take (10) or (10') to be the defining equations for $\delta^{(k)}(P)$ when $k < \tfrac{1}{2}(p + q - 2)$. If, on the other hand, $k \geqslant \tfrac{1}{2}(p + q - 2)$, we shall define $(\delta_1^{(k)}(P), \varphi)$ and $(\delta_2^{(k)}(P), \varphi)$ as the regularizations of (10) and (10') in a sense to be described below.

Collecting the above considerations we shall say that for $p > 1$ and $q > 1$ the generalized functions $\delta_1^{(k)}(P)$ and $\delta_2^{(k)}(P)$ are defined by

$$(\delta_1^{(k)}(P), \varphi) = \int_0^\infty \left[\left(\frac{\partial}{2s\,\partial s} \right)^k \left\{ s^{q-2} \frac{\psi(r, s)}{2} \right\} \right]_{s=r} r^{p-1}\,dr, \qquad (11)$$

$$(\delta_2^{(k)}(P), \varphi) = (-1)^k \int_0^\infty \left[\left(\frac{\partial}{2r\,\partial r} \right)^k \left\{ r^{p-2} \frac{\psi(r, s)}{2} \right\} \right]_{r=s} s^{q-1}\,ds, \qquad (11')$$

where $\psi(r, s)$ is $r^{1-p}s^{1-q}$ multiplied by the integral of φ over the surface $x_1^2 + \ldots + x_p^2 = r^2$, $x_{p+1}^2 + \ldots + x_{p+q}^2 = s^2$. The integrals converge and coincide for

$$k < \frac{p + q - 2}{2}.$$

If, on the other hand,

$$k \geqslant \frac{p + q - 2}{2},$$

these integrals must be understood in the sense of their regularizations.

The regularization of (11) and (11') we shall understand as follows. Note first that $\psi(r, s)$ as defined by (9) is an infinitely differentiable function of r^2 and s^2 with bounded support. Let us make the formal change of variables $r^2 = u$, $s^2 = v$ in (11), writing

$$\psi(r, s) = \psi_1(u, v),$$

and obtaining

$$(\delta_1^{(k)}(P), \varphi) = \frac{1}{4} \int_0^\infty \left[\frac{\partial^k}{\partial v^k} \{ v^{\frac{1}{2}(q-2)} \psi_1(u, v) \} \right]_{v=u} u^{\frac{1}{2}(p-2)}\,du.$$

Now $\psi_1(u, v)$ is an infinitely differentiable function of u and v with bounded support. But then we may write

$$\left[\frac{\partial^k}{\partial v^k} \{ v^{\frac{1}{2}(q-2)} \psi_1(u, v) \} \right]_{v=u} = u^{\frac{1}{2}(q-2)-k} \Psi(u),$$

where $\Psi(u)$ is again an infinitely differentiable function with bounded support. We thus have

$$(\delta_1^{(k)}(P), \varphi) = \frac{1}{4} \int_0^\infty u^{\frac{1}{2}(p+q)-2-k} \Psi(u)\,du. \qquad (12)$$

Now such integrals have been regularized in Section 3 of Chapter I. The right-hand side of (12) can be written $\frac{1}{4}(u_+^\lambda, \psi(u))$ with

$\lambda = \frac{1}{2}(p + q) - 2 - k$, where u_+^λ is, of course, the functional equal to u^λ for $u > 0$ and to zero for $u < 0$. Its regularization is the generalized function we have also denoted by u_+^λ, which is obtained for all $\lambda \neq -n$, where n is a positive integer, by analytic continuation of u_+^λ from the region Re $\lambda > -1$. At $\lambda = -n$ this generalized analytic function has simple poles, and we have defined u_+^{-n} as the constant term in the Laurent expansion for u_+^λ about $\lambda = -n$.

Thus if the dimension $n = p + q$ of the space is odd, the regularization of (12) is defined as the analytic continuation of

$$f(\lambda) = \frac{1}{4}\int_0^\infty u^\lambda \, \Psi(u) \, du$$

to $\lambda = \frac{1}{2}(p + q) - 2 - k$ (in accordance with the equations given in Section 3 of Chapter I), whereas if n is even it is defined as the constant term in the Laurent series for $f(\lambda)$ about $\lambda = \frac{1}{2}(p + q) - 2 - k$.

The regularization of (11') is defined similarly. Now in general $\delta_1^{(k)}(P)$ and $\delta_2^{(k)}(P)$ may not be the same generalized function. In Section 2.3 we shall show that in a space of odd dimension these always coincide, while if the dimension n is even, for $k \geqslant \frac{1}{2}n - 1$ the difference between them is a generalized function concentrated at the vertex of the $P = 0$ cone. In the same section we shall give a more natural definition of homogeneous functions concentrated on the surface of the $P = 0$ cone.

Note that the definition of these generalized functions implies that in any case

$$\delta_2^{(k)}(P) = (-1)^k \, \delta_1^{(k)}(-P).$$

We have been assuming that $p > 1$ and $q > 1$. The case in which either p or q is equal to unity is a special case, since in this case the transition to bipolar coordinates loses its meaning. Let us, define $\delta_1^{(k)}(P)$ and $\delta_2^{(k)}(P)$ for this special case.

Assume first that $p = q = 1$, or that $P = x^2 - y^2$. Then we may choose our local coordinates to be P and x, and proceeding as in deriving Eq. (10), we obtain

$$(\delta_1^{(k)}(x^2 - y^2), \varphi) = \int_{-\infty}^{+\infty} \left(\frac{\partial}{2y\,\partial y}\right)^k \left[\frac{\varphi(x, y)}{2y}\right] dx,$$

where the integration is over the two lines given by $x^2 - y^2 = 0$. Thus

$(\delta_1^{(k)}(x^2 - y^2), \varphi)$

$$= \int_{-\infty}^{+\infty} \left[\left(\frac{\partial}{2y\,\partial y}\right)^k \left\{\frac{\varphi(x, y)}{2y}\right\}\right]_{y=x} dx + \int_{-\infty}^{+\infty} \left[\left(\frac{\partial}{2y\,\partial y}\right)^k \left\{\frac{\varphi(x, y)}{2y}\right\}\right]_{y=-x} dx.$$

The integrals in this expression are understood in the sense of their regularizations (again there is no problem in using the results of Chapter I Section 3).

In particular, for $k = 0$ we have

$$(\delta_1(x^2 - y^2), \varphi) = \int_{-\infty}^{+\infty} \left\{ \left[\frac{\varphi(x, y)}{2y} \right]_{y=x} + \left[\frac{\varphi(x, y)}{2y} \right]_{y=-x} \right\} dx,$$

or

$$\delta_1(x^2 - y^2) = \frac{1}{2y}\, \delta(x - y) + \frac{1}{2y}\, \delta(x + y). \tag{13}$$

Similarly, if $p = 1$ while $q = n - 1 > 1$, we arrive at the definitions

$$(\delta_1^{(k)}(P), \varphi) = \int_{-\infty}^{+\infty} \left[\left(\frac{\partial}{2s\, \partial s} \right)^k \left\{ s^{n-3}\, \frac{\psi(x_1, s)}{2} \right\} \right]_{s=|x_1|} dx_1, \tag{14}$$

$$(\delta_2^{(k)}(P), \varphi) = (-1)^k \int_0^\infty \left[\left(\frac{\partial}{2x_1\, \partial x_1} \right)^k \left(\frac{\psi(x_1, s)}{2x_1} \right) \right]_{x_1 = s} s^{n-2}\, ds$$

$$+ (-1)^k \int_0^\infty \left[\left(\frac{\partial}{2x_1\, \partial x_1} \right)^k \left(\frac{\psi(x_1, s)}{2x_1} \right) \right]_{x_1 = -s} s^{n-2}\, ds, \tag{14'}$$

where $\psi(x_1, s)$ is s^{1-q} times the integral of φ over the manifold $x_1 = \text{const}$, $x_2^2 + \ldots + x_n^2 = s^2$.

If, on the other hand, $q = 1$ while $p = n - 1 > 1$, we arrive at the definitions

$$(\delta_2^{(k)}(P), \varphi) = (-1)^k \int_{-\infty}^{+\infty} \left[\left(\frac{\partial}{2r\, \partial r} \right)^k \left\{ r^{n-3}\, \frac{\psi(r, x_n)}{2} \right\} \right]_{r=|x_n|} dx_n, \tag{15}$$

$$(\delta_1^{(k)}(P), \varphi) = \int_0^\infty \left[\left(\frac{\partial}{2x_n\, \partial x_n} \right)^k \left(\frac{\psi(r, x_n)}{2x_n} \right) \right]_{x_n = r} r^{n-2}\, dr$$

$$+ \int_0^\infty \left[\left(\frac{\partial}{2x_n\, \partial x_n} \right)^k \left(\frac{\psi(r, x_n)}{2x_n} \right) \right]_{x_n = -r} r^{n-2}\, dr, \tag{15'}$$

where $\psi(r, x_n)$ is r^{1-p} times the integral of φ over the manifold $x_1^2 + \ldots + x_{n-1}^2 = r^2$, $x_n = \text{const}$. All the integrals in these equations should be understood in the sense of their regularizations.[1]

---

[1] Note that it is sometimes possible to avoid averaging φ over a sphere when $p = 1$ or $q = 1$. Equations (14') and (15') can also be written in the following forms. If $p = 1$,

$$(\delta_2^{(k)}(P), \varphi) = (-1)^k \int \left(\frac{\partial}{2x_1\, \partial x_1} \right)^k \left(\frac{\varphi}{2x_1} \right) dx_2 \ldots dx_n,$$

where the integral is over the surface of the $P = 0$ cone. If $q = 1$,

$$(\delta_1^{(k)}(P), \varphi) = \int \left(\frac{\partial}{2x_n\, \partial x_n} \right)^k \left(\frac{\varphi}{2x_n} \right) dx_1 \ldots dx_{n-1},$$

where again the integral is over the $P = 0$ cone.

2.2. The Generalized Function P_+^λ

Again consider

$$P(x) = x_1^2 + \dots + x_p^2 - x_{p+1}^2 - \dots - x_{p+q}^2 \qquad (1)$$

with $p + q = n$. We define the generalized function P_+^λ, where λ is a complex number, by

$$(P_+^\lambda, \varphi) = \int_{P>0} P^\lambda(x)\, \varphi(x)\, dx, \qquad (2)$$

where $x = (x_1, \dots, x_n)$ and $dx = dx_1 \dots dx_n$. For Re $\lambda \geqslant 0$, this integral converges and is an analytic function of λ. Analytic continuation to Re $\lambda < 0$ can be used to extend the definition of (P_+^λ, φ).

Let us find the singularities of (P_+^λ, φ). For this purpose we transform to bipolar coordinates

$$x_1 = r\omega_1, \dots, x_p = r\omega_p,\, x_{p+1} = s\omega_{p+1}, \dots, x_{p+q} = s\omega_{p+q} \qquad (3)$$

in Eq. (2), where[2]

$$r = \sqrt{x_1^2 + \dots + x_p^2}, \qquad s = \sqrt{x_{p+1}^2 + \dots + x_{p+q}^2} \,. \qquad (4)$$

Then Eq. (2) becomes

$$(P_+^\lambda, \varphi) = \int_{P>0} (r^2 - s^2)^\lambda\, \varphi r^{p-1}\, s^{q-1}\, dr\, ds\, d\Omega^{(p)}\, d\Omega^{(q)}. \qquad (5)$$

Now proceeding as in Section 2.1 to write

$$\psi(r, s) = \int \varphi\, d\Omega^{(p)}\, d\Omega^{(q)}, \qquad (6)$$

we obtain

$$(P_+^\lambda, \varphi) = \int_0^\infty \int_0^r (r^2 - s^2)^\lambda\, \psi(r, s)\, r^{p-1}\, s^{q-1}\, ds\, dr. \qquad (7)$$

Since $\varphi(x)$ is in K, $\psi(r, s)$ as defined in Eq. (6) is an infinitely differentiable function of r^2 and s^2 with bounded support. We now make the change of variables $u = r^2$, $v = s^2$ in the integrand of (7), writing

$$\psi(r, s) = \psi_1(u, v), \qquad (8)$$

---

[2] For simplicity we shall assume throughout what follows that $p > 1$ and $q > 1$. The results we shall obtain, however, are valid also for the special case $p = 1$ or $q = 1$.

to obtain

$$(P_+^\lambda, \varphi) = \frac{1}{4} \int_0^\infty \int_0^u (u - v)^\lambda \, \psi_1(u, v) \, u^{\frac{1}{2}(p-2)} \, v^{\frac{1}{2}(q-2)} \, dv \, du. \tag{9}$$

Finally, we write $v = ut$, which transforms (9) to the form

$$(P_+^\lambda, \varphi) = \frac{1}{4} \int_0^\infty u^{\lambda + \frac{1}{2}(p+q) - 1} \, du \int_0^1 (1 - t)^\lambda \, t^{\frac{1}{2}(q-2)} \, \psi_1(u, tu) \, dt. \tag{10}$$

This equation shows that (P_+^λ, φ) has two sets of poles. The first of these consists of the poles of

$$\Phi(\lambda, u) = \frac{1}{4} \int_0^1 (1 - t)^\lambda \, t^{\frac{1}{2}(q-2)} \, \psi_1(u, tu) \, dt. \tag{11}$$

Like the function (x_+^λ, φ) studied in Chapter I, Section 3, this function is regular for all λ except

$$\lambda = -1, -2, ..., -k, ...,$$

where it has simple poles. At these poles we have

$$\operatorname*{res}_{\lambda = -k} \Phi(\lambda, u) = \frac{(-1)^{k-1}}{4(k-1)!} \left[\frac{\partial^{k-1}}{\partial t^{k-1}} \{ t^{\frac{1}{2}(q-2)} \, \psi_1(u, tu) \} \right]_{t=1}. \tag{12}$$

Thus $\operatorname{res}_{\lambda = -k} \Phi(\lambda, u)$ is a functional concentrated on the surface of the $P = 0$ cone.

On the other hand, even at regular points of $\Phi(\lambda, u)$ the integral

$$(P_+^\lambda, \varphi) = \int_0^\infty u^{\lambda + \frac{1}{2}(p+q) - 1} \, \Phi(\lambda, u) \, du \tag{13}$$

may also have poles. This occurs at

$$\lambda = -\frac{n}{2}, -\frac{n}{2} - 1, ..., -\frac{n}{2} - k, ...,$$

where $n = p + q$ is the dimension of the underlying space. At these points

$$\operatorname*{res}_{\lambda = -\frac{1}{2}n - k} (P_+^\lambda, \varphi) = \frac{1}{k!} \left[\frac{\partial^k}{\partial u^k} \Phi \left(-\frac{n}{2} - k, u \right) \right]_{u=0}. \tag{14}$$

Thus the residue of (P_+^λ, φ) at $\lambda = -\frac{1}{2}u - k$ is a functional concentrated on the vertex of the cone.

Consequently, (P^λ_+, φ) has two sets of singularities, namely

$$\lambda = -1, -2, ..., -k, ... \tag{15}$$

and

$$\lambda = -\frac{n}{2}, -\frac{n}{2} - 1, ..., -\frac{n}{2} - k, ... \tag{15'}$$

The residue of (P^λ_+, φ) at a singular λ is a functional concentrated on the surface of the $P = 0$ cone if λ is a point in the first set, and on the vertex of this cone if λ is in the second set. When λ is in both sets simultaneously, the picture, of course, becomes rather more complicated.

Let us now study each case separately.

Case 1. The singular point $\lambda = -k$ belongs to the first set, but not to the second. This is always the case when the dimension $n = p + q$ is odd, but is also true if n is even and $\lambda > -\frac{1}{2} n$.

Let us write (11) in the neighborhood of $\lambda = -k$ in the form

$$\Phi(\lambda, u) = \frac{\Phi_0(u)}{\lambda + k} + \Phi_1(\lambda, u), \tag{16}$$

where $\Phi_0(u) = \operatorname{res}_{\lambda = -k} \Phi(\lambda, u)$, and $\Phi_1(\lambda, u)$ is regular at $\lambda = -k$. Inserting this into (13), we obtain

$$(P^\lambda_+, \varphi) = \frac{1}{\lambda + k} \int_0^\infty u^{\lambda + \frac{1}{2}(p+q)-1} \Phi_0(u) \, du + \int_0^\infty u^{\lambda + \frac{1}{2}(p+q)-1} \Phi_1(\lambda, u) \, du. \tag{17}$$

Under the assumptions we have made concerning λ, the integrals in (17) are regular functions of λ at $\lambda = -k$. Therefore (P^λ_+, φ) has a simple pole at such a point, and

$$\operatorname*{res}_{\lambda = -k} (P^\lambda_+, \varphi) = \int_0^\infty u^{-k + \frac{1}{2}(p+q)-1} \Phi_0(u) \, du,$$

where for $k \geqslant \frac{1}{2}(p + q)$ the integral is understood in the sense of its regularization. Inserting Eq. (12) for $\Phi_0(u)$, we arrive at

$$\operatorname*{res}_{\lambda = -k} (P^\lambda_+, \varphi)$$
$$= \frac{(-1)^{k-1}}{4(k-1)!} \int_0^\infty \left[\frac{\partial^{k-1}}{\partial t^{k-1}} \{ t^{\frac{1}{2}(q-2)} \psi_1(u, tu) \} \right]_{t=1} u^{-k+\frac{1}{2}(p+q)-1} \, du. \tag{18}$$

Consequently the residue of P^λ_+ at $\lambda = -k$ is a generalized function concentrated on the $P = 0$ cone.

We wish to establish the relation between this function and $\delta_1^{(k)}(P)$ of Section 2.1.

Note that if we write $tu = v$, we obtain

$$\left[\frac{\partial^{k-1}}{\partial t^{k-1}}\{t^{\frac{1}{2}(q-2)}\,\psi_1(u, tu)\}\right]_{t=1} = \left[\frac{\partial^{k-1}}{\partial v^{k-1}}\{v^{\frac{1}{2}(q-2)}\,\psi_1(u, v)\}\right]_{v=u} u^{-\frac{1}{2}q+k}.$$

so that we may rewrite (18) in the form

$$\operatorname*{res}_{\lambda=-k}(P_+^\lambda, \varphi) = \frac{(-1)^{k-1}}{4(k-1)!}\int_0^\infty \left[\frac{\partial^{k-1}}{\partial v^{k-1}}\{v^{\frac{1}{2}(q-2)}\,\psi_1(u, v)\}\right]_{v=u} u^{\frac{1}{2}p-2}\,du,$$

where the integral is to be understood in the sense of its regularization for $k \geqslant \frac{1}{2}n$. According to Section 2.1, on the other hand,

$$(\delta_1^{(k-1)}(P), \varphi) = \frac{1}{4}\int_0^\infty \left[\frac{\partial^{k-1}}{\partial v^{k-1}}\{v^{\frac{1}{2}(q-2)}\,\psi_1(u, v)\}\right]_{v=u} u^{\frac{1}{2}(p-2)}\,du,$$

with the same interpretation of the integral. But when $n = p + q$ is odd, the regularization is defined by analytic continuation. Hence

$$\operatorname*{res}_{\lambda=-k}(P_+^\lambda, \varphi) = \frac{(-1)^{k-1}}{(k-1)!}(\delta_1^{(k-1)}(P), \varphi).$$

Summarizing, we have the following. For *odd n* and for *even n if* $k < \frac{1}{2}n$, the generalized function P_+^λ has simple poles at $\lambda = -k$ for positive integral values of k, where the residues a are

$$\operatorname*{res}_{\lambda=-k} P_+^\lambda = \frac{(-1)^{k-1}}{(k-1)!}\delta_1^{(k-1)}(P), \tag{19}$$

This result is not surprising. Indeed, we establish in Chapter I that

$$\operatorname*{res}_{\lambda=-k} x_+^\lambda = \frac{(-1)^{k-1}}{(k-1)!}\delta^{(k-1)}(x).$$

Equation (19) implies also that for a space of odd dimension and for a space of even dimension if $k < \frac{1}{2}n$, the generalized function $\delta_1^{(k-1)}(P)$ is uniquely determined by the quadratic form P.

In Section 2.1 it was pointed out that $\delta_1^{(k)}(P)$ can be defined by

$$(\delta_1^{(k)}(P), \varphi) = (-1)^k \int_{P=0} \omega_k(\varphi),$$

and that the integral here converges for $k < \frac{1}{2}(n - 2)$, whereas if $k \geqslant \frac{1}{2}(n - 2)$, it is to be understood in the sense of its regularization. We may now write

$$\operatorname*{res}_{\lambda=-k}(P_+^\lambda, \varphi) = \frac{1}{(k-1)!}\int_{P=0} \omega_{k-1}(\varphi).$$

In particular, for $k = 1$ we have

$$\operatorname*{res}_{\lambda=-1} (P_+^\lambda, \varphi) = \int \frac{\varphi(x)\, d\sigma}{|\operatorname{grad} P|},$$

where $d\sigma$ is the element of area on the $P = 0$ cone.

Case 2. The singular point λ is in the second set, but not in the first. This occurs when $\lambda = -\frac{1}{2}n - k$, for k a nonnegative integer and dimension n odd.

In this case $\Phi(\lambda, u)$ is regular at the singular point, so that (P_+^λ, φ) as defined by Eq. (13) will have a simple pole with residue given by

$$\operatorname*{res}_{\lambda=-\frac{1}{2}n-k} (P_+^\lambda, \varphi) = \operatorname*{res}_{\lambda=-\frac{1}{2}n-k} \int_0^\infty u^{\lambda+\frac{1}{2}n-1} \Phi\left(-\frac{n}{2} - k, u\right) du,$$

or

$$\operatorname*{res}_{\lambda=-\frac{1}{2}n-k} (P_+^\lambda, \varphi) = \frac{1}{k!} \left[\frac{\partial^k \Phi\left(-\dfrac{n}{2} - k, u\right)}{\partial u^k} \right]_{u=0} \tag{20}$$

Thus the residue of (P_+^λ, φ) at $\lambda = -\frac{1}{2}n - k$ is a functional concentrated at the origin.

Let us express this functional directly in terms of the derivatives of $\varphi(x)$ at the origin.

For $\lambda = -\frac{1}{2}n$, Eq. (20) gives

$$\operatorname*{res}_{\lambda=-\frac{1}{2}n} (P_+^\lambda, \varphi) = \Phi\left(-\frac{n}{2}, 0\right).$$

We insert Eq. (11) for Φ to obtain

$$\operatorname*{res}_{\lambda=-\frac{1}{2}n} (P_+^\lambda, \varphi) = \frac{1}{4} \int_0^1 (1-t)^{-\frac{1}{2}n} \, t^{\frac{1}{2}(q-2)} \, dt \, \psi_1(0,0).$$

But

$$\int_0^1 (1-t)^{-\frac{1}{2}n} \, t^{\frac{1}{2}(q-2)} \, dt = \frac{\Gamma\left(\dfrac{q}{2}\right) \Gamma\left(-\dfrac{n}{2} + 1\right)}{\Gamma\left(-\dfrac{p}{2} + 1\right)}.$$

If p is even, $\Gamma(-\frac{1}{2}p + 1) = \infty$, so that

$$\operatorname*{res}_{\lambda=-\frac{1}{2}n} (P_+^\lambda, \varphi) = 0. \tag{21}$$

Now assume that p is odd and q is even. From (8) and (6) we have

$$\psi_1(0,0) = \psi(0,0) = \int \varphi(0)\, d\Omega^{(p)}\, d\Omega^{(q)},$$

whence

$$\psi_1(0, 0) = \Omega_p \Omega_q \, \varphi(0),$$

where Ω_p and Ω_q are the hypersurface areas of the unit spheres in R_p and R_q, respectively. We thus have

$$\operatorname*{res}_{\lambda=-\frac{1}{2}n} (P_+^\lambda, \varphi) = \frac{\Gamma\left(\frac{q}{2}\right) \Gamma\left(1 - \frac{n}{2}\right)}{4\Gamma\left(1 - \frac{p}{2}\right)} \Omega_p \Omega_q \varphi(0),$$

or

$$\operatorname*{res}_{\lambda=-\frac{1}{2}n} P_+^\lambda = \frac{\Gamma\left(\frac{q}{2}\right) \Gamma\left(1 - \frac{n}{2}\right)}{4\Gamma\left(1 - \frac{p}{2}\right)} \Omega_p \Omega_q \delta(x). \tag{22}$$

Now the area Ω_s of the unit sphere in R_s is given by

$$\Omega_s = \frac{2\pi^{\frac{1}{2}s}}{\Gamma\left(\frac{s}{2}\right)}.$$

If we insert the expressions for Ω_p and Ω_q into (22) and perform some elementary manipulations involving the properties of the gamma function, we arrive at

$$\operatorname*{res}_{\lambda=-\frac{1}{2}n} P_+^\lambda = \frac{(-1)^{\frac{1}{2}q} \, \pi^{\frac{1}{2}n}}{\Gamma\left(\frac{n}{2}\right)} \delta(x). \tag{23}$$

To find the residues of P_+^λ at $\lambda = -\frac{1}{2}n - k$, where k is a positive integer, consider the homogeneous differential operator

$$L = P\left(\frac{\partial}{\partial x_1}, \ldots, \frac{\partial}{\partial x_n}\right),$$

by which we mean

$$L = \frac{\partial^2}{\partial x_1^2} + \ldots + \frac{\partial^2}{\partial x_p^2} - \frac{\partial^2}{\partial x_{p+1}^2} - \ldots - \frac{\partial^2}{\partial x_{p+q}^2}. \tag{24}$$

A simple calculation shows that

$$LP^{\lambda+1} = 2(\lambda + 1)(2\lambda + n) P^\lambda, \tag{25}$$

where $n = p + q$.

On the other hand, for any $\varphi(x)$ in K we may write[3]

$$\int_{P>0} [\varphi(LP^{\lambda+1}) - P^{\lambda+1}(L\varphi)]\, dx = 0 \, . \tag{26}$$

We now use Eq. (25) for $LP^{\lambda+1}$ where it appears in the above integral, obtaining

$$\int_{P>0} P^\lambda \varphi\, dx = \frac{1}{2^2(\lambda+1)\left(\lambda+\dfrac{n}{2}\right)} \int_{P>0} P^{\lambda+1}(L\varphi)\, dx,$$

or

$$(P_+^\lambda, \varphi) = \frac{1}{2^2(\lambda+1)\left(\lambda+\dfrac{n}{2}\right)} (P_+^{\lambda+1}, L\varphi) \, . \tag{27}$$

Then k-fold iteration of (27) leads to

$$(P_+^\lambda, \varphi) = \frac{1}{2^{2k}(\lambda+1)\dots(\lambda+k)\left(\lambda+\dfrac{n}{2}\right)\dots\left(\lambda+\dfrac{n}{2}+k-1\right)} (P_+^{\lambda+k}, L^k\varphi). \tag{28}$$

Consequently

$$\operatorname*{res}_{\lambda=-\frac{1}{2}n-k} (P_+^\lambda, \varphi)$$

$$= \frac{1}{2^{2k}(\lambda+1)\dots(\lambda+k)\left(\lambda+\dfrac{n}{2}\right)\dots\left(\lambda+\dfrac{n}{2}+k-1\right)}\Bigg|_{\lambda=-\frac{1}{2}n-k}$$

$$\times \operatorname*{res}_{\lambda=-\frac{1}{2}n-k} (P_+^{\lambda+k}, L^k\varphi). \tag{29}$$

But

$$\operatorname*{res}_{\lambda=-\frac{1}{2}n-k} (P^{\lambda+k}, L^k\varphi) = \operatorname*{res}_{\lambda=-\frac{1}{2}n} (P^\lambda, L^k\varphi).$$

Therefore if p is even this residue, from (21), vanishes. If p is odd, on the other hand, (23) gives

$$\operatorname*{res}_{\lambda=-\frac{1}{2}n-k} (P^{\lambda+k}, L^k\varphi) = \frac{(-1)^{\frac{1}{2}q}\, \pi^{\frac{1}{2}n}}{\Gamma\left(\dfrac{n}{2}\right)} (\delta(x), L^k\varphi).$$

---

[3] This follows from the fact that for Re $\lambda > 0$ this expression can be reduced, by Green's theorem, to the integral over the $P = 0$ surface. But since $P^{\lambda+1}$ and all its partial derivatives vanish on this surface, the integral must vanish. Then by analytic continuation Eq. (26) remains valid for Re $\lambda < 0$.

Inserting this into (29) and using the fact that

$$(\delta(x), L^k\varphi) = (L^k\delta(x), \varphi),$$

we arrive, after some elementary operations, at

$$\operatorname*{res}_{\lambda=-\frac{1}{2}n-k} (P_+^\lambda, \varphi) = \frac{(-1)^{\frac{1}{2}q}\,\pi^{\frac{1}{2}n}}{2^{2k}k!\,\Gamma\left(\frac{n}{2}+k\right)} (L^k\delta(x), \varphi).$$

Summing up we then have the following. If the dimension of the underlying space is *odd*, then for p *odd* and q *even* the generalized function P_+^λ has simple poles at $\lambda = -\frac{1}{2}n - k$, for k a nonnegative integer, where the residues are given by

$$\operatorname*{res}_{\lambda=-\frac{1}{2}n-k} P_+^\lambda = \frac{(-1)^{\frac{1}{2}q}\,\pi^{\frac{1}{2}n}}{2^{2k}k!\,\Gamma\left(\frac{n}{2}+k\right)}$$

$$\times \left(\frac{\partial^2}{\partial x_1^2} + \dots + \frac{\partial^2}{\partial x_p^2} - \frac{\partial^2}{\partial x_{p+1}^2} - \dots - \frac{\partial^2}{\partial x_n^2}\right)^k \delta(x). \tag{30}$$

If, on the other hand, p is *even* and q is *odd*, P_+^λ is regular at these points.

Case 3. The singular point λ is in both sets. This can occur for even n for all

$$\lambda = -\frac{n}{2} - k \qquad (k = 0, 1, \dots).$$

As in Case 1, we first write (P_+^λ, φ) in the form

$$(P_+^\lambda, \varphi) = \frac{1}{\lambda + \frac{n}{2} + k} \int_0^\infty u^{\lambda+\frac{1}{2}n-1}\,\Phi_0(u)\,du$$

$$+ \int_0^\infty u^{\lambda+\frac{1}{2}(p+q)-1}\,\Phi_1(\lambda, u)\,du, \tag{31}$$

where $\Phi_0(u) = \operatorname{res}_{\lambda=-\frac{1}{2}n-k}\Phi(\lambda, u)$, and $\Phi(\lambda, u)$ is a function which is regular at $\lambda = -\frac{1}{2}n - k$. By assumption, each of the integrals in (31) may have a simple pole at this value of λ.

Therefore (P_+^λ, φ) may have a pole of order two at $\lambda = -\frac{1}{2}n - k$.

In the neighborhood of such a point we may expand P_+^λ in the Laurent series

$$P_+^\lambda = \frac{c_{-2}^{(k)}}{\left(\lambda + \frac{n}{2} + k\right)^2} + \frac{c_{-1}^{(k)}}{\lambda + \frac{n}{2} + k} + \dots.$$

Let us find $c_{-2}^{(k)}$ and $c_{-1}^{(k)}$.

First, according to (31)

$$(c_{-2}^{(k)}, \varphi) = \operatorname*{res}_{\lambda=-\frac{1}{2}n-k} \int_0^\infty u^{\lambda+\frac{1}{2}n-1} \Phi_0(u)\, du = \frac{1}{k!} \Phi_0^{(n)}(0). \tag{32}$$

Thus, the generalized function $c_{-2}^{(k)}$ is concentrated on the vertex of the $P = 0$ cone.

Let us express $c_{-2}^{(k)}$ in terms of the derivatives of $\delta(x)$. When we set $k = 0$, Eq. (32) becomes

$$c_{-2}^{(0)} = \Phi_0(0).$$

Now by definition

$$\Phi_0(u) = \frac{1}{4} \operatorname*{res}_{\lambda=-\frac{1}{2}n} \int_0^1 (1-t)^\lambda t^{\frac{1}{2}(q-2)} \psi_1(u, tu)\, dt,$$

from which we obtain

$$\Phi_0(0) = \frac{1}{4} \psi_1(0,0) \operatorname*{res}_{\lambda=-\frac{1}{2}n} \int_0^1 (1-t)^\lambda t^{\frac{1}{2}(q-2)}\, dt$$

$$= \psi_1(0,0) \operatorname*{res}_{\lambda=-\frac{1}{2}n} \frac{\Gamma\left(\frac{q}{2}\right)\Gamma(\lambda+1)}{4\Gamma\left(\lambda+\frac{q}{2}+1\right)}.$$

According to the known relation

$$\Gamma(1-x)\,\Gamma(x) = \frac{\pi}{\sin \pi x}$$

for the gamma function, we may write

$$\frac{\Gamma\left(\frac{q}{2}\right)\Gamma(\lambda+1)}{\Gamma\left(\lambda+\frac{q}{2}+1\right)} = \frac{\sin\pi\left(-\frac{q}{2}-\lambda\right)}{\sin\pi(-\lambda)} \frac{\Gamma\left(\frac{q}{2}\right)\Gamma\left(-\lambda-\frac{q}{2}\right)}{\Gamma(-\lambda)}.$$

Now it is easily seen, on the other hand, that

$$\lim_{\lambda=-\frac{1}{2}n} \frac{\lambda+\frac{n}{2}}{\sin(-\pi\lambda)} = \frac{(-1)^{\frac{1}{2}n+1}}{\pi},$$

so that for the residue we have

$$
\operatorname*{res}_{\lambda=-\frac{1}{2}n} \frac{\Gamma\left(\frac{q}{2}\right)\Gamma(\lambda+1)}{\Gamma\left(\lambda+\frac{q}{2}+1\right)}
$$
$$
= (-1)^{\frac{1}{2}n+1}\left[\sin\pi\left(-\frac{q}{2}-\lambda\right)\frac{\Gamma\left(\frac{q}{2}\right)\Gamma\left(-\lambda-\frac{q}{2}\right)}{\pi\Gamma(-\lambda)}\right]_{\lambda=-\frac{1}{2}n}.
$$

Now putting $\lambda = -\frac{1}{2}n$, we have

$$
\operatorname*{res}_{\lambda=-\frac{1}{2}n} \frac{\Gamma\left(\frac{q}{2}\right)\Gamma(\lambda+1)}{\Gamma\left(\lambda+\frac{q}{2}+1\right)} = (-1)^{\frac{1}{2}n+1}\sin\frac{\pi p}{2}\frac{\Gamma\left(\frac{q}{2}\right)\Gamma\left(\frac{p}{2}\right)}{\pi\Gamma\left(\frac{n}{2}\right)}. \tag{33}
$$

Now recall that

$$
\psi_1(0,0) = \Omega_p\Omega_q\,\varphi(0),
$$

so that we may write

$$
(c^{(0)}_{-2}, \varphi) = (-1)^{\frac{1}{2}n+1}\sin\frac{\pi p}{2}\frac{\Gamma\left(\frac{q}{2}\right)\Gamma\left(\frac{p}{2}\right)}{4\pi\Gamma\left(\frac{n}{2}\right)}\Omega_p\Omega_q\varphi(0). \tag{34}
$$

If p is even, $c^{(0)}_{-2} = 0$, since then $\sin(\pi p/2) = 0$. Thus if p and q are even, P^λ_+ has just a simple pole at $\lambda = -\frac{1}{2}n$. If, on the other hand, p and q are odd, Eq. (34) implies, using

$$
\sin\frac{\pi p}{2} = (-1)^{\frac{1}{2}(p-1)} \qquad \text{and} \qquad \Omega_s = \frac{2\pi^{\frac{1}{2}s}}{\Gamma\left(\frac{s}{2}\right)}
$$

that

$$
(c^{(0)}_{-2}, \varphi) = (-1)^{\frac{1}{2}(q-1)}\frac{\pi^{\frac{1}{2}n-1}}{\Gamma\left(\frac{n}{2}\right)}\varphi(0),
$$

or in other words, for p and q odd we have

$$
c^{(0)}_{-2} = (-1)^{\frac{1}{2}(q-1)}\frac{\pi^{\frac{1}{2}n-1}}{\Gamma\left(\frac{n}{2}\right)}\delta(x). \tag{35}
$$

We now use (28) and go step by step through the same considerations as in Case 2 to find for even p and q that $c^{(k)}_2 = 0$, or that at $\lambda = -\frac{1}{2}n - k$, the function P_+^λ has a simple pole. If, on the other hand, p and q are odd, then

$$c_{-2}^{(k)} = (-1)^{\frac{1}{2}(q-1)} \frac{\pi^{\frac{1}{2}n-1}}{2^{2k}k!\,\Gamma\left(\frac{n}{2}+k\right)}$$

$$\times \left(\frac{\partial^2}{\partial x_1^2} + \cdots + \frac{\partial^2}{\partial x_p^2} - \frac{\partial^2}{\partial x_{p+1}^2} - \cdots - \frac{\partial^2}{\partial x_{p+q}^2}\right)^k \delta(x). \tag{36}$$

Let us now determine $c_{-1}^{(k)}$. Equation (31) gives

$$(c_{-1}^{(k)}, \varphi) = \int_0^\infty u^{-k-1}\,\Phi_0(u)\,du$$

$$+ \operatorname*{res}_{\lambda=-\frac{1}{2}n-k} \int_0^\infty u^{\lambda+\frac{1}{2}n-1}\,\Phi_1\left(-\frac{n}{2}-k, u\right) du. \tag{37}$$

The first of these integrals is to be understood as the constant term of the Laurent expansion of

$$f(\lambda) = \int_0^\infty u^{\lambda+\frac{1}{2}n-1}\,\Phi_0(u)\,du$$

at $\lambda = -\frac{1}{2}n - k$. This integral defines a functional concentrated on the $P = 0$ cone, which we now proceed to determine. Since

$$\Phi_0(u) = \operatorname*{res}_{\lambda=-k}\,\Phi(\lambda, u),$$

we may conclude from (12) that

$$\Phi_0(u) = \frac{(-1)^{\frac{1}{2}n+k-1}}{4\left(\frac{n}{2}+k-1\right)!} \left[\frac{\partial^{\frac{1}{2}n+k-1}}{\partial t^{\frac{1}{2}n+k-1}}\{t^{\frac{1}{2}(q-2)}\,\psi_1(u, tu)\}\right]_{t=1}$$

$$= \frac{(-1)^{\frac{1}{2}n+k-1}}{4\left(\frac{n}{2}+k-1\right)!} \left[\frac{\partial^{\frac{1}{2}n+k-1}}{\partial v^{\frac{1}{2}n+k-1}}\{v^{\frac{1}{2}(q-2)}\,\psi_1(u, v)\}\right]_{v=u} u^{\frac{1}{2}n+k-\frac{1}{2}q}.$$

From this we obtain

$$\int_0^\infty u^{-k-1}\,\Phi_0(u)\,du$$

$$= \frac{(-1)^{\frac{1}{2}n+k-1}}{4\left(\frac{n}{2}+k-1\right)!} \int_0^\infty \left[\frac{\partial^{\frac{1}{2}n+k-1}}{\partial v^{\frac{1}{2}n+k-1}}\{v^{\frac{1}{2}(q-2)}\,\psi_1(u, v)\}\right]_{v=u} u^{\frac{1}{2}(p-2)}\,du.$$

On the other hand, according to the definition in Section 2.1,

$$(\delta_1^{(\frac{1}{2}n+k-1)}(P), \varphi) = \frac{1}{4} \int_0^\infty \left[\frac{\partial^{\frac{1}{2}n+k-1}}{\partial v^{\frac{1}{2}n+k-1}} \{ v^{\frac{1}{2}(q-2)} \psi_1(u_1, v) \} \right]_{v=u} u^{\frac{1}{2}(p-2)} \, du.$$

In that section we defined the regularization of the integral in exactly the same way as we have done for $\int_0^\infty u^{-k-1} \Phi_0(u) \, du$. Therefore

$$\int_0^\infty u^{-k-1} \Phi_0(u) \, du = \frac{(-1)^{\frac{1}{2}n+k-1}}{\left(\frac{n}{2}+k-1\right)!} (\delta_1^{(\frac{1}{2}n+k-1)}(P), \varphi)). \tag{38}$$

From this, we may deduce, in particular, that in an even-dimensional space if $k \geqslant \frac{1}{2}n$, the $\delta_1^{(k-1)}(P)$ function is uniquely determined by the quadratic form P.

Let us now turn to the second term in (37). We have

$$\operatorname*{res}_{\lambda=-\frac{1}{2}n-k} \int_0^\infty u^{\lambda+\frac{1}{2}n-1} \Phi_1\left(-\frac{n}{2}-k, u\right) du$$

$$= \frac{1}{k!} \left[\frac{\partial^k \Phi_1\left(-\frac{n}{2}-k, u\right)}{\partial u^k} \right]_{u=0}. \tag{39}$$

The functional defined by this expression is concentrated, therefore, on the vertex of the $P = 0$ cone.

Thus

$$c_{-1}^{(k)} = \frac{(-1)^{\frac{1}{2}n+k-1}}{\left(\frac{n}{2}+k-1\right)!} \delta_1^{(\frac{1}{2}n+k-1)}(P) + \alpha^{(k)}, \tag{40}$$

where $\alpha^{(k)}$ is the generalized function concentrated at the vertex of $P = 0$ which is defined by [see (39)]

$$(\alpha^{(k)}, \varphi) = \frac{1}{k!} \left[\frac{\partial^k \Phi_1\left(-\frac{n}{2}-k, u\right)}{\partial u^k} \right]_{u=0}. \tag{41}$$

Let us obtain an explicit expression for $\alpha^{(k)}$ in terms of the derivatives of $\delta(x)$. Consider first the case $k = 0$. For this case (41) becomes

$$(\alpha^{(0)}, \varphi) = \Phi_1\left(-\frac{n}{2}, 0\right).$$

But $\Phi_1(-\tfrac{1}{2}n, 0)$ is the constant term of the Laurent expansion of $\Phi(\lambda, 0)$ at $\lambda = -\tfrac{1}{2}n$. For this function we have

$$\Phi(\lambda, 0) = \tfrac{1}{4} \int_0^1 (1-t)^\lambda t^{\frac{1}{2}(q-2)}\, dt\psi_1(0,0) = \frac{\Gamma(\lambda+1)\,\Gamma\left(\frac{q}{2}\right)}{4\Gamma\left(\lambda+\frac{q}{2}+1\right)} \Omega_p\Omega_q\varphi(0).$$

Again using the gamma-function relation $\Gamma(1-x)\,\Gamma(x) = \pi/\sin \pi x$ and the explicit expressions for Ω_p and Ω_q, we may rewrite this expression in the form

$$\Phi(\lambda, 0) = \frac{\sin \pi \left(\lambda + \frac{q}{2}\right)}{\sin \pi\lambda} \frac{\Gamma\left(-\lambda - \frac{q}{2}\right) \pi^{\frac{1}{2}n}}{\Gamma\left(\frac{p}{2}\right)\Gamma(-\lambda)} \varphi(0).$$

If p and q are even, this expression represents a regular function at $\lambda = -\tfrac{1}{2}n$. Therefore, $\Phi_1(-\tfrac{1}{2}n, 0) = \Phi(-\tfrac{1}{2}n, 0)$, which means that

$$(\alpha^{(0)}, \varphi) = (-1)^{\frac{1}{2}q} \frac{\pi^{\frac{1}{2}n}}{\Gamma\left(\frac{n}{2}\right)} \varphi(0).$$

If, on the other hand, p and q are odd, $\Phi(\lambda, 0)$ has a pole at $\lambda = -\tfrac{1}{2}n$. Then

$$(\alpha^{(0)}, \varphi) = \Phi_1\left(-\frac{n}{2}, 0\right) = \kappa\varphi(0),$$

where κ is the constant term in the Laurent series for

$$\frac{\sin \pi \left(\lambda + \frac{q}{2}\right)}{\sin \pi\lambda} \frac{\Gamma\left(-\lambda - \frac{q}{2}\right)}{\Gamma\left(\frac{p}{2}\right)\Gamma(-\lambda)} \pi^{\frac{1}{2}n}$$

at $\lambda = -\tfrac{1}{2}n$. Elementary calculations which we shall not go into here lead to

$$\kappa = (-1)^{\frac{1}{2}(q+1)} \frac{\Gamma'\left(\frac{p}{2}\right)\Gamma\left(\frac{n}{2}\right) - \Gamma\left(\frac{p}{2}\right)\Gamma'\left(\frac{n}{2}\right)}{\Gamma\left(\frac{p}{2}\right)\Gamma^2\left(\frac{n}{2}\right)} \pi^{\frac{1}{2}n-1},$$

or

$$\kappa = \frac{(-1)^{\frac{1}{2}(q+1)} \pi^{\frac{1}{2}n-1}}{\Gamma\left(\frac{n}{2}\right)} \left[\psi\left(\frac{p}{2}\right) - \psi\left(\frac{n}{2}\right)\right], \tag{42}$$

where $\psi(x)$ is defined as[4]

$$\psi(x) = \frac{\Gamma'(x)}{\Gamma(x)}.$$

Inserting the expressions we have obtained for $\alpha^{(0)}$ into (40), we find that for $k = 0$

$$c_{-1}^{(0)} = \frac{1}{\Gamma\left(\frac{n}{2}\right)} [(-1)^{\frac{1}{2}n-1}\delta_1^{(\frac{1}{2}n-1)}(P) + \theta\pi^{\frac{1}{2}n}\delta(x)], \tag{43}$$

where

$$\theta = (-1)^{\frac{1}{2}q} \qquad\qquad \text{if } p \text{ and } q \text{ are even, and}$$

$$\theta = (-1)^{\frac{1}{2}(q+1)}\frac{1}{\pi}\left[\psi\left(\frac{p}{2}\right) - \psi\left(\frac{n}{2}\right)\right] \qquad \text{if } p \text{ and } q \text{ are odd.}$$

Finally, in order to obtain $c_{-1}^{(k)}$ for arbitrary k, we again use the "lowering" formula (28). This gives

$$c_{-1}^{(k)} = \frac{1}{\Gamma\left(\frac{n}{2}+k\right)}\left[(-1)^{\frac{1}{2}n+k-1}\delta_1^{(\frac{1}{2}n+k-1)}(P)\right.$$

$$\left. + \theta_{p,q}\frac{\pi^{\frac{1}{2}n}}{2^{2k}k!}P^k\left(\frac{\partial}{\partial x_1}, ..., \frac{\partial}{\partial x_n}\right)\delta(x)\right]. \tag{44}$$

Here the numerical coefficient $\theta_{p,q}$ is equal to $(-1)^{\frac{1}{2}q}$ when p and q are even. Expressions for the $\alpha^{(k)}$ can also be obtained in the following way. Since $c_{-1}^{(k)} = \text{res}_{\lambda=\frac{1}{2}n-k}P_+^\lambda$ is a homogeneous function of degree $-n-2k$, so is $\alpha^{(k)}$. But $\alpha^{(k)}$ is concentrated at the origin, and must therefore be of the form[5]

$$\alpha^{(k)} = Q_{2k}\left(\frac{\partial}{\partial x_1}, ..., \frac{\partial}{\partial x_n}\right)\delta(x),$$

where Q_{2k} is a homogeneous polynomial of degree $2k$.

[4] For integral and half-integral values of the argument, $\psi(x)$ is given by

$$\psi(k) = -C + 1 + \frac{1}{2} + ... + \frac{1}{k-1},$$

$$\psi\left(k + \frac{1}{2}\right) = -C - 2\ln 2 + 2\left(1 + \frac{1}{3} + ... + \frac{1}{2k-1}\right),$$

where C is Euler's constant (see Ryshik and Gradstein, "Tables," Section 6.35). See also E. Jahnke and F. Emde, "Tables of Functions," p. 19. Dover, New York, 1945.
[5] In Volume II, Chapter II, Section 4 we shall prove that a generalized function concentrated at a point is a linear combination of the delta function and its derivatives.

On the other hand, $\alpha^{(k)}$, like P_+^λ for all values of λ, is invariant under linear transformations which preserve the quadratic form

$$P(x_1, ..., x_{p+q}) = x_1^2 + ... + x_p^2 - x_{p+1}^2 - ... - x_{p+q}^2.$$

Therefore the operator $Q_{2k}(\partial/\partial x_1, ..., \partial/\partial x_{p+q})$ must also be invariant under such transformations. Any operator satisfying these requirements must be of the form

$$Q_{2k} = c_{k,p,q}\, P^k\left(\frac{\partial}{\partial x_1}, ..., \frac{\partial}{\partial x_{p+q}}\right).$$

Thus we may write

$$\alpha^{(k)} = c_{k,p,q} P^k\left(\frac{\partial}{\partial x_1}, ..., \frac{\partial}{\partial x_{p+q}}\right) \delta(x) = c_{k,p,q} L^k \delta(x).$$

We can find $c_{k,p,q}$ by using Eq. (41) with some fixed φ in K.

We shall choose φ to be a function which near the origin is

$$\varphi(x) = (x_1^2 + ... + x_p^2 - x_{p+1}^2 - ... - x_{p+q}^2)^k.$$

Then, as is seen from Eqs. (6) and (8), in the neighborhood of $u = 0$ the function $\psi_1(u, tu)$ is of the form

$$\psi_1(u, tu) = \Omega_p \Omega_q u^k (1 - t)^k.$$

Therefore according to (11)

$$\Phi(\lambda, u) = \left[\tfrac{1}{4} \int_0^1 (1 - t)^{\lambda+k}\, t^{\frac{1}{2}(q-2)}\, dt\, \Omega_p \Omega_q\right] u^k,$$

or

$$\Phi(\lambda, u) = \frac{\Gamma(\lambda + k + 1)\, \Gamma\left(\frac{q}{2}\right)}{4\Gamma\left(\lambda + k + \frac{q}{2} + 1\right)} \Omega_p \Omega_q u^k.$$

Now $\Phi_1(-\tfrac{1}{2}n - k, u)$, which appears in the expression for $(\alpha^{(k)}, \varphi)$, is the constant term of the Laurent expansion of $\Phi(\lambda, u)$ about $\lambda = -\tfrac{1}{2}n - k$.

To be specific, consider the case in which p and q are odd. Since the coefficients of the Laurent series for

$$\frac{\Gamma(\lambda + k + 1)\, \Gamma\left(\frac{q}{2}\right)}{\Gamma\left(\lambda + k + \frac{q}{2} + 1\right)}$$

about $\lambda = -\frac{1}{2} n - k$ are the same as those of the Laurent series for

$$\frac{\Gamma(\lambda + 1) \, \Gamma\left(\frac{q}{2}\right)}{\Gamma\left(\lambda + \frac{q}{2} + 1\right)}$$

about $\lambda = -\frac{1}{2} n$, we may write

$$\Phi_1\left(-\frac{n}{2} - k, u\right) = \kappa u^k,$$

where κ is given by (42). Then according to (41) we arrive at

$$(\alpha^{(k)}, \varphi) = \kappa = \frac{(-1)^{\frac{1}{2}(q+1)} \, \pi^{\frac{1}{2}n-1}}{\Gamma\left(\frac{n}{2}\right)} \left[\psi\left(\frac{p}{2}\right) - \psi\left(\frac{n}{2}\right)\right]. \tag{45}$$

But on the other hand

$$(\alpha^{(k)}, \varphi) = c_{k,p,q} L^k \varphi(x) \big|_{x=0} .$$

Now by iterating (25) k times we obtain (recall that $\varphi = P^k$)

$$L^k \varphi(x) = 2^{2k} k! \left(\frac{n}{2}\right) \left(\frac{n}{2} + 1\right) \cdots \left(\frac{n}{2} + k - 1\right). \tag{46}$$

Comparison of (45) and (46) gives

$$c_{k,p,q} = \frac{(-1)^{\frac{1}{2}(q+1)} \, \pi^{\frac{1}{2}n-1}}{2^{2k} k! \, \Gamma\left(\frac{n}{2} + k\right)} \left[\psi\left(\frac{p}{2}\right) - \psi\left(\frac{n}{2}\right)\right]$$

for odd p and q.

Thus for odd p and q the $\theta_{p,q}$ coefficient in (44) becomes

$$\theta_{p,q} = \frac{(-1)^{\frac{1}{2}(q+1)}}{\pi} \left[\psi\left(\frac{p}{2}\right) - \psi\left(\frac{n}{2}\right)\right].$$

Summing up, we have the following. If the dimension n of the space is *even* and p and q are *even*, P_+^λ has simple poles at $\lambda = -\frac{1}{2} n - k$, where k is a nonnegative integer, where the residues are given by

$$\operatorname*{res}_{\lambda=-\frac{1}{2}n-k} P_+^\lambda = \frac{(-1)^{\frac{1}{2}n+k-1}}{\Gamma\left(\frac{n}{2} + k\right)} \delta_1^{(\frac{1}{2}n+k-1)}(P) + \frac{(-1)^{\frac{1}{2}q} \, \pi^{\frac{1}{2}n}}{2^{2k} k! \, \Gamma\left(\frac{n}{2} + k\right)} L^k \delta(x),$$

where

$$L = \frac{\partial^2}{\partial x_1^2} + \cdots + \frac{\partial^2}{\partial x_p^2} - \frac{\partial^2}{\partial x_{p+1}^2} - \cdots - \frac{\partial^2}{\partial x_{p+q}^2}.$$

If, on the other hand, p and q are *odd*, P_+^λ has poles of order 2 at $\lambda = -\frac{1}{2}n - k$. Let the Laurent expansion of P_+^λ about this point be

$$P_+^\lambda = \frac{c_{-2}^{(k)}}{\left(\lambda + \dfrac{n}{2} + k\right)^2} + \frac{c_{-1}^{(k)}}{\lambda + \dfrac{n}{2} + k} + \cdots.$$

Then the coefficients are given by

$$c_{-2}^{(k)} = \frac{(-1)^{\frac{1}{2}(q+1)} \pi^{\frac{1}{2}n-1}}{2^{2k}k!\,\Gamma\left(\dfrac{n}{2} + k\right)} L^k \delta(x)$$

and

$$c_{-1}^{(k)} = \frac{(-1)^{\frac{1}{2}n+k-1}}{\Gamma\left(\dfrac{n}{2} + k\right)} \delta_1^{(\frac{1}{2}n+k-1)}(P)$$

$$+ \frac{(-1)^{\frac{1}{2}(q+1)} \pi^{\frac{1}{2}n-1} \left[\psi\left(\dfrac{p}{2}\right) - \psi\left(\dfrac{n}{2}\right)\right]}{2^{2k}k!\,\Gamma\left(\dfrac{n}{2} + k\right)} L^k \delta(x),$$

where $\psi(x) = \Gamma'(x)/\Gamma(x)$.

In addition to P_+^λ, we can also define the generalized function P_-^λ by

$$(P_-^\lambda, \varphi) = \int_{-P>0} (-P)^\lambda \varphi \, dx. \tag{47}$$

All that we have said above about P_+^λ remains true also for P_-^λ except that p and q must be interchanged, and in all the formulas $\delta_1^{(k)}(P)$ must be replaced by $\delta_1^{(k)}(-P) = (-1)^k \delta_2^{(k)}(P)$.

2.3. The Generalized Function $\mathscr{P}^\lambda$ Associated with a Quadratic Form with Complex Coefficients

We have so far dealt exclusively with quadratic forms whose coefficients are real. Let us now turn to the space of all quadratic forms

$$\mathscr{P} = \sum_{r,s=1}^{n} g_{rs} x_r x_s \tag{1}$$

whose coefficients may be complex.

We set ourselves the task of defining the generalized function $\mathscr{P}^\lambda$, where λ is a complex number. In general, however, $\mathscr{P}^\lambda$ is not a single-valued analytic function of λ. But any quadratic form may be written in the form

$$\mathscr{P} = P_1 + iP_2,$$

and so as to remove ambiguity in λ we shall deal with the "upper half-plane" in the space of such quadratic forms, namely with *quadratic forms whose imaginary part is positive definite*, and it is for these that we shall define $\mathscr{P}^\lambda$. If a quadratic form $\mathscr{P}$ belongs to this "half-plane," we shall write

$$\mathscr{P}^\lambda = \exp \lambda(\ln |\mathscr{P}| + i \arg \mathscr{P}),$$

where $0 < \arg \mathscr{P} < \pi$. Such a function is a single-valued analytic function of λ.

Now with $\mathscr{P}^\lambda$ we shall associate the generalized function, also denoted by $\mathscr{P}^\lambda$, defined as

$$(\mathscr{P}^\lambda, \varphi) = \int \mathscr{P}^\lambda \varphi \, dx, \tag{2}$$

where the integral is taken over all space. This integral converges for Re $\lambda > 0$, and for these values it is an analytic function of λ. Then by analytic continuation we can define the generalized function for other values of λ.

We now proceed to find the singular points of $\mathscr{P}^\lambda$ for quadratic forms on the "upper half-plane," and to calculate the residues at these singularities. The calculations can be simplified considerably by a trick we shall often find very useful.

The generalized function $\mathscr{P}^\lambda$ is analytic not only in λ but also in the coefficients of the quadratic form $\mathscr{P}$. This means that $\mathscr{P}^\lambda$ is analytic on the upper "half-plane" of all quadratic forms $\mathscr{P} = P_1 + iP_2$ (or where P_2 is positive definite). But this, in turn, means that $\mathscr{P}^\lambda$ is uniquely determined by its values on the "positive imaginary axis," that is, on the set of all $\mathscr{P} = iP_2$, where P_2 is positive definite. Therefore it is sufficient for our purposes to consider generalized functions of the form $(iP_2)^\lambda$. We have, however, already solved this problem in Chapter I, Section 3, since it is always possible to transform a positive definite form into a simple sum of squares by a nonsingular linear transformation.

Let us therefore start by assuming that $\mathscr{P} = \sum_{r,s=1}^n g_{rs} x_r x_s$ lies on the "positive imaginary axis." This means that $g_{rs} = ia_{rs}$, $r, s = 1, ..., n$, where the a_{rs} are real and where $\sum_{r,s=1}^n a_{rs} x_r x_s$ is positive definite.

Then

$$(\mathscr{P}^{\lambda}, \varphi) = e^{\frac{1}{2}\pi\lambda i} \int \left(\sum a_{rs} x_r x_s\right)^{\lambda} \varphi\, dx.$$

Now there exists a linear transformation $x_r = \sum_{s=1}^{n} \alpha_{rs} x'_s$ which will transform $\sum_{r,s=1}^{n} a_{rs} x_r x_s$ into $r^2 = x_1'^2 + \ldots + x_n'^2$. The Jacobian of this transformation is $1/\sqrt{|a|}$, where $|a|$ is the determinant

$$|a| = \begin{vmatrix} a_{11} & \cdots & a_{1n} \\ \cdots & \cdots & \cdots \\ a_{n1} & \cdots & a_{nn} \end{vmatrix}.$$

We may thus write

$$(\mathscr{P}^{\lambda}, \varphi) = \frac{e^{\frac{1}{2}\pi\lambda i}}{\sqrt{|a|}} \int r^{2\lambda} \varphi\, dx',$$

or

$$(\mathscr{P}^{\lambda}, \varphi) = \frac{e^{\frac{1}{2}\pi\lambda i}}{\sqrt{(-i)^n\,|g|}} \int r^{2\lambda} \varphi\, dx', \qquad (3)$$

where $|g|$ is the determinant of the coefficients of $\mathscr{P}$, and $\sqrt{(-i)^n\,|g|}$ is merely the square root of the positive real number $(-i)^n\,|g|$.

Now we have already dealt with $(r^{2\lambda}, \varphi)$ in Chapter I, Section 3.9. We found there that the only singularities of this functional, and therefore also of $(\mathscr{P}^{\lambda}, \varphi)$, are simple poles at $\lambda = -\frac{1}{2}n - k$, where k is a nonnegative integer. For $k = 0$ we found

$$\operatorname*{res}_{\lambda=-\frac{1}{2}n} r^{2\lambda} = \frac{\pi^{\frac{1}{2}n}}{\Gamma\left(\dfrac{n}{2}\right)} \delta(x).$$

Therefore

$$\operatorname*{res}_{\lambda=-\frac{1}{2}n} \mathscr{P}^{\lambda} = \frac{e^{-\frac{1}{4}\pi n i}\,\pi^{\frac{1}{2}n}}{\sqrt{(-i)^n\,|g|}\,\Gamma\left(\dfrac{n}{2}\right)} \delta(x). \qquad (4)$$

Let us find the residues of $\mathscr{P}^{\lambda}$ at the other singular points. For this purpose we consider the differential operator

$$L_{\mathscr{P}} = \sum g^{rs} \frac{\partial^2}{\partial x_r\, \partial x_s}, \qquad (5)$$

where the g^{rs} are defined by

$$\sum_{s=1}^{n} g^{rs} g_{st} = \delta_t^r$$

($\delta_t^r = 1$ for $r = t$ and $\delta_t^r = 0$ for $r \neq t$). Thus $\| g^{rs} \|$, the matrix of the coefficients of $L_{\mathscr{P}}$, is the inverse of $\| g_{st} \|$. Using this fact, we may obtain

$$L_{\mathscr{P}} \mathscr{P}^{\lambda+1} = 4(\lambda + 1) \left(\lambda + \frac{n}{2}\right) \mathscr{P}^\lambda.$$

This relation is easily verified by direct calculation. Iterating it k times, we arrive at

$$L_{\mathscr{P}}^k \mathscr{P}^{\lambda+k} = 4^k(\lambda + 1) \ldots (\lambda + k) \left(\lambda + \frac{n}{2}\right) \ldots \left(\lambda + \frac{n}{2} + k - 1\right) \mathscr{P}^\lambda,$$

so that

$$\mathscr{P}^\lambda = \frac{1}{4^k(\lambda + 1) \ldots (\lambda + k) \left(\lambda + \frac{n}{2}\right) \ldots \left(\lambda + \frac{n}{2} + k - 1\right)} L_{\mathscr{P}}^k \mathscr{P}^{\lambda+k}. \tag{6}$$

Consequently

$$\operatorname*{res}_{\lambda=-\frac{1}{2}n-k} \mathscr{P}^\lambda$$

$$= \frac{1}{4^k(\lambda + 1) \ldots (\lambda + k) \left(\lambda + \frac{n}{2}\right) \ldots \left(\lambda + \frac{n}{2} + k - 1\right)} \Bigg|_{\lambda=-\frac{1}{2}n-k} \operatorname*{res}_{\lambda=-\frac{1}{2}n} L_{\mathscr{P}}^k \mathscr{P}^\lambda,$$

so that, inserting (4), we arrive at

$$\operatorname*{res}_{\lambda=-\frac{1}{2}n-k} \mathscr{P}^\lambda = \frac{e^{-\frac{1}{4}\pi n i} \pi^{\frac{1}{2}n}}{4^k k! \Gamma \left(\frac{n}{2} + k\right) \sqrt{(-i)^n \, | g |}} L_{\mathscr{P}}^k \delta(x). \tag{7}$$

Now this equation was obtained under the assumption that $\mathscr{P}$ lies on the "positive imaginary axis." We must now continue it analytically to the entire "upper half-plane." The analytic continuation of $L_{\mathscr{P}}$ is known, however, since the coefficients of this operator are analytic functions of the coefficients of $\mathscr{P}$. All that remains then is to study the analytic continuation of $\sqrt{(-i)^n \, | g |}$. The problem will therefore be completely solved when we are able to express this as a single-valued analytic function on the "upper half-plane" of quadratic forms.

To do this, we again write $\mathscr{P}$ in the form

$$\mathscr{P} = P_1 + iP_2,$$

where P_1 and P_2 are quadratic forms whose coefficients are real, and where P_2 is positive definite. Now there exists a nonsingular linear transformation

$$x_r = \sum_{s=1}^{n} b_{rs} y_s$$

with real coefficients such that P_1 and P_2 are transformed to

$$P_1 = \lambda_1 y_1^2 + \ldots + \lambda_n y_n^2,$$

$$P_2 = y_1^2 + \ldots + y_n^2.$$

The λ_i in the expression for P_1 are real and independent of the particular choice of transformation we have made. Thus they are invariants of $\mathscr{P}$ itself. We may thus write

$$|g| = |b|^2 (\lambda_1 + i) \ldots (\lambda_n + i),$$

where $|b|$ is the determinant of the $\|b_{ij}\|$ matrix, so that

$$(-i)^n |g| = |b|^2 (1 - \lambda_1 i) \ldots (1 - \lambda_n i).$$

Then the function we wish to define may be written

$$\sqrt{(-i)^n |g|} = \sqrt{|b|^2} (1 - \lambda_1 i)^{\frac{1}{2}} \ldots (1 - \lambda_n i)^{\frac{1}{2}}, \tag{8}$$

where the square roots are defined by

$$\sqrt{z} = |z|^{\frac{1}{2}} e^{\frac{1}{2} i \cdot \arg z}, \qquad -\pi < \arg z < \pi.$$

The function defined by (8) will be the desired single-valued analytic function on the "upper half-plane" of complex quadratic forms.

Thus, if

$$\mathscr{P} = \sum_{r,s=1}^{n} g_{rs} x_r x_s$$

is any quadratic form with positive definite imaginary part, the generalized function $\mathscr{P}^\lambda$ is a regular analytic function of λ everywhere except at $\lambda = -\frac{1}{2}n - k$, where k is a nonnegative integer, and at these points this function has simple poles with residues given by

$$\operatorname*{res}_{\lambda=-\frac{1}{2}n-k} \mathscr{P}^{\lambda} = \frac{e^{-\frac{1}{4}\pi n i}\, \pi^{\frac{1}{2}n}}{4^k k!\, \Gamma\left(\dfrac{n}{2}+k\right) \sqrt{(-i)^n\,|\,g\,|}} \left(\sum_{r,s=1}^{n} g^{rs}\, \frac{\partial^2}{\partial x_r\, \partial x_s}\right)^k \delta(x), \qquad (9)$$

where $|\,g\,|$ is the determinant of the coefficients of $\mathscr{P}$, and $\sqrt{(-i)^n\,|\,g\,|}$ is defined by (8).

The lower half-plane" of quadratic forms, that is, quadratic forms with negative definite imaginary parts, may be subjected to similar analysis. If we were to do this, we would obtain the following result.

If the quadratic form

$$\mathscr{P} = P_1 - iP_2 = \sum_{r,s=1}^{n} g_{rs} x_r x_s$$

belongs to the "lower half-plane," the generalized function $\mathscr{P}^{\lambda}$ is again a regular analytic function of λ everywhere except at $\lambda = -\frac{1}{2}n - k$, where k is a nonnegative integer, and at these points this function has simple poles with residue given by

$$\operatorname*{res}_{\lambda=-\frac{1}{2}n-k} \mathscr{P}^{\lambda} = \frac{e^{-\frac{1}{4}\pi n i}\, \pi^{\frac{1}{2}n}}{4^k k!\, \Gamma\left(\dfrac{n}{2}+k\right) \sqrt{i^n\,|\,g\,|}} \left(\sum_{r,s=1}^{n} g^{rs}\, \frac{\partial^2}{\partial x_r\, \partial x_s}\right)^k \delta(x), \qquad (9')$$

where, as before, $|\,g\,|$ is the determinant of the coefficients of $\mathscr{P}$, and $\sqrt{i^n\,|\,g\,|}$ is given by the analog of (8), namely

$$\sqrt{i^n\,|\,g\,|} = \sqrt{|\,b\,|^2}(1 + \lambda_1 i)^{\frac{1}{2}} \ldots (1 + \lambda_n i)^{\frac{1}{2}}.$$

2.4. The Generalized Functions $(P + i0)^{\lambda}$ and $(P - i0)^{\lambda}$

We can now use the results of Section 2.3 to study any *real* quadratic form raised to some power λ.

Let

$$P = \sum_{r,s=1}^{n} g_{rs} x_r x_s$$

be a nondegenerate quadratic form with real coefficients. Then we may define the analogs of $(x + i0)^{\lambda}$ and $(x - i0)^{\lambda}$ of Chapter I, Section 3, which we shall call $(P + i0)^{\lambda}$ and $(P - i0)^{\lambda}$.

For this purpose consider the quadratic form

$$\mathscr{P} = P + iP',$$

where P' is a positive definite quadratic form (with real coefficients). It is easily shown that as the coefficients of P' converge to zero, the generalized function $(P + iP')^\lambda$ converges to a well-defined limit. This limit we shall call $(P + i0)^\lambda$.

Indeed, this assertion is obvious for Re $\lambda > 0$, since in this case the limit can be taken under the integral sign in $\int \mathscr{P}^\lambda \varphi \, dx$. But then according to Eq. (6) of Section 2.3 the assertion remains valid at all points of analyticity of $\mathscr{P}^\lambda$.[6]

Similarly, we shall define the generalized function $(P - i0)^\lambda$ as the limit of the generalized function $(P - iP')^\lambda$ as the coefficients of P' converge to zero, where P' is a positive definite quadratic form.

From the definition of $(P + i0)^\lambda$ and $(P - i0)^\lambda$ we may deduce that they are analytic in λ everywhere except at $\lambda = -\frac{1}{2}n - k$, where k is a nonnegative integer.

It is easily shown also by using Eq. (6) of Section 2.3 that at these points our functions have simple poles with residues given by

$$\operatorname*{res}_{\lambda=-\frac{1}{2}n-k} (P + i0)^\lambda = \lim_{P' \to 0} \operatorname*{res}_{\lambda=-\frac{1}{2}n-k} (P + iP')^\lambda,$$

$$\operatorname*{res}_{\lambda=-\frac{1}{2}n-k} (P - i0)^\lambda = \lim_{P' \to 0} \operatorname*{res}_{\lambda=-\frac{1}{2}n-k} (P - iP')^\lambda.$$

In order to obtain these residues, therefore, we need only find the limits, as P' converges to zero, of $\sqrt{(-i)^n \,|\, g \,|}$ and $\sqrt{i^n \,|\, g \,|}$, where $|\, g \,|$ is the determinant of the coefficients of the complex quadratic form $P + iP'$.

Without loss of generality we may assume that P' is of the form

$$P' = \epsilon(x_1^2 + ... + x_n^2), \qquad \epsilon > 0.$$

Then we can use Eq. (8) of. Section 2.3 to write

$$\sqrt{(-i)^n \,|\, g \,|} = (\epsilon - i\lambda_1)^{\frac{1}{2}} ... (\epsilon - i\lambda_n)^{\frac{1}{2}},$$

where the λ_i are the eigenvalues of P. Let us assume that p of these eigenvalues are positive, and q negative. Then as $\epsilon \to 0$, we arrive at

$$\lim_{\epsilon \to 0} \sqrt{(-i)^n \,|\, g \,|} = \sqrt{|\, \lambda_1 ... \lambda_n \,|} \, (-i)^{\frac{1}{2}p} \, i^{\frac{1}{2}q},$$

---

[6] We shall not study in detail the special case in which λ is a negative integer which is not a singularity of $\mathscr{P}^\lambda$. We state without proof merely that at such points $(P + i0)^\lambda$ has no singularities.

or

$$\lim_{\epsilon \to 0} \sqrt{(-i)^n \, | \, g \, |} = e^{-\frac{1}{4}\pi(p-q)i} \sqrt{| \, \Delta \, |},$$

where Δ is the determinant of the coefficients of P. We now use Eq. (9) of Section 2.3 to express the residue of $\mathscr{P}^\lambda$ at $\lambda = - \frac{1}{2} n - k$, which yields

$$\operatorname*{res}_{\lambda=-\frac{1}{2}n-k} (P + i0)^\lambda = \frac{e^{-\frac{1}{2}\pi q i} \, \pi^{\frac{1}{2}n}}{4^k k! \, \Gamma \left(\dfrac{n}{2} + k \right) \sqrt{| \, \Delta \, |}} \left(\sum_{r,s=1}^{n} g^{rs} \frac{\partial^2}{\partial x_r \, \partial x_s} \right)^k \delta(x). \qquad (1)$$

In a similar way

$$\operatorname*{res}_{\lambda=-\frac{1}{2}n-k} (P - i0)^\lambda = \frac{e^{\frac{1}{2}\pi q i} \, \pi^{\frac{1}{2}n}}{4^k k! \, \Gamma \left(\dfrac{n}{2} + k \right) \sqrt{| \, \Delta \, |}} \left(\sum_{r,s=1}^{n} g^{rs} \frac{\partial^2}{\partial x_r \partial x_s} \right)^k \delta(x). \qquad (1')$$

Thus the residues of $(P + i0)^\lambda$ and $(P - i0)^\lambda$ at $\lambda = - \frac{1}{2} n - k$ are generalized functions concentrated on the vertex of the $P = 0$ cone.

These new generalized functions can be expressed in terms of P_+^λ and P_-^λ defined in Section 2.2 by

$$(P + i0)^\lambda = P_+^\lambda + e^{\pi \lambda i} \, P_-^\lambda, \qquad (2)$$

$$(P - i0)^\lambda = P_+^\lambda + e^{-\pi \lambda i} \, P_-^\lambda. \qquad (2')$$

Indeed, for Re $\lambda > 0$, the functionals (P_+^λ, φ) and (P_-^λ, φ) correspond to the functions

$$P_+^\lambda = \begin{cases} P^\lambda, & \text{where } P \geqslant 0, \\ 0, & \text{where } P \leqslant 0; \end{cases}$$

$$P_-^\lambda = \begin{cases} 0, & \text{where } P \geqslant 0, \\ (-P)^\lambda, & \text{where } P \leqslant 0. \end{cases}$$

For this case (2) and (2') follow directly from the definition of $(P + i0)^\lambda$ and $(P - i0)^\lambda$. But then by analytic continuation Eqs. (2) and (2') must remain valid also for other λ.

We mention incidentally that it follows from (2) and (2') that when λ is a nonnegative integer the functions $(P + i0)^\lambda$, $(P - i0)^\lambda$, and P^λ coincide.

We may now use Eqs. (2) and (2') to establish the relation between the residues of P_+^λ and P_-^λ at $\lambda = -\frac{1}{2}n - k$, $k = 0, 1, \ldots$. The Laurent expansions of these functions about such a point are

$$P_+^\lambda = \frac{c_{-2}^{(k)}}{\left(\lambda + \dfrac{n}{2} + k\right)^2} + \frac{c_{-1}^{(k)}}{\lambda + \dfrac{n}{2} + k} + \cdots,$$

$$P_-^\lambda = \frac{c_{-2}'^{(k)}}{\left(\lambda + \dfrac{n}{2} + k\right)^2} + \frac{c_{-1}'^{(k)}}{\lambda + \dfrac{n}{2} + k} + \cdots.$$

According to (2) and (2')

$$P_+^\lambda = \frac{e^{-\pi\lambda i}(P + i0)^\lambda - e^{\pi\lambda i}(P - i0)^\lambda}{-2i \sin \pi\lambda},$$

$$P_-^\lambda = \frac{(P + i0)^\lambda - (P - i0)^\lambda}{2i \sin \pi\lambda}. \tag{3}$$

Using Eqs. (1) and (1') for the residues of $(P + i0)^\lambda$ and $(P - i0)^\lambda$, we find that if the dimension n is *odd*, as well as if n is *even* and p and q are *even*, $c_{-2}^{(k)} = c_{-2}'^{(k)} = 0$. If, on the other hand, p and q are *odd*,

$$c_{-2}^{(k)} = (-1)^{\frac{1}{2}n+k+1} c_{-2}'^{(k)} = \frac{(-1)^{\frac{1}{2}(q-1)} \pi^{\frac{1}{2}n-1}}{4^k k! \, \Gamma\left(\dfrac{n}{2} + k\right) \sqrt{|\varDelta|}} L_P^k \delta(x),$$

where

$$L_P = \sum_{r,s=1}^{n} g^{rs} \frac{\partial^2}{\partial x_r \, \partial x_s}.$$

We have already obtained these results in a different way in Section 2.2.

Further, according to (2), we may write

$$\operatorname*{res}_{\lambda=-\frac{1}{2}n-k} (P + i0)^\lambda = \operatorname*{res}_{\lambda=-\frac{1}{2}n-k} P_+^\lambda + \operatorname*{res}_{\lambda=-\frac{1}{2}n-k} e^{\pi\lambda i} P_-^\lambda ,$$

so that

$$\operatorname*{res}_{\lambda=-\frac{1}{2}n-k} (P + i0)^\lambda = c_{-1}^{(k)} + e^{-\pi(\frac{1}{2}n+k)i} c_{-1}'^{(k)} + \pi i e^{-\pi(\frac{1}{2}n+k)i} c_{-2}'^{(k)}.$$

Inserting the expressions for the residue of $(P + i0)^\lambda$ at $\lambda = -\frac{1}{2}n - k$ and $c_{-2}'^{(k)}$, we find that if the dimension n is *odd*, as well as if n is *even* and p and q are *even*,

$$\operatorname*{res}_{\lambda=-\frac{1}{2}n-k} P_+^\lambda + e^{-\pi(\frac{1}{2}n+k)i} \operatorname*{res}_{\lambda=-\frac{1}{2}n-k} P_-^\lambda = \frac{e^{-\frac{1}{2}\pi q i} \pi^{\frac{1}{2}n}}{4^k k! \, \Gamma\left(\dfrac{n}{2} + k\right) \sqrt{|\varDelta|}} L_P^k \delta(x) \tag{4}$$

while if p and q are *odd*,

$$\operatorname*{res}_{\lambda=-\frac{1}{2}n-k} P_+^\lambda + e^{-\pi(\frac{1}{2}n+k)i} \operatorname*{res}_{\lambda=-\frac{1}{2}n-k} P_-^\lambda = 0. \tag{5}$$

We may note also that if n is *odd*, as well as if n is *even* and $k < \frac{1}{2}n$, then according to (2) we have

$$\operatorname*{res}_{\lambda=-k} P_+^\lambda + (-1)^k \operatorname*{res}_{\lambda=-k} P_-^\lambda = \operatorname*{res}_{\lambda=-k} (P+i0)^\lambda = 0. \tag{6}$$

Let us assume from now on that P is in its canonical form

$$P = x_1^2 + \dots + x_p^2 - x_{p+1}^2 - \dots - x_{p+q}^2.$$

In Section 2.2 we obtained explicit expressions for the residues of P_+^λ and P_-^λ in terms of $\delta_1^{(k)}(P)$, $\delta_1^{(k)}(-P) = (-1)^k \delta_2^{(k)}(P)$, and $\delta(x)$. Inserting these expressions into (4), (5), and (6) we obtain

$$\delta_1^{(k)}(P) - \delta_2^{(k)}(P) = c_{p,q,k} L^{k-\frac{1}{2}n+1} \delta(x),$$

where

$$L = \frac{\partial^2}{\partial x_1^2} + \dots + \frac{\partial^2}{\partial x_p^2} - \frac{\partial^2}{\partial x_{p+1}^2} - \dots - \frac{\partial^2}{\partial x_{p+q}^2}$$

and $c_{p,q,k} = 0$ if the dimension is odd, as well as if n is even and $k < \frac{1}{2}n - 1$. In all other cases $c_{p,q,k}$ is easily obtained from the equations of Section 2.2.

Summing up, we find that if the dimension n is *odd*, as well as if n is *even* and $k < \frac{1}{2}n - 1$, then $\delta_1^{(k)}(P) = \delta_2^{(k)}(P)$. If, on the other hand, n is *even*, $\delta_1^{(k)}(P) - \delta_2^{(k)}(P)$ for $k \geq \frac{1}{2}n - 1$ is a generalized function concentrated on the vertex of the $P = 0$ cone.

According to the general theory which we shall treat in Section 4, a more natural definition of homogeneous generalized functions concentrated on the $P = 0$ surface is the following (see Section 4.5):

$$\delta^{(k)}(P_+) = (-1)^k k! \operatorname*{res}_{\lambda=-k-1} P_+^\lambda$$

and similarly

$$\delta^{(k)}(P_-) = (-1)^k k! \operatorname*{res}_{\lambda=-k-1} P_-^\lambda.$$

For odd n, as well as for even n and $k < \frac{1}{2}n - 1$ we have

$$\delta^{(k)}(P_+) = \delta_1^{(k)}(P), \qquad \delta^{(k)}(P_-) = \delta_1^{(k)}(-P)$$

while in the case of even dimension and $k \geqslant \frac{1}{2} n - 1$,

$$\delta^{(k)}(P_+) - \delta_1^{(k)}(P) \quad \text{and} \quad \delta^{(k)}(P_-) - \delta_1^{(k)}(-P)$$

are generalized functions concentrated at the vertex of the $P = 0$ cone.

Equations (4)–(6) imply that if p and q are both even and if $k \geqslant \frac{1}{2} n - 1$, then

$$(-1)^k \, \delta^{(k)}(P_+) - \delta^{(k)}(P_-) = \frac{(-1)^{\frac{1}{2}q} \, \pi^{\frac{1}{2}n}}{4^{k+1-\frac{1}{2}n} \left(k + 1 - \dfrac{n}{2} \right)!} L^{k+1-\frac{1}{2}n} \, \delta(x),$$

while in all other cases

$$\delta^{(k)}(P_-) = (-1)^k \delta^{(k)}(P_+).$$

2.5. Elementary Solutions of Linear Differential Equations

Let us now apply the results of Sections 2.3 and 2.4 to elementary solutions of equations of the form

$$L^k u = f(x), \tag{1}$$

where L is a linear homogeneous differential operator of the form

$$L = \frac{\partial^2}{\partial x_1^2} + \cdots + \frac{\partial^2}{\partial x_p^2} - \frac{\partial^2}{\partial x_{p+1}^2} - \cdots - \frac{\partial^2}{\partial x_{p+q}^2}$$

and where k is a positive integer.

Recall that an elementary solution of (1) is a generalized function K such that[7]

$$L^k K = \delta(x). \tag{2}$$

Any linear differential equation of second order with constant coefficients and only the derivatives of highest order, for instance, can be written in the form of Eq. (1) with $k = 1$. Such an equation is often called *ultrahyperbolic*. For either p or q equal to zero, such an equation becomes Laplace's equation, and for p or q equal to one, it becomes the wave equation.

One might naturally try to solve Eq. (2) with a homogeneous function (or an associated function; see Chapter I, Section 4.1), since L and $\delta(x)$

---

[7] In this section K will stand for a generalized function, rather than the space of test functions. In later sections (see Section 2.7) the same symbol with a subscript will be used to denote modified Bessel functions.

are both homogeneous. Since a derivative of order $2k$ of K gives the delta function, which is homogeneous of degree $-n$ (here $n = p + q$ is the dimension of the space), K would have to be homogeneous of degree $-n + 2k$.

In addition, Eq. (2) is invariant under linear transformations that preserve the quadratic form

$$P = x_1^2 + \ldots + x_p^2 - x_{p+1}^2 - \ldots - x_{p+q}^2 \, .$$

We shall seek its solution in the form

$$K = f(P) \, .$$

In Section 2.4 we studied $(P + i0)^\lambda$ and $(P - i0)^\lambda$, which are homogeneous generalized functions of P. In addition, for $\lambda = -\frac{1}{2}n + k$, these functions have the desired degree $-n + 2k$. We shall now show that unless n is even and $k \geqslant \frac{1}{2}n$ both $(P + i0)^{-\frac{1}{2}n+k}$ and $(P - i0)^{-\frac{1}{2}n+k}$ are, to within a constant factor, elementary solutions of $L^k u = f(x)$.

From the results of Sections 2.3 and 2.4 we have

$$L^k(P + i0)^{\lambda+k}$$
$$= 4^k(\lambda + 1) \ldots (\lambda + k) \left(\lambda + \frac{n}{2}\right) \ldots \left(\lambda + \frac{n}{2} + k - 1\right) (P + i0)^\lambda. \quad (3)$$

Setting $\lambda = -\frac{1}{2}n$, we obtain

$$L^k(P + i0)^{-\frac{1}{2}n+k} = 4^k \left(1 - \frac{n}{2}\right) \ldots \left(k - \frac{n}{2}\right)(k - 1)! \operatorname*{res}_{\lambda=-\frac{1}{2}n} (P + i0)^\lambda. \quad (4)$$

Thus if n is *even* and $k \geqslant \frac{1}{2}n$, the right-hand side of (4) vanishes and we obtain

$$L^k (P + i0)^{-\frac{1}{2}n+k} = 0,$$

so that $(P + i0)^{-\frac{1}{2}n+k}$ is a solution of the homogeneous equation corresponding to (1).

In all other cases, Eq. (4) becomes

$$L^k(P + i0)^{-\frac{1}{2}n+k} = 4^k \left(1 - \frac{n}{2}\right) \ldots \left(k - \frac{n}{2}\right)(k - 1)! \, \frac{e^{-\frac{1}{2}nqi} \, \pi^{\frac{1}{2}n}}{\Gamma\left(\frac{n}{2}\right)} \, \delta(x).$$

Therefore *except* when n is *even* and $k \geqslant \frac{1}{2}n$, both

$$K_1 = (-1)^k \, \frac{e^{\frac{1}{2}nqi} \, \Gamma\left(\frac{n}{2} - k\right)}{4^k(k - 1)! \, \pi^{\frac{1}{2}n}} \, (P + i0)^{-\frac{1}{2}n+k} \quad (5)$$

and, similarly,

$$K_2 = (-1)^k \frac{e^{-\frac{1}{2}n q i} \, \Gamma\left(\frac{n}{2} - k\right)}{4^k (k-1)! \, \pi^{\frac{1}{2}n}} (P - i0)^{-\frac{1}{2}n + k} \tag{5'}$$

are elementary solutions of $L^k u = f(x)$. If, on the other hand, *n is even* and $k \geqslant \frac{1}{2}n$, then $(P + i0)^{-\frac{1}{2}n+k} = (P - i0)^{-\frac{1}{2}n+k}$ is a solution of the corresponding homogeneous equation $L^k u = 0$.

Note that K_1 and K_2 are complex conjugates.

Equations (5) and (5') are the most convenient forms for the elementary solutions of our differential equation. We could also attempt to find real solutions to our problem by appropriately combining the real and imaginary parts of K_1 and K_2. These new solutions, however, would be quite different depending on whether n, p, and q are even or odd.

The formulas for K_1 and K_2 can also be written directly in terms of the generalized functions P_+^λ and P_-^λ.

To do this we use the equations of Section 2.4 which give $(P + i0)^\lambda$ and $(P - i0)^\lambda$ in terms of P_+^λ and P_-^λ. If n is odd, P_+^λ and P_-^λ are regular at $\lambda = -\frac{1}{2}n + k$, and we obtain

$$K_1 = \bar{K}_2 = (-1)^k \frac{e^{\frac{1}{2}n q i} \, \Gamma\left(\frac{n}{2} - k\right)}{4^k (k-1)! \, \pi^{\frac{1}{2}n}} (P_+^{-\frac{1}{2}n+k} + e^{\pi(-\frac{1}{2}n+k)i} \, P_-^{-\frac{1}{2}n+k}). \tag{6}$$

If, on the other hand, n is even and $k < \frac{1}{2}n$, both P_+^λ and P_-^λ have simple poles at $\lambda = -\frac{1}{2}n + k$, with residues given by

$$\operatorname*{res}_{\lambda = -\frac{1}{2}n+k} P_+^\lambda = (-1)^{\frac{1}{2}n-k-1} \operatorname*{res}_{\lambda = -\frac{1}{2}n+k} P_-^\lambda = \frac{(-1)^{\frac{1}{2}n-k-1}}{\left(\frac{n}{2} - k - 1\right)!} \delta^{(\frac{1}{2}n-k-1)}(P_+).$$

Let $P_+^{-\frac{1}{2}n+k}$ and $P_-^{-\frac{1}{2}n+k}$ denote the constant terms of the Laurent expansions for P_+^λ and P_-^λ about this point. Then we obtain

$$K_1 = \bar{K}_2 = (-1)^k \frac{e^{\frac{1}{2}n q i} \, \Gamma\left(\frac{n}{2} - k\right)}{4^k (k-1)! \, \pi^{\frac{1}{2}n}}$$

$$\times \left[P_+^{-\frac{1}{2}n+k} + (-1)^{-\frac{1}{2}n+k} P_-^{-\frac{1}{2}n+k} + \frac{(-1)^{-\frac{1}{2}n+k} \pi i}{\left(\frac{n}{2} - k - 1\right)!} \delta^{(\frac{1}{2}n-k-1)}(P_+) \right]. \tag{7}$$

From Eqs. (6) and (7) we find, in particular, that K_1 and K_2 are linearly independent. The generalized function $K_1 - K_2$ is a solution of $L^k u = 0$.

Finally consider the case in which n is even and $k \geqslant \frac{1}{2} n$. We expand $(P + i0)^{\lambda+k}$ in a Taylor's series about $\lambda = -\frac{1}{2} n$, obtaining

$$(P + i0)^{\lambda+k} = (P + i0)^{-\frac{1}{2}n+k} + \left(\lambda + \frac{n}{2}\right) (P + i0)^{-\frac{1}{2}n+k} \ln (P + i0) + \dots.$$

Now we insert this into (3) and compare the coefficients of $\lambda + \frac{1}{2} n$ on both sides of the equation. This gives

$$L^k \left[(P + i0)^{-\frac{1}{2}n+k} \ln (P + i0)\right]$$

$$= 4^k (-1)^{\frac{1}{2}n-1} \left(\frac{n}{2} - 1\right)! \left(k - \frac{n}{2}\right)! (k - 1)! \operatorname*{res}_{\lambda=-\frac{1}{2}n} (P + i0)^{\lambda}.$$

We now insert the expression for the residue $(P + i0)^{\lambda}$ at $\lambda = -\frac{1}{2} n$, and find that when n is *even* and $k \geqslant \frac{1}{2} n$, an elementary solution of $L^k u = f(x)$ is the associated function

$$K_1 = (-1)^{\frac{1}{2}n-1} \frac{e^{\frac{1}{2}n q i}}{4^k \left(k - \frac{n}{2}\right)! (k - 1)!} (P + i0)^{-\frac{1}{2}n+k} \ln (P + i0).$$

Similarly, another elementary solution of this equation for this case will be

$$K_2 = (-1)^{\frac{1}{2}n-1} \frac{e^{-\frac{1}{2}n q i}}{4^k \left(k - \frac{n}{2}\right)! (k - 1)!} (P - i0)^{-\frac{1}{2}n+k} \ln (P - i0).$$

It should be mentioned that these equations for the elementary solutions of $L^k u = f(x)$ remain true also when L is *any* linear homogeneous differential operator of order two, which we may write

$$L = \sum_{r,s=1}^{n} g^{rs} \frac{\partial^2}{\partial x_r \, \partial x_s}.$$

In this case P is the quadratic form

$$P = \sum_{r,s=1}^{n} g_{rs} x_r x_s,$$

where $\sum_{s=1}^{n} g_{rs} g^{st} = \delta_r^t$ (for $r, t = 1, \dots, n$). The expressions for the factor multiplying $(P + i0)^{-\frac{1}{2}n+k}$ and $(P - i0)^{-\frac{1}{2}n+k}$ should now be multiplied by $\sqrt{|\Delta|}$, where Δ is the determinant of the coefficients of P.

For instance, elementary solutions of

$$\left(2\,\frac{\partial^2}{\partial x_1\,\partial x_3} + \frac{\partial^2}{\partial x_2^2}\right) u = f(x_1, x_2, x_3)$$

are the functions

$$K_1 = \frac{1}{4\pi i}\,(2x_1 x_3 + x_2^2 + i0)^{-\frac{1}{2}}$$

and

$$K_2 = \bar{K}_1.$$

2.6. Fourier Transforms of $(P + i0)^\lambda$ and $(P - i0)^\lambda$

To obtain the Fourier transforms of $(P + i0)^\lambda$ and $(P - i0)^\lambda$, we shall use analytic continuation of the quadratic form, as described in Section 2.3.

Let

$$\mathscr{P} = \alpha_1 x_1^2 + \dots + \alpha_n x_n^2 \qquad (1)$$

be a quadratic form with complex coefficients, and let Im $\tilde{\mathscr{P}}$ be a positive definite quadratic form (in other words, Im $\alpha_s > 0$, $s = 1, \dots, n$). The generalized function $\mathscr{P}^\lambda$, and therefore also its Fourier transform $\tilde{\mathscr{P}}^\lambda$, are analytic functions of the α_s in the region Im $\alpha_s > 0$. Therefore in order to find $\tilde{\mathscr{P}}^\lambda$ we need only treat the case in which all the α_s are imaginary, and we shall write $\alpha_s = ib_s$, with $b_s > 0$. For this case we have

$$\tilde{\mathscr{P}}^\lambda = e^{\frac{1}{2}\pi\lambda i}\int (b_1 x_1^2 + \dots + b_n x_n^2)^\lambda \exp\left[i(x_1 s_1 + \dots + x_n s_n)\right] dx.$$

A suitable change of variables in the integrand transforms this to the form

$$\tilde{\mathscr{P}}^\lambda = \frac{e^{\frac{1}{2}\pi\lambda i}}{\sqrt{b_1}\dots\sqrt{b_n}}\int r^{2\lambda} \exp\left[i\left(x_1\,\frac{s_1}{\sqrt{b_1}} + \dots + x_n\,\frac{s_n}{\sqrt{b_n}}\right)\right] dx,$$

where $r^2 = x_1^2 + \dots + x_n^2$.

The Fourier transform of the generalized function r^λ has already been calculated in Section 3.3 of Chapter II. Using the results of that section, we have

$$\tilde{\mathscr{P}}^\lambda = \frac{e^{\frac{1}{2}\pi\lambda i}\,2^{2\lambda+n}\,\pi^{\frac{1}{2}n}\,\Gamma\left(\lambda + \dfrac{n}{2}\right)}{\sqrt{b_1}\dots\sqrt{b_n}\,\Gamma(-\lambda)}\left(\frac{s_1^2}{b_1} + \dots + \frac{s_n^2}{b_n}\right)^{-\lambda-\frac{1}{2}n},$$

or

$$\tilde{\mathscr{P}}^\lambda = \frac{e^{-\frac{1}{4}\pi n i}\, 2^{2\lambda+n}\, \pi^{\frac{1}{2}n}\, \Gamma\left(\lambda + \frac{n}{2}\right)}{\sqrt{-i\alpha_1}\, \cdots\, \sqrt{-i\alpha_n}\, \Gamma(-\lambda)} \left(\frac{s_1^2}{\alpha_1} + \cdots + \frac{s_n^2}{\alpha_n}\right)^{-\lambda-\frac{1}{2}n} \tag{2}$$

Now the uniqueness of analytic continuation implies that (2) remains valid also for any quadratic form whose imaginary part is positive definite. To use this expression we write the square roots appearing in it in the form $\sqrt{z} = |z|^{\frac{1}{2}} \exp\left(\frac{1}{2}i\arg z\right)$. Note that the quadratic form $s_1^2/\alpha_1 + \cdots + s_n^2/\alpha_n$ has *negative definite* imaginary part.

Now let us write

$$P = x_1^2 + \cdots + x_p^2 - x_{p+1}^2 - \cdots - x_{p+q}^2.$$

$$Q = s_1^2 + \cdots + s_p^2 - s_{p+1}^2 - \cdots - s_{p+q}^2.$$

By letting the imaginary part of the quadratic form now approach zero in Eq. (2) and setting $\alpha_1 = \cdots = \alpha_p = 1$, $\alpha_{p+1} = \cdots = \alpha_{p+q} = -1$, we obtain

$$F[(P + i0)^\lambda] = \frac{e^{-\frac{1}{2}\pi q i}\, 2^{2\lambda+n}\, \pi^{\frac{1}{2}n}\, \Gamma\left(\lambda + \frac{n}{2}\right)}{\Gamma(-\lambda)} (Q - i0)^{-\lambda-\frac{1}{2}n} \tag{3}$$

and in a similar way

$$F[(P - i0)^\lambda] = \frac{e^{\frac{1}{2}\pi q i}\, 2^{2\lambda+n}\, \pi^{\frac{1}{2}n}\, \Gamma\left(\lambda + \frac{n}{2}\right)}{\Gamma(-\lambda)} (Q + i0)^{-\lambda-\frac{1}{2}n}. \tag{3'}$$

We may now use Eq. (3) of Section 2.4 to arrive, after some elementary operations, at

$$\tilde{P}_+^\lambda = 2^{2\lambda+n}\, \pi^{\frac{1}{2}n-1}\, \Gamma(\lambda + 1)\, \Gamma\left(\lambda + \frac{n}{2}\right)$$

$$\times \frac{1}{2i}\, [e^{-\pi(\lambda+\frac{1}{2}q)i}\, (Q - i0)^{-\lambda-\frac{1}{2}n} - e^{\pi(\lambda+\frac{1}{2}q)i}\, (Q + i0)^{-\lambda-\frac{1}{2}n}], \tag{4}$$

$$\tilde{P}_-^\lambda = -2^{2\lambda+n}\, \pi^{\frac{1}{2}n-1}\, \Gamma(\lambda + 1)\, \Gamma\left(\lambda + \frac{n}{2}\right)$$

$$\times \frac{1}{2i}\, [e^{-\frac{1}{2}\pi q i}\, (Q - i0)^{-\lambda-\frac{1}{2}n} - e^{\frac{1}{2}\pi q i}\, (Q + i0)^{-\lambda-\frac{1}{2}n}]. \tag{4'}$$

It should be noted that these equations for the Fourier transforms remain valid also if P is any arbitrary real nondegenerate quadratic form

$$P = \sum_{\alpha,\beta=1}^{n} g_{\alpha\beta} x_\alpha x_\beta.$$

In this case, we replace Q by

$$Q = \sum_{\alpha,\beta=1}^{n} g^{\alpha\beta} s_\alpha s_\beta,$$

where $\sum_{\beta=1}^{n} g_{\alpha\beta} g^{\beta\gamma} = \delta_\alpha^\gamma$ $(\alpha, \gamma = 1, ..., n)$. In addition, all the formulas have to be multiplied on the right-hand side by the factor $1/\sqrt{|\varDelta|}$ where $\varDelta$ is the determinant of the coefficients of P.

2.7. Generalized Functions Associated with Bessel Functions

We shall now turn to a class of functions $\mathscr{P}^\lambda f(\mathscr{P},\lambda)$, where $f(z, \lambda)$ is an entire function of z and λ, and $\mathscr{P}$ is a complex quadratic form with positive definite imaginary part. These generalized functions are defined for Re $\lambda > 0$ by

$$(\mathscr{P}^\lambda f(\mathscr{P}, \lambda), \varphi(x)) = \int \mathscr{P}^\lambda f(\mathscr{P}, \lambda)\, \varphi(x)\, dx. \tag{1}$$

For Re $\lambda > -1$, obviously, $\mathscr{P}^\lambda f(\mathscr{P}, \lambda)$ is an analytic function of λ. Its definition can be extended by analytic continuation to other values of λ. The functions $\mathscr{P}^\lambda \ln^m \mathscr{P}f(\mathscr{P}, \lambda)$ may be defined analogously.

It is easily shown [starting, for instance, by expanding $f(z, \lambda)$ in a power series in z] that for every real quadratic form P the limit

$$(P + i0)^\lambda f(P + i0, \lambda) = \lim_{P_1 \to 0} (P + iP_1)^\lambda f(P + iP_1, \lambda) \tag{2}$$

exists, where P_1 is a positive definite quadratic form.

In complete analogy one can define the generalized function

$$(P - i0)^\lambda f(P - i0, \lambda)$$

for any real quadratic form P.

Obviously if P is positive definite, then

$$(P + i0)^\lambda f(P + i0, \lambda) = (P - i0)^\lambda f(P - i0, \lambda) = P^\lambda f(P, \lambda).$$

Moreover, since for all positive integral values of n we have

$$(P + i0)^n = (P - i0)^n = P^n,$$

it follows that

$$f(P + i0, n) = f(P - i0, n) = f(P, n).$$

Now let us write the generalized functions $(P + i0)^\lambda f(P, \lambda)$ and $(P - i0)^\lambda f(P, \lambda)$ in terms of P_+^λ and P_-^λ. For this purpose we use the relations, developed in Section 2.4,

$$(P + i0)^\lambda = P_+^\lambda + e^{i\lambda\pi}P_-^\lambda,$$

$$(P - i0)^\lambda = P_+^\lambda + e^{-i\lambda\pi}P_-^\lambda.$$

Since

$$P_+^\lambda f(P, \lambda) = P_+^\lambda f(P_+, \lambda) \qquad \text{and} \qquad P_-^\lambda f(P, \lambda) = P_-^\lambda f(P_-, \lambda),$$

we obtain

$$(P + i0)^\lambda f(P, \lambda) = P_+^\lambda f(P_+, \lambda) + e^{i\lambda\pi}P_-^\lambda f(P_-, \lambda) \tag{3}$$

and

$$(P - i0)^\lambda f(P, \lambda) = P_+^\lambda f(P_+, \lambda) + e^{-i\lambda\pi}P_-^\lambda f(P_-, \lambda). \tag{4}$$

The class of generalized functions we have defined is quite large. In it, in particular, are generalized functions such as $\mathscr{P}^{\frac12\lambda}J_\lambda(\mathscr{P}^{\frac12})$ and $\mathscr{P}^{-\frac12\lambda}J_\lambda(\mathscr{P}^{\frac12})$, where $J_\lambda(z)$ is a Bessel function. This is easily seen by making use of the power series expansion

$$J_\lambda(z) = \left(\frac{z}{2}\right)^\lambda \sum_{m=0}^\infty \frac{(-1)^m \left(\frac{z}{2}\right)^{2m}}{m!\,\Gamma(\lambda + m + 1)} \tag{5}$$

for the Bessel function.

In addition to $J_\lambda(z)$ we shall consider also $N_\lambda(z)$, $H_\lambda^{(1)}(z)$, $H_\lambda^{(2)}(z)$, $I_\lambda(z)$, and $K_\lambda(z)$, which we define for nonintegral λ by

$$N_\lambda(z) = \frac{1}{\sin\lambda\pi}[\cos\lambda\pi I_\lambda(z) - J_{-\lambda}(z)],$$

$$H_\lambda^{(1)}(z) = J_\lambda(z) + iN_\lambda(z),$$

$$H_\lambda^{(2)}(z) = J_\lambda(z) - iN_\lambda(z),$$

$$I_\lambda(z) = e^{-\frac12\pi\lambda i} J_\lambda(iz),$$

$$K_\lambda(z) = \frac{\pi}{2\sin\lambda\pi}[I_{-\lambda}(z) - I_\lambda(z)].$$

For integral values of λ these functions are defined by convergence in λ.

Power series expansions for $N_\lambda(z^{\frac{1}{2}})$, $H_\lambda^{(1)}(z^{\frac{1}{2}})$, $H_\lambda^{(2)}(z^{\frac{1}{2}})$, $I_\lambda(z^{\frac{1}{2}})$, and $K_\lambda(z^{\frac{1}{2}})$ for nonintegral values of λ are easily obtained from Eq. (5) and the definition of these functions. It is then found that all these functions belong to the class we have defined at the beginning of this section, which makes it possible for us to define generalized functions such as $K_\lambda[(P + i0)^{\frac{1}{2}}]$, $(P + i0)^{-\frac{1}{2}\lambda} K_\lambda[(P + i0)^{\frac{1}{2}}]$, and others.

2.8. Fourier Transforms of $(c^2 + P + i0)^\lambda$ and $(c^2 + P - i0)^\lambda$

In Section 2 of Chapter II we already saw that the Fourier transforms of the generalized functions $(x^2 - 1)^\lambda$, $(1 - x^2)^\lambda$, and $(1 + x^2)^\lambda$ can be written in terms of Bessel functions. Here and in the next section we shall show that this is also true of the Fourier transforms of their n-dimensional analogs $(c^2 + P)_+^\lambda$ and $(c^2 + P)_-^\lambda$.

We start by considering the generalized function $(c^2 + P)^\lambda$ for a positive definite quadratic form P. The Fourier transform of this generalized function for $\operatorname{Re} \lambda < -\frac{1}{2} n$ is given by

$$F[(c^2 + P)^\lambda] = \int (c^2 + P)^\lambda \, e^{i(x,s)} \, dx, \tag{1}$$

where $(x, s) = x_1 s_1 + \ldots + x_n s_n$.

Let us first consider the case in which the canonical form of P is $\sum_{k=1}^n x_k^2$. Clearly in this case the generalized function $F[(c^2 + P)^\lambda]$ depends only on $|s|$, the magnitude of the vector s. Therefore without loss of generality we may assume that the components of s are given by $s = (|s|, 0, 0, \ldots, 0)$, so that the integral in (1) becomes

$$\int (c^2 + |x|^2)^\lambda \, e^{ix_1 |s|} \, dx, \tag{2}$$

where $\operatorname{Re} \lambda < -\frac{1}{2} n$.

We shall perform the integration in (2) by going to polar coordinates. After integrating over the angles $\varphi_2, \ldots, \varphi_{n-1}$ and using the fact that

$$\Omega_{n-1} = \frac{2(\sqrt{\pi})^{n-1}}{\Gamma\left(\dfrac{n-1}{2}\right)}$$

we arrive at

$$\frac{2(\sqrt{\pi})^{n-1}}{\Gamma\left(\dfrac{n-1}{2}\right)} \int_0^\infty \int_0^\pi (c^2 + r^2)^\lambda \, e^{ir\,|s|\cos\varphi_1} \sin^{n-2}\varphi_1 \, r^{n-1} \, d\varphi_1 \, dr.$$

Now it is known that

$$\int_0^\pi e^{ir|s|\cos\varphi_1} \sin^{n-2}\varphi_1 \, d\varphi_1 = \frac{\Gamma\left(\frac{n-1}{2}\right)\sqrt{\pi}}{\left(\frac{r|s|}{2}\right)^{\frac{1}{2}n-1}} J_{\frac{1}{2}n-1}(r|s|),$$

and that

$$\int_0^\infty r^{\frac{1}{2}n}(c^2 + r^2)^\lambda J_{\frac{1}{2}n-1}(r|s|) \, dr = \left(\frac{2}{|s|}\right)^{\lambda+1} \frac{c^{\frac{1}{2}n+\lambda}}{\Gamma(-\lambda)} K_{\frac{1}{2}n+\lambda}(c|s|).$$

Therefore for Re $\lambda < -\frac{1}{2}n$, the integral of (2) becomes

$$\int (c^2 + |x|^2)^\lambda e^{ix_1|s|} \, dx = \frac{2^{\lambda+1}(\sqrt{2\pi})^n}{\Gamma(-\lambda)} \left(\frac{c}{|s|}\right)^{\frac{1}{2}n+\lambda} K_{\frac{1}{2}n+\lambda}(c|s|). \tag{3}$$

For other values of λ Eq. (3) remains valid by analytic continuation in λ.

In order to obtain the expression for the Fourier transform of *any* positive definite quadratic form, we need only rewrite (3) in the form

$$\sqrt{\Delta} \int (c^2 + |x|^2)^\lambda e^{i(x,s)} \, dx = \frac{2^{\lambda+1}(\sqrt{2\pi})^n}{\Gamma(-\lambda)} \left(\frac{c}{|s|}\right)^{\frac{1}{2}n+\lambda} K_{\frac{1}{2}n+\lambda}(c|s|) \tag{4}$$

(note that the determinant of the coefficients $\Delta = 1$ for the quadratic form $|x|^2 = \Sigma_{k=1}^n x_k^2$). Now $\sqrt{\Delta} \, dx$ remains invariant under a coordinate transformation that carries $|x|^2$ into $P = \Sigma_{k,r=1}^n g_{kr}x_k x_r$, and the square of the length of the dual vector s becomes $\Sigma_{k,r=1}^n g^{kr}s_k s_r$ ($Q = \Sigma_{k,r=1}^n g^{kr}s_k s_r$ is the quadratic form dual to $P = \Sigma_{k,r=1}^n g_{kr}x_k x_r$). Thus under such a coordinate transformation Eq. (4) becomes

$$\sqrt{\Delta} \int (c^2 + P)^\lambda e^{i(x,s)} \, dx = \frac{2^{\lambda+1}(\sqrt{2\pi})^n c^{\frac{1}{2}n+\lambda}}{\Gamma(-\lambda)} \frac{K_{\frac{1}{2}n+\lambda}(cQ^{1/2})}{Q^{\frac{1}{2}(\frac{1}{2}n+\lambda)}}.$$

Our result may be summarized as follows.

Let P be a positive definite quadratic form, and let Q be its dual. Then the Fourier transform of the generalized function $(c^2 + P)^\lambda$ is

$$F[(c^2 + P)^\lambda] = \frac{2^{\lambda+1}(\sqrt{2\pi})^n c^{\frac{1}{2}n+\lambda}}{\Gamma(-\lambda)\sqrt{\Delta}} \frac{K_{\frac{1}{2}n+\lambda}(cQ^{\frac{1}{2}})}{Q^{\frac{1}{2}(\frac{1}{2}n+\lambda)}} \tag{5}$$

where Δ is the determinant of the coefficients of P.

Now let P be any real quadratic form. We wish to consider the generalized functions $(c^2 + P + i0)^\lambda$ and $(c^2 + P - i0)^\lambda$ defined by

$$(c^2 + P + i0)^\lambda = \lim_{\epsilon \to 0} (c^2 + P + i\epsilon P_1)^\lambda \qquad (6)$$

and

$$(c^2 + P - i0)^\lambda = \lim_{\epsilon \to 0} (c^2 + P - i\epsilon P_1)^\lambda , \qquad (7)$$

where $\epsilon > 0$ and P_1 is a positive definite quadratic form. The existence of the limits in (6) and (7) was established in Section 2.4 for $c = 0$, and the existence for $c \neq 0$ follows from the absence of singular points on the $c^2 + P = 0$ hypersurface.

If the quadratic form $\mathscr{P}$ lies in the "upper half-plane," its dual $\mathscr{Q}$ lies in the "lower half-plane." Therefore according to the uniqueness of analytic continuation, Eq. (5) implies that

$$F[(c^2 + P + i0)^\lambda] = \frac{2^{\lambda+1}(\sqrt{2\pi})^n c^{\frac{1}{2}n+\lambda}}{\Gamma(-\lambda)\sqrt{\Delta}} \frac{K_{\frac{1}{2}n+\lambda}[c(Q - i0)^{\frac{1}{2}}]}{(Q - i0)^{\frac{1}{2}(\frac{1}{2}n+\lambda)}} . \qquad (8)$$

Here by $\sqrt{\Delta}$ we denote the analytic continuation from the sheet on which this function is positive for positive definite quadratic forms.[8] We mention that $\sqrt{\Delta} = \sqrt{|\Delta|}\, e^{\frac{1}{2}q\pi i}$ if the canonical form of P has q negative terms,

Similarly, it can be shown that

$$F[(c^2 + P - i0)^\lambda] = \frac{2^{\lambda+1}(\sqrt{2\pi})^n c^{\frac{1}{2}n+\lambda}}{\Gamma(-\lambda)\sqrt{\Delta}} \frac{K_{\frac{1}{2}n+\lambda}[c(Q + i0)^{\frac{1}{2}}]}{(Q + i0)^{\frac{1}{2}(\frac{1}{2}n+\lambda)}} . \qquad (9)$$

In this case $\sqrt{\Delta} = \sqrt{|\Delta|}\, e^{-q\pi i}$, where again q is the number of negative terms in the canonical form of P.

Let us now express $F[(c^2 + P + i0)^\lambda]$ and $F[(c^2 + P - i0)^\lambda]$ in terms of Q_+ and Q_-. By using the power series expansion for $K_\lambda(z)$ and Eq. (4) of Section 2.7, we arrive after some simple operations at

$$F[(c^2 + P + i0)^\lambda]$$

$$= \frac{2^{\lambda+\frac{1}{2}n+1}\pi^{\frac{1}{2}n} e^{-\frac{1}{2}q\pi i} c^{\lambda+\frac{1}{2}n}}{\Gamma(-\lambda)\sqrt{|\Delta|}} \cdot \left[\frac{K_{\lambda+\frac{1}{2}n}(cQ_+^{\frac{1}{2}})}{Q_+^{\frac{1}{2}(\lambda+\frac{1}{2}n)}} + \frac{\pi i}{2} \frac{H^{(1)}_{-\lambda-\frac{1}{2}n}(cQ_-^{\frac{1}{2}})}{Q_-^{\frac{1}{2}(\lambda+\frac{1}{2}n)}} \right] \qquad (10)$$

---

[8] Similar analytic continuation was studied in detail in Section 2.3, and we shall not go into it again here.

and

$$F[(c^2 + P - i0)^\lambda]$$

$$= \frac{2^{\lambda + \frac{1}{2}n+1} \pi^{\frac{1}{2}n} e^{\frac{1}{2}q\pi i} c^{\lambda + \frac{1}{2}n}}{\Gamma(-\lambda) \sqrt{|\varDelta|}} \cdot \left[\frac{K_{\lambda + \frac{1}{2}n}(cQ^{\frac{1}{2}}_+)}{Q^{\frac{1}{2}(\lambda + \frac{1}{2}n)}_+} - \frac{\pi i}{2} \frac{H^{(2)}_{-\lambda - \frac{1}{2}n}(cQ^{\frac{1}{2}}_-)}{Q^{\frac{1}{2}(\lambda + \frac{1}{2}n)}_-} \right] \qquad (11)$$

Equations (10) and (11) can be considerably simplified if P is definite. If P is positive definite only the first term remains in the square brackets, while if it is negative definite, only the second term remains.

2.9. Fourier Transforms of $(c^2 + P)^\lambda_+$ and $(c^2 + P)^\lambda_-$

Let us now turn to the generalized functions $(c^2 + P)^\lambda_+$ and $(c^2 + P)^\lambda_-$. They are linear combinations of $(c^2 + P + i0)^\lambda$ and $(c^2 + P - i0)^\lambda$ of the previous section. Their Fourier transforms are therefore linear combinations of the Fourier transforms of the latter, namely,

$$F[(c^2 + P)^\lambda_+] = \frac{i}{2 \sin \lambda \pi} \{ e^{-i\lambda \pi} F[(c^2 + P + i0)^\lambda] - e^{i\lambda \pi} F[(c^2 + P - i0)^\lambda] \} \qquad (1)$$

and

$$F[(c^2 + P)^\lambda_-] = - \frac{i}{2 \sin \lambda \pi} \{ F[(c^2 + P + i0)^\lambda] - F[(c^2 + P - i0)^\lambda] \}. \qquad (2)$$

Inserting Eqs. (8) and (9) of Section 2.8, we obtain

$$\frac{F[(c^2 + P)^\lambda_+]}{\Gamma(\lambda + 1)} = - \frac{2^{\lambda + \frac{1}{2}n} i \pi^{\frac{1}{2}n - 1} c^{\frac{1}{2}n + \lambda}}{\sqrt{|\varDelta|}}$$

$$\times \left\{ e^{-i(\lambda + \frac{1}{2}q)\pi} \frac{K_{\frac{1}{2}n + \lambda}[c(Q - i0)^{\frac{1}{2}}]}{(Q - i0)^{\frac{1}{2}(\lambda + \frac{1}{2}n)}} - e^{i(\lambda + \frac{1}{2}q)\pi} \frac{K_{\frac{1}{2}n + \lambda}[c(Q + i0)^{\frac{1}{2}}]}{(Q + i0)^{\frac{1}{2}(\lambda + \frac{1}{2}n)}} \right\} \qquad (3)$$

and

$$\frac{F[(c^2 + P)^\lambda_-]}{\Gamma(\lambda + 1)} = \frac{2^{\lambda + \frac{1}{2}n} i \pi^{\frac{1}{2}n - 1} c^{\frac{1}{2}n + \lambda}}{\sqrt{|\varDelta|}}$$

$$\times \left\{ e^{-\frac{1}{2}q\pi i} \frac{K_{\frac{1}{2}n + \lambda}[c(Q - i0)^{\frac{1}{2}}]}{(Q - i0)^{\frac{1}{2}(\lambda + \frac{1}{2}n)}} - e^{\frac{1}{2}q\pi i} \frac{K_{\frac{1}{2}n + \lambda}[c(Q + i0)^{\frac{1}{2}}]}{(Q + i0)^{\frac{1}{2}(\lambda + \frac{1}{2}n)}} \right\}. \qquad (4)$$

The generalized functions on the left-hand sides of these equations can be expressed in terms of Q_+ and Q_- by using Eqs. (10) and (11) of Section 2.8 in Eqs. (1) and (2) of the present section. This leads then to

$$\frac{F[(c^2 + P)_+^\lambda]}{\Gamma(\lambda + 1)} = \frac{2^{\lambda + \frac{1}{2}n + 1}\,\pi^{\frac{1}{2}n - 1}\,c^{\frac{1}{2}n + \lambda}}{\sqrt{|\varDelta|}}\left\{-\sin\left(\lambda + \tfrac{1}{2}q\right)\pi\,\frac{K_{\lambda + \frac{1}{2}n}(cQ_+^{\frac{1}{2}})}{Q_+^{\frac{1}{2}(\lambda + \frac{1}{2}n)}}\right.$$

$$+\frac{\pi}{2\sin(\lambda + \frac{1}{2}n)\,\pi}\left[\sin\left(\lambda + \tfrac{1}{2}q\right)\pi\,\frac{J_{\lambda + \frac{1}{2}n}(cQ_-^{\frac{1}{2}})}{Q_-^{\frac{1}{2}(\lambda + \frac{1}{2}n)}} + \sin\tfrac{1}{2}p\pi\,\frac{J_{-\lambda - \frac{1}{2}n}(cQ_-^{\frac{1}{2}})}{Q_-^{\frac{1}{2}(\lambda + \frac{1}{2}n)}}\right]\right\}. \quad (5)$$

In this formula p denotes the number of positive terms in the canonical form of P, and $p + q = n$. The expression for $F[(c^2 + P)_-^\lambda]/\Gamma(\lambda + 1)$ can be obtained from that for $F[(c^2 + P)_+^\lambda]/\Gamma(\lambda + 1)$ by replacing $\sin(\lambda + \frac{1}{2}q)\pi$ by $-\sin\frac{1}{2}q\pi$, and $\sin\frac{1}{2}p\pi$ by $-\sin(\lambda + \frac{1}{2}p)\pi$. We then get

$$\frac{F[(c^2 + P)_-^\lambda]}{\Gamma(\lambda + 1)} = \frac{2^{\lambda + \frac{1}{2}n + 1}\,\pi^{\frac{1}{2}n - 1}\,c^{\frac{1}{2}n + \lambda}}{\sqrt{|\varDelta|}}\left\{\sin\tfrac{1}{2}q\pi\,\frac{K_{\lambda + \frac{1}{2}n}(cQ_+^{\frac{1}{2}})}{Q_+^{\frac{1}{2}(\lambda + \frac{1}{2}n)}}\right.$$

$$-\frac{\pi}{2\sin(\lambda + \frac{1}{2}n)\,\pi}\left[\sin\tfrac{1}{2}q\pi\,\frac{J_{\lambda + \frac{1}{2}n}(cQ_-^{\frac{1}{2}})}{Q_-^{\frac{1}{2}(\lambda + \frac{1}{2}n)}} + \sin\left(\lambda + \tfrac{1}{2}p\right)\pi\,\frac{J_{-\lambda - \frac{1}{2}n}(cQ_-^{\frac{1}{2}})}{Q_-^{\frac{1}{2}(\lambda + \frac{1}{2}n)}}\right]\right\}.$$

$$(6)$$

By setting $c = 0$ in Eqs. (3) and (4) we obtain Eqs. (5) and (6) of Section 2.6 for $\tilde{P}_+^\lambda$ and $\tilde{P}_-^\lambda$. Special cases of all these formulas are the ones we obtained in Chapter II, Section 2.5 for the Fourier transforms of $(1 - x^2)^\lambda$, $(1 + x^2)^\lambda$, and $(x^2 - 1)^\lambda$.

2.10. Fourier Transforms of $\dfrac{(c^2 + P)_+^\lambda}{\Gamma(\lambda + 1)}$ and $\dfrac{(c^2 + P)_-^\lambda}{\Gamma(\lambda + 1)}$ for Integral λ.

Fourier Transforms of $\delta(c^2 + P)$ and Its Derivatives

The Fourier transforms deduced in Section 2.9 for $(c^2 + P)_+^\lambda/\Gamma(\lambda + 1)$ and $(c^2 + P)_-^\lambda/\Gamma(\lambda + 1)$ require further study for integral values of λ. This is because

$$\frac{K_{\frac{1}{2}n + \lambda}[c(Q + i0)^{\frac{1}{2}}]}{(Q + i0)^{\frac{1}{2}(\frac{1}{2}n + \lambda)}} \quad \text{and} \quad \frac{K_{\frac{1}{2}n + \lambda}[c(Q - i0)^{\frac{1}{2}}]}{(Q - i0)^{\frac{1}{2}(\frac{1}{2}n + \lambda)}}$$

have poles at such points.

Now the first of these is defined by the expansion

$$
\frac{K_{\frac{1}{2}n+\lambda}\left[c(Q+i0)^{\frac{1}{2}}\right]}{(Q+i0)^{\frac{1}{2}(\frac{1}{2}n+\lambda)}} = \frac{\pi(c/2)^{\frac{1}{2}n+\lambda}}{2\sin\left(\frac{1}{2}n+\lambda\right)\pi}
$$

$$
\times \left[\sum_{m=0}^{\infty} \frac{(c/2)^{-2\lambda-n+2m}(Q+i0)^{-\lambda-\frac{1}{2}n+m}}{m!\,\Gamma(-\lambda-\frac{1}{2}n+m+1)} \right.
$$

$$
\left. - \sum_{m=0}^{\infty} \frac{(c/2)^{2m}(Q+i0)^{m}}{m!\,\Gamma(\lambda+\frac{1}{2}n+m+1)} \right]. \qquad (1)
$$

But according to Section 2.4, $(Q+i0)^{\lambda}$ has a pole at $\lambda = -\frac{1}{2}n - k$ with residue[9]

$$
\operatorname*{res}_{\lambda=-\frac{1}{2}n-k} (Q+i0)^{\lambda} = \frac{e^{-\frac{1}{2}n\pi i}\,\pi^{\frac{1}{2}n}\,\sqrt{|\varDelta|}}{4^{k}k!\Gamma(\frac{1}{2}n+k)} \left(\sum_{r,t=1}^{n} g_{rt}\frac{\partial^{2}}{\partial s_{r}\,\partial s_{t}} \right)^{k} \delta(s).
$$

Therefore the generalized function of Eq. (1) has poles at $\lambda = t$, $t > 0$ with residues given by

$$
\operatorname*{res}_{\lambda=t} \frac{K_{\frac{1}{2}n+\lambda}[c(Q+i0)^{\frac{1}{2}}]}{(Q+i0)^{\frac{1}{2}(\frac{1}{2}n+\lambda)}}
$$

$$
= \frac{(-1)^{t}\,\pi^{\frac{1}{2}n}(c/2)^{t-\frac{1}{2}n}\,e^{-\frac{1}{2}n\pi i}\,\sqrt{|\varDelta|}}{2} \sum_{m=0}^{t} \frac{(-1)^{m}(c/2)^{-2m}}{4^{m}m!(t-m)!} L^{m}\delta(s), \qquad (2)
$$

where we have written

$$
L = \sum_{r,t=1}^{n} g_{rt}\frac{\partial^{2}}{\partial s_{r}\,\partial s_{t}}.
$$

(in calculating the residue we again use $\Gamma(x)\Gamma(1-x) = \pi/\sin\pi x$).

Let us now find the regular part of the generalized function of Eq. (1) at $\lambda = t$. This is easily done for odd n. In this case the regular part at $\lambda = t$ is the sum of two terms, the first of which is given by the series in Eq. (1) for $\lambda = t$, in which $(Q+i0)^{-t-\frac{1}{2}n+m}$ for $m \leqslant t$ is understood as the regular part of $(Q+i0)^{\lambda}$ at $\lambda = -t - \frac{1}{2}n + m$. The second term is of the form $\sum_{m=0}^{t} \alpha_{m}\beta_{m}$, where

$$
\alpha_{m} = \operatorname*{res}_{\lambda=-t-\frac{1}{2}n+m} (Q+i0)^{\lambda},
$$

$$
\beta_{m} = \frac{(-1)^{m}}{2(m!)}(c/2)^{2m-\frac{1}{2}n}\frac{d}{d\lambda}\left[\frac{(c/2)^{-\lambda}}{\Gamma(\lambda+\frac{1}{2}n-m)} \right]_{\lambda=t}.
$$

---

[9] The determinant of the coefficients of Q is $1/\varDelta$, where $\varDelta$ is the determinant of the coefficients of P.

We shall denote the total regular part by

$$\frac{K_{\frac{1}{2}n+t}[c(Q+i0)^{\frac{1}{2}}]}{(Q+i0)^{\frac{1}{2}(\frac{1}{2}n+t)}}.$$

The situation is somewhat more complicated for even n.[10] In this case we shall define the generalized function of Eq. (1) at $\lambda = t$ as the sum of two terms, the first of which is again $\sum_{m=0}^{t} \alpha_m \beta_m$, but the second of which is now

$$(-1)^{\frac{1}{2}n+t} \left(\ln \frac{c(Q+i0)}{2} + \gamma \right) \frac{I_{\frac{1}{2}n+t}^{-}[c(Q+i0)^{\frac{1}{2}}]}{(Q+i0)^{\frac{1}{2}(\frac{1}{2}n+t)}}$$

$$+ \frac{(-1)^{\frac{1}{2}n+t}}{2} \left(\frac{c}{2} \right)^{\frac{1}{2}n+t} \sum_{m=1}^{\frac{1}{2}n+t} \frac{(-1)^m (m-1)!}{(\frac{1}{2}n+t-m)!} \left(\frac{c}{2} \right)^{-2m} (Q+i0)^{-m}$$

$$+ \frac{(-1)^{\frac{1}{2}n+t}}{2} \left(\frac{c}{2} \right)^{\frac{1}{2}n+t} \sum_{m=0}^{\infty} \frac{h_{m+\frac{1}{2}n+t} + h_m}{m!(\frac{1}{2}n+t-m)!} \left(\frac{c}{2} \right)^{2m} (Q+i0)^m, \qquad (3)$$

where γ is Euler's constant, and $h_m = \sum_{r=1}^{m} r^{-1}$.

As above, by $(Q+i0)^{-m}$ for $m \geqslant \frac{1}{2}n$ we here mean the regular part of $(Q+i0)^\lambda$ at $\lambda = -m$. The generalized function

$$\frac{K_{\frac{1}{2}n+t}[c(Q+i0)^{\frac{1}{2}}]}{(Q+i0)^{\frac{1}{2}(\frac{1}{2}n+t)}}$$

is now equal to the regular part of

$$\frac{K_{\frac{1}{2}n+\lambda}[c(Q+i0)^{\frac{1}{2}}]}{(Q+i0)^{\frac{1}{2}(\frac{1}{2}n+\lambda)}} \qquad (4)$$

at $\lambda = t$.

Let us now consider the same generalized function for negative integral $\lambda = -t$. In this case it is defined by Eq. (1) with $\lambda = -t$ for odd n, and by Eq. (3) for even n.

Similar results are obtained for

$$\frac{K_{\frac{1}{2}n+\lambda}[c(Q-i0)^{\frac{1}{2}}]}{(Q-i0)^{\frac{1}{2}(\frac{1}{2}n+\lambda)}},$$

---

[10] For even n the expansion given in Eq. (1) to define our generalized function loses meaning.

except that the residue at $\lambda = t$ is

$$\operatorname*{res}_{\lambda=t} \frac{K_{\frac{1}{2}n+\lambda}\left[c(Q-i0)^{\frac{1}{2}}\right]}{(Q-i0)^{\frac{1}{2}(\frac{1}{2}n+\lambda)}}$$

$$= \frac{(-1)^t \pi^{\frac{1}{2}n}(c/2)^{t-\frac{1}{2}n}\, e^{\frac{1}{2}\pi q i}\, \sqrt{|\varDelta|}}{2} \sum_{m=0}^{t} \frac{(-1)^m (c/2)^{-2m}}{4^m m!(t-m)!}. \qquad (5)$$

We can now proceed to a consideration of the generalized functions

$$\frac{F[(c^2+P)_+^\lambda]}{\Gamma(\lambda+1)} \quad \text{and} \quad \frac{F[(c^2+P)_-^\lambda]}{\Gamma(\lambda+1)}$$

for integral λ. If $\lambda = -t$, where t is a positive integer, the left-hand side of Eq. (3) of Section 2.9 becomes $\delta^{(t-1)}(c^2+P)$. This means that

$$F[\delta^{(t-1)}(c^2+P)] = (-1)^{t+1}\frac{i}{\sqrt{|\varDelta|}}\, 2^{\frac{1}{2}n-t}\, \pi^{\frac{1}{2}n-1}\, c^{\frac{1}{2}n-t}$$

$$\times \left[e^{-\frac{1}{2}\pi q i}\frac{K_{\frac{1}{2}n-t}\left[c(Q-i0)^{\frac{1}{2}}\right]}{(Q-i0)^{\frac{1}{2}(\frac{1}{2}n-t)}} - e^{\frac{1}{2}\pi q i}\frac{K_{\frac{1}{2}n-t}\left[c(Q+i0)^{\frac{1}{2}}\right]}{(Q+i0)^{\frac{1}{2}(\frac{1}{2}n-t)}} \right]. \qquad (6)$$

In particular,

$$F[\delta(c^2+P)] = -\frac{i}{\sqrt{|\varDelta|}}\,(2\pi c)^{\frac{1}{2}n-1}$$

$$\times \left[-e^{-\frac{1}{2}\pi q i}\frac{K_{\frac{1}{2}n-1}\left[c(Q-i0)^{\frac{1}{2}}\right]}{(Q-i0)^{\frac{1}{2}(\frac{1}{2}n-1)}} + e^{\frac{1}{2}\pi q i}\frac{K_{\frac{1}{2}n-1}\left[c(Q+i0)^{\frac{1}{2}}\right]}{(Q+i0)^{\frac{1}{2}(\frac{1}{2}n-1)}} \right]. \qquad (7)$$

As for positive integers t, we have

$$\frac{K_{\frac{1}{2}n+\lambda}\left[c(Q+i0)^{\frac{1}{2}}\right]}{(Q+i0)^{\frac{1}{2}(\frac{1}{2}n+\lambda)}} = \frac{c_1}{\lambda-t} + \frac{K_{\frac{1}{2}n+t}\left[c(Q+i0)^{\frac{1}{2}}\right]}{(Q+i0)^{\frac{1}{2}(\frac{1}{2}n+t)}} + \dots,$$

where c_1 denotes the residue of Eq. (4) at $\lambda = t$, and the omitted terms converge to zero as $\lambda \to t$. Similarly,

$$\frac{K_{\frac{1}{2}n+\lambda}\left[c(Q-i0)^{\frac{1}{2}}\right]}{(Q-i0)^{\frac{1}{2}(\frac{1}{2}n+1)}} = \frac{c_2}{\lambda-t} + \frac{K_{\frac{1}{2}n+t}\left[c(Q-i0)^{\frac{1}{2}}\right]}{(Q-i0)^{\frac{1}{2}(\frac{1}{2}n+t)}} + \dots,$$

where c_2 is the residue at $\lambda = t$.

Inserting these expressions into Eqs. (3) and (4) of Section 2.9 and going to the limit as $\lambda \to t$, we arrive at

$$\frac{F[(c^2 + P)^t_-]}{\Gamma(t+1)} = \frac{i \cdot 2^{t+\frac{1}{2}n}\, \pi^{\frac{1}{2}n-1}\, c^{\frac{1}{2}n+t}}{\sqrt{|\Delta|}}$$

$$\times \left[e^{-\frac{1}{2}q\pi i}\, \frac{K_{\frac{1}{2}n+t}\,[c(Q-i0)^{\frac{1}{2}}]}{(Q-i0)^{\frac{1}{2}(\frac{1}{2}n+t)}} - e^{\frac{1}{2}q\pi i}\, \frac{K_{\frac{1}{2}n+t}\,[c(Q+i0)^{\frac{1}{2}}]}{(Q+i0)^{\frac{1}{2}(\frac{1}{2}n+t)}} \right] \qquad (8)$$

and

$$\frac{F[(c^2 + P)^t_+]}{\Gamma(t+1)} = (-1)^{t+1}i2^{t+\frac{1}{2}n}\pi^{\frac{1}{2}n-1}c^{\frac{1}{2}n+t}$$

$$\times \left[e^{-\frac{1}{2}q\pi i}\, \frac{K_{\frac{1}{2}n+t}\,[c(Q-i0)^{\frac{1}{2}}]}{(Q-i0)^{\frac{1}{2}(\frac{1}{2}n+t)}} - e^{\frac{1}{2}q\pi i}\, \frac{K_{\frac{1}{2}n+t}\,[c(Q+i0)^{\frac{1}{2}}]}{(Q+i0)^{\frac{1}{2}(\frac{1}{2}n+t)}} \right]$$

$$+ (2\pi)^n \sum_{m=0}^{t} \frac{(-1)^m (c/2)^{2t-2m}}{4^m m!(t-m)!} L^m \delta(s). \qquad (9)$$

According to these equations,

$$\frac{F[(c^2 + P)^t]}{\Gamma(t+1)} = (2\pi)^n \sum_{m=0}^{t} \frac{(-1)^m (c/2)^{2t-2m}}{4^m m!(t-m)!} L^m \delta(s). \qquad (10)$$

3. Homogeneous Functions

3.1. Introduction

We have already dealt with some types of homogeneous generalized functions in Chapter I (Sections 3 and 4), as well as in Section 2 of the present chapter. We now wish to consider arbitrary homogeneous generalized functions of any degree in n dimensions. We recall that a generalized function $f(x) = f(x_1, \ldots, x_n)$ is called homogeneous of degree λ if for any $\alpha > 0$ we have

$$f(\alpha x_1, \ldots, \alpha x_n) = \alpha^\lambda f(x_1, \ldots, x_n) \qquad (1)$$

or, which is the same,

$$\left(f, \varphi\left(\frac{x_1}{\alpha}, \ldots, \frac{x_n}{\alpha} \right) \right) = \alpha^{\lambda+n}(f, \varphi(x_1, \ldots, x_n)). \qquad (2)$$

In particular, to every ordinary homogeneous function $f(x)$ of degree λ with Re $\lambda > - n$, continuous for $x \neq 0$, corresponds the generalized function

$$(f, \varphi) = \int f(x) \, \varphi(x) \, dx, \cdot$$

which is also homogeneous of degree λ. If, on the other hand, $f(x)$ is an ordinary homogeneous function of degree λ with Re $\lambda \leqslant - n$, its singularity at the origin is nonsummable, and it is not clear that there exists a regularization which is also homogeneous and of degree λ. We shall therefore call such an ordinary function *formally homogeneous*.

We shall later need to make use of the following property of homogeneous generalized functions.

Theorem. A generalized function f is homogeneous of degree λ if and only if it satisfies the Euler equation

$$\sum_{i=1}^{n} x_i \frac{\partial f}{\partial x_i} = \lambda f. \tag{3}$$

Proof. Note first that when applied to any $\varphi(x)$ in K this equation becomes

$$-\left(f, \sum_{i=1}^{n} \frac{\partial(x_i \varphi)}{\partial x_i}\right) = \lambda(f, \varphi),$$

or

$$-\left(f, \sum_{i=1}^{n} x_i \frac{\partial \varphi}{\partial x_i}\right) = (\lambda + n)(f, \varphi). \tag{4}$$

Let us assume that f is a generalized homogeneous function of degree λ, and thus satisfies Eq. (2), which we differentiate with respect to α. It is clear that on the left-hand side the differential operator may be applied to φ, and we obtain

$$-\left(f, \sum_{i=1}^{n} \frac{x_i}{\alpha^2} \frac{\partial}{\partial x_i} \varphi\left(\frac{x_1}{\alpha}, ..., \frac{x_n}{\alpha}\right)\right) = (\lambda + n) \, \alpha^{\lambda+n-1}(f, \varphi(x_1, ..., x_n)).$$

In order to obtain Eq. (4) now we need only set $\alpha = 1$.

Conversely, let f be a generalized function satisfying Eq. (4). Consider the fraction

$$\frac{(f, \varphi(x_1/\alpha, ..., x_n/\alpha))}{\alpha^{\lambda+n}}$$

for $\alpha > 0$. Differentiating this with respect to α and using Eq. (4), we find that the derivative vanishes. This means that

$$\frac{(f, \varphi(x_1/\alpha, ..., x_n/\alpha))}{\alpha^{\lambda+n}} = \text{const} = \frac{(f, \varphi(x_1, ..., x_n))}{1},$$

or f is homogeneous of degree λ.

3.2. Positive Homogeneous Functions of Several Independent Variables

Consider a continuous homogeneous function of the first degree in the variables x_1, x_2, ..., x_n, and assume that this function is positive everywhere with the possible exception of the origin. An example of such a function is $r = \sqrt{x_1^2 + x_2^2 + ... + x_n^2}$ or more generally $P^{1/2m}(x_1, x_2, ..., x_n)$ where P is a positive definite form of degree $2m$. We shall denote this function by $f(x_1, x_2, ..., x_n) = f(x)$, and consider the generalized function

$$(f^\lambda(x), \varphi(x)) = \int f^\lambda(x)\, \varphi(x)\, dx. \tag{1}$$

for Re $\lambda > - n$. Here $\varphi(x) = \varphi(x_1, x_2, ..., x_n)$ is, as usual, a function in K. It is easily shown that for those values of λ for which the integral converges, Eq. (1) defines a homogeneous generalized function of degree λ analytic in λ. We shall show now that f^λ can be analytically continued to the entire complex λ plane except for the points $\lambda = - n - k$, where k is a nonnegative integer, where f^λ may have simple poles.

The generalized function so obtained will be the regularization of $\int f^\lambda \varphi\, dx$ (see Chapter I, Section 1.7), and since for $\lambda > - n$ this regularization coincides with the ordinary integral, we shall maintain the notation $\int f^\lambda \varphi\, dx$.

To prove the assertion in n dimensions, let G be any region containing the origin, let Γ be its (smooth) boundary, and for Re $\lambda > - n$ write the integral of (1) in the form

$$(f^\lambda, \varphi) = \int_G f^\lambda(x)\, [\varphi(x) - \varphi(0)]\, dx$$
$$+ \int_{R-G} f^\lambda(x)\, \varphi(x)\, dx + \varphi(0) \int_G f^\lambda(x)\, dx \tag{2}$$

(here $R - G$ is the complement of G). Now we may rewrite the last integral on the right: since $f^\lambda(x)$ is homogeneous of degree λ, we have

$$\sum x_k \frac{\partial f^\lambda(x)}{\partial x_k} = \lambda f^\lambda(x)$$

so that

$$\int_G f^\lambda(x)\, dx = \frac{1}{\lambda} \sum_{k=1}^{n} \int_G x_k \frac{\partial f^\lambda(x)}{\partial x_k}\, dx. \qquad (3)$$

Now we may integrate each of the terms on the right-hand side (that is, the kth term with respect to the kth variable) by parts to rewrite it in the form

$$\frac{1}{\lambda} \int_\Gamma f^\lambda(x)[x_1\, dx_2 \ldots dx_n - x_2\, dx_1\, dx_3 \ldots dx_n$$

$$+ \ldots \pm x_n\, dx_1 \ldots dx_{n-1}] - \frac{n}{\lambda} \int_G f^\lambda(x)\, dx\,,$$

so that

$$\int_G f^\lambda(x)\, dx = \frac{1}{\lambda+n} \int_\Gamma f^\lambda(x)[x_1\, dx_2 \ldots dx_n - \ldots \pm x_n\, dx_1 \ldots dx_{n-1}]. \qquad (4)$$

Inserting this into Eq. (2) we have

$$(f^\lambda, \varphi) = \int_G f^\lambda(x)[\varphi(x) - \varphi(0)]\, dx + \int_{R-G} f^\lambda(x)\, \varphi(x)\, dx$$

$$+ \frac{\varphi(0)}{\lambda+n} \int_\Gamma f^\lambda(x)[x_1\, dx_2 \ldots dx_n - \ldots \pm x_n\, dx_1 \ldots dx_{n-1}]. \qquad (5)$$

Now $f^\lambda(x)$ is homogeneous of degree λ, and $\varphi(x) - \varphi(0)$ has a zero of first order at $x = 0$, so that the first integral on the right-hand side converges for $\mathrm{Re}\,\lambda > -n-1$, while the second and third converge for all λ, since the integrals are taken outside a neighborhood of the origin and $\varphi(x)$ has bounded support. Thus (5) is meaningful for $\mathrm{Re}\,\lambda > -n-1$ and defines a generalized function analytic in λ whose first pole is located at $\lambda = -n$. Note that the location of the first singularity depends on the dimension.

Thus Eq. (2) is an explicit expression for the analytic continuation of $\int f^\lambda(x)\varphi(x)dx$ into $\mathrm{Re}\,\lambda > -n-1$, $\lambda \neq -n$. Before extending this analytic continuation, let us study the residue at $\lambda = -n$, which is of some importance. From Eq. (5) it is clear that this residue is

$$\varphi(0) \int_\Gamma \frac{x_1\, dx_2 \ldots dx_n - \ldots \pm x_n\, dx_1 \ldots dx_{n-1}}{f^n(x)}. \qquad (6)$$

Henceforth we shall briefly denote the differential form

$$x_1\, dx_2 \ldots dx_n - x_2\, dx_1\, dx_3 \ldots dx_n + \ldots \pm x_n\, dx_1 \ldots dx_{n-1}$$

by ω. It is easily verified that if σ is a hypersurface, then $n^{-1} \int_\sigma \omega$ is the volume of the cone whose vertex is at the origin and whose base is σ.

Remark.   Let us assume that Γ is given by an equation of the form $P(x) = 1$, where $P(x)$ is a supplementary homogeneous function of degree one; we may then show that ω is related to the surface $P(x) - 1 = 0$ in the sense of Section 1.2. To prove this we need only verify that at points on Γ,

$$dP \cdot \omega = dx_1 \ldots dx_n.$$

Thus we multiply $dP = \partial P/\partial x_1 \, dx_1 + \ldots + \partial P/\partial x_n \, dx_n$ by the expression for ω. Using the anticommutation rule $dx_i dx_j = - dx_j dx_i$, the left-hand side becomes

$$\left(\frac{\partial P}{\partial x_1} x_1 + \ldots + \frac{\partial P}{\partial x_n} x_n \right) dx_1 \ldots dx_n.$$

Now by Euler's theorem the expression in the brackets is just P, and by definition $P = 1$ on Γ.

Returning to the residue of $\int f^\lambda(x)\varphi(x)dx$ at $\lambda = -n$, we may write it briefly in the form

$$\varphi(0) \int_\Gamma \frac{\omega}{f^n(x)} . \tag{7}$$

Since this is the residue of an analytic function, it cannot depend on Γ.[1] Therefore the integral in (7) is determined by the values of $f^{-n}(x)$ in any neighborhood of the origin.

We shall call $\int_\Gamma \omega/f^n(x)$ the *residue at the origin of the ordinary homogeneous function* $f^{-n}(x)$ of degree $-n$. [We emphasize that this is not the same as the residue of the analytic generalized function f^λ at $\lambda = -n$, which is this integral multiplied by $\delta(x)$.] If this residue vanishes, the analytic generalized function $f^\lambda(x)$ has no pole at $\lambda = -n$. Then according to the fifth property of homogeneous generalized functions listed in Chapter I, Section 3.11, Eq. (5) defines a generalized function which is homogeneous of degree $-n$ at $\lambda = -n$. We shall make use of this fact in the next section.

The residue at the origin of a homogeneous function of degree $-n$ has a relatively simple geometric meaning. This may be seen by choosing Γ to be the closed surface whose equation is $f(x) = 1$. We may then write the residue in the form $\int_\Gamma \omega$. Since $\int_\sigma \omega$ is n times the volume of the cone whose base is σ, it follows that the residue of $f^{-n}(x)$ is nV, where V is the volume of the region $f(x) \leqslant 1$.

Let us now proceed to extend the analytic continuation of the integral. As before, by adding to and subtracting from $\varphi(x)$ higher terms in its

---

[1] It is left to the reader to prove this statement.

Taylor's series, we arrive at a formula analogous to (5) which will give the analytic continuation into $\operatorname{Re} \lambda > -n - k - 1$. To do this we need only use the homogeneity of $f^\lambda(x) x_1^{\alpha_1} \ldots x_n^{\alpha_n}$ to transform integrals of the form

$$\int_G f^\lambda(x)\, x_1^{\alpha_1} \ldots x_n^{\alpha_n}\, dx,$$

into surface integrals over the boundary Γ, as we have already done for

$$\int_G f^\lambda(x)\, dx.$$

The final expression so obtained will be

$$\int_G f^\lambda(x)\, \varphi(x)\, dx$$

$$= \int_G f^\lambda(x) \left[\varphi(x) - \varphi(0) - \ldots - \frac{1}{k!} \sum_{\alpha_1 + \ldots + \alpha_n = k} x_1^{\alpha_1} \ldots x_n^{\alpha_n} \frac{\partial^k \varphi(0)}{\partial x_1^{\alpha_1} \ldots \partial x_n^{\alpha_n}} \right] dx$$

$$+ \int_{R-G} f^\lambda(x)\, \varphi(x)\, dx$$

$$+ \sum_{m=0}^{k} \frac{1}{m!(\lambda + n + m)} \sum_{\alpha_1 + \ldots + \alpha_n = m} \frac{\partial^m \varphi(0)}{\partial x_1^{\alpha_1} \ldots \partial x_n^{\alpha_n}} \int_\Gamma f^\lambda(x)\, x_1^{\alpha_1} \ldots x_n^{\alpha_n}\, \omega. \qquad (8)$$

This equation shows that $\int f^\lambda(x)\varphi(x)dx$ is an analytic function for $\operatorname{Re} \lambda > -n - k - 1$ except at $\lambda = -n - m$, where $m = 0, 1, \ldots, k$ (at $\lambda = -n - k - 1$ the first integral fails to converge), where it has simple poles whose residues are obvious from Eq. (8). In particular, the residue of the generalized function $f^\lambda(x)$ at $\lambda = -n - m$ is

$$\frac{(-1)^m}{m!} \sum_{\alpha_1 + \ldots + \alpha_n = m} \frac{\partial^m \delta(x)}{\partial x_1^{\alpha_1} \ldots \partial x_n^{\alpha_n}} \int_\Gamma \frac{x_1^{\alpha_1} \ldots x_n^{\alpha_n}}{f^{n+m}(x)}\, \omega. \qquad (9)$$

Thus, for instance, we obtain the following results:

$$\lim_{\lambda \to -n} (\lambda + n) f^\lambda(x) = \delta(x) \int_\Gamma \frac{\omega}{f^n(x)},$$

$$\lim_{\lambda \to -n-1} (\lambda + n + 1) f^\lambda(x) = -\frac{\partial \delta(x)}{\partial x_1} \int_\Gamma \frac{x_1 \omega}{f^{n+1}(x)} - \ldots - \frac{\partial \delta(x)}{\partial x_n} \int_\Gamma \frac{x_n \omega}{f^{n+1}(x)},$$

and similar expressions for the other singular points.

Now the integral

$$\int_\Gamma \frac{x_1^{\alpha_1} \dots x_n^{\alpha_n}}{f^{n+m}(x)}\, \omega$$

is independent of the choice of Γ. If we choose this surface to be $f(x) = 1$ again, the integral becomes $\int_{f=1} x_1^{\alpha_1} \dots x_n^{\alpha_n} \omega$, and this is equal, up to a factor, to one of the mth moments of the region bounded by $f = 1$.

This may be proved, for instance, as follows. Consider the moment

$$I = \int_{f \leqslant 1} x_1^{\alpha_1} \dots x_n^{\alpha_n}\, dx$$

of the region $f \leqslant 1$. On this region we introduce the new coordinates $u_1 = f$, u_2, ..., u_n with Jacobian $D\left(\frac{x}{u}\right) \neq 0$. In these coordinates

$$dx = D\left(\frac{x}{u}\right) du_1\, du_2 \dots du_n.$$

Let us write

$$\omega = D\left(\frac{x}{u}\right) du_2 \dots du_n.$$

On $f = 1$ this differential form is the same as the ω already defined. We may then write I as

$$I = \int_0^1 \int_{f=c} x_1^{\alpha_1} \dots x_n^{\alpha_n}\, \omega\, dc.$$

Obviously the inner integral is a homogeneous function of its argument c of degree $\alpha_1 + \dots + \alpha_n - n + 1$. This means that

$$\int_{f=c} x_1^{\alpha_1} \dots x_n^{\alpha_n} \omega = c^{\alpha_1 + \dots + \alpha_n - n + 1} \int_{f=1} x_1^{\alpha_1} \dots x_n^{\alpha_n} \omega$$

and

$$I = \int_{f=1} x_1^{\alpha_1} \dots x_n^{\alpha_n} \omega \int_0^1 c^{\alpha_1 + \dots + \alpha_n - n + 1}\, dc,$$

so that

$$\int_{f \leqslant 1} x_1^{\alpha_1} \dots x_n^{\alpha_n}\, dx = \frac{1}{\alpha_1 + \dots + \alpha_n - n + 2} \int_{f=1} x_1^{\alpha_1} \dots x_n^{\alpha_n} \omega.$$

Thus the residue of the generalized analytic function $f^\lambda(x)$ at $\lambda = -n - m$ is a linear differential operator of order m with constant coefficients applied to $\delta(x)$. The coefficients of this operator are, up to a factor independent of f, all the mth moments of the region bounded by the $f = 1$ surface.

In analogy with what we have done before, we shall call the intergals

$$\int_\Gamma \frac{x_1^{\alpha_1} \dots x_n^{\alpha_n}}{f^{n+m}(x)} \, \omega, \qquad \alpha_1 + \dots + \alpha_n = m,$$

the *residues at the origin of the ordinary homogeneous function* $f^{-n-m}(x)$ of degree $-n-m$. The residue of the generalized analytic function $f^\lambda(x)$ at $\lambda = -n-m$ is a linear combination of the above residues with coefficients

$$\frac{(-1)^m}{m!} \frac{\partial^m \delta(x)}{\partial x_1^{\alpha_1} \dots \partial x_n^{\alpha_n}}.$$

If all the residues of the ordinary function $f^{-n-m}(x)$ vanish, then according to Eq. (8) with $\lambda = -n-m$, to this function corresponds a homogeneous generalized function of degree $-n-m$. We shall make use of this fact in Section 3.5.

To calculate the generalized function f^λ for Re $\lambda < -n$, Eq. (8) can be replaced by a more convenient and symmetric expression. For this purpose, note that for Re $\lambda < -n-k$, an integral of the form

$$\int_{R-G} f^\lambda(x) \, x_1^{\alpha_1} \dots x_n^{\alpha_n} \, dx, \qquad \alpha_1 + \dots + \alpha_n = m < k, \tag{10}$$

converges. Now the boundary of $R - G$ is the same surface Γ that bounds G, except that the orientation is opposite. If we therefore transform Eq. (10) to a surface integral, we obtain

$$\int_{R-G} f^\lambda(x) \, x_1^{\alpha_1} \dots x_n^{\alpha_n} \, dx = - \frac{1}{\lambda + m + n} \int_\Gamma f^\lambda(x) \, x_1^{\alpha_1} \dots x_n^{\alpha_n} \omega.$$

Let us use this expression to replace the surface integrals over Γ in Eq. (8) by volume integrals over $R - G$. Uniting all the volume integrals thus obtained, we arrive at the representation

$$\int f^\lambda(x) \, \varphi(x) \, dx$$

$$= \int f^\lambda(x) \left[\varphi(x) - \varphi(0) - \dots - \frac{1}{k!} \sum_{\alpha_1 + \dots + \alpha_n = k} x_1^{\alpha_1} \dots x_n^{\alpha_n} \frac{\partial^k \varphi(0)}{\partial x_1^{\alpha_1} \dots \partial x_n^{\alpha_n}} \right] dx \tag{11}$$

for our functional, a representation valid in the strip $-n-k-1 <$ Re $\lambda < -n-k$.

It may be noted that given any ordinary homogeneous function $\Phi(x)$ of degree λ (not necessarily of definite sign) with a singularity only at the origin, Eq. (11) can be used to regularize the integral

$$\int \Phi(x)\, \varphi(x)\, dx$$

for $-n - k - 1 < \text{Re } \lambda < -n - k$.

The functional obtained in this way is a homogeneous generalized function of degree λ. Thus every formally homogeneous (ordinary) function $\Phi(x)$ of degree λ with $\lambda \neq -n - k$ can be put in correspondence, using Eq. (11), with a generalized homogeneous function of the same degree, the regularization of $\Phi(x)$. Now a homogeneous function of any degree with a singularity only at the origin is determined by its values on any closed surface Γ which intersects avery ray from the origin at only one point. We may thus conclude that every continuous function on such a closed surface can be put in correspondence with a generalized homogeneous function of degree λ for $\lambda \neq -n - k$.

3.3. Generalized Homogeneous Functions of Degree $- n$

Consider the (ordinary) formally homogeneous function $\Phi(x) \equiv \Phi(x_1, x_2, ..., x_n)$, not necessarily of definite sign, and of degree $- n$. Let this function have a singularity (that is, a point of local nonsummability) only at the origin. We shall define the corresponding generalized function, and thereby regularize the divergent integral

$$\int \Phi(x)\, \varphi(x)\, dx,$$

by choosing an arbitrary region G containing the origin and writing

$$\int \Phi(x)\, \varphi(x)\, dx = \int_G \Phi(x)[\varphi(x) - \varphi(0)]\, dx + \int_{R-G} \Phi(x)\, \varphi(x)\, dx \ . \qquad (1)$$

The regularization of our integral defined in this way depends, of course, on the choice of G. Let us denote the generalized function so obtained by $\Phi \mid_G$, and study the way it behaves when G is replaced by some other region G_1. It is immediately obvious that if G is replaced by $G_1 \subset G$, the functional we obtain will differ by

$$\varphi(0) \int_{G-G_1} \Phi(x)\, dx$$

from the original. Thus

$$\Phi \mid_G - \Phi \mid_{G_1} = \delta(x) \int_{G-G_1} \Phi(x) \, dx. \tag{2}$$

Since Eq. (2) implies that the difference between $\Phi \mid_G$ and $\Phi \mid_{G_1}$ is a homogeneous generalized function of degree $-n$, it follows that the homogeneity or inhomogeneity of the generalized function defined by (1) is independent of the choice of G.

We may now ask under what circumstances the generalized function defined by (1) will be homogeneous of degree $-n$, that is, under what circumstances we have

$$\left(\Phi, \varphi \left(\frac{x}{\alpha} \right) \right) = (\Phi, \varphi).$$

Before proceeding with the calculation, let us compare our Eq. (1) with Eq. (5) of Section 3.2. If we write $\Phi = f^{-n}$ in the present Eq. (1), where $f(x)$ is a positive homogeneous function of the first degree, we shall obtain Eq. (5) of Section 3.2, except that $\int_\Gamma f^{-n}(x)\omega$, the residue of $f^{-n}(x)$, is replaced by zero. But the vanishing of this residue is necessary and sufficient for the generalized function $f^\lambda(x)$ to have no pole at $\lambda = -n$, or, as mentioned in Section 3.2, for Eq. (5) of that section to define a generalized homogeneous function of degree $-n$ at $\lambda = -n$.

We shall show now that in the more general case a similar condition is necessary and sufficient for the generalized function of Eq. (1) to be homogeneous of degree $-n$.

Consider the integral

$$\int_G \Phi(x) \left[\varphi \left(\frac{x}{\alpha} \right) - \varphi(0) \right] dx + \int_{R-G} \Phi(x) \, \varphi \left(\frac{x}{\alpha} \right) dx$$

in which we transform the independent variables according to $x_k/\alpha = x_k'$, and let αG denote the region obtained from G by this similarity transformation. We then have

$$\int_G \Phi(x) \left[\varphi \left(\frac{x}{\alpha} \right) - \varphi(0) \right] dx + \int_{R-G} \Phi(x) \, \varphi \left(\frac{x}{\alpha} \right) dx$$

$$= \int_G \Phi(x)[\varphi(x) - \varphi(0)] \, dx + \int_{R-G} \Phi(x) \, \varphi(x) \, dx + \varphi(0) \int_{G-\alpha G} \Phi(x) \, dx.$$

It is clear that $\int \Phi(x)\varphi(x)dx$ will be homogeneous if and only if

$$\int_{G-\alpha G} \Phi(x) \, dx = 0$$

for all α.

Let us restate this condition. We do this by introducing into $G - \alpha G$ the coordinate system $\rho, u_1, \ldots, u_{n-1}$, where $\rho = 1$ and $\rho = \alpha$ are the equations of Γ and $\alpha \Gamma$, respectively, and where the u_i are coordinates on the $\rho =$ const surfaces.

Now write $x_k = \rho x_k'$. Then

$$dx_1 \ldots dx_n = D \begin{pmatrix} x_1\ x_2 \ldots x_n \\ \rho\ \ u_1 \ldots u_{n-1} \end{pmatrix} d\rho\, du_1 \ldots du_{n-1}$$

$$= \rho^{n-1}\, d\rho(x_1'\, dx_2' \ldots dx_n' - x_2'\, dx_1'\, dx_3' \ldots dx_n' + \ldots \pm x_n'\, dx_1' \ldots dx_{n-1}') = \rho^{n-1}\, d\rho\omega,$$

where the differential form ω is on the $\rho = 1$ surface. On the other hand, $\Phi(x) = \rho^{-n}\Phi(x')$. Thus

$$\int_{G-\alpha G} \Phi(x)\, dx = \int_1^\alpha \frac{d\rho}{\rho} \int_\Gamma \Phi(x)\, \omega = \ln \alpha \int_\Gamma \Phi(x)\, \omega. \qquad (3)$$

Consequently the regularization of Eq. (1) for the integral of the formally homogeneous function $\Phi(x_1, \ldots, x_n)$, namely, the generalized function $\Phi \mid_G$, is a homogeneous generalized function of degree $- n$ if and only if

$$\int_\Gamma \Phi\, \omega = 0 \qquad (4)$$

If this condition is not fulfilled, $\Phi \mid_G$ is not homogeneous, but will be seen to be an associated generalized function.

We shall call the integral of Eq. (4) the *residue of $\Phi(x)$ at the origin* (this is now a generalization of the definition of Section 3.2). Consequently the generalized function (1) will be homogeneous of degree $- n$ if and only if the residue of $\Phi(x)$ vanishes at the origin.

DIFFERENTIATION OF HOMOGENEOUS FUNCTIONS OF DEGREE $-n + 1$

In Chapter I we encountered more than once situations in which an ordinary locally summable function f is differentiable everywhere except at isolated points and such that the derivative is not a locally summable function (so that it would be more correct to call it the *formal* derivative). We did not at the time have a recipe for calculating derivatives of arbitrary order in the generalized sense, and every time we dealt with a different function we had to go through a separate procedure. For instance, in Chapter I, Section 2.3 we calculated the Laplacian of $1/r$ in three dimensions essentially by going through the usual classical considerations involving eliminating a neighborhood of the origin and applying Green's theorem. Such a procedure is in a certain sense reproduced by the following equation, which we shall first state and then prove. Let Φ be an ordinary locally summable function of degree $-n+1$. Then

$$\frac{\partial \Phi}{\partial x_i} = \left(\frac{\partial \Phi}{\partial x_i}\right)\Big|_G + (-1)^{i-1}\delta(x_1, ..., x_n)\int_\Gamma \Phi\, dx_1 ... dx_{i-1}\, dx_{i+1} ... dx_n. \qquad (5)$$

Here $\partial \Phi/\partial x_i$ on the left-hand side is the derivative of Φ in the sense of generalized functions, and $(\partial \Phi/\partial x_i)\,|_G$ on the right-hand side denotes the generalized function corresponding to the formal, i.e. ordinary, derivative of Φ. Here Γ is the boundary of G.

The proof of Eq. (5) follows. We have

$$\left(\frac{\partial \Phi}{\partial x_i}, \varphi\right) = -\left(\Phi, \frac{\partial \varphi}{\partial x_i}\right) = -\int \Phi(x_1, ..., x_n)\frac{\partial \varphi(x_1, ..., x_n)}{\partial x_i}\, dx_1 ... dx_n$$

$$= -\int_G \Phi(x_1, ..., x_n)\frac{\partial}{\partial x_i}[\varphi(x_1, ..., x_n) - \varphi(0, ..., 0)]\, dx_1 ... dx_n$$

$$-\int_{R-G} \Phi(x_1, ..., x_n)\frac{\partial \varphi(0, ..., 0)}{\partial x_i}\, dx_1 ... dx_n.$$

Here $R-G$ is the complement of G.

By integrating by parts we obtain the functional $(\,(\partial \Phi/\partial x_i)\,|_G,\, \varphi)$, and all the integrals over Γ will cancel except

$$(-1)^{i-1}\varphi(0)\int_\Gamma \Phi\, dx_1 ... dx_{i-1}\, dx_{i+1} ... dx_n.$$

But this then proves Eq. (5). Now the left-hand side of that equation is independent of G, so the right-hand side must be also. The reader may verify this directly.

Example 1.   We assert that

$$\left(\frac{\partial^2}{\partial x^2} + \frac{\partial^2}{\partial y^2}\right)\ln(x^2 + y^2)^{-\frac{1}{2}} = 2\pi\delta(x, y). \qquad (6)$$

We prove it by noting first that

$$\frac{\partial}{\partial x}\ln(x^2 + y^2)^{-\frac{1}{2}} = \frac{-x}{x^2 + y^2}.$$

This is a valid equation both in the sense of ordinary and generalized functions, for $x(x^2 + y^2)^{-1}$ is locally summable. Now proceeding by Eq. (5), we obtain

$$\frac{\partial^2}{\partial x^2}\ln(x^2 + y^2)^{-\frac{1}{2}} = \frac{\partial}{\partial x}\frac{-x}{x^2 + y^2} = \left(\frac{x^2 - y^2}{(x^2 + y^2)^2}\right)\Big|_G - \delta(x, y)\int_\Gamma \frac{x\, dy}{x^2 + y^2}.$$

Similarly,

$$\frac{\partial^2}{\partial y^2} \ln (x^2 + y^2)^{-\frac{1}{2}} = \left(\frac{y^2 - x^2}{(x^2 + y^2)^2}\right)\bigg|_G + \delta(x, y) \int_\Gamma \frac{y\, dx}{x^2 + y^2}.$$

Addition of the last two equations gives

$$\left(\frac{\partial^2}{\partial x^2} + \frac{\partial^2}{\partial y^2}\right) \ln (x^2 + y^2)^{-\frac{1}{2}} = -\delta(x, y) \int_\Gamma \frac{x\, dy - y\, dx}{x^2 + y^2}.$$

Now let Γ be a circle with center at the origin; it is then easily shown that

$$-\int_\Gamma \frac{x\, dy - y\, dx}{x^2 + y^2} = 2\pi.$$

This completes the proof of (6). The result can also be stated differently, namely

$$\Delta \ln \frac{1}{r} = 2\pi\delta(x, y); \qquad r^2 = x^2 + y^2, \qquad \Delta = \frac{\partial^2}{\partial x^2} + \frac{\partial^2}{\partial y^2}.$$

Example 2. Let us calculate $\Delta(1/r)$ in three dimensions (that is, for $r^2 = x^2 + y^2 + z^2$). The first derivative can be taken in the ordinary sense:

$$\frac{\partial}{\partial x} \frac{1}{r} = -\frac{x}{r^3}, \qquad \frac{\partial}{\partial y} \frac{1}{r} = -\frac{y}{r^3}, \qquad \frac{\partial}{\partial z} \frac{1}{r} = -\frac{z}{r^3}.$$

The formal derivatives of these are, respectively,

$$\frac{3x^2 - r^2}{r^5}, \qquad \frac{3y^2 - r^2}{r^5}, \qquad \frac{3z^2 - r^2}{r^5},$$

and their sum vanishes. We now use Eq. (5) for each of the second derivatives, add the results, and obtain

$$\left(\frac{\partial^2}{\partial x^2} + \frac{\partial^2}{\partial y^2} + \frac{\partial^2}{\partial z^2}\right) \frac{1}{r} = \delta(x, y, z) \int_\Gamma \frac{x\, dy\, dz - y\, dx\, dz + z\, dx\, dy}{r^3}.$$

This time we choose Γ to be a sphere and then the integral is easily shown (for instance, in polar coordinates) to be -4π. We thus find that in three dimensions

$$\Delta \left(\frac{1}{r}\right) = -4\pi\delta(x, y, z). \tag{7}$$

A similar procedure can be used to calculate Δr^{2-n} in n dimensions ($n \geqslant 3$), which will give the result presented toward the end of Chapter I, Section 2.3.

The concept of the residue may be looked at from a different point of view that may help to clarify the reason that an ordinary homogeneous function $\Phi(x)$ of degree $- n$ with vanishing residue corresponds to a homogeneous generalized function.

Since $\Phi(x_1, ..., x_n)$ is a formally homogeneous function of degree $- n$, Euler's theorem implies that

$$\sum_{k=1}^{n} x_k \frac{\partial \Phi}{\partial x_k} = -n\Phi,$$

or, in the form of a divergence,

$$\sum_{k=1}^{n} \frac{\partial(x_k \Phi)}{\partial x_k} = 0.$$

This is a formal equation in the sense that it is satisfied at all points except the origin, where Φ has a singularity. On the other hand, we know that Euler's theorem as applied to generalized functions (see Section 3.1) characterizes the homogeneous ones of the appropriate degree.

It can be shown that if we consider $x_k \Phi$ to be a generalized function, then

$$\sum_{k=1}^{n} \frac{\partial(x_k \Phi)}{\partial x_k} = c\delta(x), \tag{8}$$

where c is the residue of the ordinary function Φ. Indeed, this follows from Eq. (5), according to which

$$\frac{\partial(x_k \Phi)}{\partial x_k} = \left(\frac{\partial(x_k \Phi)}{\partial x_k} \right) \Big|_G$$
$$+ \delta(x_1, ..., x_n) \int_{\Gamma} \Phi(-1)^{k-1} x_k \, dx_1 ... dx_{k-1} \, dx_{k+1} ... dx_n.$$

Summation over k gives

$$\sum_k \frac{\partial(x_k \Phi)}{\partial x_k} = \left(\sum_k \frac{\partial(x_k \Phi)}{\partial x_k} \right) \Big|_G + \delta(x_1, ..., x_n) \int_{\Gamma} \Phi \, \omega,$$

so that, since the divergence of $x_k \Phi$ vanishes and causes the first term on the right-hand side to vanish, we obtain Eq. (8).

This means that if $c = \int_{\Gamma} \Phi \omega \neq 0$, the homogeneity of $\Phi(x)$ breaks down at the origin. If, for instance, $\Phi(x)$ is the potential of a field, such breakdown of homogeneity reflects the presence of point sources at the singularities.

We have shown, thus, that if a formally homogeneous function $\Phi(x)$ of degree $-n$ is to define a homogeneous generalized function (by regularization of integrals of the form $\int \Phi \varphi dx$), then its residue at the origin must vanish.

Now if $\Phi(x)$ is of definite sign, of course, its residue cannot vanish. Thus a homogeneous function of degree $-n$ and of definite sign will not define a homogeneous generalized function. For instance, the generalized function $1/|x|$ on the line (see Chapter I, Section 3) is inhomogeneous. The generalized function $(x^2 + y^2)^{-1}$ [or more accurately, $(x^2 + y^2)^{-1}|_G$] in the plane, defined by

$$\int \frac{\varphi(x, y)}{x^2 + y^2} \, dx \, dy = \int_G \frac{\varphi(x, y) - \varphi(0, 0)}{x^2 + y^2} \, dx \, dy + \int_{R-G} \frac{\varphi(x, y)}{x^2 + y^2} \, dx \, dy,$$

is inhomogeneous. On the other hand, $(x^2 - y^2)(x^2 + y^2)^{-2}$, defined similarly by

$$\int \frac{x^2 - y^2}{(x^2 + y^2)^2} \varphi(x, y) \, dx \, dy$$

$$= \int_G \frac{x^2 - y^2}{(x^2 + y^2)^2} [\varphi(x, y) - \varphi(0, 0)] \, dx \, dy + \int_{R-G} \frac{x^2 - y^2}{(x^2 + y^2)^2} \varphi(x, y) \, dx \, dy,$$

is homogeneous, since the ordinary function $(x^2 - y^2)(x^2 + y^2)^{-2}$ has residue zero (as can be seen from symmetry considerations).

3.4. Generalized Homogeneous Functions of Degree $-n-m$

Now let $\Phi(x_1, ..., x_n)$ be an ordinary homogeneous function of degree $-n-m$, where m is an integer. Again let us choose some arbitrary region G containing the origin, and let us define the regularization of $\int \Phi(x)\varphi(x)dx$ by

$$\int \Phi(x)\,\varphi(x)\,dx$$

$$= \int_G \Phi(x) \left[\varphi(x) - \varphi(0) - ... - \frac{1}{m!} \sum_{\Sigma \alpha_j = m} x_1^{\alpha_1} ... x_n^{\alpha_n} \frac{\partial^m \varphi(0)}{\partial x_1^{\alpha_1} ... \partial x_n^{\alpha_n}} \right] dx$$

$$+ \int_{R-G} \Phi(x) \left[\varphi(x) - \varphi(0) - ... - \frac{1}{(m-1)!} \sum_{\Sigma \alpha_j = m-1} x_1^{\alpha_1} ... x_n^{\alpha_n} \frac{\partial^{m-1} \varphi(0)}{\partial x_1^{\alpha_1} ... \partial x_n^{\alpha_n}} \right] dx.$$

$$(1)$$

In order for this generalized function to be homogeneous, it is necessary and sufficient that

$$\int_{G-\alpha G} \Phi(x)\, x_1^{\alpha_1} \dots x_n^{\alpha_n}\, dx = 0, \qquad \alpha_1 + \dots + \alpha_n = m,$$

for all α or equivalently (see Section 3.3) that

$$\int_{\Gamma} \Phi(x)\, x_1^{\alpha_1} \dots x_n^{\alpha_n}\, \omega = 0, \qquad \alpha_1 + \dots + \alpha_n = m. \tag{2}$$

We shall call the integrals on the left-hand side of (2), as before, the residues of the ordinary function $\Phi(x)$.

Consequently an ordinary homogeneous function $\Phi(x)$ of integral degree $-n-m$ will correspond through Eq. (1) to a homogeneous generalized function of the same degree if and only if all of its residues vanish.

In order to clarify the meaning of this result, compare Eq. (1) to Eq. (8) of Section 3.2. By setting $\Phi(x) = f^{-n-m}(x)$ into Eq. (1), where $f(x)$ is a positive homogeneous function of degree one, we obtain Eq. (8) of Section 3.2, except that the integrals $\int_{\Gamma} f^{-n-m}(x) x_1^{\alpha_1} \dots x_n^{\alpha_n}\, \omega$, which represent the residues of $f^{-n-m}(x)$, do not occur and are thus set equal to zero. On the other hand, the vanishing of these residues is necessary and sufficient for the generalized analytic function $f^{\lambda}(x)$ to have no pole at $\lambda = -n-m$, in other words for Eq. (8) to define a homogeneous generalized function of degree $-n-m$ for this value of λ. If even one of the residues of $f^{-n-m}(x)$ fails to vanish, the integral

$$\int f^{-n-k}(x)\, \varphi(x)\, dx$$

regularized according to Eq. (1) is an associated homogeneous function of degree $-n-m$.

We may note that in addition to generalized homogeneous functions defined by ordinary homogeneous functions of integral degree $\lambda \leqslant -n$, we are familiar with generalized homogeneous functions related to no ordinary functions: these are $\delta(x)$ and its derivatives. The kth derivative of $\delta(x)$ is a homogeneous generalized function of degree $-n-k$. The number of different kth derivatives is exactly equal to the number of different equations in (2). In this sense the set of all homogeneous functions of degree $-n-m$, as is true for any nonintegral value of the degree λ, is the same as the set of all continuous functions on a closed surface enclosing the origin.

3.5. Generalized Functions of the Form $r^\lambda f$, Where f Is a Generalized Function on the Unit Sphere

Every ordinary homogeneous function of degree λ can be written in the form

$$\Phi(x_1, ..., x_n) = \Phi(r\omega_1, r\omega_2, ..., r\omega_n) = r^\lambda f(\omega_1, ..., \omega_n), \tag{1}$$

where $f(\omega_1, ..., \omega_n)$ is a function on the unit sphere Ω whose equation is $\Sigma \omega_i^2 = 1$. For simplicity let us choose $\operatorname{Re} \lambda > -n$. The corresponding regular functional can be written in the form

$$(\Phi(x_1, ..., x_n), \varphi(x_1, ..., x_n)) = \int \Phi\varphi \, dx_1 \, ... \, dx_n$$

$$= \int_0^\infty r^{\lambda+n-1} \, dr \int_\Omega f(\omega_1, ..., \omega_n) \, \varphi(r\omega_1, ..., r\omega_n) \, d\omega = \int_0^\infty r^{\lambda+n-1} u(r) \, dr,$$

where $\varphi(x_1, ..., x_n)$ is any function in K, and

$$u(r) = \int_\Omega ... \int f(\omega_1, ..., \omega_n) \, \varphi(r\omega_1, ..., r\omega_n) \, d\omega.$$

As usual, $d\omega$ here represents the element of area on Ω.

Now let $f(\omega_1, ..., \omega_n)$ be any generalized function on Ω. Then we may define the generalized function

$$\Phi_\lambda(x_1, ..., x_n) = r^\lambda f(\omega_1, ..., \omega_n) \tag{2}$$

by the relation

$$(\Phi_\lambda, \varphi) = \int_0^\infty r^{\lambda+n-1} u(r) \, dr, \tag{3}$$

where

$$u(r) = (f, \varphi(r\omega_1, ..., r\omega_n)). \tag{4}$$

Now $u(r)$ is infinitely differentiable and has bounded support, since $\varphi(r\omega_1, ..., r\omega_n)$ is infinitely differentiable and has bounded support with respect to r. This can be seen, for instance, by noting that

$$\frac{\partial \varphi(r\omega_1, ..., r\omega_n)}{\partial r} = \frac{\partial}{\partial r} \varphi(r\omega_1, ..., r\omega_n) = \sum_{i=1}^n \omega_i \frac{\partial \varphi}{\partial x_i} .$$

Equation (2) shows that the functional Φ_λ is analytic in λ for $\lambda \neq -n - k$, where k is a nonnegative integer. At $\lambda = -n - k$ it has simple poles.

Let us find the residue at $\lambda = -n$. We have

$$(\Phi_\lambda, \varphi) = \int_0^\infty r^{\lambda+n-1}(f, \varphi(r\omega_1, ..., r\omega_n))\, dr = \int_0^\infty r^{\lambda+n-1}u(r)\, dr$$

$$= \int_0^1 r^{\lambda+n-1}[u(r) - u(0)]\, dr + \int_1^\infty r^{\lambda+n-1}u(r)\, dr + \frac{u(0)}{\lambda + n}.$$

The first and second integrals in this expression converge for $\mathrm{Re}\,\lambda > -n - 1$. Thus the residue of (Φ_λ, φ) at $\lambda = -n$ is

$$u(0) = c\varphi(0)\,,$$

where $c = (f, 1)$ is the value obtained when the functional f operates on the function equal to unity on Ω. Thus the residue of Φ_λ at $\lambda = -n$ is $c\delta(x_1, ..., x_n)$.

It is easily shown that if $f = f_\lambda$ is a generalized function analytic in λ and if it is regular at $\lambda = -n$, then the generalized function $\Phi_\lambda = r^\lambda f_\lambda$ has a residue at $\lambda = -n$ equal to $c_{-n}\delta(x_1, ..., x_n)$, where $c_{-n} = (f_\lambda, 1)$ at $\lambda = -n$.

Let us now find the residue of $\Phi_\lambda = r^\lambda f$ at $\lambda = -n - k$. The residue of the ordinary analytic function (Φ_λ, φ) at $\lambda = -n - k$ is

$$\frac{1}{(k-1)!}\left[\frac{\partial^{k-1}}{\partial r^{k-1}}u(r)\right]_{r=0},$$

where $u(r)$ is defined by Eq. (4). Therefore

$$\left[\frac{\partial^{k-1}}{\partial r^{k-1}}u(r)\right]_{r=0} = \left(f, \frac{\partial^{k-1}}{\partial r^{k-1}}\varphi(r\omega_1, ..., r\omega_n)\right)_{r=0}$$

$$= \sum_{\Sigma\alpha_j=k-1}\left(f, \omega_1^{\alpha_1}...\omega_n^{\alpha_n}\frac{\partial^{k-1}}{\partial x_1^{\alpha_1}...\partial x_n^{\alpha_n}}\varphi(x_1, ..., x_n)\right)\bigg|_{x_1=...=x_n=0}$$

$$= \sum_{\Sigma\alpha_j=k-1}\frac{\partial^{k-1}}{\partial x_1^{\alpha_1}...\partial x_n^{\alpha_n}}\varphi(0, ..., 0)\,(f, \omega_1^{\alpha_1}, ..., \omega_n^{\alpha_n}).$$

As $f(\omega_1, ..., \omega_n)$ is a functional on the $r = 1$ surface, where $\omega_i = x_i$, we may conclude that the residue of Φ_λ at $\lambda = -n - k$ is

$$\frac{(-1)^{k-1}}{(k-1)!}\sum_{\Sigma\alpha_j=k-1}c_{\alpha_1...\alpha_n}\frac{\partial^{k-1}}{\partial x_1^{\alpha_1}...\partial x_n^{\alpha_n}}\delta(x_1, ..., x_n),$$

where $c_{\alpha_1...\alpha_n} = (f, x_1^{\alpha_1}...x_n^{\alpha_n})$.

4. Arbitrary Functions Raised to the Power λ

4.1. Reducible Singular Points

Let $G(x_1, ..., x_n)$ be any infinitely differentiable function. In this section we shall study $G^\lambda(x_1, ..., x_n)$, or actually the functional

$$(G^\lambda, \varphi) = \int_{G>0} G^\lambda(x_1, ..., x_n) \, \varphi(x_1, ..., x_n) \, dx_1 \, ... \, dx_n \tag{1}$$

as an analytic function of λ. The singularities of this analytic function are intimately related to the nature of the $G = 0$ hypersurface. We shall not consider all types of possible $G = 0$ hypersurfaces, restricting our considerations to those surfaces, most important in all applications, which consist of points that we shall call *reducible*.

A function $G(x_1, ..., x_n)$ is called *equivalent to a homogeneous function in the neighborhood of some point M* if in this neighborhood there exists a local coordinate system $\xi_1, ..., \xi_n$, in which G becomes a homogeneous function. Obviously equivalence to a homogeneous function can be defined not only in an affine space, but on any analytic manifold, for instance on a sphere in such a space.

We shall define a *reducible point* on a manifold by induction on the dimension of the space in which the manifold is imbedded or of the manifold itself. Assume that reducible points have been defined for manifolds of dimension less than n or for manifolds in a space of dimension less than n. We shall call a point M of the $G(x_1, ..., x_n) = 0$ manifold reducible if:

(1) $G(x_1, ..., x_n)$ is equivalent to a homogeneous function in a neighborhood of M, and

(2) the intersection of the $G = 0$ manifold with a sufficiently small sphere centered on M is a manifold each of whose points is reducible on the sphere.

For the case of one dimension, that is for a function $G(x)$ on the line, the second requirement is clearly unnecessary. In this one-dimensional case assume $G(x_0) = 0$; then x_0 is called reducible if in its neighborhood there exists a nonsingular infinitely differentiable function $\xi = X(x)$ such that $X(x_0) = 0$ and

$$G[X^{-1}(\xi)] \equiv \xi^m.$$

In the neighborhood of a reducible point M we can introduce local coordinates $\xi_1, ..., \xi_n$ in which G becomes a homogeneous function of degree m.

If these local coordinates can be chosen so that G depends on k variables, but cannot be chosen so that it depends on fewer, M is called *a point of order k and of degree m.*

In this way we can associate with every point of a manifold two numbers: its order k and its degree m. If, in particular, grad $G \neq 0$ on a hypersurface $G = 0$, every point of this hypersurface will be of order one and of degree one. Indeed, at any such point one of the coordinates can be chosen as $\xi_1 = G(x_1, \ldots, x_n)$ itself, and the rest can be chosen arbitrarily. In this coordinate system, obviously, the G function becomes ξ_1, which is a homogeneous function of the first degree depending on a single variable. Thus an ordinary point of a surface is a point of order one and of degree one. But if the $G = 0$ surface, for instance, consists of the three coordinate planes in three dimensions, or $G = x_1 x_2 x_3$, the origin is a point of order three, the coordinate axes are points of order two, and all other points are of order one. It is left as an exercise to the reader to find the degrees of these various points.

Because the reader may find some difficulty in analyzing Eq. (1) in general, we shall start by considering the case in which the $G = 0$ manifold consists only of first order points (Section 4.2). In Section 4.3 we deal with points of orders one and two, and in Section 4.4 we consider the general case.

We shall also need the following theorem, which we state without proof.

Theorem. Let $G(x_1, \ldots, x_n)$ be a polynomial. If the $G = 0$ hypersurface consists only of reducible points, it can be decomposed into a finite number of connected components (or submanifolds) each of which consists of points of a given order and a given degree. In particular, the $G = 0$ hypersurface has a finite number of points of order n, a finite number of lines (manifolds of dimension one) of order $(n - 1)$, etc., terminating finally with a finite number of $(n - 1)$-dimensional components of order one.

Now recall that in the neighborhood of a point of order one we can introduce local coordinates in which G depends on just a single variable, in the neighborhood of a point of order two G will depend on two variables in suitable local coordinates, etc. This means that in considering functions with reducible points of order one, we need only study generalized functions of a single independent variable; similarly, in considering functions with reducible points of order two we shall be studying generalized functions of two independent variables, etc.

Before proceeding we wish to recall that it was shown in Appendix 1.2 of Chapter I, that if there exists a covering by a countable system of open neighborhoods D_i such that every ball intersects only a finite number of these neighborhoods, every $\varphi(x)$ in K can be written as a finite sum $\varphi(x) = \Sigma \varphi_i(x)$ of functions in K such that each φ_i vanishes outside of D_i. For this reason we may always assume whatever function $\varphi(x_1, \ldots, x_n)$ we are dealing with to have support in an arbitrarily small neighborhood of the point in question.

The integral of Eq. (1), which depends on a single complex parameter, will be treated by considering a certain auxiliary function of two complex variables. This method shall be described for the simplest case in Section 4.3.

4.2. The Generalized Function G^λ When $G = 0$ Consists Entirely of First Order Points

Let $G(x_1, \ldots, x_n)$ be an infinitely differentiable function such that the $G = 0$ surface is bounded and consists of reducible points of order one. Consider the functional

$$(G^\lambda, \varphi) = \int \ldots \int_{G>0} G^\lambda(x_1, \ldots, x_n) \, \varphi(x_1, \ldots, x_n) \, dx_1 \ldots dx_n, \qquad (1)$$

which depends on the parameter λ.

For Re $\lambda > 0$ this integral converges and is an analytic function of λ. We shall prove the following:

Theorem. The generalized function G is meromorphic in λ. To every connected component (of the $G = 0$ manifold) consisting of points of degree l corresponds a sequence of simple poles at

$$\lambda = -\frac{1}{l}, -\frac{2}{l}, \ldots, -\frac{n}{l}, \ldots. \qquad (2)$$

In particular, if the $G = 0$ surface has no singular points, G has poles only at

$$\lambda = -1, -2, \ldots, -n, \ldots. \qquad (3)$$

Proof. Let M be any point on the $G = 0$ surface. According to what we have said, we may without loss of generality assume that $\varphi(x_1, \ldots, x_n) = 0$ outside some fixed small neighborhood D of M. Let D_1 be the intersection of D with the region $G \geqslant 0$. Then

$$(G^\lambda, \varphi) = \int \ldots \int_{G>0} G^\lambda(x_1, \ldots, x_n) \, \varphi(x_1, \ldots, x_n) \, dx_1 \ldots dx_n$$

$$= \int \ldots \int_{D_1} G^\lambda(x_1, \ldots, x_n) \, \varphi(x_1, \ldots, x_n) \, dx_1 \ldots dx_n.$$

Since M is a point of order one (and degree l), we can introduce into D_1 a coordinate system in which $G(x_1, ..., x_n) = \xi_1^l$, while the remaining $n - 1$ of the ξ_i are chosen as arbitrary infinitely differentiable functions of the x_i, except that we require that in D_1 the Jacobian

$$D\begin{pmatrix} x_1 \cdots x_n \\ \xi_1 \cdots \xi_n \end{pmatrix} > 0.$$

In these new coordinates let us write

$$\varphi(x_1, ..., x_n) = \varphi_1(\xi_1, ..., \xi_n)$$

and then

$$\int_{D_1} \cdots \int G^\lambda(x_1, ..., x_n)\, \varphi(x_1, ..., x_n)\, dx_1 ... dx_n$$

$$= \int_{D_1} \cdots \int \xi_1^{\lambda l}\, \varphi_1(\xi_1, ..., \xi_n)\, D\begin{pmatrix} x_1 \cdots x_n \\ \xi_1 \cdots \xi_n \end{pmatrix} d\xi_1 ... d\xi_n.$$

Let

$$\psi(\xi_1) = \int \cdots \int \varphi_1(\xi_1, ..., \xi_n)\, D\begin{pmatrix} x_1 \cdots x_n \\ \xi_1 \cdots \xi_n \end{pmatrix} d\xi_2 ... d\xi_n, \qquad (4)$$

where the integral is taken over the intersection of D_1 with the surface whose equation is $G(x_1, ..., x_n) = \xi_1 = \text{const.}$ This function $\psi(\xi_1)$ is defined in D_1, is infinitely differentiable and has bounded support; using it we can rewrite Eq. (2) in the form

$$(G^\lambda, \varphi) = \int_0^\infty \xi_1^{\lambda l}\, \psi(\xi_1)\, d\xi_1. \qquad (5)$$

Let us now use the results of Chapter I, Section 3.2 concerning the poles and residues of $\int_0^\infty x^\lambda \psi(x)\, dx$. We then find that for every first order, lth degree point M of the $G = 0$ manifold, the generalized function $G^\lambda(x_1, ..., x_n)$ has a sequence of poles at

$$\lambda = -\frac{1}{l}, -\frac{2}{l}, ..., -\frac{n}{l}, ...$$

Further, the residue of the integral in Eq. (5) at $\lambda l = -n$ (that is, at $\lambda = -n/l$) is easily expressed in terms of ψ: it is, in fact, $\psi^{(n-1)}(0)/(n - 1)!$.

Let us obtain explicit expressions for the residues of G^λ.

The residue of (G^λ, φ) at $\lambda = -1/l$ is $\psi(0)$, or by Eq. (4),

$$\int \cdots \int \varphi_1(0, \xi_2, ..., \xi_n)\, D\begin{pmatrix} x_1 \cdots x_n \\ \xi_1 \cdots \xi_n \end{pmatrix}\bigg|_{\xi_1=0} d\xi_2 ... d\xi_n.$$

Let ω be the differential form multiplying φ_1 in the integral of Eq. (4):

$$\omega = D \begin{pmatrix} x_1 \cdots x_n \\ \xi_1 \cdots \xi_n \end{pmatrix} d\xi_2 \ldots d\xi_n. \tag{6}$$

Then the residue can be written

$$\int_G \varphi(x_1, \ldots, x_n) \, \omega. \tag{7}$$

But we also have

$$dx_1 \ldots dx_n = dv = d\xi_1 \omega,$$

and by writing

$$G^{1/l}(x_1, \ldots, x_n) = P(x_1, \ldots, x_n),$$

we will obtain

$$dv = dP \, \omega.$$

Now recall the definition of $\delta(P)$ in Section 1.3 of this chapter. We see that the residue given by (7) is simply $(\delta(P), \varphi)$. Thus we have found that the residue of G^λ at $\lambda = -1/l$ is $\delta(G^{1/l})$.

Let us now look for the other residues. For this purpose we remark that Eq. (4) can be differentiated with respect to ξ_1 by taking the derivative of the integrand. Again referring to Chapter I, we find that the reside of (G^λ, φ) at $\lambda = -(k+1)/l$ is

$$\frac{1}{k!} \int \cdots \int \frac{\partial^k}{\partial \xi_1^k} \left[\varphi_1(\xi_1, \ldots, \xi_n) D \begin{pmatrix} x_1 \cdots x_n \\ \xi_1 \cdots \xi_n \end{pmatrix} \right] \Big|_{\xi_1=0} d\xi_2 \ldots d\xi_n$$

$$= \frac{1}{k!} \int_{P=0} \frac{\partial^k}{\partial \xi_1^k} \left[\varphi \cdot D \begin{pmatrix} x \\ \xi \end{pmatrix} \right] d\xi_2 \ldots d\xi_n.$$

This may be written according to the result of Section 1.5 in the form

$$\frac{1}{k!} \int_{P=0} \omega_k(\varphi) = \left(\frac{(-1)^k}{k!} \delta^{(k)}(P), \varphi \right). \tag{8}$$

Thus the residue of the generalized function G^λ at $\lambda = -(k+1)/l$ is

$$\frac{(-1)^k}{k!} \delta^{(k)}(G^{1/l}). \tag{9}$$

Now in accordance with our earlier remarks, we may assume $\varphi(x_1, \ldots, x_n)$ to have support in any finite region. In calculating the residue at $\lambda = -(k+1)/l$, the integral of (8) must be taken over those components of the $G = 0$ surface which consists of points of degree l. Of course, if λ belongs to more than one of the sequences of Eq. (2), the residue at λ is the sum of the residues obtained for the various l.

We remark that if the $G = 0$ surface consists of points of order one and degree l, the analytic functional G^λ has poles at the same values of λ as does the generalized function $(x^l)_+^\lambda$ of a single variable (see Chapter I, Section 3). At $\lambda = -(k+1)/l$ the residue of $(x^l)_+^\lambda$ is $(-1)^k \delta^{(k)}(x)/k!$, while the residue of G^λ is, $(-1)^k \delta^{(k)}(G^{1/l})/k!$. Thus the case in which the $G(x_1, ..., x_n) = 0$ surface consists of points of order one is quite analogous to the case of one independent variable.

4.3. The Generalized Function G^λ When $G = 0$ Has No Points of Order Higher Than Two

Let us now assume that the $G(x_1, ..., x_n) = 0$ surface consists of reducible points of order no higher than two. Each sufficiently small neighborhood of a point of order one generates in G^λ a sequence of poles of the type already considered in Section 4.2. As before, the generalized function $G^\lambda(x_1, ..., x_n)$ is defined by

$$(G^\lambda, \varphi) = \int_{G>0} ... \int G^\lambda(x_1, ..., x_n) \, \varphi(x_1, ..., x_n) \, dx_1 ... dx_n.$$

What remains, therefore, is to understand how the reducible second-order points (which are, of course, singular points of the $G = 0$ surface) affect the λ-dependence of this functional. We shall prove the following:

Theorem. The neighborhood of every point M of order two and of degree m generates in the generalized function $G^\lambda(x_1, ..., x_n)$ the sequence of poles

$$\lambda = -\frac{2}{m}, \ -\frac{3}{m}, \ ..., \ -\frac{k}{m}, \ .... \tag{1}$$

Further, every arbitrarily small neighborhood of a point of second order will, in general, contain points of first order and perhaps of different degrees. If some $\lambda = \lambda_0$ belongs simultaneously to a sequence such as (1) and to one or several sequences generated by first-order points included in an arbitrary neighborhood of M (that is, to first-order points incident on M), then the generalized function $G^\lambda(x_1, ..., x_n)$ has a pole of second order at $\lambda = \lambda_0$.

Proof. Let us choose φ to be nonzero only in a small neighborhood of M, and let $\xi_1, ..., \xi_n$ be a local coordinate system in this neighborhood.

Since M is a point of second order, this choice can be made so that in terms of the new coordinates G becomes a homogeneous function P of ξ_1 and ξ_2, so that the defining integral for G^λ becomes

$$\int_{P>0} \cdots \int P^\lambda(\xi_1, \xi_2)\, \varphi_1(\xi_1, ..., \xi_n)\, D \begin{pmatrix} x_1 \cdots x_n \\ \xi_1 \cdots \xi_n \end{pmatrix} d\xi_1\, d\xi_2 \cdots d\xi_n, \qquad (2)$$

where $\varphi_1(\xi_1, ..., \xi_n) \equiv \varphi(x_1, ..., x_n)$. Let us write

$$\psi(\xi_1, \xi_2) = \int \cdots \int \varphi_1(\xi_1, ..., \xi_n)\, D \begin{pmatrix} x_1 \cdots x_n \\ \xi_1 \cdots \xi_n \end{pmatrix} d\xi_3 \cdots d\xi_n. \qquad (3)$$

Then we can write (2) in the simpler form

$$(G^\lambda, \varphi) = \int\int_{P>0} P^\lambda(\xi_1, \xi_2)\, \psi(\xi_1, \xi_2)\, d\xi_1\, d\xi_2, \qquad (4)$$

where $P(\xi_1, \xi_2)$ is a homogeneous function of degree m in two variables, and $\psi(\xi_1, \xi_2)$ is a test function (i.e., in K) different from zero only in a small neighborhood of the origin.

In analogy, therefore, with the case of points of order one, for which the problem reduced to studying functions of one independent variable, the behavior of G^λ in the neighborhood of a point of order two can be understood by studying arbitrary homogeneous functions of two variables. We shall now consider, therefore, the ξ_1, ξ_2 plane.

A reducible point of second order of the $P(\xi_1, \xi_2) = 0$ curve is an isolated singular point on this curve or a point of intersection of different branches of it with distinct tangents (Fig. 7). In studying the integral

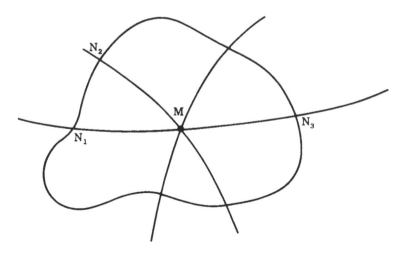

FIG. 7.

of Eq. (4) at those values of λ at which it is sure to converge, we shall use polar coordinates[1] r, u and homogeneity to rewrite Eq. (4) in the form

$$(G^\lambda, \varphi) = \int_0^\infty r^{\lambda m+1} \, dr \int P^\lambda(\bar{\xi}_1, \bar{\xi}_2) \, \psi(\xi_1, \xi_2) \, du, \tag{5}$$

where $\xi_1 = r\bar{\xi}_1$, $\xi_2 = r\bar{\xi}_2$, and $\bar{\xi}_k = \bar{\xi}_k(u)$ are the coordinates of a point on the $r = $ const curve.

Consider the auxiliary functional

$$I_{\lambda,\mu}[\psi] = \int r^\mu \, dr \int P^\lambda(\bar{\xi}_1, \bar{\xi}_2) \, \psi(\xi_1, \xi_2) \, du. \tag{6}$$

depending on two independent complex parameters λ and μ.

If we set $\mu = \lambda m + 1$, obviously, (6) is transformed into (5); it is therefore sufficient for our purposes to study Eq. (6). We assert the following.

Lemma. Let $F(\lambda, \mu)$ be a meromorphic function of two complex variables, and consider two given sequences of complex numbers

$$\lambda_1, \lambda_2, ..., \lambda_n, ..., \tag{7}$$

$$\mu_1, \mu_2, ..., \mu_n, ... \tag{8}$$

Consider $F(\lambda, \mu)$ as a function of λ, and assume that for fixed $\mu \neq \mu_n$ it has simple poles at points of the sequence in (7) and that the Laurent expansion

$$F(\lambda, \mu) = \frac{c_{-1}(\mu)}{\lambda - \lambda_n} + c_0(\mu) + c_1(\mu)(\lambda - \lambda_n) + ...$$

about every such pole has coefficients $c_i(\mu)$ which are meromorphic functions of μ with simple poles at the points of the sequence in (8).

Then the analytic function of a single variable $F(\lambda, \lambda)$ has poles at points given by both sequences (7) and (8). If, further, some point λ_0 belongs simultaneously to both sequences, then $F(\lambda, \lambda)$ has a second-order pole at $\lambda = \lambda_0$ such that the coefficient of $(\lambda - \lambda_0)^{-2}$ in the Laurent expansion about this point is equal to the residue of $c_{-1}(\mu)$ at $\mu = \lambda_0$.

The proof of this lemma follows immediately from the Laurent expansion of $F(\lambda, \mu)$ in two variables in the neighborhood of (λ_n, μ_m).

---

[1] We could use, instead of polar coordinates, any curvilinear coordinate system in which the $r = $ const. lines enclose the origin.

Let us now find the poles of the function given by (6). Let us write

$$f_\lambda(r) = \int_\Gamma P^\lambda(\xi_1, \xi_2)\, \psi(\xi_1, \xi_2)\, du \qquad (9)$$

[where Γ denotes that part of the $r = 1$ curve on which $P(\xi_1, \xi_2) > 0$]. The singularities with respect to λ of $f_\lambda(r)$ can arise only from those points $N_1, N_2, \ldots$ on the curve at which $P(\xi_1, \xi_2)$ vanishes. But these points of the $P(\xi_1, \xi_2) = 0$ curve are reducible on Γ.[2] Let N_i be such a point.

Reducibility on Γ means simply that in a neighborhood of N_i on Γ we can introduce a local coordinate ξ such that P becomes homogeneous and can thus be written ξ^{l_i}.

Now we choose ψ to be nonzero only in a neighborhood of N_i, which reduces the study of the singularities of $f_\lambda(r)$ to the study of the singularities of integrals such as $\int \xi^{\lambda l_i} \psi_1(r, \xi)\, d\xi$, where $\psi_1(r, \xi) = \psi(\xi_1, \xi_2)$.

But we know about the singularities of such integrals from Chapter I, Section 3. In particular, such an integral has simple poles at

$$\lambda = -\frac{1}{l_i}, \ -\frac{2}{l_i}, \ \ldots, \ -\frac{k}{l_i}, \ \ldots. \qquad (10)$$

We thus find that the integral of Eq. (9) for $f_\lambda(r)$ has simple poles at points of the sequence given by (10), where the l_i are the degrees of the N_i (see Fig. 7) on the branches of the curve that pass through the point M under question.

Now in terms of $f_\lambda(r)$ Eq. (6) becomes

$$I_{\lambda,\mu}[\psi] = \int_0^\infty r^\mu f_\lambda(r)\, dr. \qquad (11)$$

We may again use the results of Chapter I, Section 3, and then we see that for fixed λ belonging to none of the sequences of Eq. (10), $I_{\lambda,\mu}[\psi]$ can be analytically continued into the entire complex μ plane with the exception of the points

$$\mu = -1, -2, \ldots, -k-1, \ldots,$$

at which it has simple poles. Setting $\mu = \lambda m + 1$ and making use of our lemma, we find that

$$\int_0^\infty r^{\lambda m+1}\, dr \int P^\lambda(\xi_1, \xi_2)\, \psi(\xi_1, \xi_2)\, du = I_{\lambda, m\lambda+1}[\psi]$$

---

[2] It can be shown that if a hypersurface $F(x) = 0$ is reducible in n dimensions and F depends, say, only on x_1 and x_2, then the line $F(x) = 0$ in the x_1, x_2 plane is also reducible.

has poles at the points sequences of (10) as well as on the sequence

$$\lambda = -\frac{2}{m}, -\frac{3}{m}, ..., -\frac{k+2}{m}, ...\tag{12}$$

If $\lambda = \lambda_0$ belongs simultaneously to (12) and any one of the sequences of (10), the integral will in general have a pole of second order.

Since in the local coordinate system

$$\int_{G>0} ... \int G^\lambda(x_1, ..., x_n)\,\varphi(x_1, ..., x_n)\,dx_1\,...\,dx_n$$

$$= \int_{P>0}\int P^\lambda(\xi_1, \xi_2)\,\psi(\xi_1, \xi_2)\,d\xi_1\,d\xi_2$$

$$= \int_0^\infty r^{\lambda m+1}\,dr \int P^\lambda\left(\frac{\xi_1}{r}, \frac{\xi_2}{r}\right)\psi(\xi_1, \xi_2)\,du,$$

the theorem asserted at the beginning of this section is proven.

Let us calculate the residues of the generalized function $G^\lambda(x_1, ..., x_n)$ at its simple poles.

First let $\lambda_0 = -(k+1)/l_i$ be a pole of $G^\lambda(x_1, ..., x_n)$ arising from the vanishing of $G(\dot{x}_1, ..., x_n)$ at points of order one. We have shown in Section 4.2 that the residue of (G^λ, φ) at such a pole is $1/k!\int \omega_k(\varphi)$, where $\omega_k(\varphi)$ is a differential form defined in the neighborhood of any first-order point of the $G = 0$ surface. If the part of the surface over which $\omega_k(\varphi)$ is being integrated has second-order points of the $G = 0$ surface on its boundary, $\int \omega_k(\varphi)$ may diverge. In this case the residue is equal to the regularization of the corresponding integral.[3]

Now consider a simple pole $\lambda_0 = -(k+2)/m$ of $G^\lambda(x_1, ..., x_n)$ which arises because $G(x_1, ..., x_n)$ vanishes at points of second order. In the neighborhood of such a point the defining integral for (G^λ, φ) can be written

$$\int_0^\infty r^{\lambda m+1} f_\lambda(r)\,dr,$$

where $f_\lambda(r)$ is given by (9). For those values of λ for which the integral diverges, $f_\lambda(r)$ is defined as its regularization [the analytic continuation of $f_\lambda(r)$ from positive values of λ].

Now the residue of $\int_0^\infty r^{\lambda m+1} f_\lambda(r)\,dr$ at $\lambda = \lambda_0$ is $(-1)^k f_\lambda^{(k)}(0)/k!$.

[3] Since the pole λ_0 we are now discussing is simple,

$$\lim_{\lambda \to \lambda_0}(\lambda - \lambda_0)\int_{G>0} ... \int G^\lambda(x_1, ..., x_n)\,\varphi(x_1, ..., x_n)\,dx_1\,...\,dx_n$$

exists. Then the regularization of $(1/k!)\int \omega_k(\varphi)$ may be defined, for instance, as the limit.

We differentiate $f_{\lambda_0}(r)$ with respect to r under the integral sign in Eq. (9), with $\xi_i = r\tilde{\xi}_i$ (this is permissible for positive λ and therefore also for the regularization in general), and set $r = 0$. This leads to the following expression for the residue:

$$\frac{1}{k!} \sum_{\alpha, \beta} \frac{\partial^k \psi(0, 0)}{\partial \xi_1^\alpha \, \partial \xi_2^\beta} \int_\Gamma \xi_1^\alpha \xi_2^\beta \, P^{-(k+2)/m}(\xi_1, \xi_2) \, du$$

$$= \frac{1}{k!} \sum_{\alpha, \beta} \frac{\partial^k \psi(0, 0)}{\partial \xi_1^\alpha \, \partial \xi_2^\beta} \int_\Gamma \frac{\xi_1^\alpha \xi_2^\beta (\xi_1 \, d\xi_2 - \xi_2 \, d\xi_1)}{P^{(k+2)/m}(\xi_1, \xi_2)}. \tag{13}$$

In Section 3.2 we obtained a formula for the residue of the analytic generalized function f^λ, where f is a positive homogeneous function of the first degree [Eq. (8) of that section]. Comparing that expression with Eq. (13), we find that the residue of the generalized function $G^\lambda(x_1, ..., x_n)$ is written in terms of local coordinates quite similarly to the residue of a positive homogeneous function, except that the integral over Γ is understood here in the sense of its regularization (this integral always converges for a function which is positive in the neighborhood of the second order point).

4.4. The Generalized Function G^λ in General

Let $G^\lambda(x_1, ..., x_n)$ be a generalized function whose $G = 0$ surface consists of reducible points of arbitrary orders from 1 to n inclusive.

For simplicity we shall assume that G is a polynomial. We shall prove the following theorem.

Theorem. The generalized function $G^\lambda(x_1, ..., x_n)$ defined by

$$(G^\lambda, \varphi) = \int \cdots \int_{G>0} G^\lambda(x_1, ..., x_n) \, \varphi(x_1, ..., x_n) \, dx_1 \ldots dx_n$$

for Re $\lambda > 0$ and by its analytic continuation for other λ, is meromorphic in λ, and its poles are given by a finite number of arithmetic progressions. Each connected component consisting of points of order r and degree m on the $G = 0$ surface (see Section 4.1) generates a set of poles in the functional $G^\lambda[\varphi]$ at

$$\lambda = -\frac{r}{m}, \ -\frac{r+1}{m}, \ ..., \ -\frac{r+k}{m}, \ ... \ . \tag{1}$$

Further, assume that there exists a sequence of two, three, or more incident connected components of the $G = 0$ surface, each consisting of points of a different fixed order, and that $\lambda = \lambda_0$ occurs in two, three, or more sets (1), each generated by one of the components in the sequence. Then at $\lambda = \lambda_0$ the generalized function $G^\lambda(x_1, ..., x_n)$ has a pole of order two, three, or more, depending on the number of sets in which it occurs.

Proof.   Let us assume that we already know the poles (in the λ plane) generated in

$$\int_{G>0} ... \int G^\lambda(x_1, ..., x_n) \, \varphi(x_1, ..., x_n) \, dx_1 ... dx_n, \tag{2}$$

by points of order less than or equal to $n - 1$. We then move on to study (2) in an arbitrarily small neighborhood of a point M of order n.

In such a neighborhood we introduce a local coordinate system $\xi_1, ..., \xi_n$ in which $P(\xi_1, ..., \xi_n) = G(x_1, ..., x_n)$ is a homogeneous function of degree m, and choose $\varphi(x_1, ..., x_n)$ to vanish outside of this neighborhood. Then (2) becomes

$$I_\lambda[\psi] = \int_D ... \int P^\lambda(\xi_1, ..., \xi_n) \, \psi(\xi_1, ..., \xi_n) \, d\xi_1 ... d\xi_n, \tag{3}$$

where

$$\psi(\xi_1, ..., \xi_n) \equiv \varphi(x_1, ..., x_n) \, D \begin{pmatrix} x_1 ... x_n \\ \xi_1 ... \xi_n \end{pmatrix},$$

and D is the intersection of the neighborhood in which $\psi \neq 0$ with the region $P(\xi_1, ..., \xi_n) > 0$.

We now go over the spherical coordinates for those λ for which the integral in (3) converges. Because P is homogeneous in the ξ_i we obtain

$$I_\lambda[\varphi] = \int_0^\infty r^{\lambda m + n - 1} \, dr \int_\Gamma P^\lambda(\tilde{\xi}_1, ..., \tilde{\xi}_n) \, \psi(\xi_1, ..., \xi_n) \, d\omega, \tag{4}$$

where Γ is the intersection of the unit sphere with the $P(\tilde{\xi}_1, ..., \tilde{\xi}_n) > 0$ region, a point on Γ is represented by $(\tilde{\xi}_1, ..., \tilde{\xi}_n)$ with $\xi_k = r\tilde{\xi}_k$, and $d\omega$ is the element of area on the unit sphere.

The integral over Γ is an integral over the $(n - 1)$-sphere, on which the $P = 0$ surface has by assumption only reducible points of order no greater than $n - 1$.

By assumption we know the poles of such an integral as well as their orders.

Again introducing, as in Section 4.3, the functional

$$I_{\lambda,\mu}[\psi_1] = \int r^\mu \, dr \int_\Gamma P^\lambda(\tilde{\xi}_1, ..., \tilde{\xi}_n) \, \psi(\xi_1, ..., \xi_n) \, d\omega \tag{5}$$

depending on two complex variables, and denoting the integral over Γ by $f_\lambda(r)$, we have

$$f_\lambda(r) = \int_\Gamma P^\lambda(\xi_1, ..., \xi_n)\, \psi(\xi_1, ..., \xi_n)\, d\omega, \tag{6}$$

which reduces the problem to the situation already treated in Section 4.3. In particular, for fixed $\mu \neq -n$, the functional $I_{\lambda,\mu}[\psi]$ has poles in the λ plane at a finite number of sequences of points, and for any fixed λ which fails to occur in any of these sequences, it has poles at $\mu = -1$, $-2, ..., -n, ...$. The only difference from the case of Section 4.3, where the $G = 0$ surface contained no points of order higher than two, is that the poles in λ may be multiple rather than simple.

It is easily verified that the lemma of Section 4.3 can be used here.

Setting $\mu = \lambda m + n - 1$, we find that $I_\lambda[\psi]$, or the defining integral (2) for (G^λ, φ) has poles (in λ) other than those generated in $f_\lambda(r)$ by points of order less than n on the $G = 0$ surface; these additional poles occur at the points

$$\lambda = -\frac{n}{m}, -\frac{n+1}{m}, ..., -\frac{n+k}{m}, ... \tag{7}$$

If $f_\lambda(r)$ has a pole of order p at $\lambda = \lambda_0$, and if λ_0 also belongs to the sequence of (7), the generalized function $G^\lambda(x_1, ..., x_n)$ will have a pole of order $p + 1$ at $\lambda = \lambda_0$. This completes the proof of our theorem.

We have thus located all of the poles of the generalized function defined by Eq. (2).

We have seen in Section 3 that in the simplest case, namely when G is a homogeneous nonnegative function, it is convenient for many reasons to introduce what we called residues at the origin of a formally homogeneous function, and that these residues characterize the singularity of this function at the origin in the same way that the residues of an analytic function characterize its isolated singularities. When dealing, as now, with an arbitrary function G with reducible singular points, we can define its residues analogously.

We proceed as follows. Let D be a maximal connected component of order r and degree m belonging to the $G = 0$ manifold; we shall say that the *residue* of the ordinary function $G^{-(r+k)/m}(x_1, ..., x_n)$ on this component is the residue in the λ plane of the analytic functional $G^\lambda(x_1, ..., x_n)$ at $\lambda = -(r+k)/m$. This residue is then a generalized function.

For instance, if $G(x_1, ..., x_n)$ is a homogeneous function of degree m, then the residue of $G^{-(n+k)/m}$ at the origin is a linear combination of the

kth partial derivatives of the delta function, with coefficients given by integrals over closed surfaces surrounding the origin.

If $G(x_1, ..., x_n)$ is any function with a reducible singular point of order $n - 1$ and degree m at the origin, the residue of $G^{-(n+k)/m} (x_1, ..., x_n)$ at the origin is a linear combination of the kth and lower derivatives of the delta function with coefficients determined by G. The residue not only at a point, but on any connected component of degree m and order r of the $G = 0$ manifold, can be defined similarly.

4.5. Integrals of an Infinitely Differentiable Function over a Surface Given by $G = c$

Let us apply the above results to the study of integrals of infinitely differentiable functions of n variables over a surface given by a polynomial $G(x_1, ..., x_n)$. We shall restrict our considerations to the case in which the $G = 0$ surface consists only of reducible points, and the $G = c$ surfaces with $c > 0$ have no singular points at all. Consider the integral

$$I(c) = \int_{G=c} \varphi(x_1, ..., x_n)\, \omega, \tag{1}$$

where ω is the differential form on the $G = c$ surface defined by the equation $dv = dG\omega$. We have the obvious identity (in λ)

$$\int_0^\infty I(c)\, c^\lambda\, dc = \int \cdots \int_{G>0} G^\lambda(x_1, ..., x_n)\, \varphi(x_1, ..., x_n)\, dx_1 ... dx_n. \tag{2}$$

We have already established that the integral on the right-hand side of this equation, and therefore also that on the left-hand side, is a meromorphic function of λ, and we have found how its poles depend on the types of singularities of the $G = 0$ surface.

Now the behavior of $I(c)$ for $c > \epsilon > 0$ has no influence on the singularities of $\int_0^\infty I(c) c^\lambda dc$, and therefore if we know these singularities we can write out an asymptotic expansion of $I(c)$ for small c.

Specifically, let $F(\lambda)$ be the function defined by $\int_0^\infty I(c) c^\lambda dc$ and its analytic continuation. Then it can be shown that if the poles of $F(\lambda)$ are arranged in decreasing order

$$-\lambda_1, -\lambda_2, ..., -\lambda_k, ...; \qquad 0 < \lambda_1 < \lambda_2 < ... < \lambda_k < ... ,$$

and if m_k is the multiplicity of the kth pole, $I(c)$ has the following asymptotic expansion for small c:

$$I(c) \approx \sum_{k=1}^\infty \sum_{m=1}^{m_k} a_{k.m}\, c^{\lambda_k - 1}\, \ln^{m-1} c. \tag{3}$$

It is easily shown that each term in this expansion gives a pole of order m (see Chapter I, Section 3). Here $a_{k,m}$ is $(-1)^{m-1}/(m-1)!$ times the coefficient of $(\lambda + \lambda_k)^{-m}$ in the Laurent expansion of $F(\lambda)$ about $\lambda = -\lambda_k$. Recalling the definition of $\delta(G)$, we can rewrite $I(c)$ in the form

$$I(c) = \int_{G=c} \varphi \, \omega = (\delta(G-c), \varphi).$$

It follows from this that the asymptotic expansion for $I(c)$ in powers of c is at the same time an expansion for $(\delta(G-c), \varphi)$ in powers of c. From Eq. (3) we obtain in this way an asymptotic expansion for $\delta(G-c)$ for small c.

Example 1. Let $G(x, y) = xy$. Then the first pole of the generalized function $(xy)^\lambda$ defined by

$$\int\int_{xy>0} (xy)^\lambda \, \varphi(x, y) \, dx \, dy,$$

will occur at $\lambda = -1$. The expansion of this integral in powers of $\lambda + 1$ is

$$\int\int_{xy>0} (xy)^\lambda \, \varphi(x, y) \, dx \, dy$$

$$= \frac{2\varphi(0, 0)}{(\lambda + 1)^2} + \left[\int_{-\infty}^{\infty} \frac{\varphi(x, 0)}{x} \, dx + \int_{-\infty}^{\infty} \frac{\varphi(0, y)}{y} \, dy \right] \frac{1}{\lambda + 1} + \cdots .$$

Therefore the integral[4] $I(c) = \int_{xy=c} \varphi(x, y)\omega$ has the asymptotic expansion

$$I(c) = -2\varphi(0, 0) \ln c + \int_{-\infty}^{\infty} \frac{\varphi(x, 0)}{x} \, dx + \int_{-\infty}^{\infty} \frac{\varphi(0, y)}{y} \, dy + \cdots ,$$

where we have omitted terms converging to zero as $c \to 0$.

From what has been said above, this can be used to obtain an expansion of $\delta(xy - c)$ in powers of c. Such an expansion will be

$$\delta(xy - c) = -2\delta(x, y) \ln c + \left(\frac{\delta(y)}{x} + \frac{\delta(x)}{y} \right) + o(c). \qquad (4)$$

---

[4] Here we may write either $\omega = y^{-1} \, dy$ or $\omega = -x^{-1} \, dx$, so that we may be considering either

$$\int_{-\infty}^{\infty} \frac{\varphi(x, y) \, dy}{y} \qquad \text{or} \qquad -\int_{+\infty}^{-\infty} \frac{\varphi(x, y) \, dx}{x} .$$

Example 2. $G = x^2 + y^2 + z^2 - t^2$. In this case the integral

$$\iint_{G>0} (x^2 + y^2 + z^2 - t^2)^\lambda\, \varphi(x, y, z, t)\, dv$$

has a simple pole with residue $\delta(x^2 + y^2 + z^2 - t^2)$ at $\lambda = -1$, and a pole of order two at $\lambda = -2$, where the integral has the expansion

$$\iint_{G>0} G^\lambda \varphi\, dv = -\frac{1}{8} \frac{\varphi(0, 0, 0, 0)}{(\lambda + 2)^2} + \frac{1}{(\lambda + 2)} \left\{ \varphi(0, 0, 0, 0) \ln 2 \right.$$

$$+ \frac{8\pi}{3} \left[\int_0^1 \frac{\varphi(0, 0) - \varphi(r, r)}{r}\, dr - \int_1^\infty \frac{\varphi(r, r)}{r}\, dr + \int_0^\infty \frac{\partial \varphi(r, \rho)}{\partial \rho} \bigg|_{\rho = r}\, dr \right] \bigg\} + \dots.$$

Thus we can now write $\delta(G - c)$ in the form

$$\delta(G - c) = \delta(G) + c \ln c\, \frac{\delta(x, y, z, t)}{8} + c\delta'(G) + \dots, \tag{5}$$

We have omitted terms of order $o(c)$, and $\delta'(G)$ represents the divergent integral $\int_{G=0} \omega_1$ regularized according to the expression in braces.

In general if G is a function whose $G = 0$ surface has only reducible singularities, the low-c expansion of the generalized function $\delta(G - c)$ can be used to define $\delta(G)$, $\delta'(G)$, and the higher derivatives when the integrals of the corresponding differential forms diverge.

More specifically, if

$$\delta(G - c) = \sum_{k=1}^{\infty} \sum_{m=1}^{m_k} T_{k,m}\, c^{\lambda_k - 1} \ln^{m-1} c,$$

where the $T_{k,m}$ are generalized functions such that

$$(T_{k,m}, \varphi) = a_{k,m}$$

[where the $a_{k,m}$ are the coefficients of Eq. (3)] then $\delta(G)$ will denote the constant term in this expansion, $\delta'(G)$ the negative of the coefficient of the first power of c, etc. In particular, Eq. (4) may be written

$$\delta(xy - c) = -2 \ln c\delta(x, y) + \delta(xy) + \dots, \tag{4'}$$

so that[5]

$$\delta(xy) = \frac{\delta(x)}{y} + \frac{\delta(y)}{x}.\tag{6}$$

When the $G = 0$ surface has no singular points this definition will of course agree with the previous one. Indeed, the integral $\int_{G>0} G^\lambda \varphi \, dv$ then has only simple poles at $\lambda = -k$ (where k is a positive integer) with residues

$$\frac{(-1)^{k-1}}{(k-1)!}\,\delta^{(k-1)}(G).$$

Therefore,

$$\delta(G - c) = \delta(G) - c\delta'(G) + \frac{c^2}{2}\,\delta''(G) + \dots + \frac{(-1)^k}{k!}\,\delta^{(k)}(G) + \dots.$$

There is, incidentally, an interesting consequence of Eq. (4). Differentiating that equation with respect to c, we obtain

$$-\delta'(xy - c) = -\frac{2}{c}\,\delta(x, y) + \dots.$$

Now multiplying by $-c$ and allowing c to converge to zero, we find that

$$c\delta'(xy - c) \to 2\delta(x, y).$$

Similarly, we may differentiate Eq. (5) twice with respect to c, and again multiply by c and allow it to converge to zero. This gives

$$c\delta''(x^2 + y^2 + z^2 - t^2 - c) \to \frac{\delta(x, y, z, t)}{8}.$$

By applying these expressions to a function φ in K, we will obtain the value of φ at the origin if we know its integrals over the hyperbolas $xy = c$ or the hyperboloids $x^2 + y^2 + z^2 - t^2 = c$.

---

[5] This equation could not have been obtained in Section 1. When we derived

$$\delta(PQ) = \frac{\delta(P)}{Q} + \frac{\delta(Q)}{P}$$

in Section 1.7, we assumed that the $P = 0$ and $Q = 0$ surfaces did not intersect. We can now, however, derive this result immediately from Eq. (6) for surfaces that intersect so that there exists a coordinate system $u_1, \dots, u_n$ in which $u_1 = P$ and $u_2 = Q$.

SUMMARY OF FUNDAMENTAL DEFINITIONS AND EQUATIONS OF VOLUME I

Chapter I, Section 1

1. A *test function* $\varphi(x) = \varphi(x_1, ..., x_n)$ is an infinitely differentiable function that vanishes outside a bounded region in the n-dimensional space R_n (has bounded support).

2. The *test-function space* K is the set of all test functions. Linear operations in K are defined in the usual way. A sequence $\varphi_\nu(x) \in K$ is said to converge to zero if the $\varphi_\nu(x)$ converge to zero uniformly with all their derivatives and all vanish outside a common bounded region.

3. A *generalized function* is a continuous linear functional on K.

4. A *regular* generalized function is a functional on K which can be written

$$(f, \varphi) = \int f(x)\, \varphi(x)\, dx,$$

where $f(x)$ is a locally summable function.

5. All other generalized functions are called *singular*.

6. $\theta(x)$ denotes the *theta function* equal to unity for $x > 0$ and to zero for $x < 0$ (defined for $n = 1$).

7. The *delta function* $\delta(x - x_0)$ is the singular functional defined by

$$(\delta(x - x_0), \varphi(x)) = \varphi(x_0).$$

8. The functional $f(x)$ is said to *vanish in a region* G if $(f, \varphi) = 0$ for every φ in K vanishing on some region G_1 such that G and G_1 cover R_n.

9. The *support* of a functional f is the closed set on whose complement f vanishes.

10. A sequence f_ν of generalized functions is said to converge to the generalized function f if for every $\varphi \in K$

$$\lim_{\nu \to \infty} (f_\nu, \varphi) = (f, \varphi).$$

11. *Complex* test functions and generalized functions combine according to

$$\alpha(f, \varphi) = (f, \alpha\varphi) = (\bar{\alpha}f, \varphi).$$

A *regular functional f* corresponding to a function $f(x)$ is defined by

$$(f, \varphi) = \int \overline{f(x)}\, \varphi(x)\, dx.$$

12. The generalized function $\bar{f}$ complex conjugate to f is defined by

$$(\bar{f}, \varphi) = \overline{(f, \bar{\varphi})}.$$

13. The space S of *rapidly decreasing functions* consists of infinitely differentiable functions $\varphi(x)$ such that

$$|\, x^k\, \varphi^{(q)}(x)\,| \leqslant C_{kq}, \qquad n = 1;\, k, q = 0, 1, 2, \ldots$$

or, in the case of several variables,

$$\left|\, x_1^{k_1} \ldots x_n^{k_n}\, \frac{\partial^{q_1+\cdots+q_n}\, \varphi(x_1, \ldots, x_n)}{\partial x_1^{q_1} \ldots \partial x_n^{q_n}}\, \right| \leqslant C_{k_1 \cdots q_n}, \qquad k_1, \ldots, q_n = 0, 1, 2, \ldots$$

with convergence defined accordingly.

Chapter I, Section 2

1. The *derivative* of the generalized function with respect to x_j is defined by

$$\left(\frac{\partial f}{\partial x_j}, \varphi\right) = \left(f, -\frac{\partial \varphi}{\partial x_j}\right).$$

2. $\theta'(x) = \delta(x).$

3. $\dfrac{d}{dx} \ln(x + i0) = \dfrac{1}{x} - i\pi\delta(x),$

where

$$\left(\frac{1}{x}, \varphi(x)\right) = \lim_{\epsilon \to 0} \int_{|x| > \epsilon} \frac{\varphi(x)}{x}\, dx.$$

4. $\varDelta\, \dfrac{1}{r^{n-2}} = -(n-2)\, \Omega_n\, \delta(x)$ where Ω_n is the hypersurface area of the unit sphere imbedded in R_n ($n > 2$).

$$\varDelta \ln \frac{1}{r} = -2\pi\delta(x), \qquad n = 2.$$

5. If a sequence f_ν of generalized functions converges to f, the sequence $\partial f_\nu/\partial x_j$ converges to $\partial f/\partial x_j$.

6. $\displaystyle\sum_{n=1}^{\infty} \cos nx = -\tfrac{1}{2} + \pi \sum_{-\infty}^{\infty} \delta(x - 2\pi n).$

7. $\displaystyle\sum_{n=1}^{\infty} n \sin nx = -\pi \sum_{-\infty}^{\infty} \delta'(x - 2\pi n).$

8. $1 + e^{ix} + e^{2ix} + \dots + e^{-ix} + e^{-2ix} + \dots = 2\pi \displaystyle\sum_{-\infty}^{\infty} \delta(x - 2\pi n).$

9. $\displaystyle\sum_{n=1}^{\infty} \sin nx = \frac{1}{2} \cot \frac{x}{2}.$

10. $\dfrac{1}{\pi} \dfrac{\epsilon}{x^2 + \epsilon^2} \to \delta(x)$ as $\epsilon \to 0.$

11. $\dfrac{1}{2\sqrt{\pi t}} \exp\left(-\dfrac{x^2}{4t}\right) \to \delta(x)$ as $t \to 0.$

12. $\dfrac{1}{\pi} \dfrac{\sin \nu x}{x} \to \delta(x)$ as $\nu \to \infty.$

Chapter I, Section 3

1. The generalized function x_+^λ is defined as follows. For Re $\lambda > -1$

$$(x_+^\lambda, \varphi) = \int_0^\infty x^\lambda \varphi(x)\, dx; \tag{1}$$

for Re $\lambda > -n-1$, $\lambda \neq -1, -2, \dots, -n,$

$$(x_+^\lambda, \varphi) = \int_0^1 x^\lambda \left[\varphi(x) - \varphi(0) - x\varphi'(0) - \dots - \frac{x^{n-1}}{(n-1)!}\varphi^{(n-1)}(0)\right] dx$$

$$+ \int_1^\infty x^\lambda \varphi(x)\, dx + \sum_{k=1}^{n} \frac{\varphi^{(k-1)}(0)}{(k-1)!\,(\lambda + k)}; \tag{2}$$

for $-n-1 < \text{Re } \lambda < -n,$

$$(x_+^\lambda, \varphi) = \int_0^\infty x^\lambda \left[\varphi(x) - \varphi(0) - x\varphi'(0) - \dots - \frac{x^{n-1}}{(n-1)!}\varphi^{(n-1)}(0)\right] dx. \tag{3}$$

This generalized function coincides with the ordinary function x^λ for $x > 0$, and with zero for $x < 0$. It is analytic for all λ except $\lambda = -k$ where k is a positive integer, where it has simple poles.

We may differentiate with respect to x:

$$\frac{d}{dx} x_+^\lambda = \lambda x_+^{\lambda-1}, \qquad \lambda \neq -1, -2, \dots. \tag{4}$$

2. The generalized function x_-^λ is defined as follows. For Re $\lambda > -1$

$$(x_-^\lambda, \varphi) = \int_{-\infty}^0 |x|^\lambda \varphi(x)\, dx = \int_0^\infty x^\lambda \varphi(-x)\, dx; \tag{5}$$

for Re $\lambda > -n-1$, $\lambda \neq -1, -2, \dots, -n$,

$$(x_-^\lambda, \varphi) = \int_0^1 x^\lambda \left[\varphi(-x) - \varphi(0) + x\varphi'(0) - \dots - (-1)^{n-1} \frac{x^{n-1}}{(n-1)!} \varphi^{(n-1)}(0)\right] dx$$

$$+ \int_1^\infty x^\lambda \varphi(-x)\, dx + \sum_{k=1}^n \frac{(-1)^{k-1} \varphi^{(k-1)}(0)}{(k-1)!\,(\lambda+k)}; \tag{6}$$

for $-n-1 < \text{Re} < -n$,

$$(x_-^\lambda, \varphi) = \int_0^\infty x^\lambda \left[\varphi(-x) - \varphi(0) + x\varphi'(0) - \dots - (-1)^{n-1} \frac{x^{n-1}}{(n-1)!}\varphi^{(n-1)}(0)\right] dx \tag{7}$$

This generalized function coincides with the ordinary function $|x|^\lambda$ for $x < 0$ and to zero for $x > 0$. It is analytic for all λ except $\lambda = -k$, where k is a positive integer, where it has simple poles.

We may differentiate with respect to x:

$$\frac{d}{dx} x_-^\lambda = -\lambda x_-^{\lambda-1}, \qquad \lambda = -1, -2, \dots. \tag{8}$$

3. The generalized function $|x|^\lambda$ is defined as follows. For Re $\lambda > -1$

$$(|x|^\lambda, \varphi) = \int_{-\infty}^\infty |x|^\lambda \varphi(x)\, dx = \int_0^\infty x^\lambda[\varphi(x) + \varphi(-x)]\, dx; \tag{9}$$

for Re $\lambda > -2m-1$, $\lambda \neq -1, -3, \dots, -2m+1$,

$$(|x|^\lambda, \varphi) = \int_0^1 x^\lambda \Big\{\varphi(x) + \varphi(-x)$$

$$- 2\left[\varphi(0) + \frac{x^2}{2!}\varphi''(0) + \dots + \frac{x^{2m-2}}{(2m-2)!}\varphi^{(2m-2)}(0)\right]\Big\} dx$$

$$+ \int_1^\infty x^\lambda[\varphi(x) + \varphi(-x)]\, dx + \sum_{k=0}^{m-1} \frac{\varphi^{(2k)}(0)}{(2k)!(\lambda+2k+1)}; \tag{10}$$

for $-2m - 1 < \operatorname{Re} \lambda < -2m + 1$,

$$(|x|^{\lambda}, \varphi) = \int_0^{\infty} x^{\lambda} \bigg\{ \varphi(x) + \varphi(-x)$$

$$- 2 \bigg[\varphi(0) + \frac{x^2}{2!} \varphi''(0) + \ldots + \frac{x^{2m-2}}{(2m-2)!} \varphi^{(2m-2)}(0) \bigg] \bigg\} dx. \quad (11)$$

This generalized function correspond to the ordinary function $|x|^{\lambda}$ for $x \neq 0$. It has singularities (simple poles) at $\lambda = -2m - 1$, where m is a nonnegative integer.

At $\lambda = -2m$ we write $|x|^{-2m} = x^{-2m}$, so that

$$(x^{-2m}, \varphi) = \int_0^{\infty} x^{-2m} \bigg\{ \varphi(x) + \varphi(-x)$$

$$- 2 \bigg[\varphi(0) + \frac{x^2}{2!} \varphi''(0) + \ldots + \frac{x^{2m-2}}{(2m-2)!} \varphi^{(2m-2)}(0) \bigg] \bigg\} dx. \quad (12)$$

In particular,

$$(x^{-2}, \varphi) = \int_0^{\infty} \frac{\varphi(x) + \varphi(-x) - 2\varphi(0)}{x^2} \, dx. \quad (13)$$

4. The generalized function $|x|^{\lambda} \operatorname{sgn} x$ is defined as follows. For $\operatorname{Re} \lambda > -2$,

$$(|x|^{\lambda} \operatorname{sgn} x, \varphi) = \int_{-\infty}^{\infty} |x|^{\lambda} \operatorname{sgn} x \varphi(x) \, dx = \int_0^{\infty} x^{\lambda} [\varphi(x) - \varphi(-x)] \, dx; \quad (14)$$

for $\operatorname{Re} \lambda > -2m - 2, \lambda \neq -2, -4, \ldots, -2m$,

$$(|x|^{\lambda} \operatorname{sgn} x, \varphi) = \int_0^1 x^{\lambda} \bigg\{ \varphi(x) - \varphi(-x)$$

$$- 2 \bigg[x\varphi'(0) + \frac{x^3}{3!} \varphi'''(0) + \ldots + \frac{x^{2m-1}}{(2m-1)!} \varphi^{(2m-1)}(0) \bigg] \bigg\} dx$$

$$+ \int_1^{\infty} x^{\lambda} [\varphi(x) - \varphi(-x)] \, dx + 2 \sum_{k=0}^{m-1} \frac{\varphi^{(2k)}(0)}{(2k+1)! (\lambda + 2k + 2)}; \quad (15)$$

for $-2m - 2 < \operatorname{Re} \lambda < -2m$,

$$(|x|^{\lambda} \operatorname{sgn} x, \varphi) = \int_0^{\infty} x^{\lambda} \bigg\{ \varphi(x) - \varphi(-x)$$

$$- 2 \bigg[x\varphi'(0) + \frac{x^3}{3!} \varphi'''(0) + \ldots + \frac{x^{2m-1}}{(2m-1)!} \varphi^{(2m-1)}(0) \bigg] \bigg\} dx. \quad (16)$$

This generalized function coincides with the ordinary function $|x|^\lambda \operatorname{sgn} x$ for $x \neq 0$. It has singularities (simple poles) at $\lambda = -2m$, where m is a positive integer. At $\lambda = -2m - 1$ we write $|x|^{-2m-1} \operatorname{sgn} x = x^{-2m-1}$, so that

$$(x^{-2m-1}, \varphi) = \int_0^\infty x^{-2m-1} \left\{ \varphi(x) - \varphi(-x) \right.$$

$$\left. - 2 \left[x\varphi'(0) + \frac{x^3}{3!} \varphi'''(0) + \dots + \frac{x^{2m-1}}{(2m-1)!} \varphi^{(2m-1)}(0) \right] \right\} dx. \qquad (17)$$

In particular,

$$(x^{-1}, \varphi) = \int_0^\infty \frac{\varphi(x) - \varphi(-x)}{x} \, dx, \qquad (18)$$

$$(x^{-3}, \varphi) = \int_0^\infty \frac{\varphi(x) - \varphi(-x) - 2x\varphi'(0)}{x^3} \, dx. \qquad (19)$$

5. The q-fold indefinite integral of $|x|^\lambda$ may be written

$$\underbrace{\int \dots \int}_{q} |x|^\lambda \, d^q x = \frac{|x|^{\lambda+q}(\operatorname{sgn} x)^q}{(\lambda+1)\dots(\lambda+q)} + \sum_{k=1}^{[q/2]} \frac{x^{q-2k}}{(2k-1)!(q-2k)!} \frac{1}{\lambda+2k}. \qquad (20)$$

6. The generalized functions

$$\frac{x_+^\lambda}{\Gamma(\lambda+1)}, \qquad \frac{x_-^\lambda}{\Gamma(\lambda+1)}, \qquad \frac{|x|^\lambda}{\Gamma\left(\dfrac{\lambda+1}{2}\right)}, \qquad \frac{|x|^\lambda \operatorname{sgn} x}{\Gamma\left(\dfrac{\lambda+2}{2}\right)} \qquad (21)$$

are entire functions of λ.

In particular, as $\lambda \to -n$,

$$\frac{x_+^\lambda}{\Gamma(\lambda+1)} \to \delta^{(n-1)}(x), \qquad \frac{x_-^\lambda}{\Gamma(\lambda+1)} \to (-1)^{n-1}\delta^{(n-1)}(x); \qquad (22)$$

as $\lambda \to -2m-1$,

$$\frac{|x|^\lambda}{\Gamma\left(\dfrac{\lambda+1}{2}\right)} \to \frac{(-1)^m \delta^{(2m)}(x) \, m!}{(2m)!} \, ; \qquad (23)$$

as $\lambda \to -2m$,

$$\frac{|x|^\lambda \operatorname{sgn} x}{\Gamma\left(\dfrac{\lambda+2}{2}\right)} \to \frac{(-1)^m \delta^{(2m-1)}(x) \, (m-1)!}{(2m-1)!}. \qquad (24)$$

7. The functions $\ln(x + i0)$ and $\ln(x - i0)$ are defined as follows:

$$\ln(x + i0) = \lim_{y \to +0} \ln(x + iy) = \ln|x| + i\pi\theta(-x), \tag{25}$$

$$\ln(x - i0) = \lim_{y \to +0} \ln(x - iy) = \ln|x| - i\pi\theta(-x), \tag{26}$$

8. The functions $(x + i0)^\lambda$ and $(x - i0)^\lambda$ are defined as follows:

$$(x + i0)^\lambda = \lim_{y \to +0}(x + iy)^\lambda = \begin{cases} e^{i\lambda\pi}|x|^\lambda & \text{for} \quad x < 0, \\ x^\lambda & \text{for} \quad x > 0; \end{cases} \tag{27}$$

$$(x - i0)^\lambda = \lim_{y \to +0}(x - iy)^\lambda = \begin{cases} e^{-i\lambda\pi}|x|^\lambda & \text{for} \quad x < 0, \\ x^\lambda & \text{for} \quad x > 0. \end{cases} \tag{28}$$

These functions exist for all complex λ and define regular functionals for Re $\lambda > -1$. The corresponding generalized functions are defined in the following way. For $\lambda \neq -n$, where n is a positive integer,

$$(x + i0)^\lambda = x_+^\lambda + e^{i\lambda\pi}x_-^\lambda, \tag{29}$$

$$(x - i0)^\lambda = x_+^\lambda + e^{-i\lambda\pi}x_-^\lambda, \tag{30}$$

where the generalized functions x_+^λ and x_-^λ are defined by Eqs. (1)-(3) and (5)-(7). For $\lambda = -n$,

$$(x + i0)^{-n} = x^{-n} - \frac{i\pi(-1)^{n-1}}{(n-1)!}\delta^{(n-1)}(x), \tag{31}$$

$$(x - i0)^{-n} = x^{-n} + \frac{i\pi(-1)^{n-1}}{(n-1)!}\delta^{(n-1)}(x), \tag{32}$$

where the generalized function x^{-n} is defined by Eq. (12) or (17). Thus the generalized functions $(x + i0)^\lambda$ and $(x - i0)^\lambda$ are defined for all λ. They are entire analytic in λ.

The relations

$$(x + i0)^\lambda = \lim_{y \to +0}(x + iy)^\lambda,$$

$$(x - i0)^\lambda = \lim_{y \to +0}(x - iy)^\lambda$$

hold both in the ordinary sense and in the sense of generalized functions.

We may differentiate with respect to x:

$$\left.\begin{array}{ll} \dfrac{d}{dx}(x+i0)^{\lambda} = \lambda(x+i0)^{\lambda-1} & (\lambda \neq 0), \\[2ex] \dfrac{d}{dx}(x-i0)^{\lambda} = \lambda(x-i0)^{\lambda-1} & (\lambda \neq 0), \end{array}\right\} \tag{33}$$

$$\left.\begin{array}{l} \dfrac{d}{dx}\ln(x+i0) = \dfrac{1}{x+i0}, \\[2ex] \dfrac{d}{dx}\ln(x-i0) = \dfrac{1}{x-i0}. \end{array}\right\} \tag{34}$$

9. For Re $\lambda > -n$ the generalized function r^{λ} is given by

$$(r^{\lambda}, \varphi) = \int_{R_n} r^{\lambda}\varphi(x)\,dx = \Omega_n \int_0^{\infty} r^{\lambda+n-1} S_{\varphi}(r)\,dr, \tag{35}$$

where Ω_n is the hypersurface area of the unit sphere in n dimensions, and $S_{\varphi}(r)$ is the average of φ over the sphere of radius r.

This generalized function can be analytically continued to the entire λ plane with the exception of the points $\lambda = -n, -n-2, -n-4, \ldots$, where it has simple poles.

10. The generalized function $\dfrac{2r^{\lambda}}{\Omega_n \Gamma\left(\dfrac{\lambda+n}{2}\right)}$ is an entire analytic function of λ. At $\lambda = -n$ it becomes $\delta(x)$, and at $\lambda = -n-2k$ it becomes

$$\frac{(-1)^k \Delta^k \delta(x)}{2^k n(n+2)\ldots(n+2k-2)}. \tag{36}$$

11. The plane-wave expansion formula for r^{λ} is

$$\frac{1}{\pi^{\frac{1}{2}(n-1)}} \int_{\Omega} \frac{|\omega_1 x_1 + \ldots + \omega_n x_n|^{\lambda}\,d\omega}{\Gamma\left(\dfrac{\lambda+1}{2}\right)} = 2\frac{r^{\lambda}}{\Gamma\left(\dfrac{\lambda+n}{2}\right)}. \tag{37}$$

Special cases. For n odd we have

$$\delta(x) = \frac{(-1)^{\frac{1}{2}(n-1)}}{2(2\pi)^{n-1}} \int_{\Omega} \delta^{(n-1)}(\omega_1 x_1 + \ldots + \omega_n x_n)\,d\omega; \tag{38}$$

For n even we have

$$\delta(x) = \frac{(-1)^{\frac{1}{2}n}(n-1)!}{(2\pi)^n} \int_{\Omega} (\omega_1 x_1 + \ldots + \omega_n x_n)^{-n}\,d\omega, \tag{39}$$

For any n we have

$$\delta(x) = \frac{(n-1)!}{(2\pi i)^n} \int_\Omega (\omega_1 x_1 + \dots + \omega_n x_n - i0)^{-n}\, d\omega. \tag{40}$$

Chapter I, Section 4

1. The generalized function $x_+^\lambda \ln^m x_+$ (for $\lambda \neq -k$, where k is a positive integer) is defined as follows. For Re $\lambda > -1$,

$$(x_+^\lambda \ln^m x_+, \varphi) = \int_0^\infty x^\lambda \ln^m x\, \varphi(x)\, dx\;; \tag{1}$$

for Re $\lambda > -n-1$, $\lambda \neq -1, -2, \dots, -n$,

$$(x_+^\lambda \ln^m x_+, \varphi) = \int_0^1 x^\lambda \ln^m x \left[\varphi(x) - \varphi(0) - x\varphi'(0) - \dots - \frac{x^{n-1}}{(n-1)!}\varphi^{(n-1)}(0)\right] dx$$

$$+ \int_1^\infty x^\lambda \ln^m x\, \varphi(x)\, dx + \sum_{k=1}^n \frac{(-1)^m\, m!\varphi^{(k-1)}(0)}{(k-1)!\,(\lambda+k)^{m+1}} \tag{2}$$

for $-n-1 < $ Re $\lambda < -n$,

$$(x_+^\lambda \ln^m x_+, \varphi)$$

$$= \int_0^\infty x^\lambda \ln^m x \left[\varphi(x) - \varphi(0) - x\varphi'(0) - \dots - \frac{x^{n-1}}{(n-1)!}\varphi^{(n-1)}(0)\right] dx. \tag{3}$$

2. We may differentiate with respect to λ:

$$\frac{\partial^m}{\partial\lambda^m} x_+^\lambda = x_+^\lambda \ln^m x_+, \qquad \lambda \neq -1, -2, \dots. \tag{4}$$

The Taylor's series expansion of x_+^λ about a point of regularity λ_0 is

$$x_+^\lambda = x_+^{\lambda_0} + (\lambda-\lambda_0) x_+^{\lambda_0} \ln x_+ + \tfrac{1}{2}(\lambda-\lambda_0)^2 x_+^{\lambda_0} \ln^2 x_+ + \dots. \tag{5}$$

3. The generalized function x_+^{-n} is defined by

$$(x_+^{-n}, \varphi) = \int_0^\infty x^{-n} \left[\varphi(x) - \varphi(0) - x\varphi'(0) - \dots \right.$$

$$\left. - \frac{x^{n-2}}{(n-2)!}\varphi^{(n-2)}(0) - \frac{x^{n-1}}{(n-1)!}\varphi^{(n-1)}(0)\, \theta(1-x)\right] dx. \tag{6}$$

This generalized function coincides with the ordinary function x^{-n} for $x > 0$, and with zero for $x < 0$. It is *not* equal to x_+^λ at $\lambda = -n$.

4. The generalized function $x_+^{-n} \ln^m x_+$ is defined by

$$(x_+^{-n} \ln^m x_+, \varphi) = \int_0^\infty x^{-n} \ln^m x \left[\varphi(x) - \varphi(0) - x\varphi'(0) - \ldots \right.$$

$$\left. - \frac{x^{n-2}}{(n-2)!} \varphi^{(n-2)}(0) - \frac{x^{n-1}}{(n-1)!} \varphi^{(n-1)}(0) \, \theta(1-x) \right] dx. \qquad (7)$$

5. The Laurent series expansion of x_+^λ about the pole at $\lambda = -n$ is

$$x_+^\lambda = \frac{(-1)^{n-1} \delta^{(n-1)}(x)}{(n-1)!} \frac{1}{\lambda+n} + x_+^{-n} + (\lambda+n) x_+^{-n} \ln x_+$$

$$+ \frac{1}{2!} (\lambda+n)^2 x_+^{-n} \ln^2 x_+ + \ldots. \qquad (8)$$

6. The generalized function $x_-^\lambda \ln^m x_-$, $\lambda \neq -1, -2, \ldots$, is defined as follows. For Re $\lambda > -1$,

$$(x_-^\lambda \ln^m x_-, \varphi) = \int_{-\infty}^0 |x|^\lambda \ln^m |x| \, \varphi(x) \, dx = \int_0^\infty x^\lambda \ln^m x \, \varphi(-x) \, dx \,; \qquad (9)$$

For Re $\lambda > -n-1$, $\lambda \neq -1, -2, \ldots, -n$,

$$(x_-^\lambda \ln x_-^m, \varphi)$$

$$= \int_0^1 x^\lambda \ln^m x \left[\varphi(-x) - \varphi(0) + x\varphi'(0) - \ldots - (-1)^{n-1} \frac{x^{n-1}}{(n-1)!} \varphi^{(n-1)}(0) \right] dx$$

$$+ \int_1^\infty x^\lambda \ln^m x \, \varphi(-x) \, dx + \sum_{k=1}^n \frac{(-1)^{m+k-1} m! \, \varphi^{(k-1)}(0)}{(k-1)!(\lambda+k)^{m+1}} \,; \qquad (10)$$

for $-n-1 < $ Re $\lambda < -n$,

$$(x_-^\lambda \ln x_-^m, \varphi)$$

$$= \int_0^\infty x^\lambda \ln^m x \left[\varphi(-x) - \varphi(0) + x\varphi'(0) - \ldots - (-1)^{n-1} \frac{x^{n-1}}{(n-1)!} \varphi^{(n-1)}(0) \right] dx. \qquad (11)$$

7. We may differentiate with respect to λ:

$$\frac{\partial^m}{\partial \lambda^m} x_-^\lambda = x_-^\lambda \ln^m x_-, \qquad \lambda \neq -1, -2, \ldots. \qquad (12)$$

The Taylor's series for x_-^λ about a point of regularity λ_0 is

$$x_-^\lambda = x_-^{\lambda_0} + (\lambda - \lambda_0)\, x_-^{\lambda_0} \ln x_- + \tfrac{1}{2}(\lambda - \lambda_0)^2\, x_-^{\lambda_0} \ln^2 x_- + \dots . \tag{13}$$

8. The generalized function x_-^{-n} is defined by

$$(x_-^{-n}, \varphi) = \int_0^\infty x^{-n} \Big[\varphi(-x) - \varphi(0) + x\varphi'(0) - \dots$$
$$- (-1)^{n-1} \frac{x^{n-1}}{(n-1)!}\, \varphi^{(n-1)}(0)\, \theta(1-x)\Big]\, dx. \tag{14}$$

This generalized function coincides with the ordinary function $|x|^{-n}$ for $x < 0$ and with zero for $x > 0$. It is *not* equal to x_-^λ at $\lambda = -n$.

9. The generalized function $x_-^{-n} \ln^m x_-$ is defined by

$$(x_-^{-n} \ln^m x_-, \varphi) = \int_0^\infty x^{-n} \ln^m x \Big[\varphi(-x) - \varphi(0) + x\varphi'(0) - \dots$$
$$- (-1)^{n-1} \frac{x^{n-1}}{(n-1)!}\, \varphi^{(n-1)}(0)\, \theta(1-x)\Big]\, dx. \tag{15}$$

10. The Laurent series expansion of x_-^λ about the pole at $\lambda = -n$ is

$$x_-^\lambda = \frac{\delta^{(n-1)}(x)}{(n-1)!}\frac{1}{\lambda+n} + x_-^{-n} + (\lambda+n)\, x_-^{-n} \ln x_- + \frac{1}{2!}(\lambda+n)^2\, x_-^{-n} \ln^2 x_- + \dots . \tag{16}$$

11. The generalized function $|x|^\lambda \ln^k |x|$ is defined by formulas similar to (9)–(11) of the summary of Chapter I, Section 3, except that x^λ is replaced everywhere by $x^\lambda \ln^k x$. The Taylor's series for $|x|^\lambda$ about a point of regularity λ_0 is

$$|x|^\lambda = |x|^{\lambda_0} + (\lambda - \lambda_0)\, |x|^{\lambda_0} \ln |x| + \frac{1}{2!}(\lambda - \lambda_0)^2\, |x|^{\lambda_0} \ln^2 |x| + \dots . \tag{17}$$

The Laurent series expansion of $|x|^\lambda$ about the pole at $\lambda_0 = -2m-1$ is

$$|x|^\lambda = \frac{2\delta^{(2m)}(x)}{(2m)!}\frac{1}{\lambda+2m+1} + |x|^{-2m-1}$$
$$+ (\lambda+2m+1)\, |x|^{-2m-1} \ln |x| + \dots , \tag{18}$$

where

$$|x|^{-2m-1} = x_+^{-2m-1} + x_-^{-2m-1}; \tag{19}$$

$$|x|^{-2m-1} \ln^k |x| = x_+^{-2m-1} \ln^k x_+ + x_-^{-2m-1} \ln^k x_-. \tag{19a}$$

In particular,

$$(|x|^{-2m-1}, \varphi) = \int_0^\infty x^{-2m-1} \left\{ \varphi(x) + \varphi(-x) \right.$$

$$\left. - 2 \left[\varphi(0) + \frac{x^2}{2!} \varphi''(0) + \dots + \frac{x^{2m}}{(2m)!} \varphi^{(2m)}(0) \, \theta(1-x) \right] \right\} dx. \qquad (20)$$

The generalized function $|x|^{-2m-1}$ coincides with the ordinary function $|x|^{-2m-1}$ for $x \neq 0$.

12. The generalized function $|x|^\lambda \operatorname{sgn} x \ln^k |x|$ is defined by equations similar to (14)–(16) of the summary of Chapter I, Section 3, except that x^λ is replaced everywhere by $x^\lambda \ln {}^k x$.

The Taylor's series for $|x|^\lambda \operatorname{sgn} x$ about a point of regularity λ_0 is

$$|x|^\lambda \operatorname{sgn} x = |x|^{\lambda_0} \operatorname{sgn} x + (\lambda - \lambda_0) |x|^{\lambda_0} \ln |x| \operatorname{sgn} x$$

$$+ \frac{1}{2!} (\lambda - \lambda_0)^2 |x|^{\lambda_0} \ln |x| \operatorname{sgn} x + \dots. \qquad (21)$$

The Laurent series for $|x|^\lambda \operatorname{sgn} x$ about the pole at $\lambda_0 = -2m$ is

$$|x|^\lambda \operatorname{sgn} x = -2 \frac{\delta^{(2m-1)}(x)}{(2m-1)!} \frac{1}{\lambda + 2m} + |x|^{-2m} \operatorname{sgn} x$$

$$+ (\lambda + 2m) |x|^{-2m} \ln |x| \operatorname{sgn} x + \dots, \qquad (22)$$

where

$$|x|^{-2m} \operatorname{sgn} x = x_+^{-2m} - x_-^{-2m}; \qquad (23)$$

$$|x|^{-2m} \ln^k |x| \operatorname{sgn} x = x_+^{-2m} \ln^k x_+ - x_-^{-2m} \ln^k x_-. \qquad (23a)$$

In particular,

$$(|x|^{-2m} \operatorname{sgn} x, \varphi) = \int_0^\infty x^{-2m} \left\{ \varphi(x) - \varphi(-x) \right.$$

$$\left. - 2 \left[x\varphi'(0) + \frac{x^3}{3!} \varphi'''(0) + \dots + \frac{x^{2m-1}}{(2m-1)!} \varphi^{(2m-1)}(0) \, \theta(1-x) \right] \right\} dx. \qquad (24)$$

It should be emphasized that $|x|^{-2m-1}$ is *not* the value of $|x|^\lambda$ at $\lambda = -2m-1$, and that $|x|^{-2m} \operatorname{sgn} x$ is *not* the value of $|x|^\lambda \operatorname{sgn} x$ at $\lambda = -2m$.

The generalized function $|x|^{-2m} \operatorname{sgn} x$ coincides with the ordinary function $|x|^{-2m} \operatorname{sgn} x$ for $x \neq 0$.

13. The derivatives of the generalized functions $(x + i0)^\lambda$ and $(x - i0)^\lambda$ with respect to λ are denoted, respectively, by

$$(x + i0)^\lambda \ln (x + i0) \qquad \text{and} \qquad (x + i0)^\lambda \ln (x - i0).$$

These are entire functions of λ and may be written

$$(x + i0)^\lambda \ln (x + i0)$$

$$= \begin{cases} x_+^\lambda \ln x_+ + i\pi e^{i\lambda\pi} x_-^\lambda + e^{i\lambda\pi} x_-^\lambda \ln x_- & \text{for} \quad \lambda \neq -n, \\ (-1)^n i\pi x_-^{-n} + (-1)^{n-1} \dfrac{\pi^2}{2} \dfrac{\delta^{(n-1)}(x)}{(n-1)!} + x^{-n} \ln |x| & \text{for} \quad \lambda = -n; \end{cases} \tag{25}$$

$$(x - i0)^\lambda \ln (x - i0)$$

$$= \begin{cases} x_+^\lambda \ln x_+ - i\pi e^{-i\lambda\pi} x_-^\lambda + e^{-i\lambda\pi} x_-^\lambda \ln x_- & \text{for} \quad \lambda \neq -n, \\ (-1)^{n-1} i\pi x_-^{-n} + (-1)^{n-1} \dfrac{\pi^2}{2} \dfrac{\delta^{(n-1)}(x)}{(n-1)!} + x^{-n} \ln |x| & \text{for} \quad \lambda = -n. \end{cases} \tag{26}$$

14. The Laurent series expansion of r^λ about the pole at $\lambda = -n - 2k$ (here n is the dimension) is

$$r^\lambda = \Omega_n \frac{\delta^{(2k)}(r)}{(2k)!} \frac{1}{\lambda + n + 2k} + \Omega_n r^{-n-2k}$$
$$+ \Omega_n (\lambda + n + 2k) r^{-n-2k} \ln r + \dots. \tag{27}$$

Here the functionals $\delta^{(2k)}(r)$, r^{-n-2k}, and $r^{-n-2k} \ln^m r$ on the right act on the $S_\varphi(r)$ functions according to

$$(\delta^{(2k)}(r), S_\varphi(r)) = S_\varphi^{(2k)}(0), \tag{28}$$

$$(r^{-n-2k}, S_\psi(r)) = \int_0^\infty r^{-2k-n} \left[S_\varphi(r) - \varphi(0) - \dots \right.$$
$$\left. - \frac{r^{2k-2}}{(2k-2)!} S_\varphi^{(2k-2)}(0) - \frac{r^{2k}}{(2k)!} S_\varphi^{(2k)}(0) \theta(1 - r) \right] dr, \tag{29}$$

$$(r^{-n-2k} \ln^m r, S_\varphi(r)) = \int_0^\infty r^{-2k-n} \ln^m r \left[S_\varphi(r) - \varphi(0) - \dots \right.$$
$$\left. - \frac{r^{2k}}{(2k)!} S_\varphi^{(2k)}(0) \theta(1 - r) \right] dr. \tag{30}$$

The number $S_\varphi^{(2k)}(0)$ is given explicitly in terms of $\varphi(x)$ and its derivatives by

$$S_\varphi^{(2k)}(0) = \frac{(2k)! \Delta^n \varphi(0)}{2^k k! n(n + 2) \dots (n + 2k - 2)}. \tag{31}$$

Chapter I, Section 5

1. The *direct product* $f(x) \times g(y)$ of the two functionals $f(x)$ and $g(y)$ is the functional on the space of functions (infinitely differentiable and with bounded support) $\varphi(x, y)$ defined by

$$(f(x) \times g(y), \varphi(x, y)) = (f(x), (g(y), \varphi(x, y))) \ .$$

2. The *convolution* $f(x) * g(x)$ of the functionals f and g is the functional defined on K (the space on which f and g are defined) by

$$(f * g, \varphi) = (f(x), (g(y), \varphi(x + y))) \ .$$

The convolution exists if one of the following conditions holds:

(a) either f or g has bounded support;

(b) the supports of f and g are bounded on the same side.

When these conditions are fulfilled the convolution is both commutative and associative.

3. $\delta * f = f$ for all f.

4. $\dfrac{\partial \delta}{\partial x_j} * f = \dfrac{\partial f}{\partial x_j}$.

5. $\dfrac{\partial}{\partial x_j} (f * g) = \dfrac{\partial f}{\partial x_j} * g = f * \dfrac{\partial g}{\partial x_j}$.

6. $f_\nu \to f$ implies $f_\nu * g \to f * g$ under any of the following conditions:

(a) all the f_ν are concentrated on a given bounded set;

(b) the functional g has bounded support;

(c) the supports of f_ν and g are bounded on the same side, and the bound is independent of ν.

7. An *elementary solution of the equation*

$$P\left(\frac{\partial}{\partial x}\right) u = g \tag{1}$$

is a generalized function $E(x)$ such that

$$P\left(\frac{\partial}{\partial x}\right) E = \delta(x).$$

The solution of (1) can be written

$$u = E * g$$

if this convolution exists. See also the summaries of Chapter I, Section 6, paragraphs 1–3; Chapter II, paragraph 8; Chapter III, Section 2, paragraph 5.

8. An *elementary solution of Cauchy's problem* for an equation of the form

$$\frac{\partial u}{\partial t} = P\left(\frac{\partial}{\partial x}\right) u \tag{2}$$

is a generalized function $E(x, t)$ which depends continuously on the parameter t, is a solution of Eq. (2) for $t > 0$, and converges to $\delta(x)$ as $t \to 0$.

The solution of Eq. (2) with the initial condition $u(x, 0) = u_0(x)$ can be written in the form

$$u(x, t) = E(x, t) * u_0(x),$$

so long as this convolution exists.

9. An *elementary solution of Cauchy's problem* for an equation of the form

$$P\left(\frac{\partial}{\partial t}, \frac{\partial}{\partial x}\right) u(x, t) = 0, \tag{3}$$

of mth order in t, is a generalized function $E(x, t)$ which depends continuously on the parameter t, is a solution of Eq. (3) for $t > 0$, and is such that

$$\lim_{t \to 0} E(x, t) = 0, \qquad \lim_{t \to 0} \frac{\partial E(x, t)}{\partial t} = 0,$$

$$\lim_{t \to 0} \frac{\partial^{m-2} E(x, t)}{\partial t^{m-2}} = 0, \qquad \lim_{t \to 0} \frac{\partial^{m-1} E(x, t)}{\partial t^{m-1}} = \delta(x).$$

The solution of Eq. (3) with the initial conditions

$$u(x, 0) = \frac{\partial u(x, 0)}{\partial t} = \cdots = \frac{\partial^{m-2} u(x, 0)}{\partial t^{m-2}} = 0, \qquad \frac{\partial^{m-1} u(x, 0)}{\partial t^{m-1}} = u_{m-1}(x)$$

can be written in the form

$$u(x, t) = E(x, t) * u_{m-1}(x)$$

so long as this convolution exists. See also the summaries of Chapter I, Section 6, paragraphs 4 and 5; Chapter II, paragraph 8; Chapter III, Section 1, paragraph 6.

10. The *integral of order* λ of a generalized function $g(x)$ which

vanishes for $x < 0$ (we are dealing here with a single variable) is defined by

$$g_\lambda(x) = g(x) * \frac{x_+^{\lambda-1}}{\Gamma(\lambda)}.$$

For Re $\lambda < 0$ this formula defines the derivative of order λ of $g(x)$.

11. The hypergeometric function

$$\Gamma(\alpha, \beta, \gamma, x) = \frac{\Gamma(\gamma)}{\Gamma(\beta)\,\Gamma(\gamma-\beta)} \int_0^1 t^{\beta-1}(1-t)^{\gamma-\beta-1}(1-tx)^{-\alpha}\, dt$$

satisfies the equation

$$\frac{x^{\gamma-1}}{\Gamma(\gamma)} F(\alpha, \beta, \gamma, x) = \frac{d^{\beta-\gamma}}{dx^{\beta-\gamma}} \left[\frac{x_+^{\beta-1}(1-x)_+^{-\alpha}}{\Gamma(\beta)} \right]$$

or

$$\frac{x^{\gamma-1}(1-x)^{\alpha+\beta-\gamma}}{\Gamma(\gamma)} F(\alpha, \beta, \gamma, x) = \frac{d^{-\beta}}{dx^{-\beta}} \left[\frac{x_+^{\gamma-\beta-1}(1-x)_+^{\alpha-\gamma}}{\Gamma(\gamma-\beta)} \right].$$

12. The Bessel function $J_\lambda(\sqrt{u})$ can be expressed in terms of fractional derivatives of elementary functions:

$$2^\lambda \sqrt{\pi}\ u^{\frac{1}{2}\lambda}\ J_\lambda(\sqrt{u}) = \frac{d^{-\lambda-\frac{1}{2}}}{du^{-\lambda-\frac{1}{2}}} \frac{\cos\sqrt{u}}{\sqrt{u}}.$$

Chapter I, Section 6

1. An elementary solution of the elliptic equation $P(\partial/\partial x)E(x) = \delta(x)$ can be written

$$E(x) = \int_\Omega v_\omega(\omega_1 x_1 + \dots + \omega_n x_n, - n)\, d\omega, \tag{1}$$

where

$$v_\omega(\xi, \lambda) = \frac{1}{\Omega_n \pi^{\frac{1}{2}(n-1)}\, \Gamma\left(\frac{\lambda+1}{2}\right)} \int_{-\infty}^\infty G(\xi - \eta, \omega)\,|\,\eta\,|^\lambda\, d\eta,$$

and $G(\xi, \omega)$ is a solution of

$$P\left(\omega_1 \frac{d}{d\xi}, \dots, \omega_n \frac{d}{d\xi}\right) G(\xi, \omega) = \delta(\xi).$$

For odd n expression (1) can be reduced to (with $\xi = \omega_1 x_1 + \ldots + \omega_n x_n$)

$$E(x) = \frac{(-1)^{\frac{1}{2}(n-1)}}{\Omega_n (2\pi)^{\frac{1}{2}(n-1)} 1 \cdot 3 \ldots (n-2)} \int_\Omega \frac{\partial^{n-1} G(\xi, \omega)}{\partial \xi^{n-1}} \, d\omega.$$

2. If $P(\partial/\partial x)$ is a homogeneous polynomial of degree m in the $\partial/\partial x_j$, then $E(x)$ is of the form

$$E(x) = \frac{(-1)^{n-1}}{4(2\pi)^{m-1}(2m-n)!} \int_\Omega |\omega_1 x_1 + \ldots + \omega_n x_n|^{2m-n} \frac{d\omega}{P(\omega_1, \ldots, \omega_n)},$$

$$2m > n, \; n \text{ odd};$$

$$= \frac{(-1)^{\frac{1}{2}(n-2)}}{(2\pi)^n (2m-n)!} \int_\Omega \frac{(\omega_1 x_1 + \ldots + \omega_n x_n)^{2m-n} \ln|\omega_1 x_1 + \ldots + \omega_n x_n|}{P(\omega_1, \ldots, \omega_n)} \, d\omega,$$

$$2m \geqslant n, \; n \text{ even};$$

$$= \frac{(-1)^{\frac{1}{2}(n-1)}}{2(2\pi)^{n-1}} \int_\Omega \delta^{(n-2m-1)}(\omega_1 x_1 + \ldots + \omega_n x_n) \frac{d\omega}{P(\omega_1, \ldots, \omega_n)},$$

$$2m < n, \; n \text{ odd};$$

$$= \frac{(-1)^{\frac{1}{2}n}(n-2m-1)!}{(2\pi)^n} \int_\Omega \frac{|\omega_1 x_1 + \ldots + \omega_n x_n|^{2m-n}}{P(\omega_1, \ldots, \omega_n)} \, d\omega,$$

$$2m < n, \; n \text{ even}.$$

3. These equations remain valid also for homogeneous operators of the form $P(\partial/\partial x)$ such that grad $P(\omega)$ does not vanish on $P(\omega) = 0$, $\omega \neq 0$, but then the integrals must be understood in the sense of their regularizations (the Cauchy principal values).

4. An elementary solution of Cauchy's problem for the equation $P(\partial/\partial t, \, \partial/\partial x) = 0$ is of the form

$$u(x, t) = \frac{1}{\Omega_n \pi^{\frac{1}{2}(n-1)}} \int_\Omega \left\{ \int_{-\infty}^{\infty} G_\omega(t, \xi - \eta) \frac{|\eta|^\lambda}{\Gamma\left(\frac{\lambda + 1}{2}\right)} \bigg|_{\lambda = -n} d\eta \right\} d\omega,$$

where $G_\omega(t, \xi)$ is an elementary solution of Cauchy's problem for the equation

$$P\left(\frac{\partial}{\partial t}, \, \omega_1 \frac{d}{d\xi}, \, \ldots, \, \omega_n \frac{d}{d\xi}\right) u(t, \xi) = 0.$$

In particular, for odd n we have

$$u(x, t) = \frac{(-1)^{\frac{1}{2}(n-1)} \left(\frac{n-1}{2}\right)!}{\Omega_n \pi^{\frac{1}{2}(n-1)}(n-1)!} \int_\Omega \frac{d^{n-1}}{d\xi^{n-1}} G_\omega(t, \xi) \, d\omega.$$

5. If $P(\partial/\partial t, \partial/\partial x)$ is a homogeneous hyperbolic polynomial of degree m, then $u(x, t)$ becomes

$$u(x, t) = \frac{(-1)^{\frac{1}{2}(n+1)}}{2(2\pi)^{n-1}(m-n-1)!}$$

$$\times \int_{P(1,\xi)=0} \left(\sum x_k \xi_k + t\right)^{m-n-1} \operatorname{sgn}^{m-1}\left(\sum x_k \xi_k + t\right) \omega,$$

$$m \geqslant n-1, n \text{ odd};$$

$$= \frac{2(-1)^{\frac{1}{2}n}}{(2\pi)^n(m-n-1)!}$$

$$\times \int_{P(1,\xi)=0} \left(\sum x_k \xi_k + t\right)^{m-n-1} \ln \left| \frac{\sum x_k \xi_k + t}{\sum x_k \xi_k} \right| \omega,$$

$$m \geqslant n-1, n \text{ even};$$

$$u(x, t) = \frac{(-1)^{\frac{1}{2}(n+1)}}{(2\pi)^{n-1}} \int_{P(1,\xi)=0} \delta^{(n-m)}\left(\sum x_k \xi_k + t\right) \omega,$$

$$m < n-1, n \text{ odd};$$

$$= \frac{(-1)^{\frac{1}{2}(n+2)}(n-m)!}{(2\pi)^n} \int_{P(1,\xi)=0} \frac{\omega}{(\sum x_k \xi_k + t)^{n-m+1}}$$

$$m < n-1, n \text{ even}.$$

Here

$$\omega = d\sigma \left\{ |\operatorname{grad} H| \operatorname{sgn}\left(\sum_{k=1}^{n} \xi_k H_{\xi_k}\right) \right\}^{-1},$$

and $H(\xi) \equiv P(1, \xi)$.

Chapter II

1. The *Fourier transform of a function* $\varphi(x)$ in K is defined as

$$\psi(s) = \int \varphi(x) \exp\left[i(x_1 s_1 + \dots + x_n s_n)\right] dx_1 \dots dx_n$$

and is an entire analytic function of the complex variables $s_1 = \sigma_1 + i\tau_1, \dots, s_n = \sigma_n + i\tau_n$. This Fourier transform satisfies inequalities of the form

$$|s_1^{q_1} \dots s_n^{q_n} \psi(s_1, \dots, s_n)| \leqslant C_{\mathfrak{p}} \exp\left(a_1 |\tau_1| + \dots + a_n |\tau_n|\right)$$

where $|x_j| \leqslant a_j$ is a region containing the support of $\varphi(x)$.

2. The set of all $\psi(s)$ functions of this form is called *the space Z of slowly increasing functions.* A sequence $\psi_\nu(s) \in Z$ is said to *converge to zero in Z* if the inverse Fourier transforms $\varphi_\nu(x) \in K$ converge to zero in K.

3. The *Fourier transform of a generalized function f* defined as a functional on K is the functional $g = F(f)$ defined on Z by the equation

$$(g, \psi) = (2\pi)^n (f, \varphi),$$

where $\psi(s) \in Z$ is the Fourier transform of $\varphi(x) \in K$.

Under multiplication and differentiation Fourier transforms behave according to

$$F\left[P\left(\frac{\partial}{\partial x}\right)f\right] = P(-is)\,F[f],$$

$$F[P(x)f] = P\left(-i\frac{\partial}{\partial s}\right)F[f].$$

5. The Fourier transform of the direct product is given by

$$F[f \times g] = F[f] \times F[g].$$

6. The Fourier transform of a functional with bounded support is a functional corresponding to the function

$$F[f] = (\overline{f(x)},\, e^{i(x.s)}) = \overline{(f(x),\, e^{-i(x.s)})}.$$

7. $F[S] = S$.

8. If $u(x, t)$ is an elementary solution of Cauchy's problem for the equation $\partial u/\partial t - P(i\,\partial/\partial x)u = 0$, then

$$E(x, t) = \begin{cases} u(x, t) & \text{for} \quad t > 0, \\ 0 & \text{for} \quad t < 0 \end{cases}$$

satisfies the equation

$$\frac{\partial E(x, t)}{\partial t} - P\left(i\frac{\partial}{\partial x}\right)E(x, t) = \delta(x, t).$$

9. The Fourier transforms of many particular generalized functions are given in the table of Fourier transforms.

Chapter III, Section 1

1. The *Leray form* ω for the hypersurface $P(x_1, \ldots, x_n) = 0$ is defined by

$$dP \cdot \omega = dx_1 \ldots dx_n.$$

At those points where $\partial P/\partial x_j \neq 0$, we may write

$$\omega = (-1)^{j-1} - \frac{dx_1 \dots dx_{j-1}\, dx_{j+1} \dots dx_n}{\partial P/\partial x_j}.$$

2. The functional $\delta(P)$ is defined by

$$(\delta(P), \varphi) = \int_{P=0} \varphi(x)\, \omega.$$

3. $\theta(P)$ is the characteristic function of the region $P(x) \geqslant 0$. Its derivative is

$$\theta'(P) = \delta(P),$$

which is understood in the sense that

$$\frac{\partial \theta(P)}{\partial x_j} = \frac{\partial P}{\partial x_j} \delta(P).$$

4. The differential forms ω_k, $k = 1, 2, \dots$ are given by the conditions

$$\omega_0(\varphi) = \varphi \cdot \omega,$$
$$d\omega_0(\varphi) = dP \cdot \omega_1(\varphi),$$
$$\dots \dots \dots \dots \dots \dots$$
$$d\omega_{k-1}(\varphi) = dP \cdot \omega_k(\varphi).$$
$$\dots \dots \dots \dots \dots \dots$$

The integral of $\omega_k(\varphi)$ over the $P = 0$ hypersurface is determined uniquely.

5. We have

$$P\delta(P) = 0,$$
$$P\delta'(P) + \delta(P) = 0,$$
$$P\delta''(P) + 2\delta(P) = 0,$$
$$\dots \dots \dots \dots \dots \dots$$
$$P\delta^{(k)}(P) + k\delta^{(k-1)}(P) = 0,$$
$$\dots \dots \dots \dots \dots \dots$$

6. An elementary solution of Cauchy's problem for the wave equation $(\Delta - \partial^2/\partial t^2)u = 0$ with $n = 2k + 3$ is

$$E(x, t) = \frac{(-1)^k}{2\pi^{k+1}} \delta^{(k)}(r^2 - t^2), \qquad r^2 = x_1^2 + \dots + x_n^2.$$

7. If the hypersurfaces $P = 0$ and $Q = 0$ have no points in common, then

$$\delta(PQ) = P^{-1}\delta(Q) + Q^{-1}\delta(P).$$

8. If $a(x)$ fails to vanish, then

$$\delta(aP) = a^{-1}\delta(P),$$

. . . .

$$\delta^{(k)}(aP) = a^{-(k+1)}\delta^{(k)}(P),$$

. . . .

9. The Leray form ω of the manifold $P_1 = P_2 = \ldots = P_k = 0$ is defined by

$$dP_1 \ldots dP_k \cdot \omega = dx_1 \ldots dx_n.$$

In particular, if the Jacobian $D\begin{pmatrix} P_1 \ldots P_k \\ x_1 \ldots x_k \end{pmatrix} \neq 0$, we may write

$$\omega = \frac{dx_{k+1} \ldots dx_n}{D\begin{pmatrix} P_1 \ldots P_k \\ x_1 \ldots x_k \end{pmatrix}}.$$

10. The generalized function $\delta(P_1, \ldots, P_k)$ is defined by

$$(\delta(P_1, \ldots, P_k), \varphi(x_1, \ldots, x_n)) = \int_{P_1 = \ldots = P_k = 0} \varphi \cdot \omega.$$

11. The differential form $\omega_{\alpha_1 \ldots \alpha_{j+1} \ldots \alpha_k}(\varphi)$ is defined in terms of the differential form $\omega_{\alpha_1 \ldots \alpha_j \ldots \alpha_k}(\varphi)$ by

$$d(dP_1 \ldots dP_{j-1} dP_{j+1} \ldots dP_k \omega_{\alpha_1 \ldots \alpha_j \ldots \alpha_k}) = (-1)^{j-1} dP_1 \ldots dP_k \omega_{\alpha_1 \ldots \alpha_{j+1} \ldots \alpha_k}.$$

12. The generalized function $\dfrac{\partial^m \delta(P_1 \ldots P_k)}{\partial P_1^{\alpha_1} \ldots \partial P_k^{\alpha_k}}$ is defined by

$$\left(\frac{\partial^m \delta(P_1 \ldots P_k)}{\partial P_1^{\alpha_1} \ldots \partial P_k^{\alpha_k}}, \varphi \right) = (-1)^m \int_{P_1 = \ldots = P_k = 0} \omega_{\alpha_1 \ldots \alpha_k}(\varphi), \qquad (m = \alpha_1 + \ldots + \alpha_k).$$

13. We have

$$\frac{\partial}{\partial x_j} \delta(P_1, \ldots, P_k) = \sum_{i=1}^{k} \frac{\partial \delta(P_1, \ldots, P_k)}{\partial P_i} \frac{\partial P_i}{\partial x_j},$$

$$P_i \delta(P_1, \ldots, P_k) = 0,$$

$$P_i P_j \delta(P_1, \ldots, P_k) = 0,$$

. . . .

$$P_1 P_2 \ldots P_k \delta(P_1, \ldots, P_k) = 0,$$

and other equations obtained from these by formal differentiation with respect to the P_i.

Chapter III, Section 2

1. *Notation*:

$$P = x_1^2 + \ldots + x_p^2 - x_{p+1}^2 - \ldots - x_{p+q}^2$$

$$Q = s_1^2 + \ldots + s_p^2 - s_{p+1}^2 - \ldots - s_{p+q}^2$$

$$L = \frac{\partial^2}{\partial x_1^2} + \ldots + \frac{\partial^2}{\partial x_p^2} - \frac{\partial^2}{\partial x_{p+1}^2} - \ldots - \frac{\partial^2}{\partial x_{p+q}^2},$$

$n = p + q$ is the dimension of the space, and $p, q > 0$.

2. *Definitions of the generalized functions* $\delta_1^{(k)}(P)$, $\delta_2^{(k)}(P)$, $\delta^{(k)}(P_+)$, $\delta^{(k)}(P_-)$, P_+^λ, P_-^λ, $(P + i0)^\lambda$, $(P - i0)^\lambda$.

2.1. For $p > 1$, $q > 1$,

$$(\delta_1^{(k)}(P), \varphi) = \frac{1}{4} \int_0^\infty \left[\frac{\partial^k}{\partial v^k} \{v^{\frac{1}{2}(q-2)} \psi_1(u, v)\} \right]_{v=u} u^{\frac{1}{2}(p-2)} \, du,$$

$$(\delta_2^{(k)}(P), \varphi) = \frac{(-1)^k}{4} \int_0^\infty \left[\frac{\partial^k}{\partial u^k} \{u^{\frac{1}{2}(p-2)} \psi_1(u, v)\} \right]_{u=v} v^{\frac{1}{2}(q-2)} \, dv,$$

where $\psi_1(u, v)$ is the integral of φ over the manifold $x_1^2 + \ldots + x_p^2 = u$, $x_{p+1}^2 + \ldots + x_{p+q}^2 = v$ divided by $u^{\frac{1}{2}(p-1)} v^{\frac{1}{2}(q-1)}$.

These integrals converge for $k < \frac{1}{2} n - 1$. For $k \geq \frac{1}{2} n - 1$ they are understood in the sense of their regularizations according to Chapter I, Section 3.

The generalized functions $\delta_1^{(k)}(P)$ and $\delta_2^{(k)}(P)$ are defined similarly for the special cases $p = 1$ or $q = 1$.

2.2 $(P_+^\lambda, \varphi) = \int_{P>0} P^\lambda \varphi \, dx_1 \ldots dx_n;$

$$(P_-^\lambda, \varphi) = \int_{P<0} (-P)^\lambda \varphi \, dx_1 \ldots dx_n.$$

These integrals converge for $\operatorname{Re} \lambda \geq 0$, and are analytic functions of λ. For $\operatorname{Re} \lambda < 0$ they are obtained by analytic continuation in λ.

2.3. $\delta^{(k)}(P_+) = (-1)^k k! \operatorname*{res}_{\lambda = -k-1} P_+^\lambda,$

$\delta^{(k)}(P_-) = (-1)^k k! \operatorname*{res}_{\lambda = -k-1} P_-^\lambda.$

2.4. $(P + i0)^\lambda = P_+^\lambda + e^{\pi i \lambda} P_-^\lambda,$

$(P - i0)^\lambda = P_+^\lambda + e^{-\pi i \lambda} P_-^\lambda.$

3. *The singularities of P_+^λ, P_-^λ, $(P + i0)^\lambda$, and $(P - i0)^\lambda$.*

3.1. *p even, q odd.* The generalized function P_+^λ has simple poles at $\lambda = -k$, where k is a positive integer;

$$\operatorname*{res}_{\lambda=-k} P_+^\lambda = \frac{(-1)^{k-1}}{(k-1)!} \delta_1^{(k-1)}(P).$$

3.2 *p odd, q even.* The generalized function P_+^λ has simple poles at $\lambda = -k$, where k is a positive integer, and at $\lambda = -\frac{1}{2}n - k$, where k is a nonnegative integer;

$$\operatorname*{res}_{\lambda=-k} P_+^\lambda = \frac{(-1)^{k-1}}{(k-1)!} \delta_1^{(k-1)}(P),$$

$$\operatorname*{res}_{\lambda=-\frac{1}{2}n-k} P_+^\lambda = \frac{(-1)^{\frac{1}{2}q} \pi^{\frac{1}{2}n}}{4^k k! \, \Gamma\left(\frac{n}{2} + k\right)} L^k \delta(x_1, ..., x_n).$$

3.3. *p and q even.* The generalized function P_+^λ has simple poles at $\lambda = -k$, where k is a positive integer;

$$\operatorname*{res}_{\lambda=-k} P_+^\lambda = \frac{(-1)^{k-1}}{(k-1)!} \delta_1^{(k-1)}(P), \quad \text{for} \quad k < \tfrac{1}{2}n;$$

$$\operatorname*{res}_{\lambda=-\frac{1}{2}n-k} P_+^\lambda = \frac{(-1)^{\frac{1}{2}n+k-1}}{\Gamma\left(\frac{n}{2} + k\right)} \delta_1^{(\frac{1}{2}n+k-1)}(P) + \frac{(-1)^{\frac{1}{2}q} \pi^{\frac{1}{2}n}}{4^k k! \, \Gamma\left(\frac{n}{2} + k\right)} L^k \delta(x_1, ..., x_n).$$

3.4. *p and q odd.* The generalized function P_+^λ has simple poles at $\lambda = -1, -2, ..., -(\frac{1}{2}n - 1)$, and poles of second order at $\lambda = -\frac{1}{2}n - k$, where k is a nonnegative integer.

For $k < \frac{1}{2}n$,

$$\operatorname*{res}_{\lambda=-k} P_+^\lambda = \frac{(-1)^{k-1}}{(k-1)!} \delta_1^{(k-1)}(P).$$

In the neighborhood of $\lambda = -\frac{1}{2}n - k$,

$$P_+^\lambda = \frac{c_{-2}^{(k)}}{\left(\lambda + \frac{n}{2} + k\right)^2} + \frac{c_{-1}^{(k)}}{\left(\lambda + \frac{n}{2} + k\right)} + ...,$$

where

$$c_{-2}^{(k)} = \frac{(-1)^{\frac{1}{2}(q+1)} \pi^{\frac{1}{2}n-1}}{4^k k! \Gamma\left(\frac{n}{2} + k\right)} L^k \delta(x_1, ..., x_n)$$

and

$$c_{-1}^{(k)} = \frac{(-1)^{\frac{1}{2}n+k-1}}{\Gamma\left(\frac{n}{2}+k\right)}\, \delta_1^{(\frac{1}{2}n+k-1)}(P)$$

$$+ \frac{(-1)^{\frac{1}{2}(q+1)}\, \pi^{\frac{1}{2}n-1}\left[\psi\left(\frac{p}{2}\right) - \psi\left(\frac{n}{2}\right)\right]}{4^k k!\, \Gamma\left(\frac{n}{2}+k\right)}\, L^k \delta(x_1, ..., x_n).$$

3.5. To obtain the results for P^λ, use paragraphs 3.1–3.4 above, interchanging the roles of p and q and replacing L by $-L$ and $\delta_1^{(k-1)}(P)$ by $\delta_1^{(k-1)}(-P)$.

3.6 The generalized functions $(P + i0)^\lambda$ and $(P - i0)^\lambda$ have only simple poles at $\lambda = -\frac{1}{2}n - k$, where k is a nonnegative integer;

$$\operatorname*{res}_{\lambda=-\frac{1}{2}n-k} (P + i0)^\lambda = \overline{\operatorname*{res}_{\lambda=-\frac{1}{2}n-k} (P - i0)^\lambda}$$

$$= \frac{e^{-\frac{1}{2}n q i}\, \pi^{\frac{1}{2}n}}{4^k k!\, \Gamma\left(\frac{n}{2}+k\right)}\, L^k \delta(x_1, ..., x_n).$$

4. *Relations between* $\delta_1^{(k)}(P)$, $\delta_2^{(k)}(P)$, $\delta^{(k)}(P_+)$, *and* $\delta^{(k)}(P_-)$.

4.1 $\delta_1^{(k)}(-P) = (-1)^k\, \delta_2^{(k)}(P)$.

4.2. If n is odd, as well as if n is even and $k < \frac{1}{2}n - 1$, then

$$\delta_1^{(k)}(P) = \delta_2^{(k)}(P) = \delta^{(k)}(P_+);$$

if n is even and $k \geqslant \frac{1}{2}n - 1$, then

$$\delta_1^{(k)}(P) - \delta_2^{(k)}(P) = c_{p,q,k} L^{k-\frac{1}{2}n+1}\delta(x_1, ..., x_n),$$
$$\delta_1^{(k)}(P) - \delta^{(k)}(P_+) = c'_{p,q,k} L^{k-\frac{1}{2}n+1}\delta(x_1, ..., x_n),$$

where $c_{p,q,k}$ and $c'_{p,q,k}$ are numerical coefficients.

4.3. If p and q are even and $k \geqslant \frac{1}{2}n - 1$, then

$$(-1)^k\delta^{(k)}(P_+) - \delta^{(k)}(P_-) = \frac{(-1)^{\frac{1}{2}q}\, \pi^{\frac{1}{2}n}}{4^{k+1-\frac{1}{2}n}\left(k+1-\frac{n}{2}\right)!}\, L^{k+1-\frac{1}{2}n}\delta(x_1, ..., x_n),$$

and for all other cases

$$\delta^{(k)}(P_-) = (-1)^k\delta^{(k)}(P_+).$$

5. *Elementary solution K of the differential equation $L^k u = f(x)$.*

5.1. If n is odd, then

$$K_1 = (-1)^k \frac{e^{\frac{1}{2}nqi}\Gamma\left(\frac{n}{2} - k\right)}{4^k(k-1)!\pi^{\frac{1}{2}n}} (P + i0)^{-\frac{1}{2}n+k}$$

$$= (-1)^k \frac{e^{\frac{1}{2}nqi}\Gamma\left(\frac{n}{2} - k\right)}{4^k(k-1)!\,\pi^{\frac{1}{2}n}} (P_+^{-\frac{1}{2}n+k} + e^{\pi(-\frac{1}{2}n+k)i}\, P_-^{-\frac{1}{2}n+k}),$$

$$K_2 = \bar{K}_1.$$

5.2. If n is even and $k < \frac{1}{2}n$, then

$$K_1 = (-1)^k \frac{e^{\frac{1}{2}nqi}\Gamma\left(\frac{n}{2} - k\right)}{4^k(k-1)!\,\pi^{\frac{1}{2}n}} (P + i0)^{-\frac{1}{2}n+k}$$

$$= (-1)^k \frac{e^{\frac{1}{2}nqi}\Gamma\left(\frac{n}{2} - k\right)}{4^k(k-1)!\,\pi^{\frac{1}{2}n}}$$

$$\times \left[P_+^{-\frac{1}{2}n+k} + (-1)^{-\frac{1}{2}n+k} P_-^{-\frac{1}{2}n+k} + \frac{(-1)^{-\frac{1}{2}n+k}\,\pi i}{\left(\frac{n}{2} - k - 1\right)!} \delta^{(\frac{1}{2}n-k-1)}(P_+) \right],$$

where $P_+^{-\frac{1}{2}n+k}$ and $P_-^{-\frac{1}{2}n+k}$ are the constant terms of the Laurent expansions for P_+^λ and P_-^λ about $\lambda = -\frac{1}{2}n + k$;

$$K_2 = \bar{K}_1.$$

5.3. If n is even and $k \geqslant \frac{1}{2}n$, then

$$K_1 = (-1)^{\frac{1}{2}n-1} \frac{e^{\frac{1}{2}nqi}}{4^k \left(k - \frac{n}{2}\right)!\,(k-1)!} (P + i0)^{-\frac{1}{2}n+k} \ln (P + i0),$$

$$K_2 = \bar{K}_1.$$

6. The equations of paragraphs 3 and 5 remain valid for any non-degenerate quadratic form

$$P = \sum_{\alpha,\beta=1}^{n} g_{\alpha\beta} x_\alpha x_\beta.$$

For this case we write

$$L = \sum_{\alpha,\beta=1}^{n} g^{\alpha\beta} \frac{\partial^2}{\partial x_\alpha \, \partial x_\beta},$$

$$Q = \sum_{\alpha,\beta=1}^{n} g^{\alpha\beta} s_\alpha s_\beta,$$

where the $g^{\alpha\beta}$ are defined by $\sum_{\beta=1}^{n} g_{\alpha\beta} g^{\beta\gamma} = \delta_\alpha^\gamma$ (here δ_α^β is the Kronecker delta symbol).

In order to use the equations of paragraph 3 in this case we must multiply on the right-hand sides by $1/\sqrt{|\Delta|}$, and in order to use those of paragraph 5 we must multiply by $\sqrt{|\Delta|}$, where Δ is the determinant of the coefficients of P.

7. The Fourier transforms of the functions of paragraphs 1–6 are given in the table.

Chapter III, Section 3

1. The generalized function $f(x_1, \ldots, x_n)$ is called a *homogeneous function of degree* λ if for any $\varphi(x)$ in K and for $\alpha > 0$ we have

$$\left(f, \varphi\left(\frac{x}{\alpha}\right) \right) = \alpha^{\lambda+n}(f, \varphi).$$

2. If $f(x_1, \ldots, x_n)$ is a positive continuous homogeneous function of the first degree, the generalized function

$$(f^\lambda, \varphi) = \int f^\lambda \, \varphi \, dx,$$

is defined by the integral for Re $\lambda > -n$ and can be analytically continued into the entire λ plane except for the points $\lambda = -n - k$, where k is a nonnegative integer. At these points it has simple poles. The analytic continuation f^λ is given for Re $\lambda > -n - k - 1$ by

$$(f^\lambda, \varphi) = \int_G f^\lambda(x) \left[\varphi(x) \dot{-} \varphi(0) - \ldots \right.$$

$$\left. \ldots - \frac{1}{k!} \sum_{\alpha_1+\ldots+\alpha_n=k} x_1^{\alpha_1} \ldots x_n^{\alpha_n} \frac{\partial^k \varphi(0)}{\partial x_1^{\alpha_1} \ldots \partial x_n^{\alpha_n}} \right] dx$$

$$+ \int_{R-G} f^\lambda(x) \, \varphi(x) \, dx$$

$$+ \sum_{m=0}^{k} \frac{1}{m!(\lambda + n + m)} \sum_{\alpha_1+\ldots+\alpha_n=m} \frac{\partial^m \varphi(0)}{\partial x_1^{\alpha_1} \ldots \partial x_n^{\alpha_n}} \int_\Gamma f^\lambda(x) \, x_1^{\alpha_1} \ldots x_n^{\alpha_n} \cdot \omega,$$

where G is a region containing the origin, Γ is its boundary, and ω is the differential form

$$x_1 \, dx_2 \dots dx_n - x_2 \, dx_1 \, dx_3 \dots dx_n + \dots + (-1)^{n+1} x_n \, dx_1 \dots dx_{n-1}.$$

In the strip $-n-k-1 < \mathrm{Re} \, \lambda < -n-k$ this same equation can be written

$$(f^\lambda, \varphi) = \int f^\lambda(x) \left[\varphi(x) - \varphi(0) - \dots - \frac{1}{k!} \sum_{\alpha_1 + \dots + \alpha_n = k} x_1^{\alpha_1} \dots x_n^{\alpha_n} \frac{\partial^k \varphi(0)}{\partial x_1^{\alpha_1} \dots \partial x_n^{\alpha_n}} \right] dx.$$

The residue of f^λ at $\lambda = -n-k$ is

$$\frac{(-1)^k}{k!} \sum_{\alpha_1 + \dots + \alpha_n = k} \frac{\partial^k \delta(x)}{\partial x_1^{\alpha_1} \dots \partial x_n^{\alpha_n}} \int_\Gamma \frac{x_1^{\alpha_1} \dots x_n^{\alpha_n}}{f^{n+k}(x)} \, \omega.$$

In particular, at $\lambda = -n$ it is

$$\delta(x) \cdot \int_\Gamma \frac{\omega}{f^n(x)} \, .$$

3. Any formally homogeneous function $\Phi(x)$ of degree $-n$ and any region G containing the origin can be used to construct a generalized function

$$\Phi \mid_G = \int_G \Phi(x) \left[\varphi(x) - \varphi(0) \right] dx + \int_{R-G} \Phi(x) \, \varphi(x) \, dx,$$

which coincides locally with $\Phi(x)$ everywhere except at the origin. This generalized function is homogeneous of degree $-n$ if and only if

$$\int_\Gamma \Phi \cdot \omega = 0,$$

where Γ is the boundary of the region G, and ω is the differential form given above.

4. Any formally homogeneous function $\Phi(x)$ of degree $-n-m$ and any region G containing the origin can be used to construct a generalized function

$$\Phi \mid_G = \int_G \Phi(x) \left[\varphi(x) - \varphi(0) - \dots - \frac{1}{m!} \sum_{\Sigma \alpha_j = m} x_1^{\alpha_1} \dots x_n^{\alpha_n} \frac{\partial^m \varphi(0)}{\partial x_1^{\alpha_1} \dots \partial x_n^{\alpha_n}} \right] dx$$

$$+ \int_{R-G} \Phi(x) \left[\varphi(x) - \varphi(0) - \dots - \frac{1}{(m-1)!} \sum_{\Sigma \alpha_j = m-1} x_1^{\alpha_1} \dots x_n^{\alpha_n} \frac{\partial^{m-1} \varphi(0)}{\partial x_1^{\alpha_1} \dots \partial x_n^{\alpha_n}} \right] dx,$$

which coincides locally with $\Phi(x)$ everywhere except at the origin. This generalized function is homogeneous of degree $-n-m$ if and only if

$$\int_\Gamma \Phi(x)\, x_1^{\alpha_1} \dots x_n^{\alpha_n} \cdot \omega = 0, \qquad \sum_j \alpha_j = m.$$

5. A homogeneous function Φ of degree $-n+1$ can be differentiated (in the sense of generalized functions) according to

$$\frac{\partial \Phi}{\partial x_j} = \frac{\partial \Phi}{\partial x_j}\bigg|_G + (-1)^{j-1}\, \delta(x_1, \dots, x_n) \int_\Gamma \Phi(x)\, dx_1 \dots dx_{j-1}\, dx_{j+1} \dots dx_n,$$

and the result is independent of the choice of G.

Chapter III, Section 4

1. A function $G(x) = G(x_1, \dots, x_n)$ is called *equivalent to a homogeneous function* in the neighborhood of a point M if in this neighborhood there exists a local coordinate system $\xi_1, \dots, \xi_n$, in terms of which $G(x)$ is a homogeneous function.

2. *Inductive definition of a reducible point.* In one dimension the point x_0 on the real axis is called reducible (with respect to a function $G(x)$ such that $G(x_0) = 0$) if $G(x)$ is equivalent to a homogeneous function in a neighborhood of x_0.

A point M on a hypersurface $G(x_1, \dots, x_n) = 0$ is called reducible if in some neighborhood of M the function G is equivalent to a homogeneous function, and the intersection of the $G = 0$ hypersurface with every sufficiently small sphere centered at M is a manifold each of whose points is reducible on this sphere.

3. If local coordinates in the neighborhood of M can be choosen so that the homogeneous function G of degree m depends on k of the variables, but cannot be chosen so that it depends on $k-1$ of the variables, M is called a *point of order k and degree m*.

4. Assume that the $G = 0$ hypersurface consists only of reducible points. The functional defined by the integral

$$(G^\lambda, \varphi) = \int_{G>0} G^\lambda(x_1, \dots, x_n)\, \varphi(x_1, \dots, x_n)\, dx_1 \dots dx_n,$$

convergent for Re $\lambda > 0$, is an analytic function of λ. This analytic function can be continued to the entire λ plane as a meromorphic function whose poles are given by a finite number of arithmetic progressions. Specifically, each connected component (submanifold) of the $G = 0$ hypersurface consisting of points of order r and degree m generates poles at $\lambda = - (r + k)/m$, $k = 0, 1, \ldots$. If there exist several connected components imbedded one in the other, each component consisting of points of a given order, and if $\lambda = \lambda_0$ belongs to more than one progression generated by these components, the multiplicity of the pole at $\lambda = \lambda_0$ increases correspondingly.

In particular, if the $G = 0$ hypersurface has no singular points (that is, all of its points are reducible of order one and degree one), the poles of (G^λ, φ) are *all* included in the set $\lambda = - 1, - 2, \ldots, - n$.

5. Assume that the $G = 0$ hypersurface consists only of reducible points and that the $G = c$ hypersurface, with $c > 0$, has no singular points. Further, let the differential form ω be defined by $dG \cdot \omega = dv$. Then the integral

$$I(c) = \int_{G=c} \varphi(x_1, \ldots, x_n) \cdot \omega = (\delta(G - c), \varphi)$$

has the asymptotic expansion

$$I(c) \cong \sum_{k=1}^{\infty} \sum_{m=1}^{m_k} a_{km} c^{\lambda_k - 1} \ln^{m-1} c,$$

where $0 > - \lambda_1 > - \lambda_2 > \ldots$ are the poles of (G^λ, φ), m_k is the multiplicity of the pole at $- \lambda_k$, and a_{km} is $(- 1)^{m-1}/(m - 1)!$ times the coefficient of $(\lambda + \lambda_k)^{-m}$ in the Laurent expansion for (G^λ, φ) about the pole at $- \lambda_k$.

6. In particular,

$$\int_{xy=c} \varphi \cdot \omega = \int_{-\infty}^{\infty} \frac{\varphi(x, y)}{y} \, dy = - \int_{-\infty}^{\infty} \frac{\varphi(x, y) \, dx}{x}$$

$$= -2\varphi(0, 0) \ln c + \int_{-\infty}^{\infty} \frac{\varphi(x, 0)}{x} \, dx + \int_{-\infty}^{\infty} \frac{\varphi(0, y)}{y} \, dy + \ldots,$$

where we have omitted terms which vanish as $c \to 0$. Equivalently,

$$\delta(xy - c) = -2\delta(x, y) \ln c + \frac{\delta(y)}{x} + \frac{\delta(x)}{y} + o(c).$$

7. For $G \equiv x^2 + y^2 + z^2 - t^2$, we have

$$\delta(G - c) = \delta(G) + c \ln c \frac{\delta(x, y, z, t)}{8} + c\delta'(G) + o(c).$$

TABLE OF FOURIER TRANSFORMS

1. Functions of a Single Variable

Entry no.	Generalized function f	Fourier transform $F[f]$
1	Ordinary summable function $f(x)$	$F[f] = \displaystyle\int_{-\infty}^{\infty} f(x)\, e^{ix\sigma}\, dx$
2	$\delta(x)$	1
3	1	$2\pi\, \delta(\sigma)$
4	Polynomial $P(x)$	$2\pi P\left(-i\dfrac{d}{d\sigma}\right)\delta(\sigma)$
5	$\delta^{(2m)}(x)$	$(-1)^m \sigma^{2m}$
6	$\delta^{(2m+1)}(x)$	$(-1)^{m+1}\, i\sigma^{2m+1}$
7	e^{bx}	$2\pi\, \delta(s - ib)$
8	$\sin bx$	$-i\pi[\delta(s+b) - \delta(s-b)]$
9	$\cos bx$	$\pi[\delta(s+b) + \delta(s-b)]$
10	$\sinh bx$	$\pi[\delta(s-ib) - \delta(s+ib)]$
11	$\cosh bx$	$\pi[\delta(s-ib) + \delta(s+ib)]$
12	$\exp\left(\dfrac{x^2}{2}\right)$	Analytic functional $i\sqrt{2\pi}\,\exp(s^2/2)$ (integration along the imaginary axis)
13	$\|x\|^{\lambda}\quad (\lambda \neq -1, -3, \dots)$	$-2\sin\dfrac{\lambda\pi}{2}\,\Gamma(\lambda+1)\,\|\sigma\|^{-\lambda-1}$
14	$f_\lambda(x) = 2^{-\frac{1}{2}\lambda}\,\dfrac{\|x\|^{\lambda}}{\Gamma\left(\dfrac{\lambda+1}{2}\right)}$	$\sqrt{2\pi}f_{-\lambda-1}(\sigma) = \sqrt{2\pi}\,\dfrac{2^{\frac{1}{2}(\lambda+1)}\,\|\sigma\|^{-\lambda-1}}{\Gamma\left(-\dfrac{\lambda}{2}\right)}$
15	$\|x\|^{\lambda}\,\mathrm{sgn}\,x$ $(\lambda \neq -2, -4, \dots)$	$2i\cos\dfrac{\lambda\pi}{2}\,\Gamma(\lambda+1)\,\|\sigma\|^{-\lambda-1}\,\mathrm{sgn}\,\sigma$
16	$g_\lambda(x) = 2^{-\frac{1}{2}\lambda}\,\dfrac{\|x\|^{\lambda}\,\mathrm{sgn}\,x}{\Gamma\left(\dfrac{\lambda+2}{2}\right)}$	$\sqrt{2\pi}\,ig_{-\lambda-1}(\sigma)$ $= \sqrt{2\pi}\,i\,\dfrac{2^{\frac{1}{2}(\lambda+1)}\,\|\sigma\|^{-\lambda-1}\,\mathrm{sgn}\,\sigma}{\Gamma\left(\dfrac{1-\lambda}{2}\right)}$

Continued

Entry no.	Generalized function f	Fourier transform $F[f]$
17	x^m	$2(-i)^m \pi\, \delta^{(m)}(\sigma)$
18	x^{-m}	$i^m \dfrac{\pi}{(m-1)!}\, \sigma^{m-1} \operatorname{sgn} \sigma$
19	x^{-1}	$i\pi \operatorname{sgn} \sigma$
20	x^{-2}	$-\pi\,\lvert\,\sigma\,\rvert$
21	$x_+^\lambda \quad (\lambda \neq -1, -2, \ldots)$	$ie^{i\lambda(\pi/2)}\, \Gamma(\lambda + 1)\,(\sigma + i0)^{-\lambda-1}$ $= i\Gamma(\lambda + 1)$ $\quad \times [e^{\lambda i(\pi/2)}\, \sigma_+^{-\lambda-1} - e^{-i\lambda(\pi/2)}\, \sigma_-^{-\lambda-1}]*$
22	x_+^n	$i^{n+1} n!\, \sigma^{-n-1} + (-i)^n \pi\, \delta^{(n)}(\sigma)$
23	$\theta(x)$	$i\sigma^{-1} + \pi\delta(\sigma)$
24	$x_-^\lambda \quad (\lambda \neq -1, -2, \ldots)$	$-ie^{-i\lambda(\pi/2)}\, \Gamma(\lambda + 1)\,(\sigma - i0)^{-\lambda-1}$ $= i\Gamma(\lambda + 1)$ $\quad \times [e^{i\lambda(\pi/2)}\, \sigma_-^{-\lambda-1} - e^{-i\lambda(\pi/2)}\, \sigma_+^{-\lambda-1}]*$
25	$(x + i0)^\lambda$	$\dfrac{2\pi e^{i\lambda(\pi/2)}}{\Gamma(-\lambda)}\, \sigma_-^{-\lambda-1}$
26	$(x - i0)^\lambda$	$\dfrac{2\pi e^{-i\lambda(\pi/2)}}{\Gamma(-\lambda)}\, \sigma_+^{-\lambda-1}$

In Entries 27–38 we write:

$$ie^{i\lambda(\pi/2)}\, \Gamma(\lambda + 1) = \frac{a_{-1}^{(n)}}{\lambda + n} + a_0^{(n)} + a_1^{(n)}(\lambda + n) + \ldots,$$

$$-ie^{-i\lambda(\pi/2)}\, \Gamma(\lambda + 1) = \frac{b_{-1}^{(n)}}{\lambda + n} + b_0^{(n)} + b_1^{(n)}(\lambda + n) + \ldots,$$

$$-2\sin\frac{\lambda\pi}{2}\, \Gamma(\lambda + 1) = \frac{c_{-1}^{(n)}}{\lambda + n} + c_0^{(n)} + c_1^{(n)}(\lambda + n) + \ldots,$$

$$2\cos\frac{\lambda\pi}{2}\, \Gamma(\lambda + 1) = \frac{d_{-1}^{(n)}}{\lambda + n} + d_0^{(n)} + d_1^{(n)}(\lambda + n).$$

* Second expression for $\lambda \neq 0,\ \pm 1,\ \pm 2,\ \pm 3, \ldots$.

Entry no.	Generalized function f	Fourier transform $F[f]$

The $a_{-1}^{(n)}$, $a_0^{(n)}$, ... are given by:

$$a_{-1}^{(n)} = \frac{i^{n-1}}{(n-1)!} \, ;$$

$$a_0^{(n)} = \frac{i^{n-1}}{(n-1)!} \left[1 + \frac{1}{2} + ... + \frac{1}{n-1} + \Gamma'(1) + i\frac{\pi}{2} \right] ;$$

$$a_1^{(n)} = \frac{i^{n-1}}{(n-1)!} \left\{ \sum_{j,k=1}^{n-1} \frac{1}{jk} - \frac{\pi^2}{8} + \left(1 + \frac{1}{2} + ... + \frac{1}{n-1} \right) \Gamma'(1) + \Gamma''(1) \right.$$

$$\left. + i\frac{\pi}{2} \left[1 + \frac{1}{2} + ... + \frac{1}{n-1} + \Gamma'(1) \right] \right\} ;$$

$$b_i^{(n)} = \bar{a}_i^{(n)}; \quad c_i^{(n)} = 2 \operatorname{Re} a_i^{(n)}; \quad d_i^{(n)} = 2 \operatorname{Im} a_i^{(n)}.$$

In particular,

$$b_{-1}^{(n)} = \frac{(-i)^{n-1}}{(n-1)!} \, ; \quad c_{-1}^{(n)} = \frac{2(-1)^{n-1}}{(n-1)!} \cos (n-1) \frac{\pi}{2} \, ;$$

$$d_{-1}^{(n)} = \frac{2(-1)^n}{(n-1)!} \sin (n-1) \frac{\pi}{2} \, .$$

Entry no.	Generalized function f	Fourier transform $F[f]$
27	$\lvert x \rvert^{-2m-1}$	$c_0^{(2m+1)}\sigma^{2m} - c_{-1}^{(2m+1)}\sigma^{2m} \ln \lvert \sigma \rvert$
28	$x^{-2m} \operatorname{sgn} x$	$i d_0^{(2m)}\sigma^{2m-1} - i d_{-1}^{(2m)}\sigma^{2m-1} \ln \lvert \sigma \rvert$
29	$x_+^\lambda \ln x_+$ $(\lambda \neq -1, -2, ...)$	$ie^{i\lambda(\pi/2)} \left\{ \left[\Gamma'(\lambda + 1) + i\frac{\pi}{2} \Gamma(\lambda + 1) \right] \right.$ $\times (\sigma + i0)^{-\lambda-1}$ $\left. -\Gamma(\lambda + 1)(\sigma + i0)^{-\lambda-1} \ln (\sigma + i0) \right\}$
30	$x_-^\lambda \ln x_-$ $(\lambda \neq -1, -2, ...)$	$-ie^{-i\lambda(\pi/2)} \left\{ \left[\Gamma'(\lambda + 1) - i\frac{\pi}{2} \Gamma(\lambda + 1) \right] \right.$ $\times (\sigma - i0)^{-\lambda-1}$ $\left. -\Gamma(\lambda + 1)(\sigma - i0)^{-\lambda-1} \ln (\sigma - i0) \right\}$

Continued

Entry no.	Generalized function f	Fourier transform $F[f]$
31	$\ln x_+$	$i\left\{\left(\Gamma'(1) + i\dfrac{\pi}{2}\right)(\sigma + i0)^{-1}\right.$
		$\left. -(\sigma + i0)^{-1}\ln(\sigma + i0)\right\}$
32	$\ln x_-$	$-i\left\{\left(\Gamma'(1) - i\dfrac{\pi}{2}\right)(\sigma - i0)^{-1}\right.$
		$\left. -(\sigma - i0)^{-1}\ln(\sigma - i0)\right.$
33	$\|x\|^{\lambda}\ln\|x\|$ $(\lambda \neq -1, -2, \ldots)$	$ie^{i\lambda(\pi/2)}\left\{\left[\Gamma'(\lambda + 1) + i\dfrac{\pi}{2}\Gamma(\lambda + 1)\right]\right.$
		$\times (\sigma + i0)^{-\lambda-1}$
		$\left. -\Gamma(\lambda + 1)(\sigma + i0)^{-\lambda-1}\ln(\sigma + i0)\right\}$
		$-ie^{-i\lambda(\pi/2)}\left\{\left[\Gamma'(\lambda + 1) - i\dfrac{\pi}{2}\Gamma(\lambda + 1)\right]\right.$
		$\times (\sigma - i0)^{-\lambda-1}$
		$\left. -\Gamma(\lambda + 1)(\sigma - i0)^{-\lambda-1}\ln(\sigma - i0)\right\}$
34	$\|x\|^{\lambda}\ln\|x\|\,\mathrm{sgn}\,x$ $(\lambda \neq -1, -2, \ldots)$	$ie^{i\lambda(\pi/2)}\left\{\left[\Gamma'(\lambda + 1) + i\dfrac{\pi}{2}\Gamma(\lambda + 1)\right]\right.$
		$\times (\sigma + i0)^{-\lambda-1}$
		$\left. -\Gamma(\lambda + 1)(\sigma + i0)^{-\lambda-1}\ln(\sigma + i0)\right\}$
		$+ie^{-i\lambda(\pi/2)}\left\{\left[\Gamma'(\lambda + 1) - i\dfrac{\pi}{2}\Gamma(\lambda + 1)\right]\right.$
		$\times (\sigma - i0)^{-\lambda-1}$
		$\left. -\Gamma(\lambda + 1)(\sigma - i0)^{-\lambda-1}\ln(\sigma - i0)\right\}$
35	$x^{-2m}\ln\|x\|$	$c_1^{(2m)}\|\sigma\|^{2m-1} - c_0^{(2m)}\|\sigma\|^{2m-1}\ln\|\sigma\|$
36	$x^{-2m-1}\ln\|x\|$	$id_1^{(2m+1)}\sigma^{2m}\,\mathrm{sgn}\,\sigma - id_0^{(2m+1)}\sigma^{2m}\ln\|\sigma\|\,\mathrm{sgn}\,\sigma$
37	$\|x\|^{-2m-1}\ln\|x\|$	$c_1^{(2m+1)}\sigma^{2m} - c_0^{(2m+1)}\sigma^{2m}\ln\|\sigma\|$
		$+ \dfrac{1}{2}c_{-1}^{(2m+1)}\sigma^{2m}\ln^2\|\sigma\|$

Continued

Entry no.	Generalized function f	Fourier transform $F[f]$
38	$\lvert x \rvert^{-2m} \ln \lvert x \rvert \, \mathrm{sgn}\, x$	$id_1^{(2m)}\sigma^{2m-1} - id_0^{(2m)}\sigma^{2m-1}\ln\lvert\sigma\rvert$ $+\dfrac{i}{2}d_{-1}^{(2m)}\sigma^{2m-1}\ln^2\lvert\sigma\rvert$
39	$(1-x^2)_+^\lambda$ $(\lambda \neq -1, -2, \ldots)$	$\sqrt{\pi}\,\Gamma(\lambda+1)\left(\dfrac{\sigma}{2}\right)^{-\lambda-\frac{1}{2}}J_{\lambda+\frac{1}{2}}(\sigma)$
40	$(1+x^2)_+^\lambda$	$\dfrac{2\sqrt{\pi}}{\Gamma(-\lambda)}\left\lvert\dfrac{\sigma}{2}\right\rvert^{-\lambda-\frac{1}{2}}K_{-\lambda-\frac{1}{2}}(\lvert\sigma\rvert)$
41	$(x^2-1)_+^\lambda$ $(\lambda \neq -1, -2, \ldots)$	$-\Gamma(\lambda+1)\sqrt{\pi}\left\lvert\dfrac{\sigma}{2}\right\rvert^{-\lambda-\frac{1}{2}}N_{-\lambda-\frac{1}{2}}(\lvert\sigma\rvert)$ $=\Gamma(\lambda+1)\sqrt{\pi}\left\lvert\dfrac{\sigma}{2}\right\rvert^{-\lambda-\frac{1}{2}}$ $\times\dfrac{\cos\pi(\lambda+\frac{1}{2})\,J_{-\lambda-\frac{1}{2}}(\lvert\sigma\rvert)-J_{\lambda+\frac{1}{2}}(\lvert\sigma\rvert)}{\sin\pi(\lambda+\frac{1}{2})}$
42	$(x^2-1)_+^n$	$(-1)^n\,2\pi\left(1+\dfrac{d^2}{d\sigma^2}\right)^n\delta(\sigma)$ $+(-1)^{n+1}\sqrt{\pi}\left(\dfrac{\sigma}{2}\right)^{-n-\frac{1}{2}}J_{n+\frac{1}{2}}(\sigma)$

2. Functions of Several Variables

1	$\delta(x_1, \ldots, x_n)$	1
2	1	$(2\pi)^n\delta(\sigma_1, \ldots, \sigma_n)$
3	Polynomial $P(x_1, \ldots, x_n)$	$(2\pi)^n P\left(-i\dfrac{\partial}{\partial\sigma_1}, \ldots, -i\dfrac{\partial}{\partial\sigma_n}\right)\delta(\sigma)$
4	$r^\lambda \quad \left(r=\sqrt{\sum x_j^2}\right)$	$2^{\lambda+n}\pi^{\frac{1}{2}n}\dfrac{\Gamma\left(\dfrac{\lambda+n}{2}\right)}{\Gamma\left(-\dfrac{\lambda}{2}\right)}\rho^{-\lambda-n}\quad\left(\rho=\sqrt{\sum\sigma_j^2}\right)$
5	$f_\lambda(r)=\dfrac{2^{-\frac{1}{2}\lambda}r^\lambda}{\Gamma\left(\dfrac{\lambda+n}{2}\right)}$	$(2\pi)^{\frac{1}{2}n}f_{-\lambda-n}(\rho)=(2\pi)^{\frac{1}{2}n}\dfrac{2^{\frac{1}{2}(\lambda+n)}r^{-\lambda-n}}{\Gamma\left(-\dfrac{\lambda}{2}\right)}$

Continued

Entry no.	Generalized function f	Fourier transform $F[f]$

In Entries 6–9 we write:

$$C_\lambda = 2^{\lambda+n}\,\pi^{\frac{1}{2}n}\,\frac{\Gamma\left(\dfrac{\lambda+n}{2}\right)}{\Gamma\left(-\dfrac{\lambda}{2}\right)}$$

$$= \frac{c_{-1}^{(n+2m)}}{\lambda+n+2m} + c_0^{(n+2m)} + c_1^{(n+2m)}(\lambda+n+2m) + \dots;$$

the right-hand side is the Laurent expansion of this function about $\lambda = -n-2m$. Further,

$$\Omega_n = \frac{2\pi^{\frac{1}{2}n}}{\Gamma\left(\dfrac{n}{2}\right)}$$

is the hypersurface area of the unit sphere in n dimensions.

Entry no.	Generalized function f	Fourier transform $F[f]$
6	$r^\lambda \ln r$ $(\lambda \neq -n, -n-2, \dots)$	$\dfrac{dC_\lambda}{d\lambda}\rho^{-\lambda-n} + C_\lambda \rho^{-\lambda-n}\ln\rho$
7	$r^\lambda \ln^2 r$ $(\lambda \neq -n, -n-2, \dots)$	$\dfrac{d^2C_\lambda}{d\lambda^2}\rho^{-\lambda-n} + 2\dfrac{dC_\lambda}{d\lambda}\rho^{-\lambda-n}$ $\times \ln\rho + C_\lambda\rho^{-\lambda-n}\ln^2\rho$
8	$\Omega_n r^{-2m-n}$	$c_{-1}^{(n+2m)}\rho^{2m}\ln\rho + c_0^{(n+2m)}\rho^{2m}$
9	$\Omega_n r^{-2m-n}\ln r$	$\tfrac{1}{2}c_{-1}^{(n+2m)}\rho^{2m}\ln^2\rho + c_0^{(n+2m)}\rho^{2m}\ln\rho$ $+ c_1^{(n+2m)}\rho^{2m}$
10	$\delta(r-a) \qquad (n \geqslant 1)$	$2^{\frac{1}{2}n-1}\,\Gamma(\tfrac{1}{2}n-\tfrac{1}{2})\,\Gamma(\tfrac{1}{2})\,\Omega_{n-1}a^{\frac{1}{2}n}\rho^{1-\frac{1}{2}n}$ $\times J_{\frac{1}{2}(n-2)}(a\rho)$
11	The same, for $n = 3$	$4\pi a\,\dfrac{\sin a\rho}{\rho}$
12	$\left(\dfrac{d}{a\,da}\right)^m \dfrac{\delta(r-a)}{a}$	$2^{\frac{1}{2}n-1}\,\Gamma(\tfrac{1}{2}n-\tfrac{1}{2})\,\Gamma(\tfrac{1}{2})\,\Omega_{n-1}\sqrt{\dfrac{2}{\pi}}\,\dfrac{\sin a\rho}{\rho}$

For the notation in Entries 13–25 see the Summary of Fundamental Definitions and Equations for Chapter III, Section 2.

Continued

Entry no.	Generalized function f	Fourier transform $F[f]$
13	$(P + i0)^\lambda$	$\dfrac{1}{\Gamma(-\lambda)}\, e^{-(\pi/2)qi}\, 2^{n+2\lambda}\, \pi^{\frac{1}{2}n}\, \Gamma(\lambda + \tfrac{1}{2} n)\, (Q - i0)^{-\lambda-\frac{1}{2}n}$
14	$(P - i0)^\lambda$	$\dfrac{e^{(\pi/2)qi}\, 2^{n+2\lambda}\, \pi^{\frac{1}{2}n}\, \Gamma(\lambda + \tfrac{1}{2} n)}{\Gamma(-\lambda)}\, (Q + i0)^{-\lambda-\frac{1}{2}n}$
15	P_+^λ	$2^{n+2\lambda}\, \pi^{\frac{1}{2}n-1}\, \Gamma(\lambda + 1)\, \Gamma(\lambda + \tfrac{1}{2} n)$ $\times \dfrac{1}{2i}\, [e^{-i(\frac{1}{2}q+\lambda)\pi} (Q - i0)^{-\lambda-\frac{1}{2}n} - e^{i(\frac{1}{2}q+\lambda)\pi} (Q + i0)^{-\lambda-\frac{1}{2}n}]$
16	P_-^λ	$-2^{n+2\lambda}\, \pi^{\frac{1}{2}n-1}\, \Gamma(\lambda + 1)\, \Gamma(\lambda + \tfrac{1}{2} n)$ $\times \dfrac{1}{2i}\, [e^{-(\pi/2)qi} (Q - i0)^{-\lambda-\frac{1}{2}n} - e^{(\pi/2)qi} (Q + i0)^{-\lambda-\frac{1}{2}n}]$
17	$(c^2 + P + i0)^\lambda$	$\dfrac{2^{\lambda+1}(\sqrt{2\pi})^n\, c^{\frac{1}{2}n+\lambda}}{\Gamma(-\lambda)\sqrt{\Delta}}\, \dfrac{K_{\frac{1}{2}n+\lambda}[c(Q - i0)^{\frac{1}{2}}]}{(Q - i0)^{\frac{1}{2}(\frac{1}{2}n+\lambda)}}$ $= \dfrac{2^{\lambda+\frac{1}{2}n+1}\, \pi^{\frac{1}{2}n}\, e^{-\frac{1}{2}q\pi i}\, c^{\lambda+\frac{1}{2}n}}{\Gamma(-\lambda)\sqrt{|\Delta|}}$ $\times \left[\dfrac{K_{\lambda+\frac{1}{2}n}(cQ_+^{\frac{1}{2}})}{Q_+^{\frac{1}{2}(\lambda+\frac{1}{2}n)}} + \dfrac{\pi i}{2}\, \dfrac{H^{(1)}_{-\lambda-\frac{1}{2}n}(cQ_-^{\frac{1}{2}})}{Q_-^{\frac{1}{2}(\lambda+\frac{1}{2}n)}} \right].$
18	$(c^2 + P - i0)^\lambda$	$\dfrac{2^{\lambda+1}(\sqrt{2\pi})^n\, c^{\frac{1}{2}n+\lambda}}{\Gamma(-\lambda)\sqrt{\Delta}}\, \dfrac{K_{\frac{1}{2}n+\lambda}[c(Q + i0)^{\frac{1}{2}}]}{(Q + i0)^{\frac{1}{2}(\frac{1}{2}n+\lambda)}}$ $= \dfrac{2^{\lambda+\frac{1}{2}n+1}\, \pi^{\frac{1}{2}n}\, e^{\frac{1}{2}q\pi i}\, c^{\lambda+\frac{1}{2}n}}{\Gamma(-\lambda)\sqrt{|\Delta|}}$ $\times \left[\dfrac{K_{\lambda+\frac{1}{2}n}(cQ_+^{\frac{1}{2}})}{Q_+^{\frac{1}{2}(\lambda+\frac{1}{2}n)}} - \dfrac{\pi i}{2}\, \dfrac{H^{(2)}_{-\lambda-\frac{1}{2}n}(cQ_-^{\frac{1}{2}})}{Q_-^{\frac{1}{2}(\lambda+\frac{1}{2}n)}} \right].$

Continued

Entry no.	Generalized function f	Fourier transform $F[f]$

19 $\dfrac{(c^2 + P)^{\lambda}_{+}}{\Gamma(\lambda + 1)}$

$$-\frac{2^{\lambda + \frac{1}{2}n}\, i\pi^{\frac{1}{2}n-1}\, c^{\frac{1}{2}n+\lambda}}{\sqrt{|\Delta|}}$$

$$\times \left\{ e^{-i(\lambda + \frac{1}{2}q)\pi}\, \frac{K_{\frac{1}{2}n+\lambda}[c(Q - i0)^{\frac{1}{2}}]}{(Q - i0)^{\frac{1}{2}(\lambda + \frac{1}{2}n)}} \right.$$

$$\left. - e^{i(\lambda + \frac{1}{2}q)\pi} \frac{K_{\frac{1}{2}n+\lambda}[c(Q + i0)^{\frac{1}{2}}]}{(Q + i0)^{\frac{1}{2}(\lambda + \frac{1}{2}n)}} \right\} = \frac{2^{\lambda + \frac{1}{2}n+1}\pi^{\frac{1}{2}n-1}c^{\frac{1}{2}n+\lambda}}{\sqrt{|\Delta|}}$$

$$\times \left\{ -\sin\left(\lambda + \tfrac{1}{2}q\right)\pi\, \frac{K_{\lambda + \frac{1}{2}n}(cQ_{+}^{\frac{1}{2}})}{Q_{+}^{\frac{1}{2}(\lambda + \frac{1}{2}n)}} + \frac{\pi}{2\sin\left(\lambda + \frac{1}{2}n\right)\pi} \right.$$

$$\times \left[\sin\left(\lambda + \tfrac{1}{2}q\right)\pi\, \frac{J_{\lambda + \frac{1}{2}n}(cQ_{-}^{\frac{1}{2}})}{Q_{-}^{\frac{1}{2}(\lambda + \frac{1}{2}n)}} \right.$$

$$\left.\left. + \sin\frac{p\pi}{2}\, \frac{J_{-\lambda - \frac{1}{2}n}(cQ_{-}^{\frac{1}{2}})}{Q_{-}^{\frac{1}{2}(\lambda + \frac{1}{2}n)}} \right] \right\}.$$

20 $\dfrac{(c^2 + P)^{\lambda}_{-}}{\Gamma(\lambda + 1)}$

$$\frac{2^{\lambda + \frac{1}{2}n}i\pi^{\frac{1}{2}n-1}c^{\frac{1}{2}n+\lambda}}{\sqrt{|\Delta|}}\left\{ e^{-\frac{1}{2}q\pi i}\, \frac{K_{\frac{1}{2}n+\lambda}[c(Q - i0)^{\frac{1}{2}}]}{(Q - i0)^{\frac{1}{2}(\lambda + \frac{1}{2}n)}} \right.$$

$$\left. - e^{\frac{1}{2}q\pi i}\, \frac{K_{\frac{1}{2}n+\lambda}[c(Q + i0)^{\frac{1}{2}}]}{(Q + i0)^{\frac{1}{2}(\lambda + \frac{1}{2}n)}} \right\} = \frac{2^{\lambda + \frac{1}{2}n+1}\pi^{\frac{1}{2}n-1}c^{\frac{1}{2}n+\lambda}}{\sqrt{|\Delta|}}$$

$$\times \left\{ \sin\frac{q\pi}{2}\, \frac{K_{\lambda + \frac{1}{2}n}(cQ_{+}^{\frac{1}{2}})}{Q_{+}^{\frac{1}{2}(\lambda + \frac{1}{2}n)}} - \frac{\pi}{2\sin\left(\lambda + \frac{1}{2}n\right)\pi} \right.$$

$$\times \left[\sin\frac{q\pi}{2}\, \frac{J_{\lambda + \frac{1}{2}n}(cQ_{-}^{\frac{1}{2}})}{Q_{-}^{\frac{1}{2}(\lambda + \frac{1}{2}n)}} + \sin\left(\lambda + \frac{p}{2}\right)\pi\, \frac{J_{-\lambda - \frac{1}{2}n}(cQ_{-}^{\frac{1}{2}})}{Q_{-}^{\frac{1}{2}(\lambda + \frac{1}{2}n)}} \right] \right\}.$$

21 $\begin{array}{l} \delta^{(t-1)} \\ \times (c^2 + P) \end{array}$

$$(-1)^{t+1}\, \frac{i}{\sqrt{|\Delta|}}\, 2^{\frac{1}{2}n-t}\pi^{\frac{1}{2}n-1}c^{\frac{1}{2}n-t}$$

$$\times \left[e^{-\frac{1}{2}\pi i q}\, \frac{K_{\frac{1}{2}n-t}[c(Q - i0)^{\frac{1}{2}}]}{(Q - i0)^{\frac{1}{2}(\frac{1}{2}n-t)}} \right.$$

$$\left. - e^{\frac{1}{2}\pi i q}\, \frac{K_{\frac{1}{2}n-t}[c(Q + i0)^{\frac{1}{2}}]}{(Q + i0)^{\frac{1}{2}(\frac{1}{2}n-t)}} \right].$$

Continued

Entry no.	Generalized function f	Fourier transform $F[f]$
22	$\delta(c^2 + P)$	$-\dfrac{i}{\sqrt{\lvert\Delta\rvert}}(2\pi c)^{\frac{1}{2}n-1}\left[-e^{-\frac{1}{2}\pi qi}\dfrac{K_{\frac{1}{2}n-1}[c(Q-i0)^{\frac{1}{2}}]}{(Q-i0)^{\frac{1}{2}(\frac{1}{2}n-1)}} + e^{\frac{1}{2}\pi qi}\dfrac{K_{\frac{1}{2}n-1}[c(Q+i0)^{\frac{1}{2}}]}{(Q+i0)^{\frac{1}{2}(\frac{1}{2}n-1)}}\right].$
23	$\dfrac{(c^2+P)_+^t}{\Gamma(t+1)}$	$(-1)^{t+1}i\,2^{t+\frac{1}{2}n}\pi^{\frac{1}{2}n-1}c^{\frac{1}{2}n+t}$ $\times\left[e^{-\frac{1}{2}qni}\dfrac{K_{\frac{1}{2}n+t}[c(Q-i0)^{\frac{1}{2}}]}{(Q-i0)^{\frac{1}{2}(\frac{1}{2}n+t)}} - e^{\frac{1}{2}qni}\dfrac{K_{\frac{1}{2}n+t}[c(Q+i0)^{\frac{1}{2}}]}{(Q+i0)^{\frac{1}{2}(\frac{1}{2}n+t)}}\right]$ $+ (2\pi)^n\sum_{m=0}^{t}\dfrac{(-1)^m(\frac{1}{2}c)^{2t-2m}}{4^m m!(t-m)!}L^m\delta(s),$
24	$\dfrac{(c^2+P)_-^t}{\Gamma(t+1)}$	$\dfrac{i\cdot 2^{t+\frac{1}{2}n}\pi^{\frac{1}{2}n-1}c^{\frac{1}{2}n+t}}{\sqrt{\lvert\Delta\rvert}}\left[e^{-\frac{1}{2}qni}\dfrac{K_{\frac{1}{2}n+t}[c(Q-i0)^{\frac{1}{2}}]}{(Q-i0)^{\frac{1}{2}(\frac{1}{2}n+t)}} - e^{\frac{1}{2}qni}\dfrac{K_{\frac{1}{2}n+t}[c(Q+i0)^{\frac{1}{2}}]}{(Q+i0)^{\frac{1}{2}(\frac{1}{2}n+t)}}\right],$
25	$\dfrac{(c^2+P)^t}{\Gamma(t+1)}$	$(2\pi)^n\sum_{m=0}^{t}\dfrac{(-1)^m(\frac{1}{2}c)^{2t-2m}}{4^m m!(t-m)!}L^m\delta(s).$

PROOF OF THE COMPLETENESS OF THE GENERALIZED-FUNCTION SPACE

In Chapter I, Section 1.8 we asserted that the generalized-function space K' is (sequentially) complete under convergence as defined there. In other words, given a sequence of functionals $f_1, f_2, ..., f_\nu, ...$ such that for every φ in K the number sequence (f_ν, φ) converges, $f(\varphi) = \lim_{\nu\to\infty} (f_\nu, \varphi)$ is again a continuous linear functional on K. We shall prove this assertion in the present appendix.

To show that $f(\varphi)$ is linear is quite simple. In fact,

$$f(\alpha_1\varphi_1 + \alpha_2\varphi_2) = \lim_{\nu\to\infty} f_\nu(\alpha_1\varphi_1 + \alpha_2\varphi_2) = \lim \{f_\nu(\alpha_1\varphi_1) + f_\nu(\alpha_2\varphi_2)\}$$
$$= \alpha_1 f(\varphi_1) + \alpha_2 f(\varphi_2).$$

What is important is to prove the continuity of $f(\varphi)$. Let $\{\varphi_\nu\}$ be a sequence of functions converging to zero in K. We must show that $f(\varphi_\nu) \to 0$.

Let us assume the contrary. Then choosing, if necessary, a subsequence, we may assume that $|f(\varphi_\nu)| \geqslant c > 0$.

Now recall that convergence of a sequence $\{\varphi_\nu\}$ to zero in K means that all the $\varphi_\nu(x)$ vanish outside some bounded region and that they all converge to zero uniformly in R_n together with all their derivatives. Again choosing a subsequence, we may assume that $|D^k\varphi_\nu(x)| \leqslant 1/4^\nu$, $k = 0, 1, ..., \nu$.

Let us now write $\psi_\nu = 2^\nu\varphi_\nu$. Then the ψ_ν also converge to zero in K, but $|f(\psi_\nu)| \to \infty$.

We now define a subsequence $\{f'_\nu\}$ and another subsequence $\{\psi'_\nu\}$ as follows.

Choose ψ'_1 so that $|f(\psi'_1)| > 1$. Now since $(f_\nu, \psi) \to f(\psi)$, we may choose f'_1 such that $|(f'_1, \psi'_1)| > 1$.

Now suppose we have chosen f'_j and ψ'_j, $j = 1, 2, ..., \nu - 1$. We now choose ψ'_ν to be one of the $\{\psi_\nu\}$ sequence with index so high that

$$|(f'_k, \psi'_\nu)| < \frac{1}{2^{\nu-k}}, \qquad k = 0, 1, ..., \nu - 1; \tag{a}$$

$$|f(\psi'_\nu)| > \sum_{j=1}^{\nu-1} |f(\psi'_j)| + \nu. \tag{b}$$

The first is possible because the ψ_ν converge to zero in K and therefore $(f_0, \psi_\nu) \to 0$ for any generalized function f_0. The second is possible because $|f(\psi_\nu)| \to \infty$. Since $(f_\nu, \psi) \to f(\psi)$, we may choose f'_ν from the $\{f_\nu\}$ sequence such that

$$|(f'_\nu, \psi'_\nu)| > \sum_{j=1}^{\nu-1} |(f'_\nu, \psi'_j)| + \nu. \tag{b'}$$

In this way we can go and construct the new infinite sequences $\{\psi'_\nu\}$ and $\{f'_\nu\}$. Now let us write

$$\psi = \sum_{\nu=1}^{\infty} \psi'_\nu.$$

By construction the series on the right converges in K, and therefore ψ is in K. Further,

$$(f'_\nu, \psi) = \sum_{j=1}^{\nu-1} (f'_\nu, \psi'_j) + (f'_\nu, \psi'_\nu) + \sum_{i=\nu+1}^{\infty} (f'_\nu, \psi'_j).$$

But from (b') and the fact that

$$\sum_{j=\nu+1}^{\infty} (f'_\nu, \psi'_j) < \sum_{j=\nu+1}^{\infty} \frac{1}{2^{j-\nu}} = 1,$$

we arrive at

$$|(f'_\nu, \psi)| > \nu - 1,$$

which means that $|(f'_\nu, \psi)| \to \infty$ as $\nu \to \infty$. But this contradicts $\lim_{\nu\to\infty} (f'_\nu, \psi) = f(\psi)$.

Consequently $f(\varphi_\nu) \to 0$, and the limit functional f is therefore continuous.

GENERALIZED FUNCTIONS
OF COMPLEX VARIABLES

This Appendix is essentially a short introduction to the theory of generalized functions of one and several complex variables. At the author's suggestion it has been moved from the fifth volume of the Russian work to the first volume of the English translation. The results of Sections B1.1–B1.7, B2.1–B2.3, B2.5, and B.2.7 are used in Chapters II–IV of Volume V.

B1. Generalized Functions of a Single Complex Variable

B1.1. The Variables z and $\bar{z}$

It is often convenient to treat a function $f(x, y)$ of two real variables x and y as a function of a single complex variable $z = x + iy$.

Then instead of using the differential operators $\partial/\partial x$ and $\partial/\partial y$, one uses derivatives with respect to z and $\bar{z}$, namely $\partial/\partial z$ and $\partial/\partial \bar{z}$. These new operators are defined with the requirement that the ordinary rules of differentiation, in particular the chain rule, hold for them, so that

$$\frac{\partial}{\partial x} = \frac{\partial z}{\partial x}\frac{\partial}{\partial z} + \frac{\partial \bar{z}}{\partial x}\frac{\partial}{\partial \bar{z}} = \frac{\partial}{\partial z} + \frac{\partial}{\partial \bar{z}}$$

and

$$\frac{\partial}{\partial y} = \frac{\partial z}{\partial y}\frac{\partial}{\partial z} + \frac{\partial \bar{z}}{\partial y}\frac{\partial}{\partial \bar{z}} = i\left(\frac{\partial}{\partial z} - \frac{\partial}{\partial \bar{z}}\right).$$

From this we find that

$$\frac{\partial}{\partial z} = \frac{1}{2}\left(\frac{\partial}{\partial x} - i\frac{\partial}{\partial y}\right), \tag{1}$$

$$\frac{\partial}{\partial \bar{z}} = \frac{1}{2}\left(\frac{\partial}{\partial x} + i\frac{\partial}{\partial y}\right). \tag{1'}$$

Note that the Laplacian $\Delta = \partial^2/\partial x^2 + \partial^2/\partial y^2$ can be written in terms of the complex differential operators in the form

$$\Delta = 4\,\frac{\partial^2}{\partial z\,\partial \bar{z}}\,.$$

There follow some examples of differentiation with respect to z and $\bar{z}$.

Example 1. $\partial|\,z\,|^2/\partial z = \partial(z\bar{z})/\partial z = \bar{z}$. Similarly, $\partial|\,z\,|^2/\partial \bar{z} = z$.

Example 2. $\partial z^n/\partial \bar{z} = 0$.

Example 3. Let $P(x, y)$ be a polynomial in x and y. By writing $x = \frac{1}{2}(z + \bar{z})$ and $y = (-i/2)\,(z - \bar{z})$, we write P as a polynomial in z and $\bar{z}$, namely $P(x, y) = P_1(z, \bar{z})$. It is easily verified that the derivative of $P_1(z, \bar{z})$ with respect to z or $\bar{z}$ is obtained as though z and $\bar{z}$ were independent variables.

When using the variables z and $\bar{z}$ we shall write a function $f(x, y)$ in the form $f_1(z, \bar{z})$, by which we shall mean

$$f_1(z, \bar{z}) = f\left(\frac{1}{2}\,[z + \bar{z}],\; -\frac{i}{2}\,[z - \bar{z}]\right).$$

In this notation an *analytic* function is written $f(z)$. This is because the Cauchy-Riemann conditions $\partial u/\partial x = \partial v/\partial y$ and $\partial u/\partial y = -\,\partial v/\partial x$ can be written for the function $f = u + iv$ in the form $\partial f/\partial \bar{z} = 0$.

Similarly, *antianalytic* functions are defined by the condition $\partial f/\partial z = 0$. It is therefore natural to write them in the form $f(\bar{z})$.[1] (We note that if f is an analytic function, $\bar{f}$ is an antianalytic.)

Let us now write the Maclaurin expansion of an arbitrary function $f(x, y) = f_1(z, \bar{z})$ in powers of z and $\bar{z}$. We have

$$f(x, y) = \sum_{n=0}^{\infty} \frac{1}{n!}\left(x\,\frac{\partial}{\partial x} + y\,\frac{\partial}{\partial y}\right)^n f(0, 0) = \sum_{j,k=0}^{\infty} \frac{\partial^{j+k} f(0, 0)}{\partial x^j\,\partial y^k}\,\frac{x^j y^k}{j!k!}\,.$$

It is easily seen that

$$x\,\frac{\partial}{\partial x} + y\,\frac{\partial}{\partial y} = z\,\frac{\partial}{\partial z} + \bar{z}\,\frac{\partial}{\partial \bar{z}}\,.$$

Therefore in terms of z and $\bar{z}$ the Maclaurin series may be written (we now write f instead of f_1)

$$f(z, \bar{z}) = \sum_{n=0}^{\infty} \frac{1}{n!}\left(z\,\frac{\partial}{\partial z} + \bar{z}\,\frac{\partial}{\partial \bar{z}}\right)^n f(0, 0) = \sum_{j,k=0}^{\infty} \frac{f^{(j,k)}(0, 0)}{j!k!}\,z^j \bar{z}^k, \qquad (3)$$

---

[1] We shall maintain this convention only for a while, until the reader becomes accustomed to the notation. Later, in the interests of conciseness, we shall write simply $f(z)$ for any arbitrary function of z and $\bar{z}$, whether or not it be analytic.

where we have introduced the notation

$$f^{(j,k)}(z, \bar{z}) \equiv \frac{\partial^{j+k} f(z, \bar{z})}{\partial z^j \, \partial \bar{z}^k}.$$

In particular, for analytic functions this series has no terms in $\bar{z}$, and for antianalytic ones none in z.

On integrating some $f(z, \bar{z})$ it is convenient to replace the differential form $dx \, dy$ by $dz \, d\bar{z}$. This new differential form is defined as the exterior product of the differential forms $dz = dx + i \, dy$ and $d\bar{z} = dx - i \, dy$, or

$$dz \, d\bar{z} = (dx + i \, dy)(dx - i \, dy) = -2i \, dx \, dy.$$

Thus

$$dx \, dy = \frac{i}{2} \, dz \, d\bar{z}. \tag{4}$$

On integrating over the complex plane, we may use the following rule for a change of variables. Let $z = \varphi(\zeta)$ be an analytic function of ζ which maps a region D_ζ of the ζ plane in a one-to-one way onto a region D_z of the z plane. Then

$$\frac{i}{2} \int_{D_z} f(z, \bar{z}) \, dz \, d\bar{z} = \frac{i}{2} \int_{D_\zeta} f[\varphi(\zeta), \bar{\varphi}(\zeta)] \, |\, \varphi'(\zeta) \,|^2 \, d\zeta \, d\bar{\zeta}. \tag{5}$$

Indeed, for this case we have $dz = \varphi'(\zeta) d\zeta$, and $d\bar{z} = \bar{\varphi}'(\zeta) d\bar{\zeta}$.

It is a simple matter also to verify the formula for integration by parts, namely

$$\frac{i}{2} \int \varphi^{(j,k)}(z, \bar{z}) f(z, \bar{z}) \, dz \, d\bar{z} = (-1)^{j+k} \frac{i}{2} \int \varphi(z, \bar{z}) f^{(j,k)}(z, \bar{z}) \, dz \, d\bar{z}, \tag{6}$$

where $\varphi(z, \bar{z})$ and $f(z, \bar{z})$ are sufficiently smooth functions of bounded support.

B1.2. Homogeneous Functions of a Complex Variable

Let λ and μ be any complex numbers such that $\lambda - \mu$ is an integer. A function $F(z, \bar{z})$ is called a *homogeneous function of degree* (λ, μ) if for every complex number $a \neq 0$ we have[2]

$$F(az, \bar{a}\bar{z}) = a^\lambda \bar{a}^\mu \, F(z, \bar{z}). \tag{1}$$

---

[2] By definition

$$a^\lambda \bar{a}^\mu \equiv |\, a \,|^{\lambda + \mu} \exp \{i(\lambda - \mu) \arg a\}$$

which, for integral $\lambda - \mu$, is a single-valued function of a.

One might have supposed that homogeneous functions could be defined by the more general condition

$$F(az, \bar{a}\bar{z}) = \alpha(a) F(z, \bar{z}), \tag{1'}$$

but this world imply that $\alpha(a)$ must satisfy the functional equation

$$\alpha(ab) = \alpha(a)\,\alpha(b).$$

The solution of this equation (obtained, for instance, by transforming to polar coordinates) is

$$\alpha(a) = a^{\lambda}\bar{a}^{\mu},$$

where λ and μ are any complex numbers such that $\lambda - \mu$ is an integer.

We shall define a *homogeneous generalized function* $F(z, \bar{z})$ of degree (λ, μ) also by Eq. (1).

To make such a definition meaningful, we rewrite it in the form in which it must appear for generalized functions. To the ordinary function $F(z, \bar{z})$ corresponds the functional

$$(F, \varphi) = \frac{i}{2} \int F(z, \bar{z})\, \varphi(z, \bar{z})\, dz\, d\bar{z}.$$

A formal change of variables leads to

$$(F, \varphi(z/a, \bar{z}/\bar{a})) = a\bar{a} \frac{i}{2} \int F(az, \bar{a}\bar{z})\, \varphi(z, \bar{z})\, dz\, d\bar{z}.$$

Therefore the condition that F be a homogeneous generalized function of degree (λ, μ) may be written

$$(F, \varphi(z/a, \bar{z}/\bar{a})) = a^{\lambda+1}\,\bar{a}^{\mu+1}(F, \varphi(z, \bar{z})). \tag{2}$$

We shall show later (Section B1.6) that to every pair of complex numbers λ, μ whose difference is an integer, there corresponds, up to a multiplicative factor, one and only one homogeneous generalized function of degree (λ, μ).

B1.3. The Homogeneous Generalized Functions $z^{\lambda}\bar{z}^{\mu}$

In this section we shall define the homogeneous generalized function $z^{\lambda}\bar{z}^{\mu}$ on K. Let us assume first that $\mathrm{Re}\,(\lambda + \mu) > -2$. We then define the generalized function $z^{\lambda}\bar{z}^{\mu}$ by the integral

$$(z^{\lambda}\bar{z}^{\mu}, \varphi) = \frac{i}{2} \int z^{\lambda}\bar{z}^{\mu}\, \varphi(z, \bar{z})\, dz\, d\bar{z}, \tag{1}$$

which converges, and in which φ is infinitely differentiable function with bounded support (i.e., φ is in K). It is clear that $z^\lambda \bar{z}^\mu$ is a homogeneous function of degree (λ, μ).

We now wish to extend the definition to $\mathrm{Re}\,(\lambda + \mu) < -2$. Equation (1) will no longer suffice, since the integral in it diverges. We shall show however, that this integral can be regularized and that its regularization yields a homogeneous generalized function.

We introduce the new variables $s = \lambda + \mu$ and $n = \lambda - \mu$ (the reader should bear in mind that n is always an integer, while s may be any complex number) and transform to polar coordinates in (1), writing $z = re^{i\alpha}$; then

$$(z^\lambda \bar{z}^\mu, \varphi) = \int_0^\infty r^{s+1} \varphi_n(r)\, dr, \tag{2}$$

where we have written

$$\varphi_n(r) = \int_0^{2\pi} \varphi(re^{i\alpha}, re^{-i\alpha})\, e^{in\alpha}\, d\alpha$$

Let us now consider a fixed function $\varphi(z, \bar{z})$ in K and a fixed integer n. Then the expression $(z^\lambda \bar{z}^\mu, \varphi)$ defined by (2) is an analytic function of s for $\mathrm{Re}\,s > -2$. We now continue this function analytically to $\mathrm{Re}\,s < -2$. This analytic continuation (at its regular points, that is, for $s \neq -2$, -3, ...) is then taken as the definition of $(z^\lambda \bar{z}^\mu, \varphi)$ for $\mathrm{Re}\,s < -2$. Obviously, the homogeneity is maintained under analytic continuation in s, so that at the regular points in s the functional $(z^\lambda \bar{z}^\mu, \varphi)$ defines a homogeneous generalized function, which we shall denote simply $z^\lambda \bar{z}^\mu$.

In other words the homogeneous generalized function $z^\lambda \bar{z}^\mu$ is defined by the integral of Eq. (1), which converges for $\mathrm{Re}\,(\lambda + \mu) > -2$, and is an analytic function of $s = \lambda + \mu$ for $\mathrm{Re}\,s > -2$. For $\mathrm{Re}\,s < -2$ this integral is to be understood in the sense of its analytic continuation in s (for fixed $n = \lambda - \mu$).

Let us now obtain an explicit expression for this new generalized function. For this purpose we rewrite the defining integral for $\mathrm{Re}\,(\lambda + \mu) > -2$ in the form

$$(z^\lambda \bar{z}^\mu, \varphi) = \frac{i}{2} \int_{|z| \leqslant 1} z^\lambda \bar{z}^\mu \left[\varphi(z, \bar{z}) - \sum_{k+l=0}^{m-1} \varphi^{(k,l)}(0,0)\frac{z^k \bar{z}^l}{k!\,l!} \right] dz\, d\bar{z}$$

$$+ \frac{i}{2} \int_{|z| > 1} z^\lambda \bar{z}^\mu\, \varphi(z, \bar{z})\, dz\, d\bar{z}$$

$$+ \sum_{k+l=0}^{m-1} \frac{\varphi^{(k,l)}(0,0)}{k!\,l!} \frac{i}{2} \int_{|z| \leqslant 1} z^{\lambda+k} \bar{z}^{\mu+l}\, dz\, d\bar{z}. \tag{3}$$

The last integral can be calculated directly. In polar coordinates, in fact, we have

$$\frac{i}{2} \int_{|z|\leqslant 1} z^{\lambda+k} \bar{z}^{\mu+l} \, dz \, d\bar{z} = \int_0^1 r^{k+l+s+1} \, dr \int_0^{2\pi} e^{i(k-l+n)\alpha} \, d\alpha$$

$$= \begin{cases} 2\pi(k+l+s+2)^{-1} & \text{if } k-l = -n, \\ 0 & \text{if } k-l \neq -n. \end{cases}$$

Thus for $\mathrm{Re}\,(\lambda + \mu) > -2$ we may write

$$(z^\lambda \bar{z}^\mu, \varphi) = \frac{i}{2} \int_{|z|\leqslant 1} z^\lambda \bar{z}^\mu \left[\varphi(z, \bar{z}) - \sum_{k+l=0}^{m-1} \varphi^{(k,l)}(0, 0) \frac{z^k \bar{z}^l}{k!l!} \right] dz \, d\bar{z} \tag{4}$$

$$+ \frac{i}{2} \int_{|z|>1} z^\lambda \bar{z}^\mu \, \varphi(z, \bar{z}) \, dz \, d\bar{z} + 2\pi \sum_{\substack{k+l=0 \\ k-l=-n}}^{m-1} \varphi^{(k,l)}(0, 0) \, [k!l!(k+l+s+2)]^{-1}.$$

Now Eq. (4) defines $z^\lambda \bar{z}^\mu$ for $\mathrm{Re}\,(\lambda + \mu) > -m - 2$. The first and second terms in this equation are analytic functions of $s = \lambda + \mu$ for $\mathrm{Re}\,s > -m - 2$, so that the only singularities in this generalized function are contributed by the third term at $s = -k - l - 2$ and $-n = k - l$, or for $\lambda = -k - 1$, $\mu = -l - 1$. Thus the generalized function $z^\lambda \bar{z}^\mu$ is analytic everywhere except at the points $\lambda, \mu = -k - 1$, where k is a nonnegative integer. At such points $z^\lambda \bar{z}^\mu$ has simple poles as a function of $s = \lambda + \mu$ (for fixed integer $n = \lambda - \mu$). At its regular points for $\mathrm{Re}\,s > -m - 2$, the functional $(z^\lambda \bar{z}^\mu, \varphi)$ is given by Eq. (4). Its residue at $\lambda = -k - 1$ and $\mu = -l - 1$ (for $k, l = 0, 1, ...$) is given by[3]

$$\mathop{\mathrm{res}}_{\substack{\lambda=-k-1 \\ \mu=-l-1}} (z^\lambda \bar{z}^\mu, \varphi) = \frac{2\pi}{k!l!} \varphi^{(k,l)}(0, 0). \tag{5}$$

This equation can also be written in the form

$$\mathop{\mathrm{res}}_{\substack{\lambda=-k-1 \\ \mu=-l-1}} z^\lambda \bar{z}^\mu = 2\pi \frac{(-1)^{k+l}}{k!l!} \, \delta^{(k,l)}(z, \bar{z}), \tag{6}$$

where the delta function $\delta(z, \bar{z})$ is defined by

$$(\delta, \varphi) = \varphi(0, 0)$$

---

[3] This is understood as the residue of the function of $s = \lambda + \mu$ for fixed $n = \lambda - \mu$.

and where[4]

$$\delta^{(k,l)}(z, \bar{z}) = \frac{\partial^{k+l} \delta(z, \bar{z})}{\partial z^k \partial \bar{z}^l}.$$

$$\left(\frac{\partial^{k+l} f}{\partial z^k \partial \bar{z}^l}, \varphi\right) = (-1)^{k+l} \left(f, \frac{\partial^{k+l} \varphi}{\partial z^k \partial \bar{z}^l}\right).$$

Note that in the strip $-m - 2 < \mathrm{Re}\,(\lambda + \mu) < -m - 1$, (4) reduces to the simpler expression

$$(z^\lambda \bar{z}^\mu, \varphi) = \frac{i}{2} \int z^\lambda \bar{z}^\mu \left[\varphi(z, \bar{z}) - \sum_{k+l=0}^{m-1} \varphi^{(k,l)}(0, 0) \frac{z^k \bar{z}^l}{k! l!}\right] dz \, d\bar{z}. \tag{7}$$

It is often convenient to consider the normalized generalized function

$$\frac{z^\lambda \bar{z}^\mu}{\Gamma(\frac{1}{2} s + \frac{1}{2} |n| + 1)},$$

where $s = \lambda + \mu$, and $n = \lambda - \mu$. The gamma function in the denominator has simple poles at $s = -k - l - 2$ and $-n = k - l$ (for k, l nonnegative integers), that is for λ and μ negative integers. These poles are at the same points as those of $z^\lambda \bar{z}^\mu$, and their residues are given by

$$\operatorname*{res}_{\substack{\lambda = -k-1 \\ \mu = -l-1}} \Gamma(\tfrac{1}{2} s + \tfrac{1}{2} |n| + 1) = 2(-1)^j / j!, \tag{8}$$

where $j = \frac{1}{2}(k + l) - \frac{1}{2}|k - l| = \min\,(k, l)$. This means that $z^\lambda \bar{z}^\mu / \Gamma(\frac{1}{2} s + \frac{1}{2} |n| + 1)$ is an entire analytic function of $s = \lambda + \mu$ for each fixed $n = \lambda - \mu$ and is homogeneous in z and $\bar{z}$ of degree (λ, μ).

It follows from (8) that at $\lambda = -k - 1$, $\mu = -l - 1$, the normalized function is given by

$$\left[\frac{z^\lambda \bar{z}^\mu}{\Gamma(\frac{1}{2} s + \frac{1}{2} |n| + 1)}\right]_{\substack{\lambda = -k-1 \\ \mu = -l-1}} = \pi \frac{(-1)^{k+l+j} \, j!}{k! l!} \delta^{(k,l)}(z, \bar{z}), \tag{9}$$

with j as defined above.

Thus for each pair of complex numbers λ and μ whose difference is an integer, we have constructed a homogeneous generalized function of degree (λ, μ). If either λ or μ differs from a nonnegative integer, this generalized function is concentrated on the entire z plane. If, however, both λ and μ are negative integers, it is concentrated at the point $z = 0$.

---

[4] The derivatives of a generalized function f are defined by

We shall show in Section B1.6 that no other homogeneous generalized functions exist.

B1.4. The Generalized Function z^{-k-1} and Its Derivatives

As we have seen in the preceding section, $z^\lambda \bar{z}^\mu$ is a generalized function analytic in λ and μ, regular at $\lambda = -1, -2, \dots$ and $\mu = 0$. Thus there exists a generalized function which we may call z^{-k-1}, where k is a nonnegative integer. It is defined as the regularization of the integral

$$(z^{-k-1}, \varphi) = \frac{i}{2} \int z^{-k-1} \varphi(z, \bar{z}) \, dz \, d\bar{z}, \tag{1}$$

which can be calculated by integrating by parts k times over z in the equation

$$(z^\lambda \bar{z}^\mu, \varphi) = \frac{i}{2} \int z^\lambda \bar{z}^\mu \varphi(z, \bar{z}) \, dz \, d\bar{z}$$

and then setting $\lambda = -k - 1$, $\mu = 0$. This gives

$$(z^{-k-1}, \varphi) = \frac{i}{2k!} \int z^{-1} \varphi^{(k,0)}(z, \bar{z}) \, dz \, d\bar{z}, \tag{2}$$

which can be written in the form

$$z^{-k-1} = (-1)^k \frac{1}{k!} \frac{\partial^k (z^{-1})}{\partial z^k}.$$

Let us now calculate the derivative of z^{-k-1} with respect to $\bar{z}$. It may seem at first that this derivative vanishes. It does not, however, since z^{-k-1} is not analytic at $z = 0$, where it has in fact a pole. Therefore its derivative with respect to $\bar{z}$ should be a generalized function concentrated at $z = 0$.

To calculate the desired derivative we first note that for $\alpha \neq 0$

$$\frac{\partial}{\partial \bar{z}} z^{-k-1+\alpha} \bar{z}^\alpha = \alpha z^{-k-1+\alpha} \bar{z}^{\alpha-1}.$$

Now let α approach zero. According to Section B1.3 we have[5]

$$\lim_{\alpha \to 0} \alpha z^{-k-1+\alpha} \bar{z}^{\alpha-1} = \frac{1}{2} \operatorname*{res}_{\substack{\lambda=k-1 \\ \mu=-1}} z^\lambda \bar{z}^\mu = (-1)^k \frac{\pi}{k!} \delta^{(k,0)}(z, \bar{z}).$$

[5] See footnote 3 in Section B.1.3.

Thus

$$\frac{\partial}{\partial \bar{z}} z^{-k-1} = (-1)^k \frac{\pi}{k!} \delta^{(k,0)}(z, \bar{z}). \tag{3}$$

B1.5. Associated Homogeneous Functions

A generalized function $F(z, \bar{z})$ is called associated homogeneous of order one and of degree (λ, μ) if for every complex number $a \neq 0$ we have

$$F(az, \bar{a}\bar{z}) = a^\lambda \bar{a}^\mu [F(z, \bar{z}) + F_0(z, \bar{z}) \ln |a|], \tag{1}$$

where $F_0(z, \bar{z})$ is some homogeneous generalized function of the same degree. Associated homogeneous generalized functions of higher orders are defined similarly.

Let us calculate some associated homogeneous generalized functions of degree (λ, μ).

Assume first that λ and μ are not both negative integers. Then the generalized function $z^{\lambda + \frac{1}{2}s}\bar{z}^{\mu + \frac{1}{2}s}$ is regular in the neighborhood of $s = 0$. Expanding it in a Taylor's series about this point, we obtain

$$z^{\lambda + \frac{1}{2}s} \bar{z}^{\mu + \frac{1}{2}s} = z^\lambda \bar{z}^\mu + s z^\lambda \bar{z}^\mu \ln |z| + \frac{s^2}{2!} z^\lambda \bar{z}^\mu \ln^2 |z| + \cdots. \tag{2}$$

It is easily verified that the coefficients of this expansion, namely the generalized functions $z^\lambda \bar{z}^\mu \ln^m |z|$ are associated functions of order m, and that they are given explicitly by

$$(z^\lambda \bar{z}^\mu \ln^m |z|, \varphi) = \frac{i}{2} \int z^\lambda \bar{z}^\mu \ln^m |z| \, \varphi(z, \bar{z}) \, dz \, d\bar{z}, \tag{3}$$

where the integral is understood in the sense of its regularization.[6]

Let us now turn to the special case in which $\lambda = -k - 1$ and $\mu = -l - 1$ (with k and l nonnegative integers). In this case $z^{\lambda + \frac{1}{2}s}\bar{z}^{\mu + \frac{1}{2}s}$ has a simple pole at $s = 0$; we then expand it in a Laurent series in powers of s. The coefficients of this series will again be associated generalized functions. The constant term of the Laurent expansion, an

---

[6] The regularization of (3) is defined in the same way as the regularization of the integral for $(z^\lambda \bar{z}^\mu, \varphi)$ in Section B1.3.

associated generalized function of order one, shall be denoted by $z^{-k-1}\bar{z}^{-l-1}$. It is defined by

$$(z^{-k-1}\bar{z}^{-l-1}, \varphi) = \frac{i}{2} \int z^{-k-1}\bar{z}^{-l-1} \left[\varphi(z, \bar{z}) - \sum_{h+j=0}^{k+l-1} \varphi^{(h,j)}(0, 0) \frac{z^h \bar{z}^j}{h!\,j!} \right.$$

$$\left. - \theta(1 - |z|) \sum_{h+j=k+l} \varphi^{(h,j)}(0, 0) \frac{z^h \bar{z}^j}{h!\,j!} \right] dz\,d\bar{z}, \qquad (4)$$

where $\theta(x) = 0$ for $x < 0$, and $\theta(x) = 1$ for $x > 0$.

This formula is obtained directly from the expression for $(z^\lambda \bar{z}^\mu, \varphi)$ [that is, Eq. (4) of Section B1.3].

B1.6. Uniqueness Theorem for Homogeneous Generalized Functions

In Section B1.3 we found that to every pair of complex numbers λ, μ whose difference is an integer there corresponds a homogeneous generalized function of degree (λ, μ) defined on K, and we constructed these generalized functions. We shall now show that these functions exhaust all the homogeneous generalized functions of z and $\bar{z}$ (up to a multiplicative factor). In other words, we shall show that up to a constant factor there exists one and only one homogeneous generalized function of each degree.

We start by deriving a differential equation. Let F be a homogeneous generalized function of degree (λ, μ), so that

$$F(az, \bar{a}\bar{z}) = a^\lambda \bar{a}^\mu F(z, \bar{z}). \qquad (1)$$

Differentiating with respect to a and setting $a = 1$, we arrive at

$$z \frac{\partial F}{\partial z} = \lambda F. \qquad (2)$$

Similarly, by differentiating (1) with respect to $\bar{a}$ and setting $\bar{a} = 1$, we obtain

$$\bar{z} \frac{\partial F}{\partial \bar{z}} = \mu F. \qquad (2')$$

Thus homogeneous generalized functions of degree (λ, μ) satisfy the system of differential equations represented by (2) and (2'). Let us now solve these equations. We transform to polar coordinates $z = re^{\alpha i}$ and $\bar{z} = re^{-\alpha i}$, transforming the equations to the form

$$\frac{1}{2} \left[r \frac{\partial F}{\partial r} - i \frac{\partial F}{\partial \alpha} \right] = \lambda F, \qquad \frac{1}{2} \left[r \frac{\partial F}{\partial r} + i \frac{\partial F}{\partial \alpha} \right] = \mu F. \qquad (3)$$

For $r \neq 0$ these equations can be integrated in the usual way. Simple operations lead to

$$F = Cr^{\lambda+\mu} e^{i(\lambda-\mu)\alpha} = Cz^\lambda \bar{z}^\mu. \tag{4}$$

This shows that if $F(z, \bar{z})$ is a homogeneous generalized function of degree (λ, μ), it coincides with the ordinary function $Cz^\lambda \bar{z}^\mu$ for $z \neq 0$.

We now prove uniqueness for all z in the nonsingular case $(\lambda, \mu) \neq (-k - 1, -l - 1)$, where $k, l = 0, 1, 2, \ldots$. In this case $z^\lambda \bar{z}^\mu$ as constructed in Section B1.3 is nonzero in the neighborhood of any point z. Let $\Phi(z, \bar{z})$ be a homogeneous generalized function of degree (λ, μ). As we have just shown, for $z \neq 0$ we may write $\Phi(z, \bar{z}) = Cz^\lambda \bar{z}^\mu$. Consequently,

$$\Phi_1(z, \bar{z}) = \Phi(z, \bar{z}) - Cz^\lambda \bar{z}^\mu$$

is a generalized function concentrated at $z = 0$. It remains to be shown that $\Phi_1(z, \bar{z}) \equiv 0$. This will be seen to follow from the fact[7] that any generalized function of two variables concentrated at the point $x = 0$, $y = 0$ is a linear combination of derivatives of $\delta(x, y)$.

Since $\Phi_1(z, \bar{z})$ is concentrated at $z = 0$, we may write

$$\Phi_1(z, \bar{z}) = \sum_{k,l=0}^{n} c_{kl} \delta^{(k,l)}(z, \bar{z}).$$

Each term of the form $c_{kl}\delta^{(k,l)}(z, \bar{z})$ in this sum is a homogeneous generalized function of degree $(-k - 1, -l - 1)$, and since $\Phi_1(z, \bar{z})$ is of degree $(\lambda, \mu) \neq (-k - 1, -l - 1)$, it must vanish. Thus for this nonsingular case we have proven that any such homogeneous generalized function may be written

$$\Phi(z, \bar{z}) = Cz^\lambda \bar{z}^\mu.$$

Let us now consider the singular case in which $\lambda = -k - 1$, $\mu = -l - 1$, where k and l are nonnegative integers. For this case $\delta^{(k,l)}(z, \bar{z})$ is a homogeneous generalized function of degree $(-k - 1, -l - 1)$ concentrated at $z = 0$. It is easily seen that there exist no other homogeneous generalized functions of this degree concentrated at $z = 0$.

We shall show further that there exist also no homogeneous generalized functions of this degree concentrated anywhere on the z plane. Indeed, assume that $\Phi(z, \bar{z})$ is such a generalized function. Then for $z \neq 0$, it coincides with the ordinary function $Cz^{-k-1}\bar{z}^{-l-1}$ for some

---

[7] See Volume II, Chapter II, Section 4.5.

$C \neq 0$. On the other hand, as we have seen in Section B1.5, the *associated* homogeneous generalized function $z^{-k-1}\bar{z}^{-l-1}$ coincides with the ordinary function $z^{-k-1}\bar{z}^{-l-1}$ for $z \neq 0$.[8] Therefore these two generalized functions can differ only at the origin, and hence

$$\Phi_1(z, \bar{z}) = \Phi(z, \bar{z}) - Cz^{-k-1}\bar{z}^{-l-1}$$

is an *associated* homogeneous generalized function concentrated at $z = 0$. But all generalized functions concentrated at $z = 0$ are linear combinations of the $\delta^{(k,l)}(z, \bar{z})$, which are homogeneous but not associated homogeneous. Thus Φ_1 vanishes identically, and Φ is *associated* homogeneous, contrary to assumption. It follows, then, that any homogeneous generalized function of degree $(-k - 1, -l - 1)$ must be concentrated at $z = 0$.

We have thus established that to every pair of complex numbers λ, μ whose difference is an integer there corresponds, up to a multiplicative factor, one and only one homogeneous generalized function of degree (λ, μ).

It is easily shown in the same way that to every such pair of complex numbers there corresponds, again up to a multiplicative factor, one and only one associated homogeneous generalized function of any order and of degree (λ, μ).

B1.7. Fourier Transforms of Test Functions and of Generalized Functions

We shall call the Fourier transform of a function $\varphi(z, \bar{z})$ in K the function $\tilde{\varphi}(w, \bar{w})$ defined by

$$\tilde{\varphi}(w, \bar{w}) = \frac{i}{2} \int \varphi(z, \bar{z}) \exp\left[i \operatorname{Re}(zw)\right] dz \, d\bar{z}$$

$$= \frac{i}{2} \int \varphi(z, \bar{z}) \exp\left[\tfrac{1}{2} i(zw + \bar{z}\bar{w})\right] dz \, d\bar{z}. \tag{1}$$

When we transform from the complex variables z, w to real ones by writing $z = x + iy$ and $w = u + iv$, Eq. (1) becomes

$$\tilde{\varphi}(u, v) = \int \varphi(x, y) \, e^{i(xu - vv)} \, dx \, dy. \tag{1'}$$

Thus to within the sign of v, our definition of $\tilde{\varphi}$ agrees with the ordinary

---

[8] See Eq. (4) of Section B1.5.

Fourier transform of φ when this latter is treated as a function of the two real variables x and y.[9]

It is known that if $\varphi(z, \bar{z})$ and $f(z, \bar{z})$ are two test functions and $\tilde{\varphi}(w, \bar{w})$ and $\tilde{f}(w, \bar{w})$ are their Fourier transforms, then

$$\frac{i}{2} \int f(z, \bar{z})\, \varphi(z, \bar{z})\, dz\, d\bar{z} = \frac{1}{4\pi^2} \frac{i}{2} \int \tilde{f}(w, \bar{w})\, \bar{\tilde{\varphi}}(w, \bar{w})\, dw\, d\bar{w}. \tag{2}$$

This relation may be interpreted in the following way. If $\tilde{f}(w, \bar{w})$ is considered a generalized function, its action on $\bar{\tilde{\varphi}}(w, \bar{w})$ is given by the formula

$$(\tilde{f}, \bar{\tilde{\varphi}}) = 4\pi^2(f, \bar{\varphi}). \tag{3}$$

It is a simple matter to show that $\bar{\tilde{\varphi}}(w, \bar{w})$ is the Fourier transform of $\bar{\varphi}(-z, -\bar{z})$. Thus if we replace $\bar{\tilde{\varphi}}$ by $\tilde{\varphi}$ in (3), we arrive at

$$(\tilde{f}, \tilde{\varphi}) = 4\pi^2(f, \varphi(-z, -\bar{z})). \tag{4}$$

Now this equation may be used to define the Fourier transform of a generalized function $F(z, \bar{z})$. We shall thus say that the Fourier transform of a generalized function $F(z, \bar{z})$ is the generalized function $\tilde{F}(w, \bar{w})$ defined by the equation

$$(\tilde{F}, \tilde{\varphi}) = 4\pi^2(F, \varphi_1), \tag{5}$$

where $\tilde{\varphi}(w, \bar{w})$ is the Fourier transform of $\varphi(z, \bar{z})$, and $\varphi_1(z, \bar{z}) = \varphi(-z, -\bar{z})$.

Note that $\tilde{F}$ is a generalized function defined not on K, but on Z, the space of Fourier transforms of functions in K. If, however, we had been dealing with generalized functions defined on S, the space of rapidly decreasing infinitely differentiable functions (recall that all their derivatives must also be rapidly decreasing), then F and $\tilde{F}$ would both have been defined on the same test-function space.

We wish to prove that the Fourier transform $\tilde{F}$ of a homogeneous generalized function F of degree (λ, μ) is homogeneous of degree $(-\lambda - 1, -\mu - 1)$. Indeed, if $\tilde{\varphi}(w, \bar{w})$ is the Fourier transform of $\varphi(z, \bar{z})$, the Fourier transform of $\varphi(z/a, \bar{z}/\bar{a})$ will be $|a|^2\tilde{\varphi}(aw, \bar{a}\bar{w})$.

[9] It is sometimes convenient to define the Fourier transform rather by the formula

$$\tilde{\varphi}(w, \bar{w}) = \frac{i}{2} \int \varphi(z, \bar{z}) \exp\left[\tfrac{1}{2} i(z\bar{w} + \bar{z}w)\right] dz\, d\bar{z}.$$

With this definition $\tilde{\varphi}$ coincides with the ordinary Fourier transform of φ treated as a function of x and y. We shall on various occasions make use of both of these definitions.

Thus if $F(z, \bar{z})$ is homogeneous of degree (λ, μ), then

$$
\begin{aligned}
| a |^2 (\tilde{F}, \tilde{\varphi}(aw, \bar{a}\bar{w})) &= 4\pi^2 (F, \varphi(-z/a, -\bar{z}/\bar{a})) \\
&= 4\pi^2 \, a^{\lambda+1} \, \bar{a}^{\mu+1} (F, \varphi(-z, -\bar{z})) \\
&= a^{\lambda+1} \, \bar{a}^{\mu+1} (\tilde{F}, \tilde{\varphi}).
\end{aligned}
$$

We now replace a by a^{-1} in this equation, arriving at

$$
(\tilde{F}, \tilde{\varphi}(w/a, \bar{w}/\bar{a})) = a^{-\lambda} \bar{a}^{-\mu} (\tilde{F}, \tilde{\varphi}), \tag{6}
$$

as asserted.

We have shown in Section B1.3 that

$$
F(z, \bar{z}) = \frac{z^\lambda \bar{z}^\mu}{\Gamma(\frac{1}{2} s + \frac{1}{2} | n | + 1)}
$$

is a homogeneous generalized function of degree (λ, μ), where $s = \lambda + \mu$ and $n = \lambda - \mu$. Thus its Fourier transform may be written

$$
\tilde{F}(w, \bar{w}) = c(\lambda, \mu) \frac{w^{-\lambda-1} \, \bar{w}^{-\mu-1}}{\Gamma(-\frac{1}{2} s + \frac{1}{2} | n |)} \tag{7}
$$

where $c(\lambda, \mu)$ is some constant. This constant can be calculated by comparing the expressions for $(F, \varphi(-z, -\bar{z}))$ with $(\tilde{F}, \tilde{\varphi})$ for some conveniently chosen fixed test function $\varphi(z, \bar{z})$. We shall show below that

$$
c(\lambda, \mu) = 2^{\lambda+\mu+2} \, \pi i^{|\lambda-\mu|}, \tag{8}
$$

and thus that the Fourier transforms of the homogeneous generalized functions are given by

$$
\frac{\widetilde{(z^\lambda \bar{z}^\mu)}}{\Gamma(\frac{1}{2} s + \frac{1}{2} | n | + 1)} = 2^{\lambda+\mu+2} \, \pi i^{|\lambda-\mu|} \frac{w^{-\lambda-1} \bar{w}^{-\mu-1}}{\Gamma(-\frac{1}{2} s + \frac{1}{2} | n |)}, \tag{9}
$$

where $s = \lambda + \mu$ and $n = \lambda - \mu$.

Let us proceed to calculate $c(\lambda, \mu)$. In order to be specific we choose $n = \lambda - \mu \geqslant 0$. For our test function we choose[10]

$$
\varphi(z, \bar{z}) = \bar{z}^{\lambda-\mu} \exp\left(-\tfrac{1}{2} z\bar{z}\right).
$$

---

[10] This function does not have bounded support. Since, however, $z^\lambda \bar{z}^\mu$ is an algebraically increasing function, the functional represented by $(z^\lambda \bar{z}^\mu, \varphi)$ can be extended to S, and in particular to the test function we have chosen, which is in S.

To calculate the Fourier transform of this function we first note, as is seen from a simple calculation, that

$$\frac{i}{2} \int \exp \left\{ \tfrac{1}{2} \left[-z\bar{z} + i(zw + \bar{z}\bar{w}) \right] \right\} dz \, d\bar{z} = 2\pi \exp \left(-\tfrac{1}{2} w\bar{w} \right).$$

Now differentiate this n times with respect to $\bar{w}$ (recall that $n = \lambda - \mu$). This yields

$$\left(\frac{i}{2} \right)^{\lambda - \mu + 1} \int \bar{z}^{\lambda - \mu} \exp \left\{ \tfrac{1}{2} \left[-z\bar{z} + i(zw + \bar{z}\bar{w}) \right] \right\} dz \, d\bar{z}$$

$$= 2\pi \left(-\tfrac{1}{2} w \right)^{\lambda - \mu} \exp \left(-\tfrac{1}{2} w\bar{w} \right).$$

Hence for the chosen test function

$$\tilde{\varphi}(w, \bar{w}) = 2\pi i^{\lambda - \mu} w^{\lambda - \mu} \exp \left(-\tfrac{1}{2} w\bar{w} \right).$$

We now insert this expression and the right-hand side of Eq. (7) into

$$(\tilde{F}(w, \bar{w}), \tilde{\varphi}(w, \bar{w})) = 4\pi^2 \left(\frac{z^\lambda \bar{z}^\mu}{\Gamma(\tfrac{1}{2} s + \tfrac{1}{2} |n| + 1)}, \, [-\bar{z}]^{\lambda - \mu} \exp \left\{ -\tfrac{1}{2} z\bar{z} \right\} \right)$$

to arrive at

$$2\pi i^{\lambda - \mu} c(\lambda, \mu) \, (w^{-\lambda - 1} \bar{w}^{-\mu - 1}, w^{\lambda - \mu} \exp \left\{ -\tfrac{1}{2} w\bar{w} \right\}) / \Gamma(-\mu)$$

$$= (-1)^{\lambda - \mu} 4\pi^2 (z^\lambda \bar{z}^\mu, \bar{z}^{\lambda - \mu} \exp \left\{ -\tfrac{1}{2} z\bar{z} \right\}) / \Gamma(\lambda + 1). \tag{10}$$

The expressions on both sides of this equation are easily calculated with the aid of the formula

$$\frac{i}{2} \int (z\bar{z})^{\alpha - 1} \exp \left(-\tfrac{1}{2} z\bar{z} \right) dz \, d\bar{z} = 2\pi \int_0^\infty r^{2\alpha - 1} \exp \left(-\tfrac{1}{2} r^2 \right) dr = 2^\alpha \, \pi \Gamma(\alpha)$$

(for $\operatorname{Re} \alpha < 0$, the integral is to be understood in the sense of its regularization). Specifically, we have

$$(z^\lambda \bar{z}^\mu, \bar{z}^{\lambda - \mu} \exp \left\{ -\tfrac{1}{2} z\bar{z} \right\}) = \frac{i}{2} \int (z\bar{z})^\lambda \exp \left(-\tfrac{1}{2} z\bar{z} \right) dz \, d\bar{z} = 2^{\lambda + 1} \, \pi \Gamma(\lambda + 1)$$

and

$$(w^{-\lambda - 1} \bar{w}^{-\mu - 1}, w^{\lambda - \mu} \exp \left\{ -\tfrac{1}{2} w\bar{w} \right\}) = 2^{-\mu} \, \pi \Gamma(-\mu).$$

When these relations are inserted into (10), we obtain $c(\lambda, \mu) = 2^{\lambda + \mu + 2} \pi i^{\lambda - \mu}$. This result has been obtained on the assumption that

$\lambda - \mu \geqslant 0$. When $\lambda - \mu < 0$, the positions of λ and μ are interchanged. Thus we arrive at Eq. (8).

In particular, when λ and μ are integers, both of the same sign, the relevant Fourier transforms are

$$\widetilde{z^k \bar{z}^l} = 4\pi^2 (-2i)^{k+l} \delta^{(k,l)}(w, \bar{w})$$

and

$$\overline{\delta^{(k,l)}(z, \bar{z})} = (2i)^{-k-l} w^k \bar{w}^l,$$

where $k, l = 0, 1, 2, \ldots$.

B1.8. The Generalized Function $f^\lambda(z) \, \bar{f}^\mu(z)$, Where $f(z)$ Is a Meromorphic Function

Let us now define the generalized function $f^\lambda(z) \bar{f}^\mu(z)$ where $f(z)$ is a meromorphic function of z, and λ and μ are complex numbers such that $\lambda - \mu$ is an integer.

We start by considering the simpler case in which $f(z)$ is an entire function. For this case we define $f^\lambda(z) \bar{f}^\mu(z)$ by

$$(f^\lambda \bar{f}^\mu, \varphi) = \frac{i}{2} \int f^\lambda(z) \bar{f}^\mu(z) \, \varphi(z, \bar{z}) \, dz \, d\bar{z}. \tag{1}$$

For given $\lambda - \mu = n$ and for $\text{Re}\,(\lambda + \mu) > 0$, the integral in this expressions converges and is an analytic function of $\lambda + \mu$. For $\text{Re}\,(\lambda + \mu) < 0$, we shall understand the integral in the sense of its analytic continuation in $\lambda + \mu$. Our problem is to give meaning to the integral when $f(z)$ is a meromorphic function. The simple definition we have given above will not work for this case, since the integral will then fail to converge in general for all λ and μ.

Any $\varphi(z, \bar{z})$ in K can be written as a linear combination of other functions in K, each with support in a sufficiently small region. We may therefore proceed by first giving meaning to the integral in (1) for $\varphi(z, \bar{z})$ with support in a small region.

Let us thus assume that $\varphi(z, \bar{z})$ has support in a region containing a single k-fold zero and no poles of $f(z)$. Then for given $\lambda - \mu$ the integral of (1) is known to converge for $\text{Re}\,(\lambda + \mu) > 0$ and is an analytic function of $\lambda + \mu$. For $\text{Re}\,(\lambda + \mu) < 0$ we proceed by analytic continuation in $\lambda + \mu$.

Let us find the singularities of $(f^\lambda \bar{f}^\mu, \varphi)$ for this case. We may without loss of generality assume that $f(z)$ has a zero at $z = 0$ and that $\varphi(z)$ has support in a neighborhood of this point. Then

$$f(z) = z^k f_1(z),$$

where $f_1(z)$ is a function regular in the neighborhood of $z = 0$, and $f_1(0) \neq 0$. Consequently

$$(f^\lambda \bar{f}^\mu, \varphi) = \frac{i}{2} \int z^{k\lambda} \bar{z}^{k\mu} f_1^\lambda(z) \bar{f}_1^\mu(z) \varphi(z, \bar{z}) \, dz \, d\bar{z}.$$

It is seen from this result that the singularities of $(f^\lambda \bar{f}^\mu, \varphi)$ coincide with the singularities of the generalized homogeneous function $z^{k\lambda} \bar{z}^{k\mu}$, considered a function of λ and μ. The results of Section B1.3 then lead to the conclusion that if $\varphi(z)$ has its support in a neighborhood of a k-fold zero of $f(z)$, the only singularities of $(f^\lambda \bar{f}^\mu, \varphi)$ (considered an analytic function of λ and μ) are simples poles at $(\lambda, \mu) = (-p/k, -q/k)$, where $p, q = 1, 2, \ldots$ (and $\lambda - \mu$ is an integer).

Assume now that $\varphi(z)$ has support in a region containing a single pole of order l and no zeros of $f(z)$. Then the integral $(f^\lambda \bar{f}^\mu, \varphi)$ will converge for $\text{Re} \, (\lambda + \mu) < 0$; for $\text{Re} \, (\lambda + \mu) > 0$, we define it by its analytic continuation in $\lambda + \mu$. Proceeding as above, we conclude easily that for this case the only singularities in $(f^\lambda \bar{f}^\mu, \varphi)$ (considered an analytic function of λ and μ) are simple poles at $(\lambda, \mu) = (p/l, q/l)$, where p, $q = 1, 2, \ldots$ (and $\lambda - \mu$ is an integer).

We have thus interpreted the integral $(f^\lambda \bar{f}^\mu, \varphi)$ for the case in which φ has its support in a region containing only a single zero or a single pole of $f(z)$.

Finally, we define $(f^\lambda \bar{f}^\mu, \varphi)$ for any arbitrary function $\varphi(z, \bar{z})$ in K. We write

$$\varphi(z, \bar{z}) = \sum_i \varphi_i(z, \bar{z}), \tag{2}$$

where the $\varphi_i(z, \bar{z})$ are functions in K each of which has support in a region containing no more than a single zero or pole of $f(z)$. Since we have already discussed the $(f^\lambda \bar{f}^\mu, \varphi_i)$ we may write

$$(f^\lambda \bar{f}^\mu, \varphi) = \sum_i (f^\lambda \bar{f}^\mu, \varphi_i). \tag{3}$$

It is easily shown that $(f^\lambda \bar{f}^\mu, \varphi)$ does not depend on the choice of the φ_i. Indeed, let us make two such choices $\varphi = \Sigma \varphi_j'$ and $\varphi = \Sigma \varphi_j''$, where we label functions corresponding to the same zero or pole of $f(z)$ with the same index. Then the $\varphi_j' - \varphi_j''$ vanish in the neighborhood of every zero and pole of $f(z)$, and therefore for all λ and μ we have

$$(f^\lambda \bar{f}^\mu, \varphi_j' - \varphi_j'') = \frac{i}{2} \int f^\lambda \bar{f}^\mu \left[\varphi_j' - \varphi_j'' \right] dz \, d\bar{z},$$

where the integral on the right-hand side converges. Since $\Sigma(\varphi_j' - \varphi_j'') = 0$, the sum over j of these integrals vanishes. Hence we arrive at

$$\sum_j (f^\lambda \bar{f}^\mu, \varphi_j) - \sum_j (f^\lambda \bar{f}^\mu, \varphi_j'') = \sum_j (f^\lambda \bar{f}^\mu, \varphi_j' - \varphi_j'') = 0.$$

We have thus defined the generalized function $f^\lambda \bar{f}^\mu$ for any meromorphic function $f(z)$. The generalized function $f^\lambda \bar{f}^\mu$ is an analytic function of λ and μ averywhere except at

$$(\lambda, \mu) = (-p/k, -q/k) \qquad \text{and} \qquad (\lambda, \mu) = (p/l, q/l),$$

where k runs through the set of multiplicities of zeros of $f(z)$, and l through the set of multiplicities of its poles ($p, q = 1, 2, ...,$ and $\lambda - \mu$ is an integer). At these points $f^\lambda \bar{f}^\mu$ has simple poles.

In particular, we see that the generalized function $f^\lambda \bar{f}^\mu$, considered a function of λ and μ, is regular at $\lambda = k$, $\mu = 0$, $k = \pm 1, \pm 2, ...$ Thus for every meromorphic function $f(z)$ we have defined the generalized function f^k, for $k = \pm 1, \pm 2, ...$

B2. Generalized Functions of *m* Complex Variables

B2.1. The Generalized Functions $\delta(P)$ and $\delta^{(k,l)}(P)$

We shall start by defining generalized functions concentrated on a manifold S of $2m - 2$ real dimensions in a space of m complex dimensions. Consider a manifold S defined by an equation of the form

$$P(z) \equiv P(z_1, ..., z_m) = 0,$$

where $P(z)$ is an infinitely differentiable function (of z and $\bar{z}$).[1] We shall assume that the differential form $dP \, d\bar{P}$ vanishes nowhere on the $P = 0$ manifold.

This assumption has a simple geometric meaning. To understand it, we consider the space of the variables z_k to be a $2m$-dimensional

---

[1] It would be more correct to denote this function by $P(z, \bar{z})$. In order not to complicate the notation, however, we shall henceforth denote functions of m complex variables simply by $P(z)$.

space of the real variables x_k, y_k (where $z_k = x_k + iy_k$), $k = 1, ..., m$. In this real space S is given by the two equations Re $P = 0$, Im $P = 0$. Then our assumption means that the surfaces Re $P = \xi$ and Im $P = \eta$ form a lattice such that in the neighborhood of every point of S it is possible to set up a local real coordinate system in which ξ and η are two of the coordinates.

In Chapter III, Section 1.9 we were able with this assumption to define the generalized function $\delta(\text{Re } P, \text{Im } P)$ concentrated on S. We shall denote this function here by $\delta(P)$. It is convenient also to define $\delta(P)$ entirely in terms of complex variables, which we shall now proceed to do.

We define the generalized function $\delta(P)$ by the equation

$$(\delta(P), \varphi) = \int_{P=0} \varphi\omega, \tag{1}$$

where ω is the differential form of degree $2m - 2$ defined by[2]

$$\left(\frac{i}{2}\right)^m dz\, d\bar{z} = \frac{i}{2} dP\, d\bar{P}\, \omega. \tag{2}$$

We have here written

$$\left(\frac{i}{2}\right)^m dz\, d\bar{z} \equiv \left(\frac{i}{2}\right)^m dz_1\, d\bar{z}_1 ... dz_m\, d\bar{z}_m \equiv dx_1\, dy_1 ... dx_m\, dy_m.$$

When changing variables in an integrand it is often convenient to replace the order of the differentials; we have

$$\left(\frac{i}{2}\right)^m dz\, d\bar{z} = \frac{i^{m^2}}{2^m} dz_1 ... dz_m\, d\bar{z}_1 ... d\bar{z}_m.$$

Further, we have written

$$dP = \sum \left[\frac{\partial P}{\partial z_k} dz_k + \frac{\partial P}{\partial \bar{z}_k} d\bar{z}_k \right].$$

We now proceed to define the derivatives

$$\delta^{(k,l)}(P) = \frac{\partial^{k+l}\, \delta(P)}{\partial P^k\, \partial \bar{P}^l}$$

of $\delta(P)$. These generalized functions will be defined as the integrals

---

[2] The real variables x_k and y_k can be used to prove the existence of ω as defined by (2) and the uniqueness of $\delta(P)$ defined by (1) (although ω itself is not uniquely defined).

over the $P = 0$ manifold of certain differential forms $\omega_{k,l}(\varphi)$ depending on P and on the test function φ and its derivatives. First, we define $\omega_{0,0}(\varphi)$ as the differential form $\varphi\omega$, where ω is defined by Eq. (2). Then the $\omega_{k,l}(\varphi)$ for all nonnegative k, l are defined by the recurrence relations[3]

$$d\,[d\bar{P}\,\omega_{k-1,l}(\varphi)] = dP\,d\bar{P}\,\omega_{k,l}(\varphi), \tag{3}$$

$$d\,[dP\,\omega_{k,l-1}(\varphi)] = -dP\,d\bar{P}\,\omega_{k,l}(\varphi). \tag{4}$$

We then define the derivatives of $\delta(P)$ by

$$(\delta^{(k,l)}(P), \varphi) = (-1)^{k+l}\int_{P=0}\omega_{k,l}(\varphi). \tag{5}$$

It can be shown that the $\omega_{k,l}(\varphi)$ exist and that (3) and (4) define them to within an additive term of the form $d\tau + \alpha dP + \beta d\bar{P}$, where τ, α, and β are differential forms of degree $2m - 3$. Then it follows from Stokes' theorem that the $\delta^{(k,l)}(P)$ are uniquely defined by (5). The proof is exactly the same as for the case of real variables (Chapter III, Section 1.9).

Some properties of the $\delta^{(k,l)}(P)$ are the following. Again, the proofs of these properties are the same as those for the real case.

(1) The $\delta^{(k,l)}(P)$ can be differentiated in accordance with the chain rule in the sense that

$$\frac{\partial}{\partial z_i}\,\delta^{(k,l)}(P) = \frac{\partial P}{\partial z_i}\,\delta^{(k+1,l)}(P) + \frac{\partial\bar{P}}{\partial z_i}\,\delta^{(k,l+1)}(P), \tag{6}$$

$$\frac{\partial}{\partial\bar{z}_i}\,\delta^{(k,l)}(P) = \frac{\partial P}{\partial\bar{z}_i}\,\delta^{(k+1,l)}(P) + \frac{\partial\bar{P}}{\partial\bar{z}_i}\,\delta^{(k,l+1)}(P). \tag{6'}$$

(2) The following identities are satisfied by $\delta(P)$ and its derivatives:

$$P\,\delta(P) = \bar{P}\,\delta(P) = 0, \tag{7}$$

$$P\,\delta^{(k,l)}(P) + k\,\delta^{(k-1,l)}(P) = 0, \tag{8}$$

$$\bar{P}\,\delta^{(k,l)}(P) + l\,\delta^{(k,l-1)}(P) = 0. \tag{8'}$$

(3) If the equations $P = 0$ and $Q = 0$ define two nonintersecting manifolds without singular points, so that $PQ = 0$ is a manifold with no singular points, then

$$\delta(PQ) = P^{-1}\bar{P}^{-1}\,\delta(Q) + Q^{-1}\bar{Q}^{-1}\,\delta(P). \tag{9}$$

---

[3] Analogous differential forms for real variables were discussed in Chapter III, Section 1.9.

In particular, if $a(z)$ is a function which fails to vanish anywhere, then

$$\delta(aP) = a^{-1}\bar{a}^{-1}\,\delta(P). \tag{10}$$

Further interesting formulas are obtained by taking the derivative of this last relation. For instance, if P is analytic, that is if $\partial P/\partial \bar{z}_i = 0$, we have

$$\delta^{(k,l)}(aP) = a^{-k-1}\,\bar{a}^{-l-1}\,\delta^{(k,l)}(P) \tag{11}$$

for any nonvanishing function $a(z)$. The proof of this result is left to the reader.

B2.2. The Generalized Functions $G^\lambda \bar{G}^\mu$

We shall now define generalized functions associated with an entire analytic function G. Let $G(z_1, ..., z_m)$ be any entire analytic function. If λ and μ are complex numbers whose difference is an integer, then

$$G^\lambda \bar{G}^\mu = |\,G\,|^{\lambda+\mu} \exp\,[i(\lambda - \mu)\,\arg G] \tag{1}$$

is a single-valued function of the z_k.

With the ordinary function $G^\lambda \bar{G}^\mu$ we may associate the generalized function $G^\lambda \bar{G}^\mu$ defined by

$$(G^\lambda \bar{G}^\mu, \varphi) = \left(\frac{i}{2}\right)^m \int G^\lambda(z)\,\bar{G}^\mu(z)\,\varphi(z)\,dz\,d\bar{z}, \tag{2}$$

where $z = (z_1, ..., z_m)$ and $dz\,d\bar{z} = dz_1\,d\bar{z}_1 ... dz_m d\bar{z}_m$ (the integral is taken over the entire complex space). The integral is known to converge for Re $(\lambda + \mu) > 0$ and, for given $\lambda - \mu$, is an analytic function of $\lambda + \mu$. We define $(G^\lambda \bar{G}^\mu, \varphi)$ for other values of λ and μ by analytic continuation in $\lambda + \mu$.

We wish to study the generalized function $G^\lambda \bar{G}^\mu$ as an analytic function of λ and μ. Its singular points are closely related to the nature of the $G(z) = 0$ manifold. In this section we shall consider the simplest case in which the $G(z) = 0$ manifold has no singular points, that is, in which the $\partial G/\partial z_k$ are not all zero simultaneously at any point of the manifold.

The general case in which the $G = 0$ manifold has singular points will be treated in Section B2.9.

If the $G(z) = 0$ manifold has no singular points, the only singularities of the generalized function $G^\lambda \bar{G}^\mu$, considered a function of λ and μ, are simple poles at the points

$$(\lambda, \mu) = (-k - 1, -l - 1), \qquad k, l = 0, 1, ...,$$

with residues[4]

$$\operatorname*{res}_{\substack{\lambda=-k-1 \\ \mu=-l-1}} G^{\lambda} \bar{G}^{\mu} = (-1)^{k+l}\, 2\pi\, \frac{\delta^{(k,l)}(G)}{k!\,l!}. \tag{3}$$

Indeed, if the $G = 0$ manifold has no singular points, then in a neighborhood U of any point of this surface we can introduce local coordinates $w_1, \ldots, w_{m-1}, \zeta$, where $\zeta = G(z)$. Now if we pick a test function $\varphi(z)$ with support in U, we have

$$\left(\frac{i}{2}\right)^m \int G^{\lambda}(z)\, \bar{G}^{\mu}(z)\, \varphi(z)\, dz\, d\bar{z} = \frac{i}{2} \int \zeta^{\lambda} \bar{\zeta}^{\mu}\, \Phi(\zeta)\, d\zeta\, d\bar{\zeta},$$

where $\Phi(\zeta)$ denotes the integral of $\varphi(z)$ over all the w_k (for fixed ζ).

But as we have seen in Section B1.3, the only singularities of the generalized function $\zeta^{\lambda} \bar{\zeta}^{\mu}$, considered an analytic function of λ and μ, are simple poles at

$$(\lambda, \mu) = (-k-1, -l-1), \qquad k, l = 0, 1, \ldots$$

with residues

$$\operatorname*{res}_{\substack{\lambda=-k-1 \\ \mu=-l-1}} \zeta^{\lambda} \bar{\zeta}^{\mu} = (-1)^{k+l}\, 2\pi\, \frac{\delta^{(k,l)}(\zeta)}{k!\,l!}.$$

From this we obtain the asserted result for $G^{\lambda} \bar{G}^{\mu}$, namely Eq. (3).

Remark. The local nature of the considerations implies that the result holds also not only for functions defined over all of z space, but even if they are defined only on some analytic manifold without singular points.

B2.3. Homogeneous Generalized Functions

We shall call an ordinary function $f(z)$ of the m complex variables $z = (z_1, \ldots, z_m)$ a homogeneous function of degree (λ, μ) if for any complex number $\alpha \neq 0$ we have

$$f(\alpha z) = \alpha^{\lambda} \bar{\alpha}^{\mu} f(z). \tag{1}$$

As before, we assume that the difference between λ and μ is an integer. When this is so, $\alpha^{\lambda} \bar{\alpha}^{\mu}$ is unique.

---

[4] Recall that the residue is defined as the residue with respect to $s = \lambda + \mu$ for fixed $\lambda - \mu$.

We shall define a homogeneous *generalized* function $f(z)$ of degree (λ, μ) by the same equation (1), which we shall rewrite in a form applicable to generalized functions.

With an ordinary function $f(z)$ we associate the functional

$$(f, \varphi) = \left(\frac{i}{2}\right)^m \int f(z)\, \varphi(z)\, dz\, d\bar{z}, \tag{2}$$

where $dz\, d\bar{z}$ is defined as before. Then obviously the homogeneity condition (1) is equivalent to

$$(f, \varphi(z/\alpha)) = \alpha^{\lambda+m}\, \bar{\alpha}^{\mu+m}(f, \varphi(z)). \tag{3}$$

As an important example of a homogeneous generalized function, consider $\delta(z)$ defined by

$$(\delta, \varphi) = \varphi(0).$$

It is clear that $\delta(z)$ is homogeneous of degree $(-m, -m)$. Its derivatives $\partial^{k+l}\delta(z)/\partial z_i^k\, \partial \bar{z}_j^l$ are homogeneous generalized functions of degree $(-m-k, -m-l)$.

Other than $\delta(z)$ there exist no generalized functions of degree $(-m, -m)$ concentrated at $z = 0$. This follows from the fact that every generalized function concentrated at $z = 0$ is a linear combination of $\delta(z)$ and its derivatives.[5] But the derivatives of $\delta(z)$ are all of degree other than $(-m, -m)$. Therefore if f is concentrated at $z = 0$ and is homogeneous of degree $(-m, -m)$, it must be some multiple of $\delta(z)$.

We present without proofs some simple properties of homogeneous generalized functions.

(1) The product of a homogeneous generalized function f of degree (λ, μ) with an infinitely differentiable homogeneous function of degree (λ', μ') is a homogeneous generalized function of degree $(\lambda + \lambda', \mu + \mu')$.

(2) If $f(z)$ is a homogeneous generalized function of degree (λ, μ), then $\partial^{k+l} f(z)/\partial z_i^k \partial \bar{z}_j^l$ is a homogeneous generalized function of degree $(\lambda - k, \mu - l)$.

(3) In order that a generalized function f be homogeneous of degree (λ, μ) it is necessary and sufficient that it satisfy the Euler equations

$$\sum_{k=1}^m z_k \frac{\partial f}{\partial z_k} = \lambda f, \qquad \sum_{k=1}^m \bar{z}_k \frac{\partial f}{\partial \bar{z}_k} = \mu f.$$

[5] See Volume II, Chapter II, Section 4.5.

B2.4. Associated Homogeneous Functions

We shall call a function $f_1(z)$ an associated homogeneous function of order one and of degree (λ, μ) if for every complex $\alpha \neq 0$

$$f_1(\alpha z) = \alpha^\lambda \bar{\alpha}^\mu [f_1(z) + f_0(z) \ln |\alpha|], \tag{1}$$

where $f_0(z) \not\equiv 0$ is a homogeneous function of degree (λ, μ).

For instance, $\ln |z_1|$ is an associated homogeneous function of order one and degree $(0, 0)$ since

$$\ln |\alpha z_1| = \ln |z_1| + \ln |\alpha|.$$

An associated homogeneous *generalized* function of order one and degree (λ, μ) is defined by the same equation (1). To write it in a form applicable to generalized functions, we consider not $f_1(z)$, but the functional (f_1, φ). Then Eq. (1) is equivalent to

$$(f_1, \varphi(z/\alpha)) = \alpha^{\lambda+m} \bar{\alpha}^{\mu+m} [(f_1, \varphi(z)) + \ln |\alpha| (f_0, \varphi(z))], \tag{2}$$

where $f_0 \not\equiv 0$ is a homogeneous generalized function of degree (λ, μ).

We define an associated function of order k inductively. We shall say that f_k is an associated homogeneous generalized function of order k and of degree (λ, μ) if for every $\alpha \neq 0$ we have

$$(f_k, \varphi(z/\alpha)) = \alpha^{\lambda+m} \bar{\alpha}^{\mu+m} [(f_k, \varphi(z)) + \ln |\alpha| (f_{k-1}, \varphi(z))], \tag{3}$$

where f_{k-1} is an associated homogeneous generalized function of order $k - 1$ and of degree (λ, μ).

Associated functions can be obtained from homogeneous ones in the following way. Let $f_{\lambda,\mu}$ be a homogeneous generalized function of degree (λ, μ), differentiable in the parameter $s = \lambda + \mu$.[6] Then the derivative $\partial f_{\lambda,\mu}/\partial s$ will be an associated generalized function of order one. Similarly, the derivative with respect to s of an associated generalized function of order k will be one of order $k + 1$.

The proof follows simply by differentiating both sides of (3) with respect to s, bearing in mind that

$$\alpha^\lambda \bar{\alpha}^\mu = |\alpha|^s \exp [i(\lambda - \mu) \arg \alpha].$$

---

[6] More exactly, for each fixed $\lambda - \mu$, we assume $f_{\lambda,\mu}$ to be differentiable with respect to $s = \lambda + \mu$.

B2.5. The Residue of a Homogeneous Function

Many results of the theory of homogeneous functions are conveniently formulated in terms of their residues.

We first recall the definition of the residue of a homogeneous function for the case of real variables (see Chapter III, Section 3.2).

Consider an ordinary homogeneous function $f(x)$, of m real variables $x = (x_1, ..., x_m)$, of degree $-m$; that is, a function such that

$$f(\alpha x) = \alpha^{-m} f(x)$$

for every $\alpha > 0$. Let us assume that $f(x)$ is everywhere continuous except at the origin. Then we define the *residue* of $f(x)$ (at the origin) by the expression

$$\operatorname{res} f(x) = \int_\Gamma f(x)\, \omega \tag{1}$$

where

$$\omega = \sum_{k=1}^m (-1)^{k-1} x_k\, dx_1 \dots dx_{k-1}\, dx_{k+1} \dots dx_m$$

and the integral is taken over any closed surface (i.e., submanifold of dimension $m - 1$) Γ enclosing the origin.

The differential form ω has a simple geometric meaning. In fact ω/m is the volume of the cone whose vertex is at the origin and whose base is an element of area. This implies that $\operatorname{res} f(x)$ is independent of the choice of Γ, and therefore that it is completely determined by $f(x)$. In particular if $f(x) > 0$, we may may pick Γ as the closed surface defined by $f(x) = 1$, obtaining

$$\operatorname{res} f(x) = \int_\Gamma \omega = mV$$

where V is the volume of the region $f(x) \geqslant 1$ containing the origin.

We may remark that the integral of any function $f(x)$ over the entire space is conveniently expressed in terms of its residue. Let Γ be an arbitrary surface which intersects every ray from the origin at a single point. Then every point in the space can be uniquely specified by writing

$$x = \alpha y,$$

where $\alpha > 0$, and y is a point on Γ, and (α, y) may be taken as generalized polar coordinates of x. In order to integrate $f(x)$ over the entire space we may integrate first over the rays passing through $x = 0$, and then integrate the expression so obtained over Γ. Thus we may write[7]

$$\int f(x)\,dx = \int_\Gamma \left(\int_0^\infty f(\alpha y)\, \alpha^{m-1}\, d\alpha \right) \omega(y) = \mathrm{res} \int_0^\infty f(\alpha x)\, \alpha^{m-1}\, d\alpha. \tag{2}$$

This equation can be considered a transformation to generalized polar coordinates in the integrand.

We shall define the residue of a homogeneous function of several complex variables by analogy with the real case. Let $f(z)$, $z = (z_1, \ldots, z_m)$, be a homogeneous function of degree $(-m, -m)$, continuous everywhere except at the origin, and consider the differential form

$$\omega = \sum_{k=1}^m (-1)^{k-1}\, z_k\, dz_1 \ldots dz_{k-1}\, dz_{k+1} \ldots dz_m. \tag{3}$$

We then define the residue of $f(z)$ by the expression

$$\mathrm{res}\, f(z) = \frac{i^{(m-1)^2}}{2^{m-1}} \int_\Gamma f(z)\, \omega\bar\omega. \tag{4}$$

The integral is taken over any $(m - 1)$-dimensional manifold that intersects every complex line passing through the origin at a single point (with the possible exception of a set of such lines of lower dimension).[8] It can be shown that no bounded closed manifold has this property, so that in general Γ is composed of a finite number of sections of smooth manifolds. We shall show how it may be constructed.

First break up the space of the complex variables into a finite number of "sufficiently narrow" cones C_i with vertex at the origin (in other words, into regions such that if C_j contains a point z, it contains the entire complex line passing through the origin and z). In each of the C_j we choose a section Γ_j, that is, a manifold of dimension $m - 1$ that intersects at one point every line in C_j passing through the origin. We then choose Γ as the set of the Γ_j, so that

$$\frac{i^{(m-1)^2}}{2^{m-1}} \int_\Gamma f(z)\, \omega\bar\omega = \frac{i^{(m-1)^2}}{2^{m-1}} \sum_j \int_{\Gamma_j} f(z)\, \omega\bar\omega. \tag{5}$$

The integral in (4) is independent of the choice of the "surface" Γ (or more accurately, it is independent of the way the space is broken up

---

[7] Equation (2) follows directly from the differential relationship

$$dx = \alpha^{m-1}\, d\alpha\, \omega(y),$$

which is trivially verified.

[8] It should be noted that in the case of complex variables the concept of a ray belonging to a line does not exist. For this reason lines rather than rays enter into the definition.

into the C_j and of the way the Γ_j are then chosen), for $f\omega\bar{\omega}$ is homogeneous of degree $(0, 0)$ and is therefore invariant under replacement of z by αz, where $\alpha \neq 0$. Therefore $\int f\omega\bar{\omega}$ will be invariant under any deformation of Γ.

Since $f(z)\omega\bar{\omega}$ is invariant under such replacement of z by αz, it may be treated as a differential form on the space of complex lines passing through the origin. In this way (4) may be through of as an integral over the projective space of all complex lines passing through the origin.

The differential form defined in (3) has the following property.

If $z = \alpha u$, where α is a complex variable and u is a point on some analytic $(m - 1)$-dimensional manifold Γ, then[9]

$$dz \equiv dz_1 \dots dz_m = \alpha^{m-1}\, d\alpha\omega(u). \tag{6}$$

From this we obtain a convenient formula for the transition from the Cartesian coordinates $z_1, \dots, z_m$ to generalized polar coordinates. We do this by considering the complex number α and the point u on Γ as the generalized polar coordinates of $z = \alpha u$. We then have

$$\left(\frac{i}{2}\right)^m \int f(z)\, dz\, d\bar{z} = \frac{i^{(m-1)^2+1}}{2^m} \int_\Gamma \left[\int\int f(\alpha u)\, \alpha^{m-1}\, \bar{\alpha}^{m-1}\, d\alpha\, d\bar{\alpha}\right] \omega\bar{\omega}$$

$$= \operatorname{res} \frac{i}{2} \int f(\alpha z)\, \alpha^{m-1}\, \bar{\alpha}^{m-1}\, d\alpha\, d\bar{\alpha}. \tag{7}$$

Remark. The concept of a homogeneous function and of its residue can be introduced not only for a function defined on the entire space, but also for a function defined on some conical manifold with the vertex at the origin. (A manifold is called conical if together with each point z it contains, it contains the entire complex line passing through z and the origin.)

B2.6. Homogeneous Generalized Functions of Degree $(-m, -m)$

Let $f(z)$ be an ordinary homogeneous function of degree $(-m, -m)$, where m is the dimension of the space, and let it be continuous everywhere except at $z = 0$. We define the corresponding generalized function and thereby the regularization of the divergent integral

$$\left(\frac{i}{2}\right)^m \int f(z)\, \varphi(z)\, dz\, d\bar{z},$$

---

[9] Indeed,

$$dz_1 \dots dz_m = \Pi\, (\alpha\, du_i + u_i\, d\alpha) = \alpha^{m-1}\, d\alpha\omega(u),$$

since $du_1 \dots du_m = 0$.

by choosing an arbitrary region G containing the origin and writing

$$(f, \varphi) = \left(\frac{i}{2}\right)^m \int f(z)\, \varphi(z)\, dz\, d\bar{z} = \left(\frac{i}{2}\right)^m \int_G f(z)\, [\varphi(z) - \varphi(0)]\, dz\, d\bar{z}$$
$$+ \left(\frac{i}{2}\right)^m \int_{C-G} f(z)\, \varphi(z)\, dz\, d\bar{z}, \qquad (1)$$

where C denotes the entire space (so that $C - G$ is the complement of G). The regularization of our integral defined in this way depends, of course, on the choice of G.

Let us denote the generalized function so obtained by $f(z)|_G$, and let us study the way this function behaves when G is replaced by some other region G_1. It is immediately obvious that if G is replaced by $G_1 \subset G$, the functional we obtain will differ by

$$\varphi(0) \left(\frac{i}{2}\right)^m \int_{G-G_1} f(z)\, dz\, d\bar{z}$$

from the original. Thus

$$f|_G - f|_{G_1} = \delta(z) \left(\frac{i}{2}\right)^m \int_{G-G_1} f(z)\, dz\, d\bar{z}. \qquad (2)$$

Since this equation shows that $f|_G - f|_{G_1}$ is a homogeneous generalized function of degree $(-m, -m)$, it follows that the homogeneity or inhomogeneity [of degree $(-m, m)$] of the generalized function defined by (1) is independent of G.

We may thus ask under what circumstances Eq. (1) defines a homogeneous generalized function of degree $(-m, -m)$, that is, under what circumstances we have

$$(f, \varphi(z/\alpha)) = (f, \varphi(z)).$$

We shall show that the necessary and sufficient condition for this is the vanishing of the residue of the (ordinary) homogeneous function f.

Proof. With the change of variables $z_k/\alpha = z_k'$, the right-hand side of Eq. (1) becomes, since $f(z)$ is homogeneous of degree $(-m, -m)$,

$$\left(f, \varphi\left(\frac{z}{\alpha}\right)\right) = \left(\frac{i}{2}\right)^m \int_{\alpha G} f(z)\, [\varphi(z) - \varphi(0)]\, dz\, d\bar{z} + \left(\frac{i}{2}\right)^m \int_{C-\alpha G} f(z)\, \varphi(z)\, dz\, d\bar{z},$$

where αG is the region obtained from G by the similarity transformation corresponding to α. It is seen from this that the necessary and sufficient condition for the homogeneity of $f|_G$ is

$$\left(\frac{i}{2}\right)^m \int_{G-\alpha G} f(z)\, dz\, d\bar{z} = 0$$

for all α. This integral can be rewritten by subdividing the entire complex space into cones C_j such that within each cone the boundary of G consists of all points of the form $e^{i\vartheta}u$, where $0 \leqslant \vartheta \leqslant 2\pi$, and u runs over a certain section Γ_j of an analytic manifold of dimension $m - 1$ which intersects at one point every complex line in C_j passing through the origin. Then in generalized polar coordinates we have

$$\left(\frac{i}{2}\right)^m \int_{G-\alpha G} f(z)\,dz\,d\bar{z}$$

$$= \sum \frac{i^{(m-1)^2+1}}{2^m} \int_{\Gamma_j} \left[\iint_{|\alpha|\leqslant|\lambda|\leqslant 1} f(\lambda z)\,\lambda^{m-1}\,\bar{\lambda}^{-m-1}\,d\lambda\,d\bar{\lambda}\right]\omega\bar{\omega},$$

where ω is the differential form of Section B2.5, Eq. (3).

Since $f(\lambda z) = \lambda^{-m}\bar{\lambda}^{-m}f(x)$, this may be written in the form

$$\left(\frac{i}{2}\right)^m \int_{G-\alpha G} f(z)\,dz\,d\bar{z} = \frac{i}{2}\left[\iint_{|\alpha|\leqslant|\lambda|\leqslant 1}|\lambda|^{-2}\,d\lambda\,d\bar{\lambda}\right]\sum\frac{i^{(m-1)^2}}{2^{m-1}}\int_{\Gamma_j}f(z)\,\bar{\omega}\omega$$

$$= 2\pi \ln|\alpha|\operatorname{res}f(z).$$

Thus the necessary and sufficient condition that to the ordinary homogeneous function $f(z)$ of degree $(-m - m)$ there correspond a generalized homogeneous function is that the residue of $f(z)$ vanish.

We see that it is strictly no more correct to speak of the residue of a homogeneous function than it is to speak of the residue of an analytic function. In other words an analytic function is said to have a residue when the corresponding generalized function is not analytic.[10] Similarly, a homogeneous function is said to have a residue when the corresponding generalized function is not homogeneous.

B2.7. The Generalized Function $P^\lambda \bar{P}^\mu$, Where P Is a Nondegenerate Quadratic Form

Consider a nondegenerate quadratic form of m complex variables $z_1, ..., z_m$, namely

$$P = \sum_{j,k=1}^m g_{jk}\,z_j z_k. \tag{1}$$

We wish to study the homogeneous generalized function $P^\lambda \bar{P}^\mu$, or the functional

$$(P^\lambda \bar{P}^\mu, \varphi) = \left(\frac{i}{2}\right)^m \int P^\lambda(z)\,\bar{P}^\mu(z)\,\varphi(z)\,dz\,d\bar{z} \tag{2}$$

as an analytic function of λ, μ (for integral $\lambda - \mu$).

---

[10] For instance, $\partial(z^{-1})/\partial\bar{z} = \pi\delta(z)$ (see Section B1.4) shows that the generalized function z^{-1} is not analytic.

Note that the $P = 0$ manifold has a singular point at $z = 0$, so that the considerations of Section B2.2 will not apply to this case.

Equation (2) serves to define $P^\lambda \bar{P}^\mu$ only for Re $(\lambda + \mu) > 0$. Let us obtain an expression for $P^\lambda \bar{P}^\mu$ for Re $(\lambda + \mu) < 0$. To do this we express $P^\lambda \bar{P}^\mu$ in terms of $P^{\lambda+k} \bar{P}^{\mu+l}$. We introduce the differential operators

$$L_P = \sum g^{jk} \frac{\partial^2}{\partial z_j \, \partial z_k} \quad \text{and} \quad \bar{L}_P = \sum \bar{g}^{jk} \frac{\partial^2}{\partial \bar{z}_j \, \partial \bar{z}_k}, \tag{3}$$

where the g^{jk} are defined by

$$\sum_k g^{jk} g_{kl} = \delta_l^j.$$

Thus the matrix $\| g^{jk} \|$ of the coefficients of L_P is the inverse of the matrix $\| g_{jk} \|$ of the quadratic form. We then have

$$L_P P^{\lambda+1} \bar{P}^\mu = 4(\lambda + 1)(\lambda + \tfrac{1}{2} m) P^\lambda \bar{P}^\mu$$

as is easily verified by direct calculation.

Applying this formula k times, we arrive at

$$(L_P)^k P^{\lambda+k} \bar{P}^\mu = 4^k (\lambda + 1) \dots (\lambda + k)(\lambda + \tfrac{1}{2} m) \dots (\lambda + \tfrac{1}{2} m + k - 1) P^\lambda \bar{P}^\mu. \tag{4}$$

Similarly,

$$(\bar{L}_P)^l P^\lambda \bar{P}^{\mu+l} = 4^l (\mu + 1) \dots (\mu + l)(\mu + \tfrac{1}{2} m) \dots (\mu + \tfrac{1}{2} m + l - 1) P^\lambda \bar{P}^\mu. \tag{5}$$

Combining (4) and (5) we arrive at the result that the generalized function $P^\lambda \bar{P}^\mu$ may be written, for $k, l = 0, 1, \dots$,

$$P^\lambda \bar{P}^\mu = c(\lambda, k) \, c(\mu, l) \, L_P^k \bar{L}_P^l \, P^{\lambda+k} \bar{P}^{\mu+l}, \tag{6}$$

where

$$c(\nu, p) = \{ 4^p (\nu + 1) \dots (\nu + p)(\nu + \tfrac{1}{2} m) \dots (\nu + \tfrac{1}{2} m + p - 1) \}^{-1}.$$

Consequently,

$$(P^\lambda \bar{P}^\mu, \varphi) = c(\lambda, k) \, c(\mu, l) \, (P^{\lambda+k} \bar{P}^{\mu+l}, L_P^k \bar{L}_P^l \varphi). \tag{7}$$

This equation gives the desired expression for $P^\lambda \bar{P}^\mu$ for Re $(\lambda + \mu) > -k - l$, where k and l are nonnegative integers.

Let us now turn to the singularities of the generalized function $P^\lambda \bar{P}^\mu$, considered a function of λ and μ. It is seen immediately from

Eq. (6) that $P^\lambda \bar{P}^\mu$ has two sequences of singular points, namely

$$(\lambda, \mu) = (-k - 1, -l - 1), \qquad k, l = 0, 1, 2, ...,$$

and

$$(\lambda, \mu) = (-\tfrac{1}{2} m - k, -\tfrac{1}{2} m - l).$$

If (λ, μ) is a point belonging to only one of these sequences, $P^\lambda \bar{P}^\mu$ has a simple pole at this point. If, on the other hand, (λ, μ) belongs simultaneously to both sequences, which is possible only if the dimension of the space is even, $P^\lambda \bar{P}^\mu$ has a pole of order two at this point.

We now consider the singularities of $\bar{P}^\lambda \bar{P}^\mu$ in each of these cases.

Case 1. The singular point $\lambda = -k - 1$, $\mu = -l - 1$ is in the first sequence but not in the second. In this case the residue of $P^\lambda \bar{P}^\mu$ is a generalized function concentrated on the $P = 0$ manifold.

We know already that if f is an analytic function such that there are no singular points on the $f = 0$ manifold, then[11]

$$\operatorname*{res}_{\substack{\lambda = -k-1 \\ \mu = -l-1}} f^\lambda f^\mu = 2\pi (-1)^{k+l} \frac{\delta^{(k,l)}(f)}{k! l!}$$

By a natural analogy we introduce the generalized functions $\delta^{(k,l)}(P)$ concentrated on $P = 0$, defined by

$$\delta^{(k,l)}(P) = (2\pi)^{-1} (-1)^{k+l} k! l! \operatorname*{res}_{\substack{\lambda = -k-1 \\ \mu = -l-1}} P^\lambda \bar{P}^\mu. \qquad (8)$$

From the recurrence relation (6) for the $P^\lambda \bar{P}^\mu$ we may derive the formula

$$\delta^{(k,l)}(P) = \frac{\Gamma(\tfrac{1}{2} m - k - 1)\, \Gamma(\tfrac{1}{2} m - l - 1)}{4^{k+l}\, \Gamma^2(\tfrac{1}{2} m - 1)} L_P^k \bar{L}_P^l\, \delta(P). \qquad (9)$$

Case 2. The singular point (λ, μ) is in the second sequence, but not in the first, and the dimension m of the space is odd. We shall show that in this case the residue of $P^\lambda \bar{P}^\mu$ is a generalized function concentrated at $z = 0$, and that it is given by

$$\operatorname*{res}_{\substack{\lambda = -\frac{1}{2}m-k \\ \mu = -\frac{1}{2}m-l}} P^\lambda \bar{P}^\mu = \frac{(-1)^{\frac{1}{2}(m-1)}\, 2^{-(m+2k+2l-1)}\, \pi^{m+1}}{k! l!\, \Gamma(\tfrac{1}{2} m + k)\, \Gamma(\tfrac{1}{2} m + l)\, |\varDelta|} L_P^k \bar{L}_P^l\, \delta(z), \qquad (10)$$

where $\varDelta$ is the discriminant of P.

---

[11] We recall again that this is the residue of the function as a function of $s = \lambda + \mu$ for fixed integer $\lambda - \mu$.

Let us first calculate the residue of $P^\lambda \bar{P}^\mu$ at $\lambda = \mu = -\frac{1}{2} m$. This is obviously a homogeneous generalized function of degree $(-m, -m)$. We may show further that it is concentrated at $z = 0$.

Indeed, $z = 0$ is the only singular point of the $P = 0$ manifold. Therefore if φ vanishes in a neighborhood of $z = 0$, then as was shown in Section B2.2, $(P^\lambda \bar{P}^\mu, \varphi)$ is a regular function of λ and μ at $\lambda = \mu = -\frac{1}{2}m$, which means that the residue of this functional evaluated for this particular test function vanishes at this value of λ, μ.

Thus the residue of $P^\lambda \bar{P}^\mu$ at $\lambda = \mu = -\frac{1}{2} m$ is a generalized function concentrated at $z = 0$. It is, in addition, homogeneous of degree $(-m, -m)$, so that

$$\operatorname*{res}_{\lambda=\mu=-\frac{1}{2}m} P^\lambda \bar{P}^\mu = c_m \delta(z). \tag{11}$$

What remains is to calculate c_m. We assert that

$$c_m = \frac{(-1)^{\frac{1}{2}(m-1)}\, 2^{-(m-1)}\, \pi^{m+1}}{\Gamma^2(\frac{1}{2}\, m)\, |\varDelta|}, \tag{12}$$

where $\varDelta$ is the discriminant of P, and proceed to prove the assertion.

Calculation of c_m. We may define c_m by

$$c_m \varphi(0) = \operatorname*{res}_{\lambda=\mu=-\frac{1}{2}m} (P^\lambda \bar{P}^\mu, \varphi) = \operatorname*{res}_{s=-m} (|P|^s, \varphi)$$

where φ is any test function. Let us now perform a linear transformation in z space such that P takes on the form $z_1^2 + \ldots + z_m^2$. We then obtain

$$c_m \varphi(0) = \frac{1}{|\varDelta|} \operatorname*{res}_{s=-m} \left(\frac{i}{2}\right)^m \int |z_1^2 + \ldots + z_m^2|^s\, \varphi(z)\, dz\, d\bar{z}.$$

Let us now choose the test function

$$\varphi = \exp(-z_1 \bar{z}_1 - \ldots - z_m \bar{z}_m).$$

Then the above expression becomes

$$c_m = \frac{1}{|\varDelta|} \operatorname*{res}_{s=-m} \left(\frac{i}{2}\right)^m \int |z_1^2 + \ldots + z_m^2|^s \exp(-z_1 \bar{z}_1 - \ldots - z_m \bar{z}_m)\, dz\, d\bar{z}. \tag{13}$$

To calculate the integral

$$f(\lambda) = \left(\frac{i}{2}\right)^m \int |z_1^2 + \ldots + z_m^2|^{2\lambda} \exp(-z_1 \bar{z}_1 - \ldots - z_m \bar{z}_m)\, dz\, d\bar{z} \tag{14}$$

(for m either odd or even) we may use the fact that $L_P P^{\lambda+1} \bar{P}^\lambda =$

$4(\lambda + 1)(\lambda + \tfrac{1}{2}m)P^{\lambda}\bar{P}^{\lambda}$. From this we easily obtain the recurrence relation

$$f(\lambda) = [4(\lambda + 1)(\lambda + \tfrac{1}{2}m)]^{-1}f(\lambda + 1). \tag{15}$$

Hence

$$f(\lambda) = 4^{\lambda}\Gamma(\lambda + 1)\,\Gamma(\lambda + \tfrac{1}{2}m)f_1(\lambda), \tag{16}$$

where $f_1(\lambda)$ is a periodic function with period one :

$$f_1(\lambda + 1) = f_1(\lambda)$$

We now assert that $f_1(\lambda)$ is a constant. To prove this we note first that we have already established that $f(\lambda)$ has the same singularities as does $\Gamma(\lambda + 1)\,\Gamma(\lambda + \tfrac{1}{2}m)$, which means that $f_1(\lambda)$ is an entire function.

Let us now calculate the asymptotic behavior of $f_1(\lambda) = f_1(\sigma + i\tau)$ as $|\tau| \to \infty$. Because f_1 is periodic we may assume that $0 \leqslant \sigma \leqslant 1$. The original function $f(\lambda) = f(\sigma + i\tau)$ is bounded for $0 \leqslant \sigma \leqslant 1$. Therefore using the well-known asymptotic formula for the Γ-function[12]

$$|\Gamma(\sigma + i\tau)| \sim \sqrt{2\pi}\,e^{-\tfrac{1}{2}\pi|\tau|}\,|\tau|^{-\tfrac{1}{2}} \qquad \text{as} \qquad |\tau| \to \infty \tag{17}$$

we find from (16) that

$$|f_1(\sigma + i\tau)| \leqslant C\,e^{\pi|\tau|} \qquad \text{as} \qquad |\tau| \to \infty.$$

For an entire function $f_1(\lambda)$ periodic with period one, however, this is possible only if it is equal to a constant. [This can be seen by expanding $f_1(\lambda)$ in a Laurent series in powers of $z = e^{2\pi i\lambda}$ and studying the behavior of $g(z) = f_1(\lambda)$ in the neighborhood of $z = 0$ and $z = \infty$.] Therefore $f_1(\lambda) = \text{const.}$

We have thus established that

$$f(\lambda) = c4^{\lambda}\,\Gamma(\lambda + 1)\,\Gamma(\lambda + \tfrac{1}{2}m) \tag{18}$$

Now let us set $\lambda = 0$ in this equation. Since

$$f(0) = \left(\frac{i}{2}\right)^m \int \exp\left(-z_1\bar{z}_1 - \ldots - z_m\bar{z}_m\right) dz\,d\bar{z} = \pi^m,$$

we arrive from (18) at

$$c = \frac{\pi^m}{\Gamma(\tfrac{1}{2}m)}.$$

---

[12] A. Erdelyi (ed.), Bateman Manuscript Project, "Higher Transcendental Functions," Vol. I, p. 47, Eq. (6). McGraw-Hill, New York, 1953.

Finally, then, we obtain the result

$$f(\lambda) = \pi^m 4^\lambda \frac{\Gamma(\lambda + 1)\, \Gamma(\lambda + \tfrac{1}{2} m)}{\Gamma(\tfrac{1}{2} m)}. \tag{19}$$

For odd *m* this result together with (13) gives the desired equation (12) for c_m.

We now proceed to calculate the residue of $P^\lambda \bar{P}^\mu$ at $\lambda = -\tfrac{1}{2}m - k$, $\mu = -\tfrac{1}{2}m - l$, by using the recurrence relation (6) on Eq. (11). In this way it is seen that if *m* is odd, the generalized function $P^\lambda \bar{P}^\mu$ has simple poles at $(\lambda, \mu) = (-\tfrac{1}{2}m - k, -\tfrac{1}{2}m - l)$ for nonnegative integers *k* and *l*, and that the residues at these points are given by

$$\mathop{\mathrm{res}}_{\substack{\lambda=-k-\frac{1}{2}m \\ \mu=-l-\frac{1}{2}m}} P^\lambda \bar{P}^\mu = \frac{(-1)^{\frac{1}{2}(m-1)}\, 2^{-(m+2k+2l-1)}\, \pi^{m+1}}{k!\,l!\, \Gamma(\tfrac{1}{2}m + k)\, \Gamma(\tfrac{1}{2}m + l)\, |\varDelta|} L_P^k \bar{L}_P^l\, \delta(z), \tag{20}$$

where L_P and $\bar{L}_P$ are defined by Eq. (3), and $\varDelta$ is the discriminant of *P*.

Case 3. The singular point (λ, μ) is simultaneously in both sequences; i.e., $\lambda = -\tfrac{1}{2}m - k$, $\mu = -\tfrac{1}{2}m - l$, and the dimension *m* of the space is even. For this case we expand $P^\lambda \bar{P}^\mu$ in a Laurent series (in $\lambda + \mu$) about the singular point, writing

$$P^\lambda \bar{P}^\mu = a_{kl}(z)\, [(\lambda + \mu) + (m + k + l)]^{-2}$$
$$+ b_{kl}(z)\, [(\lambda + \mu) + (m + k + l)]^{-1} + \dots, \tag{21}$$

where we have omitted the regular part of the expansion.

Proceeding almost word for word as in Case 2 above, we find that

$$a_{kl}(z) = \alpha_{kl} L_P^k \bar{L}_P^l\, \delta(z), \tag{22}$$

where

$$\alpha_{kl} = \frac{(-1)^{\frac{1}{2}(m-2)}\, 2^{-(m+2k+2l-2)}\, \pi^m}{k!\,l!\, \Gamma(\tfrac{1}{2}m + k)\, \Gamma(\tfrac{1}{2}m + l)\, |\varDelta|}. \tag{23}$$

The $b_{kl}(z)$ in (21) is a generalized function concentrated on the $P = 0$ manifold. By analogy with the case in which this manifold has no singular points, we introduce the generalized function $\delta^{(\frac{1}{2}m+k-1, \frac{1}{2}m+l-1)}(P)$ defined by

$$\mathop{\mathrm{res}}_{\substack{\lambda=-\frac{1}{2}m-k \\ \mu=-\frac{1}{2}m-l}} P^\lambda \bar{P}^\mu = \beta_{kl}\, \delta^{(\frac{1}{2}m+k-1, \frac{1}{2}m+l-1)}(P), \tag{24}$$

where

$$\beta_{kl} = \frac{2\pi(-1)^{k+l}}{\Gamma(\frac{1}{2}m + k)\,\Gamma(\frac{1}{2}m + l)} \tag{25}$$

(cf. Case 1, where we dealt with simple poles).

Thus if (λ, μ) belongs simultaneously to both sequences of singular points (that is if $\lambda = -\frac{1}{2}m - k$, $\mu = -\frac{1}{2}m - l$) and the dimension m is odd, the generalized function $P^\lambda \bar{P}^\mu$ has a pole of order two at this point. The Laurent expansion of $P^\lambda \bar{P}^\mu$ about this point is

$$P^\lambda \bar{P}^\mu = [(\lambda + \mu) + (m + k + l)]^{-2}\,\alpha_{kl}\, L_P^k L_P^l \delta(z)$$

$$+ [(\lambda + \mu) + (m + k + l)]^{-1}\,\beta_{kl}\,\delta^{(\frac{1}{2}m+k-1,\,\frac{1}{2}m+l-1)}(P), \tag{26}$$

where $\delta^{(\frac{1}{2}m+k-1,\,\frac{1}{2}m+l-1)}(P)$ is a generalized function concentrated on the $P = 0$ manifold, and α_{kl} and β_{kl} are defined by (23) and (25) (and we have omitted the regular part of the Laurent expansion).

B2.8. Elementary Solutions of Linear Differential Equations in the Complex Domain

We may apply the results of Section B2.7 to obtain the elementary solutions of an equation of the form

$$L^k u = f(z), \tag{1}$$

where L is a homogeneous linear differential operator of the form

$$L = \sum_{i,j=1}^{m} g^{ij}\, \frac{\partial^2}{\partial z_i\, \partial z_j} \tag{2}$$

with a nonsingular symmetric matrix $\| g^{ij} \|$, and where $k = 1, 2, \ldots$.

Recall that an elementary solution of Eq. (1) is a generalized function K such that

$$L^k K = \delta(z).$$

Consider the quadratic form

$$P = \sum g_{ij}\, z_i z_j \tag{3}$$

whose matrix $\| g_{ij} \|$ is the inverse of that of L. We shall show that except for the case in which m is even and $k < \frac{1}{2}m$, the function

$P^{-\frac{1}{2}m+k}\bar{P}^{-\frac{1}{2}m}$ (or any constant multiple of it) is an elementary solution of Eq. (1).

To prove the assertion we use the fact that [Section B2.7, Eq. (4)]

$$L^k P^{\lambda+k}\bar{P}^{\lambda} = 4^k(\lambda + 1) \ldots (\lambda + k)(\lambda + \tfrac{1}{2}m) \ldots (\lambda + \tfrac{1}{2}m + k - 1) P^{\lambda}\bar{P}^{\mu}. \tag{4}$$

If *m* is odd and $\lambda = -\tfrac{1}{2}m$, this gives

$$L^k P^{-\frac{1}{2}m+k}\bar{P}^{-\frac{1}{2}m}$$
$$= 4^k(1 - \tfrac{1}{2}m) \ldots (k - \tfrac{1}{2}m)(k - 1)! \lim_{\lambda \to -\frac{1}{2}m} (\lambda + \tfrac{1}{2}m) P^{\lambda}\bar{P}^{\mu}. \tag{5}$$

But for odd *m* we know that[13]

$$\operatorname*{res}_{\lambda=\mu=-\frac{1}{2}m} P^{\lambda}\bar{P}^{\mu} = \frac{(-1)^{\frac{1}{2}(m-1)} 2^{-(m-1)} \pi^{m+1}}{\Gamma^2(\tfrac{1}{2}m) \, |\Delta|} \delta(z),$$

where Δ is the discriminant of *P*. Consequently in a space of odd dimension the function

$$K = (-1)^{\frac{1}{2}(m-1)+k} 2^{m-2k} \Gamma(\tfrac{1}{2}m)$$
$$\times \Gamma(\tfrac{1}{2}m - k) \, |\Delta| \, [\pi^{m+1}(k - 1)!]^{-1} P^{-\frac{1}{2}m+k}\bar{P}^{-\frac{1}{2}m} \tag{6}$$

is an elementary solution of Eq. (1).

If *m* is even and $k \geqslant \tfrac{1}{2}m$, by setting $\lambda = -\tfrac{1}{2}m$ in Eq. (4) we arrive at

$$L^k P^{-\frac{1}{2}m+k}\bar{P}^{-\frac{1}{2}m} = 4^k(-1)^{\frac{1}{2}(m-2)}(\tfrac{1}{2}m - 1)!(k - \tfrac{1}{2}m)!(k - 1)!$$
$$\times \lim_{\lambda \to -\frac{1}{2}m} (\lambda + \tfrac{1}{2}m)^2 P^{\lambda}\bar{P}^{\mu}.$$

But for even *m* we have

$$\lim_{\lambda \to -\frac{1}{2}m} (\lambda + \tfrac{1}{2}m)^2 P^{\lambda}\bar{P}^{\mu} = \frac{(-1)^{\frac{1}{2}(m-2)} 2^{-m}\pi^m}{\Gamma^2(\tfrac{1}{2}m) \, |\Delta|} \delta(z).$$

Therefore for a space of even dimension and for $k \geqslant \tfrac{1}{2}m$,

$$K = \frac{2^{m-2k} \Gamma(\tfrac{1}{2}m) \, |\Delta|}{\pi^m(k - 1)! \, (k - \tfrac{1}{2}m)!} P^{-\frac{1}{2}m+k}\bar{P}^{-\frac{1}{2}m} \tag{7}$$

is an elementary solution of Eq. (1). This proves the assertion.

Let us now consider the special case in which the dimension is even

---

[13] See footnote 11 in Section B2.7.

and $k < \frac{1}{2}m$. As was shown in the previous section, in this case $P^{\lambda+k}\bar{P}^\mu$ has a simple pole at $\lambda = \mu = -\frac{1}{2}m$, with residue

$$\operatorname*{res}_{\lambda=\mu=-\frac{1}{2}m} P^{\lambda+k}\bar{P}^\mu = \frac{2\pi(-1)^k}{(\frac{1}{2}m-k-1)!\,(\frac{1}{2}m-1)!}\,\delta^{(\frac{1}{2}m-k-1,\frac{1}{2}m-1)}(P). \tag{8}$$

Multiplying both sides of (4) by $2(\lambda + \frac{1}{2}m)$ and going to the limit as $\lambda \to -\frac{1}{2}m$, we obtain

$$\frac{2\pi(-1)^k}{(\frac{1}{2}m-k-1)!\,(\frac{1}{2}m-1)!}\,L^k\,\delta^{(\frac{1}{2}m-k-1,\frac{1}{2}m-1)}(P)$$

$$= 2^{2k+1}(1-\tfrac{1}{2}m)\ldots(k-\tfrac{1}{2}m)\,(k-1)!\,\lim_{\lambda\to-\frac{1}{2}m}\,(\lambda+\tfrac{1}{2}m)^2\,P^\lambda\bar{P}^\lambda$$

$$= 2^{2k+1}(1-\tfrac{1}{2}m)\ldots(k-\tfrac{1}{2}m)\,(k-1)!\,\frac{(-1)^{\frac{1}{2}(m-2)}\,2^{-m}\,\pi^m}{\Gamma^2(\frac{1}{2}m)\,|\Delta|}\,\delta(z).$$

Thus in the special case in which the dimension m is even and $k < \frac{1}{2}m$, the generalized function

$$K = \frac{(-1)^{\frac{1}{2}(m-2)}\,2^{m-2k}\,|\Delta|}{\pi^{m-1}(k-1)!}\,\delta^{(\frac{1}{2}m-k-1,\frac{1}{2}m-1)}(P) \tag{9}$$

concentrated on the $P = 0$ manifold, is an elementary solution of Eq. (1).

Remark. Similar considerations can be used to obtain elementary solutions of the equation

$$L^k\bar{L}^l u = f(z),$$

where L is defined by Eq. (2) and

$$\bar{L} = \sum \bar{g}^{jk}\,\frac{\partial^2}{\partial\bar{z}^j\,\partial\bar{z}^k}\,,$$

and $k, l = 1, 2, \ldots$. The solution of this problem is left to the reader.

B2.9. The Generalized Function $G^\lambda\bar{G}^\mu$ (General Case)

Let $G(z_1, \ldots, z_m)$ be any entire analytic function, and consider the generalized function $G^\lambda\bar{G}^\mu$, i.e., the functional

$$(G^\lambda\bar{G}^\mu, \varphi) = \left(\frac{i}{2}\right)^m \int G^\lambda(z)\,\bar{G}^\mu(z)\,\varphi(z)\,dz\,d\bar{z}, \tag{1}$$

as an analytic function of λ and μ. We have already studied the simplest case in which the $G = 0$ manifold has no singular points (Section B2.2). We shall now consider the case in which this manifold may have singular points.

We shall not treat entirely arbitrary $G(z) = 0$ manifolds, restricting our considerations to manifolds consisting of *reducible points*. A reducible point of an analytic manifold will be defined in analogy with the real case.

We shall say that the variables $\zeta_1 = f_1(z)$, ..., $\zeta_m = f_m(z)$ form a *local coordinate system* in some neighborhood U of a point M if the following requirements are fulfilled.

(1) The $f_i(z)$ are analytic in the neighborhood of M.

(2) The Jacobian $D(\frac{\zeta}{z})$ does not vanish in U.

(3) The coordinates of M are $\zeta_1 = ... = \zeta_m = 0$.

We shall call $G(z)$ *equivalent to a homogeneous function* in a neighborhood of M if in this neighborhood there exists a local coordinate system ζ_1, ..., ζ_m in which G is a homogeneous function (a polynomial). We shall agree to choose these coordinates always so that G depends on the least number of variables, which number we shall call the *order* of M.

Obviously a function may be equivalent to a homogeneous one even if it is defined not on the entire space, but only on some analytic manifold in the space.

As before, we shall define reducible points of a manifold by induction on the dimension of the space or of the manifold. The definition will be a local one. In other words a reducible point is defined in terms of a manifold and same arbitrarily small neighborhood of the point.

Let us assume, accordingly, that reducible points have already been defined for complex spaces (or analytic manifolds) of dimension less than m. We shall call a point M of the $G(z_1, ..., z_m) = 0$ manifold reducible if there exists a sufficiently small neighborhood U of M such that the following two requirements are fulfilled.

(1) In U the function G must be equivalent to a homogeneous function (a polynomial).

(2) Let ζ_1, ..., ζ_m be coordinates in U in which G is homogeneous. Consider an analytic manifold whose intersection P with U is such that every complex line (in the ζ_k coordinates) passing through M intersects P at no more than one point. Then the intersection of any such P with the $G = 0$ manifold must be a manifold each of whose points is reducible on P.

In the neighborhood of a reducible point M of the $G = 0$ manifold we may introduce local coordinates $\zeta_1, ..., \zeta_m$ in which G is homogeneous of degree k, and in which it depends on at least $n \leqslant m$ complex variables. We then say that M is a reducible point of order n and of degree k.

Thus with every point of the $G = 0$ manifold we associate two integers, namely its order n and its degree k. In particular, if on this manifold there is no point at which the $\partial G/\partial z_i$ all vanish simultaneously, every point of the manifold is a point of first degree and first order. In fact in the neighborhood of such a point G itself may be chosen as one of the new coordinates.

We now turn to an investigation of the singularities of the generalized function $G^\lambda \bar{G}^\mu$, considered a function of λ and μ, for the case in which all the points of the $G = 0$ manifold are reducible.

For simplicity we assume that $G(z)$ is a polynomial. Then we assert (without proof) the following results.

If $G(z_1, ..., z_m)$ is a polynomial and the $G = 0$ manifold consists only of reducible points, the manifold can be decomposed into a finite number of connected components, each of which consists of points of a given order and a given degree.

Theorem. Let $G(z)$ be a polynomial such that all points on the $G(z) = 0$ manifold are reducible. Then the generalized function $G^\lambda \bar{G}^\mu$ is meromorphic in λ and μ.[14] Its poles will then lie on a finite number of sequences. Specifically, each connected component of the $G = 0$ manifold consisting of points of order r and degree k gives rise to a set of poles of the functional $G^\lambda \bar{G}^\mu$ at

$$(\lambda, \mu) = (-[r + p]/k, -[r + q]/k), \quad p, q = 0, 1, 2, ...; \quad \lambda - \mu \quad \text{an integer.} \quad (2)$$

Further, if there occurs a sequence of two, three, or more incident connected components of the $G = 0$ manifold, each consisting of points of different fixed order, and if (λ_0, μ_0) belongs to two, three, or more sequences such as those given by (2), each corresponding to one of these components, then at the point (λ_0, μ_0), the generalized function $G^\lambda \bar{G}^\mu$ has a pole of order two, three, or more, respectively.

The proof of this theorem is essentially a repetition of the proof for the analogous theorem in the real case (Chapter III, Section 4.4). We therefore give it very briefly, omitting some of the details.

Let us assume that we already know the poles in the $(\lambda + \mu)$ plane that occur in

$$(G^\lambda \bar{G}^\mu, \varphi) = \left(\frac{i}{2}\right)^m \int G^\lambda(z) \, \bar{G}^\mu(z) \, \varphi(z) \, dz \, d\bar{z} \quad (1)$$

---

[14] Recall that λ and μ are not arbitrary numbers, for their difference must be an integer. The theorem asserts that $G^\lambda \bar{G}^\mu$ is a meromorphic function of $\lambda + \mu$ for any fixed integral value of $\lambda - \mu$.

as a result of points of order less than m. We may then proceed to investigate (1) in an arbitrarily small neighborhood of a point M of order m. In such a neighborhood there exists a local coordinate system in which G is a homogeneous polynomial of degree k in m complex variables. Without loss of generality we may assume that $G(z)$ is itself a homogeneous polynomial in z (and we may take $z = 0$ to be M).

We now go over to generalized spherical coordinates (see Section B2.5). This yields

$$(G^\lambda \bar{G}^\mu, \varphi) = \sum \frac{i^{(m-1)^2+1}}{2^m} \int_{\Gamma_i} \left[\iint G^\lambda(\alpha z) \, \bar{G}^\mu(\alpha z) \, \varphi(\alpha z) \, \alpha^{m-1} \bar{\alpha}^{m-1} \, d\alpha \, d\bar{\alpha} \right] \omega \bar{\omega}$$

$$= \sum \frac{i^{(m-1)^2}}{2^{m-1}} \int_{\Gamma_i} G^\lambda(z) \, \bar{G}^\mu(z) \, \Phi_{\lambda\mu}(z) \, \omega \bar{\omega} \quad . \tag{3}$$

where we have written

$$\Phi_{\lambda\mu}(z) = \frac{i}{2} \int \alpha^{l\lambda+m-1} \, \bar{\alpha}^{k\mu+m-1} \, \varphi(\alpha z) \, d\alpha \, d\bar{\alpha}. \tag{4}$$

The integrals in (3) run over sections Γ_i of analytic manifolds intersecting each complex line passing through M at no more than one point.

The integrals in (3) and (4) converge for $\mathrm{Re}\,(\lambda + \mu) > 0$, and for $\mathrm{Re}\,(\lambda + \mu) < 0$ they may be defined by analytic continuation in $\lambda + \mu$. Let us first find the singularities of $\Phi_{\lambda\mu}(z)$, considered a function of λ and μ. Since for $z \neq 0$ the function $\varphi(\alpha z)$ in the integrand has bounded support and is infinitely differentiable with respect to α, we may conclude on the basis of Section B2.2 that the only singularities of $\Phi_{\lambda\mu}(z)$ are simple poles at $(\lambda, \mu) = (-[m + p]/k, -[m + q]/k)$, $p, q = 0, 1, 2,$ ... ($\lambda - \mu$ is an integer).

Obviously if (λ, μ) does not belong to this sequence, $\Phi_{\lambda\mu}(z)$ is a homogeneous function of z, continuous and infinitely differentiable everywhere except at $z = 0$.

Let us now find the singularities of

$$\frac{i^{(m-1)^2}}{2^{m-1}} \int_{\Gamma_i} G^\lambda(z) \, \bar{G}^\mu(z) \, \Phi_{\lambda\mu}(z) \, \omega \bar{\omega}. \tag{5}$$

For this purpose we introduce the auxiliary functional

$$I_{\lambda\mu,\lambda'\mu'}[\varphi] = \frac{i^{(m-1)^2}}{2^{m-1}} \int_{\Gamma_i} G^\lambda(z) \bar{G}^\mu(z) \, \Phi_{\lambda'\mu'}(z) \, \omega \, \bar{\omega}. \tag{6}$$

Clearly when $\lambda' = \lambda$, $\mu' = \mu$, (6) goes over to (5), so that it is sufficient for our purposes to study the singularities of this auxiliary functional.[15]

The singular points of $I_{\lambda\mu,\lambda'\mu'}$ considered as a function of λ' and μ', have already been found. These are the points $(\lambda', \mu') = (-[m + p]/k, -[m + q]/k)$, p, $q = 0, 1, 2, \ldots$. The singular points of I, considered as a function of λ and μ, can arise only from those points of the Γ_i manifold at which $G(z) = 0$. These points are, by assumption, reducible and of order no greater than $m - 1$. Consequently (that is, by the inductive assumption) we know the poles (and their multiplicities) of I, considered as a function of λ and μ.

This proves that in addition to the poles (in λ and μ) due to points of order less than m, the integral of (5) also has poles at the points $(\lambda, \mu) = (-[m + p]/k, -[m + q]/k)$.

What remains to be proven is the assertion concerning the multiplicity of the poles. This assertion is implied by the following observation.[16]

If $I_{\lambda\mu,\lambda'\mu'}$, considered as a function of λ and μ, has a pole of order j at $(\lambda, \mu) = (\lambda_0, \mu_0)$, and if, considered as a function of λ' and μ' it has a pole of order j' at $(\lambda', \mu') = (\lambda_0, \mu_0)$, then at $(\lambda, \mu) = (\lambda_0, \mu_0)$ the function $I_{\lambda\mu,\lambda\mu}$ has a pole of order no greater than $j + j'$.

We have thus discovered the singularities of the generalized function $G^\lambda \bar{G}^\mu$ for the case in which $G(z)$ is a polynomial such that all the points of the $G = 0$ manifold are reducible.

The theorem is not in general true if $G(z)$ is any arbitrary entire analytic function for which all the points of the $G = 0$ manifold are reducible.

It is possible, however, by repeating the considerations of the theorem, to show that if $G(z)$ is an entire analytic function the following weaker assertion may be made. Every singular point of the generalized function $G^\lambda \bar{G}^\mu$ is always of the form $(\lambda, \mu) = (-r_1/k, -r_2/k)$, where r_1, r_2, and k are integers. In particular, $G^\lambda \bar{G}^\mu$ will be regular at $\lambda = -1, -2, \ldots$ and $\mu = 0$.

Thus we may assert that if G is an entire analytic function such that all the points of the $G = 0$ manifold are reducible, the generalized function G^{-k} exists. This generalized function is defined by Eq. (1) evaluated at $\lambda = -k$, $\mu = 0$.

Remark. Rather than having $G(z)$ an analytic function, we could have considered arbitrary continuous infinitely differentiable functions of z and $\bar{z}$. Then the definition of a function equivalent to a homogeneous function would be somewhat different. Specifically, the new complex variables in which G would be homogeneous could be arbitrary infinitely

---

[15] This method based on "splitting" the variables was used in Section 4.4, Chapter III to prove the analogous theorems in the real case.

[16] This is the analog of similar considerations in Section 4.3 of Chapter III.

differentiable (but not necessarily analytic) functions of the original variables. The definition of a reducible point would also change accordingly. If $G(z)$ is a polynomial in z and $\bar{z}$ such that all points of the $G(z) = 0$ manifold are reducible, one can obtain a theorem analogous to the theorem of this section for the generalized function $G^{\lambda}\bar{G}^{\mu}$. (We shall not, however, give the exact statement of the theorem.)

B2.10. Generalized Functions Corresponding to Meromorphic Functions of *m* Complex Variables

Let $f(z)$, $z = (z_1, ..., z_m)$, be a meromorphic function, i.e., a function such that in every sufficiently small neighborhood of any arbitrary point it can be written in the form $f(z) = p(z)/q(z)$, where $p(z)$ and $q(z)$ are analytic in this neighborhood.

We shall show that $f(z)$ can be associated with a generalized function $F(z)$. It is natural to require that $F(z)$ satisfy the following condition: if $f(z) = p(z)/q(z)$ in some (closed) neighborhood U in which $p(z)$ and $q(z)$ are analytic, then for any $\varphi(z)$ in K with support in U we will have

$$(F, q\varphi) = \left(\frac{i}{2}\right)^m \int p(z)\,\varphi(z)\,dz\,d\bar{z}. \tag{1}$$

It will be shown that a generalized function satisfying this condition always exists. Note that $F(z)$ is defined by (1) up to an additive generalized function F_1 such that

$$(F_1, q\varphi) = 0 \tag{2}$$

where φ is any function in K with support in U. Obviously F_1 is concentrated on the set of singular points of $f(z)$.

Let us proceed to construct $F(z)$. We first construct a generalized function F_U on the subspace consisting of functions $\varphi(z)$ with support in a sufficiently small neighborhood U of a given point z_0.

Let us write f in the form

$$f(z) = \frac{p_0(z)}{q_0(z)},$$

where $p_0(z)$ and $q_0(z)$ are analytic in U. By taking a sufficiently high derivative of $q_0(z)$ with respect to the z_j, we arrive at the analytic function

$$q_1(z) = \frac{\partial^k q_0(z)}{\partial z_1^{k_1} \dots \partial z_m^{k_m}}, \qquad k = k_1 + \dots + k_m,$$

having no zeros in U (which we have assumed sufficiently small). We then define F_U by the convergent integral

$$(F_U, \varphi) = (-1)^k \left(\frac{i}{2}\right)^m \int f(z)\, \bar{q}_0(z)\, \frac{\partial^k [\varphi(z)/\bar{q}_1(z)]}{\partial \bar{z}_1^{k_1} \dots \partial \bar{z}_m^{k_m}}\, dz\, d\bar{z}. \tag{3}$$

We then use such F_U to construct F over the entire space K. For this purpose we associate with each point z_0 a sufficiently small neighborhood U. Then from the set of neighborhoods we choose a locally finite covering of the entire space.[17] Let this covering consist of the neighborhoods U_1, U_2, ..., U_n, ... Let us now write

$$1 = \sum_{i=1}^{\infty} g_i(z)$$

where $g_i(z)$ is an infinitely differentiable function with support in U_i.

Then every test function $\varphi(z)$ can be written as a sum of functions with support in the U_i according to

$$\varphi(z) = \sum_{i=1}^{\infty} \varphi(z)\, g_i(z).$$

[Note that if $\varphi(z)$ is in K, this sum contains only a finite number of nonzero terms.]

We now define the generalized function $F(z)$ by

$$(F, \varphi) = \sum_{i=1}^{\infty} (F_{U_i}, \varphi g_i).$$

Clearly F is a continuous functional. We need show only that it satisfies (2). To see this, assume again that in some (closed) neighborhood U we have

$$f(z) = \frac{p(z)}{q(z)},$$

where $p(z)$ and $q(z)$ are analytic in U. Let $\varphi(z)$ have support in U. Then the definition of F_{U_i} [Eq. (3)] implies, after integrating by parts and canceling $q(z)$, that

$$(F_{U_i}, q\varphi g_i) = \left(\frac{i}{2}\right)^m \int p(z)\, \varphi g_i\, dz\, d\bar{z}.$$

Now summing over i we arrive at

$$(F, q\varphi) = \left(\frac{i}{2}\right)^m \int p(z)\, \varphi(z)\, dz\, d\bar{z},$$

as desired.

[17] See Appendix 1 to Chapter I.

NOTES AND REFERENCES TO THE LITERATURE

This book has dealt with problems in classical analysis most of which have a relatively long history. Therefore the references which we present below are in many cases merely the customary ones and may be only approximately correct. For instance, we assign credit for the concept of the regularization of a divergent integral to Hadamard and to M. Riesz, although Cauchy had already dealt with it (in defining the Γ function outside the region of convergence of the integral), and even Euler no doubt made use of similar considerations in his calculations.

Chapter I, Section 1

The concept of a generalized function as a functional on a certain function space was formulated by S. L. Sobolev (23). It was L. Schwartz (21) who stated it in the form in which we present it.

Chapter I, Section 2

The contents of this section, are essentially an adaptation of some of Schwartz's book.

Chapter I, Sections 3 and 4

The idea of the regularization of divergent integrals in application to problems of differential equations is due to J. Hadamard. (12). The general method of regularization by analytic continuation is due to M. Riesz (20) (see also Schwartz's book). The material presented in this section is an expansion and reworking of the corresponding subject matter in an article by Gel'fand and Shapiro (10). Not included in that article were the generalized functions $(x + i0)^\lambda$ and $(x - i0)^\lambda$, and the canonical regularization problem and its solution for the case of a single variable.

For functions of two (or more) variables with algebraic singularities, the canonical regularization problem cannot be solved. It has been shown by V. Grushin and R. Ismagilov that it is impossible to associate functionals even with functions of the form $\alpha(x, y)r^{-k}$ so as to satisfy the conditions for a canonical regularization [here $\alpha(x, y)$ is an infinitely differentiable function]. For certain smaller classes of functions of several variables canonical regularizations is possible. V. Palamadov has found several such classes. For instance, canonical regularization is possible for functions of the form $\alpha(x, y)/P(x, y)$ if each of the polynomials $P(x, y)$ has roots such that the distance from those in the upper half-plane to those in the lower half-plane is greater than some positive constant in every bounded interval on the x axis.

The problem of the plane-wave expansion of the delta function (Section 3.10) was first formulated classically (to calculate the value of a function at a point when the integrals of this function over hyperplanes are known) by J. Radon in 1917 and solved by John (14) and other authors. In this connection we may mention the note of Khachaturov (15).

Chapter I, Section 5

The contents of this section are essentially an adaptation of some of Schwartz's book. N. Ya. Vilenkin participated in working out some of the last examples of Section 5.5.

Chapter I, Section 6

The contents of Sections 6.1 and 6.3 are taken from Gel'fand and Shapiro (10). The results of Section 6.1 were obtained in a similar way by John (14) and those of Section 6.3 were obtained independently and almost simultaneously by Courant and Lax (5). The first to give formulas for the general solution of a hyperbolic equation with constant coefficients were Herglotz (13) and Petrovskii (19) for those cases in which they could be expressed in the terms of classical function theory, that is, for equations of sufficiently high order ($m \geqslant n + 1$). Hyperbolic equations with constant coefficients were treated in important work of Gårding (8) by the method of Riesz, and by Leray (16) using the general Laplace transform. Section 6.2 was written by Borovik from his own results (2).

Appendix I

The contents of this appendix are essentially an adaptation of some of Schwartz's book.

Chapter II

The Fourier transform of a function increasing as some power of its argument was first defined as a generalized function by Schwartz. Another definition of the Fourier transform for such functions occurs in the work of Bochner (1) and Carleman (4). The definition of the Fourier transform of a function of any rate of increase as given in Sections 2 and 3 is due to Gel'fand and Shilov (11). Actually that paper gives a more inclusive definition, which will be discussed in Chapter III of Volume II. The space Z was introduced by Gel'fand and Shilov (11) (where it was called Z^1)* and, simultaneously and independently, by Malgrange (17) and Ehrenpreis (6). The text of Section 2.6 is by Vilenkin and Shapiro.

Chapter III, Section 1

The definitions of the differential forms ω_j (for $j > 0$) and of the functional $\delta(P)$ and its derivatives are due to Gel'fand and Shapiro (10). The differential form ω_0 was first introduced by Leray (16). Section 1.9 is taken from an article by Shapiro (22).

Chapter III, Section 2

The analysis of the poles and residues of a quadratic form raised to a power λ for the case in which the quadratic form has no more than one minus sign was first undertaken by M. Riesz (20), for whom it formed the basis of an investigation of the solutions of the wave equation. The study of the quadratic form in the general case with the intention to apply it to representation theory was undertaken by Gel'fand and Graev (9). The elementary solutions of the ultrahyperbolic equation $Lu = \delta$ were first obtained by Y. Fourès-Bruhat (7). The form in which we present them here is that discussed by Shapiro. The results for $L^k u = \delta$ are published here for the first time. Sections 2.1 and 2.2 are based on the work of M. Riesz (20) and Gel'fand and Graev (9). The idea on which Sections 2.3–2.10 are based, namely, analytic continuation into the complex plane, is due to Gel'fand. This idea is discussed in Section 3.3. The results of Sections 3.4–3.6 belong to Gel'fand and Graev, and those of Sections 3.7–3.10 to Vilenkin, Gel'fand, and Shapiro. The relevant sections were written by those who obtained the results.

* A more detailed discussion of methods based on the space Z and the relation between the analytic functionals of Fantappie and the work of Leray will appear in Volume V.

Chapter III, Section 3

The results of this section are taken essentially from the article of Gel'fand and Shapiro (10). Section 3.6 is written by Borovik.

Chapter III, Section 4

The material of this section is taken essentially from the article of Gel'fand and Shapiro (10). M. V. Fedoryuk constructed P^λ for any polynomial of two variables with a single zero at the origin.

Appendix A

The proof of the completeness of K' given in this appendix is due to M. S. Brodskii.

Appendix B

The material in this appendix is essentially an extension to the complex domain of the results of Gel'fand and Shapiro (10). This is its first publication in the form in which it is here presented. The differential form ω was first introduced by Leray (16), and the ω_{ij} were introduced independently (for a real space) by Leray and by Gel'fand and Shapiro.

BIBLIOGRAPHY

1. S. Bochner, "Vorlesungen über Fouriersche Integrale." Leipzig, 1932.
2. V. A. Borovik, Elementary solutions of linear partial differential equations with constant coefficients (in Russian), *Doklady Akad. Nauk S.S.S.R.* **119**, 3 (1958); Dissertation, Moscow State University, 1958.
3. F. Bureau, Divergent integrals and partial differential equations, *Communs. Pure and Appl. Math.* **8**, 143–202 (1955).
4. T. Carleman, "L'intégral de Fourier et questions qui s'y rattachent." Uppsala, 1944.
5. R. Courant and A. Lax, Remarks on Cauchy's problem for hyperbolic partial differential equations with constant coefficients in several independent variables, *Communs. Pure and Appl. Math.* **8**, No. 4 (1955).
6. L. Ehrenpreis, Solution of some problems of division, *Am. J. Math.* **76**, 883–903 (1954).
7. Y. Fourès-Bruhat, Solution élémentaire d'équations ultrahyperboliques, *J. math. pures appl.* **35**, 227–288 (1956).
8. L. Gårding, Linear hyperbolic partial differential equations with constant coefficients, *Acta Math.* **85**, 1–62 (1950).
9. I. M. Gel'fand and I. Graev, Analog of the Plancherel formula for classical groups (in Russian), *Communs. (Trudy) Moscow Math. Soc.* **4**, 375–404 (1955).

10. I. M. Gel'fand and Z. Ya. Shapiro, Homogeneous functions and their applications, (in Russian), *Uspekhi Mat. Nauk* 10, 3–70 (1955). English translation: *A. Math. Soc. Translations, Ser. 2,* 8, 21 (1958).

11. I. M. Gel'fand and G. E. Shilov, Fourier transforms of rapidly increasing functions and questions of uniqueness of the solution of Cauchy's problem (in Russian), *Uspekhi Mat. Nauk* 8, 3–51 (1953). English translation: *Am. Math. Soc. Translations, Ser. 2,* 5, 221 (1957).

12. J. Hadamard, "Le problème de Cauchy et les équations aux dérivées partielles linéaires hyperboliques." Paris, 1932.

13. G. Herglotz, Über die Integration linearer, partieller Differentialgleichungen mit konstanten Koeffizienten, *Berlin. Verhandl. Sächs. Acad. Wiss. Leipzig., Math. Phys. Kl.* 78, 93–126, 287–318 (1926); 80, 60–144 (1928); *Abhandl. Math. Sem. Univ. Hamburg* 6, 189–197 (1928).

14. F. John, Bestimmung einer Funktion aus ihren Integralen über gewisse Mannigfaltigkeiten, *Math. Ann.* 100, 488–520 (1934); The fundamental solution of linear elliptic differential equations with analytic coefficients, *Communs. Pure and Appl. Math.* 3, 273–304 (1950); see also this author's book "Plane Waves and Spherical Means, Applied to Partial Differential Equations." New York, 1955.

15. A. A. Khachaturov, Determination of the value of a measure on a region in a space of n dimensions from its values for all half-spaces (in Russian), *Uspekhi Mat. Nauk* 9 205–212 (1954).

16. J. Leray, Les solutions élémentaires d'une équation aux dérivées partielles à coefficients constants, *Compt. rend. acad. sci.* 234, 1112–1115 (1952). See also this author's book "Hyperbolic Differential Equations." New York, 1955.

17. B. Malgrange, Équations aux dérivées partielles à coefficients constants, 1, *Compt. rend. acad. sci.* 237, 1620–1622 (1953).

18. J. Mikusinski, "Operatorenrechnung." Berlin, 1957 (translated from Polish).

19. I. G. Petrovskii, On the diffusion of waves and the lacunas for hyperbolic equations, *Mat. Sbornik* 17, 289–370 (1945).

20. M. Riesz, L'intégrale de Riemann-Liouville et le problème de Cauchy, *Acta Math.* 81, 1–223 (1949).

21. L. Schwartz, "Théorie des distributions," Vols. I and II. Hermann, Paris, 1957–1959.

22. Z. Ya. Shapiro, On a certain class of generalized functions (in Russian), *Uspekhi Mat. Nauk* 13, 205–212 (1958).

23. S. L. Sobolev, Méthode nouvelle à résoudre le problème de Cauchy pour les équations linéaires hyperboliques normales, *Mat. Sbornik* 1, 39–72 (1936).

24. S. L. Sobolev, "Some Applications of Functional Analysis to Mathematical Physics" (in Russian). Leningrad State Univ. Press, Leningrad, 1950.

Index

419

Index of Particular Generalized Functions

ISBN: 978-1-4704-2885-3 (Set)
ISBN: 978-1-4704-2658-3 (Vol. 1)

9 781470 426583

CHEL/377.H

About this book

The first systematic theory of generalized functions (also known as distributions) was created in the early 1950s, although some aspects were developed much earlier, most notably in the definition of the Green's function in mathematics and in the work of Paul Dirac on quantum electrodynamics in physics. The six-volume collection, *Generalized Functions*, written by I. M. Gel'fand and co-authors and published in Russian between 1958 and 1966, gives an introduction to generalized functions and presents various applications to analysis, PDE, stochastic processes, and representation theory.

Volume 1 is devoted to basics of the theory of generalized functions. The first chapter contains main definitions and most important properties of generalized functions as functional on the space of smooth functions with compact support. The second chapter talks about the Fourier transform of generalized functions. In Chapter 3, definitions and properties of some important classes of generalized functions are discussed; in particular, generalized functions supported on submanifolds of lower dimension, generalized functions associated with quadratic forms, and homogeneous generalized functions are studied in detail. Many simple basic examples make this book an excellent place for a novice to get acquainted with the theory of generalized functions. A long appendix presents basics of generalized functions of complex variables.